The Collected Papers of
Albert Einstein

Volume 8

The Collected Papers of Albert Einstein

English Translations Published to Date

Volume 1: *The Early Years, 1879–1902.* Anna Beck, translator; Peter Havas, consultant (1987)

Volume 2: *The Swiss Years: Writings, 1900–1909.* Anna Beck, translator; Peter Havas, consultant (1989)

Volume 3: *The Swiss Years: Writings, 1909–1911.* Anna Beck, translator; Don Howard, consultant (1993)

Volume 4: *The Swiss Years: Writings, 1912–1914.* Anna Beck, translator; Don Howard, consultant (1996)

Volume 5: *The Swiss Years: Correspondence, 1902–1914.* Anna Beck, translator; Don Howard, consultant (1994)

Volume 6: *The Berlin Years: Writings, 1914–1917.* Alfred Engel, translator; Engelbert Schucking, consultant (1997)

Volume 8: *The Berlin Years: Correspondence, 1914–1918.* Ann M. Hentschel, translator; Klaus Hentschel, consultant (1998)

The Collected Papers of
Albert Einstein

Volume 8

The Berlin Years: Correspondence, 1914–1918

English Translation

Ann M. Hentschel, Translator
Klaus Hentschel, Consultant

Princeton University Press
Princeton, New Jersey

Published by Princeton University Press, 41 William Street, Princeton, NJ 08540
In the United Kingdom: Princeton University Press, Chichester, West Sussex

Princeton University Press books are printed on acid-free paper and meet the guidelines for permanence and durability of the Committee on Production Guidelines for Book Longevity of the Council of Library Resources.

Printed in the United States of America

Typeset by Ann M. Hentschel

ISBN: 0–691–04841–X
Library of Congress catalog card number: 87–160800
10 9 8 7 6 5 4
ISBN-13: 978-0-691-04841-3 (pbk.)

Contents

Publisher's Foreword

We are pleased to be publishing this one-volume translation of Volume 8 of *The Collected Papers of Albert Einstein*, the companion to the two-volume, annotated, original-language documentary edition. As we have stated in all earlier volumes, these translations are not intended for use without the documentary edition, which provides the extensive editorial commentary necessary for a full historical and scientific understanding of the source documents. The translations strive first for accuracy, then literariness, though we hope that both have been achieved.

For those who are curious about the status of volume 7, a scientific-writings volume: both the documentary edition and translation will be published after the present volume. On occasion, due to the nature of scientific-historical research, the volumes will appear out of sequential order.

We welcome Ann and Klaus Hentschel as the new translator and consultant, respectively, of Einstein's correspondence. Having both worked with the Einstein Papers Project in Boston in the past, we are pleased to have them join us again and thank them for their dedication to this project.

Finally, we are happy to report that the Einstein Translation Project has been awarded a five-year grant from the National Science Foundation, to be used for the preparation of volumes 7 (already in prep.) and 9–12, that is, for the five volumes that will appear *after* this volume. We are grateful to the American taxpayers for their generosity.

Princeton University Press
February 1998

Acknowledgments

The translator would like to thank the staff of the editorial offices of the Einstein Papers in Boston for their cooperation, particularly Dr. Michel Janssen for his careful review of certain letters related to his expertise and for helpful comments made by Christopher J. Smeenk, University of Pittsburgh. Giuseppe Castagnetti, Berlin, provided the translation of the Italian letters by Levi-Civita. Particular thanks also to Alice Calaprice, Senior Editor at Princeton University Press, for her ongoing support and assistance in the preparation and production of this volume.

Texts

Vol. 1, 116. From Mileva Marić[1]

[Zurich, after 7 July 1901][2]

My Darling,

Again I was unable to reply for an eternity to your dear little letter.[3] I did want to write you immediately, but you know how much I am suffering now and this afternoon I had such a severe headache that I had to lie down. I've already become an utter whiner of a little mother.[4] But now I'm feeling very well again.– So Drude also has let something be heard of him, he really is a splendid fellow.[5] How the gentlemen do pay court to one another; I mean, the way he speaks about Boltzmann, that he believed he must have done it right. Naturally, a Boltzmann![6]

So, sweetheart, you want to look for a job immediately?[7] and have me move in with you! How happy I was when I read your little letter, and how happy I still am and always will be, as well. And I'd give my head if my happiness wasn't contagious to you as well, sweetheart! But of course it really mustn't involve the worst of positions, darling, that would be too hard for me, I couldn't stand it. Our various elders are going to be amazed, now. Incidentally, my sister wrote me that I really should invite you to visit us during the holidays; my old folks are probably in a better mood now.[8] Wouldn't you like to come along just for a while? I would be pleased! And imagine the wonderful trip we'd be taking there together! We would get off every so often and go on foot a bit or make brief stops. And then at our place everything would be new to you. And when my parents see us both physically before them, all their doubts will evaporate.

Did you have a nice time in Lenzberg?[9] On Sunday there was such a terrible thunderstorm, I was constantly afraid you might still be en route. But I hope you were already safely under cover, sweetheart. I did still want to give you cherries, you know, but you wouldn't have been able to carry them to L[enzburg] anyway. Now they're waiting for you, nicely locked away.– I am being very industrious, I still have to study Weber[10] astutely now; and inbetween I'm constantly looking forward to Sunday when I can see you again and kiss you, really and truly, not just in my thoughts,[11] and deeply, as it comes from my heart, and everywhere all over.–

What are you up to, sweetheart? Is the weather there as dreadful as it is here? When you come on Sunday, do also bring your little fiddle along,[12] and in any case, just write when you're coming so that I can expect you.

But now, my greetings and kisses from the bottom of my heart, and write back soon. Yours,

D[ollie]

Miss Popova is jealous of me about Miss Engelbrecht,[13] which naturally amuses me, and whenever we 3 meet she gets annoyed.

Vol. 5, 136a. From Dmitry Mirimanoff[1]

Parade la Ferme, Cannes, 5 February 1909

Sir,

Allow me to make some remarks about the note that you kindly directed to me.[2] It is exactly right that vector Ω is none other than Minkowski's vector $\underline{m}$, which you denote by the letter $\mathfrak{H}$. I thought it was obvious. In my paper, however, the l[etter][3] $\mathfrak{H}$ denotes *Lorentz's* vector. It is clearly this that I say on p. 193: "But, vector Ω does not signify the magnet. energy" (for *Lorentz*, of course, not for Minkowski).[4]

And it is precisely in order to avoid confusion that I assigned a different letter to this vector. For I take Lorentz's point of view. I advance the following hypotheses:

(1) I *assume as true* Lorentz's differ. equats. (I do not say that I believe them to be true; my personal opinion is of little importance.)[5] I suppose that vectors $\mathfrak{E}$, $\mathfrak{H}$, etc., exist (an indep. defin. of the princ. of relativity). I apply to these equ. Lorentz's transf. and [. . .] approximately, one can find transform. forms. such that Lorentz's equats. are not modified by the transf. I give these forms.

(2) I *assume* true the ⟨principle⟩ postulate of relativity, in other words, I suppose that Lorentz's equats. are not altered by Lorentz's transforms.

And I intend to find the relations that together form a link with the fundam. quant[ities] before and after transformation. I have no other aim. I know that the problem presents no difficulty at all and that the solution is straightforward, but I thought it not unhelpful to present it. Vector Ω plays only an ancil. role in my pap[er]. I could have omitted it, just as I could h[ave] omitted vector $\underline{\mathfrak{U}}$,[6] since the transf. form[ula]s can be obtained in a more direct man[ner]. They certain[ly] can be derived from those of Mink., but actually they follow from the fund. th. of Lorentz-Poincaré (th. of relat.).

Is it necessary to add that I do not propose any new th[eory]? The equats. I–IV are those of Lorentz; the transf. forms. for $\mathfrak{P}$, $\mathfrak{M}$, etc., are the necessary consequences of these equats. and of the princ. of relat. It would thus be natural to name all these system. equats. of Lorentz as a whole. You call it *Mink.'s system.* *You probably* give this term a wider meaning. It is not for me to pronounce myself in favor of this or that electrodyn. th. I limit myself to examining the consequences following from the application of the princ. of relat. to the fund. equats. of Lorentz's electr[odynamics]. In my opinion, my little paper should serve as a *complement* to the fine researches by Lorentz and Mink.[7] I am sorry not to have been able to express my idea in a clearer man[ner].[8]

I wonder whether, at the bottom of all this misunderstanding, it were not a matter of definition, simply a question of vocabulary.

Very respectfully yours,

D. Mir . . .

P.S. Lacking a suffic. command of German, I regret not having been able to write you in that language.

Vol. 5, 136b.   To Dmitry Mirimanoff

Berne, 9 February 1909

Esteemed Sir,

Thank you very much for your letter;[1] I see with pleasure that actually our opinions are not at variance. Nonetheless, it does not seem to me redundant when I point out in my notice that your adaptation of the Lorentz equations to the principle of relativity differs only formally but not in content from Minkowski's system of equations. For not just I but other readers as well could gain the impression upon reading your paper that now a Lorentzian theory deviating from that of Minkowski existed which, applied to Wilson's setup, led to consequences that are observable in principle, which differ from those consequences that follow from Minkowski's theory.[2] The best evidence I see for my submission not having been redundant is that, at first, the editors of the *Annalen* did not want to accept my notice, which initially had been worded more concisely. Prof. Wien wrote me that he was unable to see the correctness of my assertions.[3]

Finally, I must still express my delight at having made your acquaintance through this little war of words, and I urge you to visit me if you ever come to Berne.[4] With kind regards, yours truly,

A. Einstein.

Vol. 5, 312a. To Marie Curie

Prague, 23 November 1911

Highly esteemed Mrs. Curie,[1]

Do not laugh at me for writing you without having anything sensible to say. But I am so enraged by the base manner in which the public is presently daring to concern itself with you[2] that I absolutely must give vent to this feeling. However, I am convinced that you consistently despise this rabble, whether it obsequiously lavishes respect on you or whether it attempts to satiate its lust for sensationalism! I am impelled to tell you how much I have come to admire your intellect, your drive, and your honesty, and that I consider myself lucky to have made your personal acquaintance in Brussels. Anyone who does not number among these reptiles is certainly happy, now as before, that we have such personages among us as you, and Langevin[3] too, real people with whom one feels privileged to be in contact. If the rabble continues to occupy itself with you, then simply don't read that hogwash, but rather leave it to the reptile for whom it has been fabricated.

With most amicable regards to you, Langevin, and Perrin,[4] yours very truly,

A. Einstein

P.S. I have determined the statistical law of motion of the diatomic molecule in Planck's radiation field by means of a comical witticism, naturally under the constraint that the structure's motion follows the laws of standard mechanics. My hope that this law is valid in reality is very small, though.[5]

Vol. 5, 375a. From Walther Nernst[1]

Berlin, 26a Am Karlsbad, 23 March [1912][2]

Esteemed Professor Einstein,

With these lines I just want to express my joy at the prospect of arriving at agreement in an oral discussion, at which the presence of our colleague Planck[3] is naturally not only extremely welcome but could make mine almost superfluous. Nor do I want to address today your view that publications could do no harm because one does not have to read them[4]—I do think, though, that such an attitude, if it were prevalent, would hamper the steady development of physics— rather, only to express again my delight that I can welcome you here soon. I shall be back in Berlin from 15 April on[5] but can also come into town earlier any day and would therefore be especially obliged for advance notification.   With best regards, yours most truly,

W. Nernst.

I hope Rothmund's[6] injuries are not so serious?

Vol. 5, 430. To Unknown Addressee

[Zurich, 2 March 1913]

Highly esteemed Sir,
    . . . In the stress of various obligations I had completely lost track of this matter. Now I am all the sorrier that I cannot possibly fulfill your wish. I have had to speak and write about the foundations of relativity theory so often already that I am tired to death of this topic. Added to this, my style is extremely unwieldy and indigestible.
    But I have a solution . . . letter, the shorthand notes of a lecture I held here. These notes are unpalatable in and of themselves. However, I know a gentleman who very certainly understands the subject and who might possibly agree to a loose reworking of the lecture. The caption could say perhaps: "The Principle of Relat. (using a lecture held by A. E.) authored by . . ." The address of the gentleman I have in mind is:

Dr. Ludwig Hopf
[Academic] Assistant at the Aachen Polytechnic.[1]

You should contact this gentleman. He does not need to adhere to my lecture. You can safely leave that to him. I have no doubt of his ability to present the subject in a suitable form. If he prefers to abstain from referring to my lecture, I am in full agreement with that as well.
    With all due respect, yours very truly,

A. Einstein.

Vol. 5, 500a. To Jakob Ehrat

[Zurich, 7 January 1914]

Dear Ehrat,[1]

I telephoned you yesterday for the following reason. An old acquaintance, Mr. Besso's sister-in-law, a very pleasant, good person of our age, visited me.[2] She is a widow[3] [She was married to a cantonal schoolteacher in Aarau.] and had a position here in a shop, but it was too strenuous for her. Originally she lived with the boys in Oberwil[4] but came here to allow her boys to get a proper schooling. Now she intends to start a small boardinghouse to earn a little bit.[5] Mr. Besso is providing for the boys' upkeep.

Then it occurred to me that it would be quite smart if she went to Winterthur, provided you wanted to live with her, at least to start with. I am convinced that you also would feel very comfortable with this. She is one of the most pleasant women I have become acquainted with, very modest and industrious, a suberb personality. Do come here soon and talk to her; should she not seem suitable to you or should you not be able to agree on the conditions for some reason, you can just say no (directly to her). If you were to decide to accept, you would have a comfortable home in which you would really be master and in control.

So write a postcard *as soon as possible*[6] [she has to decide quickly where she should move] to

Mrs. Rosa Bandi
49 Beder St. II. Zurich,

in which you announce the time of your visit. Treat her very considerately; she has already had so much misfortune.

With best regards, yours,

Einstein

If you prefer, I'll go with you.

1. To Mileva Einstein-Marić

[Berlin,] 2 April 1914

Dear Mama,

I've been here in Berlin (Wilmersdorfer St.) since Sunday.[1] It was extremely pleasant at the Ehrenfests. They would like to spend a few weeks together with us in Switzerland if possible. I'm now on familiar terms with Ehrenfest.[2]

It's nice here. Yesterday I was at Haber's for the first time; he sends all of you his greetings.[3] I haven't seen Mrs. Haber[4] yet. Tomorrow I'm going to be at Koppel's.[5] He gave me a beautiful grandfather's clock as a welcoming present. Gorgeously mild weather has set in here now; I hope it's the same for you and that our Tete has no more little earaches, so that he can go outside a lot.[6] Please write me immediately where the money has been deposited so that I can pay the moving expenses. According to a letter by Maag I can do no more in the Gentner affair regarding the heating because I paid the first bill without objection.[7] For God's sake! Another old tax bill arrived from Prague for about 80 kr. which I must pay.[8] The new landlord is very decent. He's having everything renovated nicely.[9] The furniture will come on Monday but must be stacked up provisionally in the dining room because the apartment repairs will take another one and a half weeks.

Wishing you all enjoyable holidays and a good rest, yours,

Papa.

2. To Paul Ehrenfest

[Berlin, before 10 April 1914][1]

Dear Ehrenfest,

Shame on me for not having written you yet. I spent an unforgettable and absolutely pleasant week with you.[2] I thank you and your wife[3] heartily for it.

I did not write for so long because I had hoped to be able to report something sensible about the rotating system. I had already written down half of the thing. The angel had already unveiled itself halfway in its magnificence; then on further unveiling a cloven hoof was exposed, and I ran away. The root of the evil is, reduced to the simplest terms, as follows:

Mass m is conducted on to an orbital path. For quantum reasons only motions at certain rotational speeds

$$0 \pm \omega_1 \pm \omega_2. \ldots$$

are possible. If we influence adiabatically with a forming magnetic field ($\perp$ to the paper's plane), all these points on the ω line experience a shift in the same sense through induction.[4] Once the field has become constant, we have asymmetrically distributed *new* possible ω values against the old shifted ones, against $\omega = 0$, even though with a constant magnetic field the statistical mechanics of the structure ought hardly to differ from those without a magnetic field. That is why I doubt that through adiabatic influence in the sense of conventional mechanics possible conditions change again into possible conditions.[5] How do you find your way out of that? Frankly speaking, I have no solution.–

Here in Berlin I am very nicely provided for. A pretty room and an interesting companion at Haber's.[6] Otherwise I have not seen any other physicists. By contrast, I really delight in my local relatives, especially in a cousin of my age,[7] with whom I am linked by an old friendship. That is primarily why I am accustoming myself very well to the large city which is otherwise odious to me.

Grossmann wrote me that now he also is succeeding in deriving the gravitation equations from the general theory of covariants. This would be a nice addition to our examination.[8]

Astronomer Freundlich has found a method to establish light refraction by Jupiter's gravitational field.[9] In addition, he has established with astounding accuracy the shift of intensity centers of solar lines toward the red, based on an American's new, very exact observational data.[10] The American's trick consisted in his trying out all sorts of artificial light sources until he finally obtained lines whose finer structure he could no longer distinguish from the relevant solar lines.

We shall return to our countryside recreation plans.[11] Cordial greetings to both of you and your little children,[12] yours,

Einstein

Amicable greetings to your wife's aunt,[13] whom I thank very much for all the kindness she has shown toward me.

3. To Mileva Einstein-Marić, Hans Albert and Eduard Einstein

[Berlin, 10 April 1914]

My Dears,

The director was quite sympathetic. He is very much to my liking. It will be better there than I thought. Lutheran religious instruction.[1] Haber invites you all until the apartment is settled.[2]

Happy Easter wishes[3] from your

Papa.

4. From Paul Ehrenfest

[Leyden, 10 April 1914 or later][1]

Dear Einstein,

In answer to your comment about rotating electrons in a magnetic field.[2]

I protest against your assertion (or assumption): *"with a constant magnetic field the statist. mechanics of the structure ought surely hardly to differ from those* <u>*without*</u> *a magnetic field."* –

With a concise argument I can force you to admit the contrary. (And then the paradoxes indicated by you disappear!)

Here right away is the crux of the problem:

When an electron rotates in the *H* field its *"moment"* is *not*

$$mr^2\omega$$

but

$$\boxed{r^2\left(m\omega + \frac{eH}{2c}\right)}$$

[You should not forget that the ether's "electrokinetic" energy contains a term of the form

$$\frac{er\omega}{c} \cdot \frac{rH}{2}.$$

(For 2 current cycles obviously always: $\frac{c_{11}}{2}i_1^2 + c_{12}i_1 i_2 + \frac{c_{22}}{2}i_2^2$)]

5.   To Joseph Petzoldt

[Berlin,] 16[14] April 1914[1]

Highly esteemed Colleague,[2]

I read your comments on relativity theory in the *Zeitschr. für posit. Philosophie* with much pleasure.[3] From it I see with astonishment that you are closer to me in your understanding of the subject, as well as with regard to the sources from which you draw your scientific convictions, than my true colleagues in the field, even as far as they are unconditional supporters of relativity theory.[4] After great exertion I have now succeeded in establishing proof that the gravitation equations formed last year have a very high degree of covariance with acceleration transformations.[5] Seen from the *physical* standpoint, rotation and acceleration prove to be entirely relative;[6] there is no distinction between a "real" gravitational field and an "apparent" gravitational field produced through the acceleration of the reference system. In both fields the same field equations of gravitation apply.–

The only point in regard to which I do not agree with your representations is the matter of the moving clock (38).[7] Relativity theory allows the strict conclusion that a clock U' moving uniformly and in a straight line relative to the "justified" reference system K travels slower (seen from K) than identically constructed clocks U at rest in K with which we measure the time in K. We know nothing though about how U' proceeds relative to K while U' is in *accelerated* motion. But the traveling speed of U' relative to K can only be influenced *finitely* by a finite acceleration. Thus, if we allow U' to describe a closed path relative to K in such a way that U''s acceleration times disappear against U''s times while moving in a straight line (all seen relative to K), we can then disregard the influence of acceleration times on the angles traveled by the hands of clock U'. Then we *must* conclude that the hands of U' advance slower while traveling along a closed polygon[al path] than the hands of an identically designed clock that was constantly at rest relative to K. On closer consideration of this case, doubtlessly you also will have to come to this result.

I would be very pleased if we were to see each other one day soon so that we can discuss this question of common interest to us both.

With regards, your colleague

Einstein
Kais. Wilh. Inst. of Phys. Chemistry. Dahlem.

6.   To Adolf Hurwitz and Family

[Berlin, 4 May 1914][1]

Dear Hurwitz family,[2]

Here you finally receive the long-promised picture.[3] The delay is due to my wife having arrived only a short while ago and my not knowing where the various things had been packed away.[4] Against expectation I am managing to settle in well; only a certain drill with regard to attire, etc., to which I must submit myself on the order of some uncles[5] so as not to be counted among the rejects of the local human race, disturbs my peace of mind a bit. In its habitude the Academy[6] entirely resembles any faculty. It seems that most of the members restrict themselves to displaying a certain peacocklike grandeur *in writing*, otherwise they are quite human. (With the exception of the fat, unctuous Hermann Amandus Schwarz, for ex. [Amandus must not be translated, because then [it] contains an outrageous impertinence].)[7] I have not found the time to play music yet, because there is always so much else going on. Now I understand the Berliners' self-satisfaction. You get so much outside stimulation that you do not feel your own hollowness so harshly as in a calmer little spot.

I feel compelled to thank you again for the many truly fine hours I have been permitted to spend at your house.[8] With cordial greetings to all of you and best wishes for your health, yours,

A. Einstein.

[. . .][9]

7.   To Pëtr Petrovich Lazarev

[Berlin,] 16 May 1914

Highly esteemed Colleague,[1]

I thank you for the kind invitation,[2] which I nonetheless cannot accept. I am reluctant to travel without necessity to a country in which my kinsmen have been persecuted so brutally.[3]

With all due respect, yours very truly,

A. Einstein.

8.   To Paul Ehrenfest

[Charlottenburg, 18 May 1914]

Dear Ehrenfest,

I would have had the devil to answer to long ago, had he done his duty, for not answering such a dear person as you for so long. The photographs are charming. I was delighted with them.

The business about rotating bodies in a magnetic field is not quite clear to me.[1] Mutual energy certainly does exist between two electrical circuits. But there is no mutual energy between a current and a magnet; should this fiction be unpermissible in principle? Visualize the state of affairs clearly!

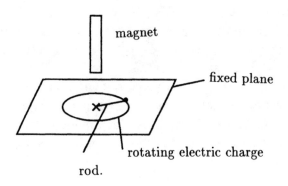

Depending on the position of the magnet, the probabilities of the various rotation velocities should be different *a priori*, even though the electrodynamic forces are constantly being compensated by the rod. This necessarily follows if you assume that the "probabilities *a priori*" are not altered by adiabatic change (infinitely slow movement of the magnet).

I have considered your last letter thoroughly. You are inclined to the view that $\lg W$ deviates from thermodynamic entropy. You prove that this is really the case if the probability function $G(q, p, a)$ depended on the a parameters.[2] I believe, however, that the assumption of such a dependency is not permissible, that it is even completely contrary to Boltzmann's conception. For if you just have a look at the states the system assumes *of its own accord* in the course of time, it is evident that then variable parameters a are altogether inconceivable *as independent entities*. Therefore I am of the opinion that G can only be dependent on p and q, but not on a. The *scale* by which the probabilities of the system states are evaluated may not itself be made dependent on the system state. Otherwise a comparison of probabilities calculated under such varying states would be futile.

With cordial greetings to you, your wife, and your dear little ones, yours,

Einstein.

My little boy is *better* but not yet *well*. They have been here for a long while.[3] The zero point energy for ideal gases peers mockingly out of Sackur's experiments.[4] Stern's analysis on gas dissociation, which you had encouraged him into doing, speaks against Nernst's theorem.[5] Planck believes he can refute my reservations regarding osm[otic] pressure.

Note on the envelope: "Forgive the unmanly paper and do not hope for greetings from the hand of the gentle sex. The content is as coarse as its writer!"

9.   From Paul Ehrenfest

Leyden, 20 May 1914

Dear Einstein,

Cordial thanks for your letter!–[1] I am triumphant! This time *I* am the brighter one. You protest that in the weighting function $G(q, p, a_1, a_2)$, a_1, a_2 offend Boltzmann's spirit—I do not want to dispute the *latter—but you yourself have been working* with such a $G(p, q, \underline{a})$ *for almost 10 years*!!!![2]

Namely with Planck's assumption of the quantization of energy [Energiestufenannahme]—this *is* a $G(p, q, \underline{a})$.–[3]

Proof: Let the energy of a resonator be

$$E(q, p, \alpha, \beta) = \frac{1}{2}(\alpha^2 q^2 + \beta^2 p^2);$$

thus the frequency:

$$\nu = \frac{\alpha\beta}{2\pi}.$$

Then, according to Planck,

$$G(q, p, \underline{\alpha}, \underline{\beta}) = \Gamma\left(\frac{E(q, p, \alpha, \beta)}{\left(\frac{\alpha\beta}{2\pi}\right)}\right)$$

where $\Gamma(i)$ is a discontinuous function *especially* with Planck (= 0 for all values except for $i = 0, h, 2h. \ldots$ for *these* value[s] = 1).

If I change α (the resonator intensity) and β (the reciprocal factor of inertia), then the "Planck" ellipses *deform* on the q, p plane.[4]

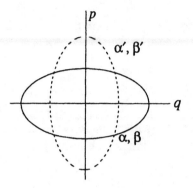

$G(q, p, a_1, a_2\ \underline{T})$ *this* I must admit is a monstrosity!! *This* I do not do! Only Herzfeld does such a thing.[5]

thus G at a specific q, p point *changes* when α, β changes, therefore, e.g., the changes in self-oscillation at compression of rigid bodies![6]

I am enormously pleased with your answer—because it documents for me that I have not done anything superfluously—now I am also certain that you will surely like my result (which is very elegant).–[7] But it is completely hopeless writing to you.

That is why I would like terribly much to come to Berlin for about 2 days.— Do you have anything against it?—I do *not* want to stay with you because you all are not organized yet.–[8] Only arrange a hotel for me—not too expensive—and make it possible for me to have concentrated conversations with you for these 2 days!

—o—

Then I'm going to defend my magnet quanta as well!

—o—

When could I visit you?– I would then like to depart somehow on a Thursday or Friday in the morning and return on Sunday.

Best regards, yours,

Ehrenfest

How happy am I to be in contact with you!!!

10.   From Paul Ehrenfest

[Leyden, 21 May 1914][1]

Dear Einstein,[2]

1. Many thanks for your nice letter![3] 2. Since I would like terribly much to discuss my things with you (correspondence certainly does not achieve this purpose), I ask you whether you are in Berlin in the period from Friday the 29th of May until Sunday the 31st of May. I would like very much to speak with you during these two days. [I would *very* certainly *not* stay with you, but clearly would like to be as nearby to you as possible!]–[4] Please reply to me (naturally absolutely frankly!!!) if possible by return post on the enclosed postcard.

—o—

Glance through the comments on the enclosed pages a bit; but you do not need to reply to them—I prefer to get my response orally.

Your remarks about

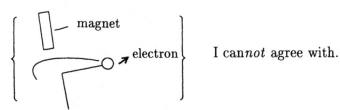

I can*not* agree with.

I shall defend my thesis to you in person, specifically with the aid of the magnet's adiabatic infinitely slow approach! Namely, this thesis:

If *without a magnetic field*[5]

$$mr^2\omega = 0, \pm\frac{h}{2\pi}, \pm 2\frac{h}{2\pi}, \ldots$$

then after the magnet's adiabatic approach it *becomes*[6]

$$mr^2\omega + \frac{eH}{c}r^2\omega = 0, \pm\frac{h}{2\pi}, \pm 2\frac{h}{2\pi}.$$

↓

perhaps a little coefficient error.

This is obtained when you consider $\int \mathfrak{E}d\sigma$ extending over the electron's orbit, while the magnet's entire period of approach is $\oint \mathfrak{E}d\sigma = \iint \frac{dH}{dt}df > 0$. [You write the *dangerous* statement "even though the electrodynamic forces are constantly being compensated by the rod"[7]—[𝔥, 𝔳] is indeed compensated, naturally.]–

On the $G(q, p, a)$ question

(1) *Einstein*:[8] "The assumption of such a dependency is not permissible; that it is even completely contrary to Boltzmann's conception, for . . . . . . *of its own accord* in the course of time . . . ."

> You can see from (9) on sheet 6, 7 that I understand you on this point[9]

(2) *I: Planck's "ellipses" hypothesis and the associated way of treating entropy and temperature works with such a $G(q, p, \underline{\alpha}, \underline{\beta})$.*

> Naturally well-known to you

Proof: Draw on the q, p plane the "Planckian ellipses"

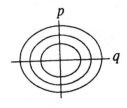

$$\frac{\frac{1}{2}(\alpha^2 q^2 + \beta^2 p^2)}{\left(\frac{\alpha\beta}{2\pi}\right)} = 0, h, 2h. \ldots \qquad (1)$$

Set these curves at weight 1, all other sections of the plane at weight 0.

$$\left.\begin{array}{l} \varepsilon = \dfrac{1}{2}\,(\alpha^2 q^2 + \beta^2 p^2) \text{ the energy} \\[2ex] \nu = \dfrac{\alpha\beta}{2\pi} \text{ the frequency} \end{array}\right\} \text{ of the resonators.}$$

You see with your own eyes that this weight placement of the q, p plane begins to dance a quadrille when you change intensity α or the resonators' reciprocal factor of inertia β infinitely slowly (e.g., collapsing a mirror cavity). This weight placement is thus of the form $G(q, p, \underline{\alpha}, \underline{\beta})$.

And what is entropy according to Planck?
This function:[10]

$$S = k \int \int dq\, dp \cdot f \lg \frac{G(q, p, \alpha, \beta)}{f},$$

where

$$f(q, p, \alpha, \beta, \mu) = N \frac{e^{-\mu\varepsilon} G(q, p, \alpha, \beta)}{\int \int dq\, dp\, e^{-\mu\varepsilon} G(q, p, \alpha, \beta)}.$$

(3) And Debye also (Wolfskehl lecture), in generalizing Planck's approach *to resonators, the energy of which has the form*

> If you develop only the data on the
> energy *line* instead of on the *q, p*
> *plane* this is not so clearly obvious!!

$$\varepsilon = \frac{\alpha^2}{2}p^2 + \left(\frac{\beta^2}{2}q^2 + cq^3 + \cdots\right)$$

has likewise worked with such a $G(q,p,\underline{a})$.[11]

(4) Whether the introduction of such an a in the *a priori* G's contradicts the
spirit of Boltzmann, since the system does *of its own accord* . . . .

Please, please, dear, dear Einstein, delay this question for a little bit longer.—
I *believe* myself entirely equal to this comment as well—but communicating this
in writing *to you* is much too hard a task.–

So simply accept *provisionally my* statement of the problem, which bases itself
on the introduction of such a $G(q,p,\underline{a})$. Then in any case the question remains,
of course:

(5) *For which $G(q,p,a)$ is the relation*

$$\underline{\mu \delta Q = \delta \lg W}$$

satisfied?

And now I could prove the following *statement:*[12]

Let $i(q,p,a)$ be the $2r$-dimensional *volume* enclosed in the $2r$-dimensional
(*molecule*) phase space of that $\varepsilon(q,p,a)$ = const. plane which goes through the
molecule phase point $q_1 \ldots p_r$.

[For Planck resonators $i(q,p,\alpha,\beta) = \dfrac{\frac{1}{2}(\alpha^2 q^2 + \beta^2 p^2)}{\left(\frac{\alpha\beta}{2\pi}\right)}$]

And let $G(q,p,a)$ be any *pure* function of $i(q,p,a)$ [such as, e.g., $= i^2$ or $\lg i$
but *not* $\underline{i^{\underline{q}}}$ or $\underline{ai} + i^2$ or $\underline{(a+q)i}$][13]

$$G(q,p,a) = \Gamma(i).$$

For *every* such

$$G(q,p,a) = \Gamma(i(q,p,a)),$$

the accompanying "most probable" distributions

$$f = N\frac{e^{-\mu\varepsilon}\Gamma(i(q,p,a))}{\int d\tau e^{-\mu\varepsilon}\Gamma(i)}$$

satisfy the relation

$$\mu \delta Q = \delta \lg W,$$

where

$$\lg W = \int d\tau f \lg \frac{\Gamma}{f}.$$

(6) However, contrary examples of such $G(q, p, a)$'s also exist in which $\mu \delta Q = \delta \lg W$ is violated.

(7) In all extensions of the quanta approach, we remain within the weight class

$$G(q, p, a) = \Gamma(i).$$

(8) Then I hope also to be able to present to you orally my reflections on why the Fokker systems satisfy the 2nd law *even though it is impossible to describe them as the "most probable" distributions* (!!!!).

(9) *As a closing remark [Schlussbemerkung]*

a) Planck, Einstein, Debye work with $G(q, p, \underline{a})$, therefore it is worthwhile to examine why these people do come up with $\mu \delta Q = \delta \lg W$ with such anti-Boltzmann spirited G's.

b) Only *once* did anyone work with $G(q, p, a, \underline{T})$: Herzfeld.[14] This displeases me.

c) Ideal gas can "of its own accord" happen to shrink to half its volume on one occasion and to a 1/3 on another, as if labeled item (a) had compelled it.—Classical Hertzian resonators at frequency ν_0 could "*by coincidence*" all be "stunned" at once. On arrival at the Planck ellipses they will belong in frequency $\nu(\alpha_1, \beta_1)$, in another instance *by chance* to Planck ellipses that belong in frequency $\nu(\alpha_2, \beta_2)$—*as if* they had been *pressed* on to these ellipses through corresponding α, β values with the aid of the quantum hypothesis lever.—*Calculate the quotients of the probabilities of both these coincidences.*

Yes sir—*this* would be the entropy calculation in Boltzmann's spirit.

You see *I understand* your comment.

But did Planck, *you*, and Debye calculate *it like this*?– No!– Rather with $G(q, p, a)$ see

e.g., *Einstein, Ann. d. Phys.* 22 (1907) p. 182 bottom.[15]

Einstein:[16] "You are inclined to the view that $\lg W$ deviates from thermodynamic entropy. You prove that this is really the case if the probability function $G(q, p, \underline{a})$ depends on the a parameters. (1) I believe, however, that the assumption of such a dependency is not permissible, that it is even completely contrary to Boltzmann's conception. (2) For if you only have a look at the states which the system assumes *of its own accord* in the course of time, it is evident that then variable parameters a are altogether inconceivable *as independent entities*. (3) The *scale* by which the probabilities of the system states are evaluated may not itself be made dependent on the system state (4) otherwise . . . . futile.

—o—

To come as quickly as possible to the objective, I do so in the following theses:
I define my problem by complete elimination of the concept of $G(q, p, a)$, *fully adopting your diction*. And then I draw up the solution to your problem using this terminology as well.

—o—

For *psychological* reasons I shall formulate everything by means of a somewhat specialized case.—If you understand me *here*, you will immediately figure out the general idea.

Einstein: The assumption of a $G(q, p, \underline{a})$ is inadmissible, completely contrary to Boltzmann's conception . . . . . system assumes of *its own accord* . . . . etc.

—o—

My problem and the solution I have found is

Leyden, 21 May 1914

Do not reply to me about this!!!—It is better we discuss it.– Dear Einstein, please do not feel it presumptuous of me that I write you again—but I am enormously interested in these things. Yours,

Ehrenfest.

11.   To Paul Ehrenfest

[Berlin, 25 May 1914]

Dear Ehrenfest!
You are entirely right, you impetuous boy. I had already noticed it before your incensed letter arrived.[1] Quantum theory requires that the G's be made

dependent on parameters or that such a dependency be permitted. Example. Electron in a fixed plane under the influence of a field.

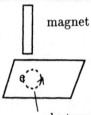

magnet

electron's path;

respective ν dependent on the magnet's position so that also the possible p's & q's [apply] in taking quantum theory as a basis.

One thing must be noted, though. The correlation between S and $\lg W$ is evident only when W is understood as the probability of a phase space *in the sense of the [relative] frequency* with which it effectuates itself. (Your W for various parameter values a cannot be perceived as such, however, as far as I can see.) It is in this sense that I believe in the exact validity of Boltzmann's principle.

I am also toiling with quanta, looking for a general formulation. But . . . . . . It is extremely stimulating here, so even if only for this reason, I recommend your coming. In this event I ask you sincerely, though, to stay with me, considering that the connections everywhere are very convenient.[2] Franck and Hertz have found that electrons in Hg atoms are reflected elastically at a velocity of up to 4.8 volts; that at this velocity upon collision they lose all their kin. energy and emit monochromatic light in such a way that within a few percentages

$$\text{Kin. energy} = h\nu.^{[3]}$$

Wonderful reversal of the photoelectric phenomenon. A striking verification of the quantum hypothesis.

Finally, I beg you not to hold against me my rash assertion in my last letters that the G's were independent of a!

To a joyful reunion! Yours,

Einstein.

Greetings to your wife and to the dear little ones.[4]

12.   To Otto Stern

[Berlin, after 4 June 1914][1]

Dear Mr. Stern,

I have read your paper many times over and enjoyed it very much.[2] It is written with exceptional clarity. I urged Haber strongly to think the matter through carefully. I continue to like it here extremely well. Intellectual stimulation abounds here, there is just too much of it. Some ideas have come to me in the area of photochemistry but nothing fully developed as yet. Yesterday I spoke with Gehrcke. If he had as much intelligence as self-esteem, it would be pleasant to discuss things with him.[3] Gravitation elicits just as much respect among my colleagues as skepticism. I am going to lecture on it in the near future at the colloquium. The Academy is amusing, actually more droll than grave. This type of thing is always subject to mass psychology.

With greetings and best wishes for the semester, yours,

Einstein.

Sackur has also just read your paper, likewise Haber.[4] Sackur wanted to maintain against your calculation that you may have perhaps assumed too high a specific heat for I_2 (due to oscillation).[5] Haber refuted him, however. From the theoretical standpoint the most doubtful aspect is that you so unhesitatingly superimposed the rotation of I_2 on to the I_2 considered to be monatomic.[6] But the deduction seems convincing to me nonetheless. Haber pointed out, though, that sensitive areas are nonsense in the case of atoms, and perhaps he is not so wrong.[7] For, the conception that the atom is an object with preferred orientations strikes as odd. But I tell myself that this unnatural construction must paralyze that of the central forces. It does not appear as if something fundamentally false would result from this approach (central forces + sensitive regions). In the end I think that Haber's argument will not capture the essentials for the following reason. In order to eliminate the entropy difference at absolute zero for chemical bonds, spatial arrangement must be excluded *in the atomic* context; states of various degrees of spatial arrangement in the *macroscopic* context (concentration!) *cannot* be excluded, however. That is why this method of inquiry is probably not the correct one at all.

13.   To Joseph Petzoldt

[Berlin, 11 June 1914]

Dear Colleague,

I have just finished reading your book with great interest, from which I gather with delight that I have long shared your convictions.–[1] I told an ailing gentleman who is a friend of mine about your relativity paper.[2] He was very interested in it. You would surprise him very nicely if you sent him an offprint. His address is

Dr. Ludwig Kraft,[3] Prinzregenten St., Berlin

With cordial greetings, yours,

A. Einstein.

I am enclosing my new paper on the covariance of the grav[itational field] equations.[4]

14.   To Wilhelm Wien

Dahlem, 15 June 1914

Dear Colleague,[1]

My colleague Haber gave me a letter today that you had written to Mr. Planck regarding the pending affairs of the German Phys. Society.[2] As Mr. Planck is very occupied with his duties as president,[3] I am writing to you as a freshman board member who has been participating in the statute consultations for a while now.[4] I do this with all the more pleasure, since you are surely convinced that I cannot have become a "Berliner" yet in such a short time.

I did not have the impression during the consultations that there was an inclination to grant fewer rights to corresponding members than to those residing in Berlin.[5] For example, at the last board meeting the proposal was accepted on Haber's petition to nominate you as first president at the first general assembly to be held according to the new statutes in April of the coming year, so that the first itinerant meeting would take place under your presidency.[6]

Haber, who after desperate resistance and after being elected despite being *in contumaciam* to assume the presidency of the society,[7] is very consternated by your letter. We do not understand in what way members do not enjoy the same rights and the same responsibilities. You would be of assistance in this matter if you could elaborate a bit on your view, so that we could take it into consideration at the meetings. I am convinced that your comments will be received most obligingly. For inst., all the essential points of suggestions that Mr. Kaufmann[8] had sent regarding the statutes were accepted at the last meeting.

Haber is exerting his best efforts to hold German physicists together within *a single* society, because in this way greater benefits can be obtained for the members. Negotiations are underway, for ex., in order to obtain journals more cheaply for members, which some journals have already agreed to do. A splintering of the society would be very detrimental also with regard to the publication of larger collected editions (e.g., Landolt & Börnstein[9]).–[10]

I ask you please not to take offense at my novice meddling in these matters; I know very well that I am inexperienced in these administrative affairs.– It is extraordinarily inspiring here in Berlin. You have certainly already seen the fine analyses by Franck and Hertz.[11] I am not doing much myself at the moment, because I must have a breather from gravitating. In Zurich I had found the proof for covariance in the gravitation equations. Now the theory of relat. really is extended to arbitrarily moving systems.[12]

With cordial greetings and asking you to send me soon a droplet of oil to pour on the troubled waters at the meeting, I remain yours very truly,

A. Einstein.

15.   From Wilhelm Wien

[Würzburg,] 19 June 1914

To Professor Einstein, Dahlem near Berlin.

Dear Colleague,

The obstacles now hindering Planck from considering closely the statutes of the Physical Society stand imposingly in my way as well, since I have to prepare for the university's festivities for the king.[1] But I do not want to withhold my involvement where such an important issue for German physics is concerned, even if the task of negotiation tends to be thankless.

The plan of holding German physicists together within a single large society is doubtlessly a very fine one. However, in a genuine German Physical Society there should be German physicists only, and not two categories, namely, Berliners and non-Berliners. Berlin should only appear in the statutes to the extent that the seat of the society should be Berlin, as Leipzig is the seat of the Society for German Scientists and Physicians; and in addition, the managing director, treasurer, and editor must reside in Berlin.

In contrast to this, the following points appear in the draft statutes:

1) Each member pays 10 M, Berlin members pay twice as much.

2) Elections are held at the general assembly, which takes place at the end of April in Berlin.

3) Of the two vice-presidents, one should have his residence within the greater Berlin area, the other outside of greater Berlin. Among the committee members, one half should reside in greater Berlin, the other outside of greater Berlin.

4) The board meetings take place in Berlin.

You will admit that these stipulations can hardly be called equal duties and equal rights for all German physicists, especially when you consider more closely the practical consequences.

As far as point (1) is concerned, one version appears possible to me that would live up to all the wishes if, as is also commonly done, a distinction is made between the contributions to the society as a whole and those to the individual sections. The former must naturally be the same for all. Each district association, thus also the Berlin one, determines for itself a special surcharge in order to cover its own costs.

Points (2) and (4) I must describe as downright *unacceptable* from the point of view of equal rights.

Elections can be held only after previous consultation with the other voters, and you cannot expect that the physicists residing outside of Berlin will travel to Berlin for the elections. If they do not do so, they forgo participation, since a paper ballot presupposes that they first contact the other physicists in a laborious exchange of correspondence. Point (2) would thus have the result that, practically, voting would be done exclusively by the Berlin physicists.

The same applies to point (4). The board members and particularly also the president could not always embark on the possibly very long trip to Berlin, especially if it happens to be, as intended, at the end of April, therefore during the semester. Thus here also the Berlin vice-president would be in charge in practice.

As with other itinerant conferences, the meetings and the elections should coincide with the society's annual assembly, and the really active members who attend the meeting would then participate in the elections.[2]

I consider point (3) less important, though also not encouraging regarding the discrimination between Berliners and non-Berliners.

If the Berlin physicists wanting to decide now on these statutes can relinquish completely all of Berlin's privileges by abandoning the 4 points I have mentioned, then nothing will stand in the way of uniting German physicists within one large society.

But far be if from me to demand or expect this. In a letter to Planck[3] I said that I completely understand if the Berlin physicists want to preserve their main influence on the society, and that is why I have suggested founding a second society which is closely associated to the Berlin society but which has the sole task of organizing the conferences of German physicists.[4]

Such an outcome has disadvantages over a single German physical society; but surely it is preferable to a division among German physicists, and this we certainly must avoid under all circumstances.

I would actually have preferred to discuss scientific problems with you. I have not yet quite penetrated the secrets of gravitation theory.

With best regards!

16.   From Walter Schottky[1]

Steglitz, 12a Fichte St., 25 June 1914

Highly esteemed Professor,

I have jotted down for you on the enclosed slip the differential equations for the space charge potential affected by an axial magnetic field so that you can see how troublesome they are.[2] In this form, in any case, [they are] still more of a "necessity" than a "virtue."

With best regards, I am very sincerely yours,

Walter Schottky.

1) $\dfrac{d}{dr}\left(r\dfrac{dV}{dr}\right) = \dfrac{2J}{u}$ (Poisson's eq.)

J current per cm filament

2) $u^2 + v^2 = 2\gamma V$ (Energy eq.)

(u, v) velocity

3) $\zeta = \dfrac{v}{u} = r\dfrac{d\phi}{dr}$

$\gamma = \dfrac{\varepsilon}{\mu}$, V potential.

⟨Difference against the filaments⟩

4) Radius of curvature $= \dfrac{\sqrt{1+\zeta^2}}{\dfrac{\zeta}{r} + \dfrac{\frac{d\zeta}{dr}}{1+\zeta^2}} = \sqrt{\dfrac{2}{\gamma}\dfrac{c}{H}}\sqrt{V}$

H magn. field strength

(Magn. deflection)

2) and 3)
$$u = \frac{\sqrt{2\gamma V}}{\sqrt{1 + \zeta^2}}$$
5)

5) and 1)
$$\frac{d}{dr}\left(r\frac{dV}{dr}\right) = \frac{2J}{\sqrt{2\gamma V}}\sqrt{1 + \zeta^2}$$
6)

6) and 4)
$$\frac{d}{dr}\left(r\frac{d}{dr}\left\{\frac{1 + \zeta^2}{\left(\frac{\zeta}{r} + \frac{\frac{d\zeta}{dr}}{1+\zeta^2}\right)^2}\right\}\right) = \frac{2J}{\sqrt{2\gamma}}\left(\sqrt{\frac{2}{\gamma}}\frac{c}{H}\right)^3\left(\frac{\zeta}{r} + \frac{\frac{d\zeta}{dr}}{1 + \zeta^2}\right)$$

With $U = \dfrac{r\zeta}{\sqrt{1 + \zeta^2}}$

$$\frac{d}{dr}r\frac{d}{dr}\left(\frac{r^2}{\frac{dU}{dr}}\right) = K\frac{1}{\sqrt{r^2 - U^2}}\frac{dU}{dr} \quad [3]$$

$$[r^2 \cdot \frac{dr}{dU} = V].\,[4]$$

Or, with $y = r$
$\quad\quad x = U$

$$y^3 y' y''' - y^3 y''^2 + 5y^2 y'^2 y'' + 4yy'^4 = \frac{K}{\sqrt{y^2 - x^2}}$$

Initial conditions:

$$\text{for } x = 0 \quad y = a \quad \text{(filament radius)}$$
$$y' = b \quad \text{(due to } V_0 = \text{const.)}$$
$$y'' = (a, b) \quad \text{(due to } \frac{dV}{dr_0} = 0)$$

Here it is assumed that the particles exit perpendicularly to the filament ($S = 0$). But ζ or $U \equiv x$ can thus also be set arbitrarily otherwise.

Marginal note by Ilse Einstein: "Arrived mid-February 1919."

17.   From Walther Nernst

Berlin, 26a Am Karlsbad, 2 July 1914

Dear Colleague,

According to Clausius-Clapeyron for the steam pressure,[1]

$$\ln p = -\frac{\lambda_0}{RT} + 2.5\,\ln T - \frac{1}{R}\int_0^T \frac{U}{T^2}dT + C; \quad \lambda = \lambda_0 - U + \frac{5}{2}RT$$

applies. U = energy content of the rigid body, C the "chemical constant" according to my heat principle, only dependent on the nature of the gas, according to Sackur[2]

$$\frac{C}{2.3026} = 1.45 + 1.5\,\log M.$$

We set according to Stern[3] $U = 3R\dfrac{\beta\nu}{e^{\frac{\beta\nu}{T}}-1}$;

thus resulting in

$$\int_0^\tau \frac{\mu}{T^2}dT = -3R\,\ln\left(e^{\frac{\beta\nu}{T}}-1\right) + \frac{3R\beta\nu}{T}.$$

Otherwise,

$$\ln p = -\frac{\lambda_0}{RT} + 2.5\,\ln T + 3\,\ln\left(e^{\frac{\beta\nu}{T}}-1\right) - 3\frac{\beta\nu}{T} + C.$$

or (T large!)

$$\ln p = -\frac{(\lambda_0 + 3R\beta\nu)}{RT} - 0.5\,\ln T + \overbrace{3\,\ln\beta\nu + C}$$

Stern commits the error of confusing

$$C \text{ with } 3\,\ln\beta\nu + C;[4]$$

his kinetic result (following the completely unnatural model) is simply nonsensical when you proceed from Sackur or from my theorem.[5]

When in the thermodynamic derivation Stern talks of "zero point energy," that is a second very gross blunder;[6] the above simple derivation shows that, as is obvious from the outset, of course, such hypotheses are not used in thermodynamic derivations.

"This was the first trick
And the second follows forthwith."[7] Namely, the diagram

Best regards, yours,

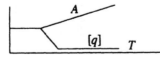

W. Nernst

N.B. It is not worth considering further the confusion, corresponding to the above, reigning in Stern's L_0 and λ_0 = model.[8]

18.   To Max Planck

Dahlem, 7 July 1914

Dear Colleague,

On the way home a little maggot popped up, about which I must inform you concerning our discussion. Take the case that the institute was established[1] and my salary is reduced accordingly by the Academy.[2] It is then revealed sooner or later that I am not the right man for the direction of the institute and for the related work arising from the endowment to be created out of it, so that I feel obliged to resign my post. Then my livelihood would be jeopardized if it is not determined in advance that the reduction in my salary by the Academy *should remain effective only as long as I hold the office concerned.* I ask you please to bring your influence to bear, to have this point taken into consideration.[3]

Then also a brief reply to a comment you made recently at the Academy in the welcoming speech.[4] There is an essential difference between the reference system restriction introduced by classical mechanics or the theory of relativity and that which I apply in the theory of gravitation. For the latter can always be adopted no matter how the $g_{\mu\nu}$'s may be selected.[5] To the contrary, the specialization introduced by the principle of the constancy of the velocity of light presupposes differential correlations between the $g_{\mu\nu}$'s, that is, correlations that ought to be very difficult to interpret physically. Satisfaction of these correlations cannot be forced by the appropriate choice of a reference system for any given manifold. According to the latter interpretation, there are two heterogenic conditions for the $g_{\mu\nu}$'s:

1) the analogy of Poisson's equation
2) the conditions that allow the introduction of a system with constant c.

With cordial greetings, yours,

A. Einstein.

Recipient's note: "seen 8 July 1914 Planck."

19.   To Paul Ehrenfest

[Berlin,] 8 July [1914][1]

Dear Ehrenfest,

Despite all the assurances lately, it is now unfortunately impossible for me to take the trip with you.[2] One of these days my mother must endure a serious operation (cancer)[3] which has upset all my plans. But even if it does not work out this year: I hope to go with all of you on my next walking excursion.

The hydrogen affair is very interesting. It seems to prove finally that quanta without zero point energy do not conform with experience.[4] The specific heat of heavily compressed gases also clamors for zero point energy. But now we must try it with quanta *and* zero p[oint] energy:[5]

$$w = (\sigma + 1)e^{-\frac{\varepsilon_\sigma}{\kappa T}}; \quad \varepsilon_\sigma = \frac{h^2}{32\pi^4 J}\left(\sigma + \frac{1}{2}\right)^2$$

$$\bar{\varepsilon} = \kappa T^2 \frac{d}{dT}\left\{ \lg \sum_0^\infty (\sigma + 1)e^{-\frac{\varepsilon_\sigma}{\kappa T}}\right\}.$$

The expressions only differ, of course, by the $\frac{1}{2}$ in $\left(\sigma + \frac{1}{2}\right)^2$ in ε_σ.

Would you not want to consider this case as well? If it also did not conform with experience, it would get the adiabatic theorem in trouble and surely also quantum theory in general. If you do, write me how the matter stands.– I have considered Nernst's theorem at length and find that it should be valid if all ⟨solid⟩ phases at absolute zero are unmixed (chemical individuals) and are crystallized in space lattices.[6] It is simply

$$S = \kappa \lg Z,$$

where Z is the number of possible realizations of the thermodynamic state. In unmixed space lattices, only atoms of the same type can be exchanged without leaving the thermodynamically defined state. If the system n_ν has ν-type atoms, it is then simply

$$Z = \prod(n_\nu!)$$

But if mixed crystals occur, then Z is larger, because states allowed by the exchange of dissimilar atoms result.

With cordial greetings to you, your wife, and your little ones, yours,

Einstein.

Amicable greetings also to your aunt.

20.   From Max Planck

Berlin W 35, 120 Potsdamer St., 12 July 1914

Highly esteemed Colleague,

In faithfully returning to you the draft sent to me, I naturally declare myself in complete agreement with the addition suggested by you.[1] Another small amendment to it has occurred to me which I have noted down.[2] But I leave it entirely to your *free discretion* whether or not you would like to include this one as well.

With best compliments, always very truly yours,

Planck.

21.   To Fritz Reiche

[Berlin, 18 July 1914]

Dear Colleague,[1]

I am expecting you ⟨tomorrow, Sunday mid-afternoon $\frac{1}{2}$ past 2 o'clock at the house of my⟩ on Monday morning, from 10 o'clock onward at the Kais. Wilh. Institute of Phys. Chemistry. Dahlem (Thielplatz). I have a little office there. Anyone at the institute will guide you to my room.

With best regards, yours,

Einstein.

22.   Memorandum to Mileva Einstein-Marić, with Comments

[Berlin, ca. 18 July 1914][1]

Conditions.

A. You make sure
1) that my clothes and laundry are kept in good order and repair
2) that I receive my three meals regularly *in my room.*
3) That my bedroom and office are always kept neat, in particular, that the desk is available *to me alone.*

B. You renounce all personal relations with me as far as maintaining them is not absolutely required for social reasons. Specifically, you do without

1) my sitting at home with you

2) my going out or traveling together with you.

C. In your relations with me you commit yourself explicitly to adhering to the following points:

1) You are neither to expect intimacy from me nor to reproach me in any way.

2) You must desist immediately from addressing me if I request it.

3) You must leave my bedroom or office immediately without protest if I so request.

D. You commit yourself not to disparage me either in word or in deed in front of my children.

I am ash[amed] for you because you let yourself be so affected by Berlin. Little boy, I only have to say, then he does it right away. Go your own way, let yourself be deceived. I really don't care. Read this slowly. It will do you good. Read it also to your family, they have nothing else to do.

Adn:[2] Ever since coming to Berlin you have become quite lazy.

M[ileva] You must also write that thing about Mrs. Haber.[3] They should also know that other people also are interested in how the famous man behaves. Nasty jokes.

23.   To Mileva Einstein-Marić

[Berlin, ca. 18 July 1914][1]

D[ear] Miza,

Yesterday Haber gave me your letter, from which I gather that you want to accept my conditions.[2] And yet I must write you again so that you are completely clear about the situation. I'm prepared to return to our apartment,[3] because I don't want to lose the children and because I don't want them to lose me, and for this reason *alone*. After all that has happened, a comradely relationship with you is out of the question. It should become a loyal business relationship; the personal aspects must be reduced to a tiny remnant. In return, however, I assure you of proper comportment on my part, such as I would exercise toward any woman as a stranger. My confidence in you suffices for this, but *only* for this. If it is impossible for you to continue living together on this basis, I shall resign myself to the necessity of a separation.

Requesting a clear reply to this, yours,

Albert.

24.   To Mileva Einstein-Marić

[Berlin, ca. 18 July 1914][1]

D[ear] Miza,

Yesterday Mr. Haber sent for me and advised me to add another comment to you regarding my letter,[2] because I had said in it that what remains of my confidence in you is just enough for a business relationship. It remains possible that I'll regain a greater degree of confidence in you through proper behavior on your part.

Let it be as he wishes. No one can see into the future. In any case, I don't consider discussions about it useful. Therefore I ask you again whether you want to live with me under the specified terms.[3] Consider it and give me a clear answer with no ifs or buts so that I know where I stand.

Kisses to my dear boys,

Albert Einstein.

25.   To Robert Heller

[Berlin, 20 July 1914]

Dear Mr. Heller,[1]

I am terribly sorry that you are afflicted with this so painful illness. If all goes well, the treatment will help you! This summer I am going to stay here in Berlin, since, as you know, I have not yet been here for long. I like it *very much* here, there is plenty of stimulation. The people are like anywhere else, but you can make a better selection because there are so very many of them. I have not done anything of real value here yet but rather have rested my nerves a bit, as befits a fellow getting on in age.

Wishing you, your wife, and your little child[2] the best of health and happiness, yours,

Einstein.

26.   To Elsa Einstein

[Berlin] Sunday. [26 July 1914][1]

Dear Else,

Today I'm already writing you the first letter, dear one, to make up for the briefness of the farewells.[2] There is already something eminently important to report. My wife requested from Haber a final meeting with me before her departure. On this occasion we determined that Miza is to remain in Zurich with both children, and all the conditions down to the specifics were laid down in writing. It lasted three hours. The way to a divorce has also been smoothed.[3] Now you have proof that I can make a sacrifice for you.[4] What you have suffered in the last few days has made such an impression on me that I couldn't act in any other way, despite the children.

Although Haber had rather discouraged than encouraged me, I do believe that he does secretly agree with me after all. I must admit that I feel a bit crushed; do you understand? Some time will have to elapse before I am again calmly in possession of myself. Such an affair is bit similar to a murder! But I came to realize that living together with the children is no blessing if the woman stands in the way. Tomorrow they will probably leave;[5] I hope I'll be able to see my boys one more time!

After the meeting at Haber's[6] I drove to your parents, who didn't come home until after 11 o'clock, though, and received the news not without a mild distaste.

Tonight I'm sleeping in your bed! It is peculiar how confusedly sentimental one is. It is just a bed like any other, as though you had not yet ever slept in it. And yet I find it comforting that I may lay myself in it, somewhat like a tender confidence.

Dear little Else, have a good rest with your children and write me soon. You will help me gradually to regain my composure and confidence after the serious operation.[7] It is really good that you are not here at the moment, because otherwise the severity of the situation would hit you very hard. For the present I cannot think of visiting you, for fear of damaging your good reputation again. Also, in the next few weeks my company is not conducive to summertime recreation. I shall attempt to overcome these difficult first weeks by working busily.

Kisses from *your*

Albert.

Best regards to Ilse and Margot.[8]

27.   To Elsa Einstein

[Berlin, after 26 July 1914][1]

Dear Darling,

Despite the severity of the situation I still had to smile a bit about your little express letter. Just don't worry, everything will turn out well. Of course you will probably have to economize a bit, but that is your passion, in which you can indulge yourself then. Dear little Else, I shall cherish you and be thankful for your love. You have proven that you felt attracted to me as a person and not to the great animal in the illustr[ated] newspapers. I know that Ilse and Margot also like coming to my home, even though leaving their grandparents' house is not easy; they will find me a very bearable stepfather.

The divorce will be initiated immediately so that everything takes its course speedily.[2] I may see my children only on neutral ground, not in our house. This is justified because it is not right to have the children see their father with a woman other than their own mother.[3] I know how to come to terms with that. Yet it is understandable that it was terribly hard at first. How can you be surprised that my parting from my children was a real blow to me? I would be a real monster if I felt any other way. I have carried these children around innumerable times day and night, taken them out in their pram, played with them, romped around and joked with them. They used to shout with joy when I came; the little one cheered even now, because he was still too small to grasp the situation instinctively. Now they are gone forever, and their mental image of their father is being spoiled systematically! But I believe they will not make very much of it, and ample care has been taken that they not be deprived of anything.

Dear darling, do not reproach yourself. The trifle that has started the ball rolling now[4] was only of significance as a release. It is not possible for an honest person to love one woman and be married to another. Aside from that, if I had not been so indifferent about my own private life, I ought to have divorced long ago before I came to know and love you.

I am not allowed to visit you. We must act very saintly during this time. See that the month of separation does not seem so very long to you; I also long for you very much already. I can visit you at your family's home, incidentally, as often as I like; so it's not so grim.

Fond kisses from your

Albert

Affectionate greetings to the children.

28.   To Elsa Einstein

[Berlin, before 30 July 1914][1]

My dearest Darling,

Now I can't even pick you up at the train and take you into my arms.[2] For my wife is returning a few hours before you to make a last attempt at preventing the divorce.[3] But I am firmly resolved to carry the matter through. I won't be able to see you until everything has been fought out. But wait until you can make me forget with a kiss the hard events that have to take place now!

With a hug from your

⟨A⟩husband.

29.   To Elsa Einstein

[Berlin, 30 July 1914][1]

Dear Else,

The last battle has been fought. Yesterday my wife left for good with the children.[2] I was at the railway station and gave them a last kiss. I cried yesterday, bawled like a little boy yesterday afternoon and yesterday evening after they had gone. Haber accompanied me to the station (9 o'clock) and then spent the evening with me. Without him I would not have managed to do it. I am glad that you are not here now;[3] you could not have withstood these harsh events. Yesterday morning I spoke with *her* for the last time; we parted rancorously.[4] She perceives my conduct as a crime against her and the children. But I know that it was the best that I could have done, even if the children are completely alienated from me. I know my wife well enough to know that with her characteristic consistency she will erect a wall between me and the boys similar to the one she had attempted to build between me and my family.

You, d[ear] little Else, will now become my wife and become convinced that it is not at all so hard to live by my side. I know that you are capable of it. After so many years you will again be able to govern and manage the house freely, and all the little people will do you honor. Your parents are pleased, and even Haber is wholeheartedly glad for your sake for getting what he had strenuously helped to gain for you.[5] He is convinced that you will do a good job. What do Ilse and Margot say to this?

Haber has impressed upon me that we must be dreadfully careful so that we, i.e., you do not become the subject of idle gossip. Do not go out alone! Haber will inform Planck so that my nearest ⟨relatives⟩ colleagues do not first hear about the matter from rumors. You will have to perform wonders of tact and restraint so that you are not looked upon as a kind of murderess; appearances are very much against us.

I am happy that soon I'll know you to be secure and soon I shall be allowed to embrace you in peace and quiet. Write soon, with fond kisses, yours,

Albert

Regards to Ilse & Margot from their stepfather!

30.   To Elsa Einstein

[Berlin,] Thursday evening [30 July 1914][1]

Dear Darling,

Already the second letter that I'm writing to you today![2] My good fortune is slowly beginning to dawn on me. How much I look forward to the quiet evenings we shall be able to spend chatting alone, and to all the tranquil shared experiences ahead of us. Now after all my ruminations and work I shall find a dear little wife at home who receives me cheerfully and contentedly. And you are no longer under your father's command but are kept on so loose a rein that no honest soul could pity you for it.

You recently reminded me of my letter which begins
. . . I am writing to you today for the last time . . .
and which ends:
     My Zurich address is . . . . .[3]
     As for you, it's like this:
Beginning: I am going on a trip for some time, in order to be away from you.
End: Come visit us at Bayrischzell.[4]

See how nicely we fit together! Today I was at my mother's. She is delighted and only says: Oh, if our poor Papa had only lived to see it![5] It is very nice of her to be so supportive of you. I would not have expected it from her at all but would rather have thought that in this case as well she would have been taken over by the illusion of an insatiable ambition. But to her credit it must be said, she is truly pleased about everything for your sake. Your parents[6] are also very happy; but they are a little resentful because I had supposedly been a bit too generous toward Miza so that our income will be a bit meager. But Haber has promised to help me get some extra income so that I can earn a little more without it being dispersed too much. So never fear, everything will turn out well! Haber is very well disposed toward you; he seems to be convinced that you are the right little woman for me. He also understands entirely that I could not live with Miza. It was not her ugliness, but her obstinacy, lack of accommodation and flexibility and tenderness which had ruled out a fusion. He does not consider me a hard inhuman person but loves me just as much as before.

My colleagues outdo one another in their friendliness toward me.[7] Nernst is extremely friendly again. Planck is so more than ever before. Surely the pain of separation from my children will ease soon and I'll be so content that I shall want to envy myself.

All three of you enjoy yourselves, with a hug to you from your prolific penman,

Albert.

I long for you very much; but I will bear it obediently and stay here. At lunchtime I stay at the institute and consume raw *milk* (!) and bread there, and this daily, alone in my room. In the evenings I am with [?].

31.   To Elsa Einstein

[Berlin,] Monday [3 August 1914][1]

My dear Else,

Since the recent events I find myself in such a harassed state that for the time being I can do nothing better than seek rest, composure, and distance in peaceful seclusion out here.[2] That's why I am avoiding meeting people as much as possible, even my mother and your parents, inasmuch as I can do so without it appearing offensive. Unfortunately even my energy for working and my ability to sleep have now diminished, and the instability of my own emotional state is hard even for me to tolerate. But this is not the first exhausting time that I have gone through; it will pass.

I received your letter. Haber also received your letter; you have probably already received his reply. I am not answering your letter now because my muddled state does not allow it. But I assure you that I love you and will never part from you. Have patience and be content! In the interim, until you are here again, you have your children and I have my work. I am associating with Haber strictly objectively again now. He feels that I must cope with my mental turmoil on my own.

Fond kisses and write soon in friendship, *yours*,

Albert.

Greet Ilse and Margot warmly from me.

32.   To Elsa Einstein

[Berlin, after 3 August 1914][1]

Dear little Darling,

I just received your detailed and dear letter and am glad not to have lost my dearest in the turbulence of the last few days. It's brutal that I had to make you so bitterly disappointed.[2] But the most important thing remains. For me there is no other female creature besides you. It is not a lack of true affection which scares me away again and again from marriage! Is it a fear of the comfortable life, of nice furniture, of the odium that I burden myself with, or even of becoming some sort of contented bourgeois? I myself don't know; but you will see that my attachment to you will endure. In this respect I am not inferior to married men. Outwardly, as well, I will always stand up for you vigorously and shall not allow any cause for pity to arise. No one can take away from us our inimitably lovely walks, either. I am never going to torment you with anything else, but when you give, I shall gratefully take. You also should be proud in front of others, and don't allow them the pleasure of sympathy at your expense. You have someone on whom you can lean just as well as a wedded husband. The little bit of distance in our external life will be sufficient to protect what has made life so wonderful for us now from becoming banal and from growing pale.

All day long I sit at the institute and enjoy working again. Presently I am conducting a difficult study which, owing to its complexity, had constantly resisted my efforts in the last few years. But now it has succeeded for the most part.[3] I have submitted to the printers the paper I had had to present on that unlucky day (Friday).[4] I see little of my mother and your parents. A considerable rift has formed which without the "fame" would be completely irreparable.

The regulations regarding mode of living are being followed strictly. Twice a day $\frac{1}{2}$ l[iter] milk; otherwise only one meal. In fulfilling these regulations, I feel like I'm performing a kind of divine service that is inextricably connected to your person.

I'm glad that our delicate relationship does not have to founder on a provincial narrowminded lifestyle. Soon you will be as cheerful and vivacious again as before. As for me, it will be even better because the living plague that had made my life so hard since my youth is gone.

Concerning Petzoldt, I have sent a very warmly phrased letter of recommendation to the minister.[5] It will not misfire.

Fond kisses from *your*

Albert.

Best regards to the ex-stepchildren! Another kiss for your dear letter.

33.   To Mileva Einstein-Marić

Dahlem, 18 August [1914][1]

Dear Wife,

The money I had given you to take along will probably run short soon. For the month of August withdraw Fr. 100 from the Kantonalbank and for each of the succeeding months Fr. 400 a month from the Kantonalbank.[2] I cannot give more now, as the difficult times exhort me to utmost caution. Should the Kantonalbank give you any trouble, contact Prof. Stern[3] who will stand by you. Ensure that I am informed about the status of our deposits at the Kantonalbank.

I beg you to notify me directly about business matters, because this indirect manner is insulting to me. I have all the more reason to expect this from you as I do not intend to demand the divorce from you, but only that you stay in Switzerland with the children.

I dismissed the maid at substantial cost & paid the bill at Gasteiger. Probably I shall have to keep the apartment[4] because renting it out at this time is hardly conceivable.

Please give me news of my dear boys every two weeks; I send them fond kisses. Albert should write me a postcard once in a while.

The keys, particularly the one for the desk, are missing. You absolutely must have these sent to me so that I can separate what's for me and what's for you. Specifically, I must assemble the bills for the moving-cost reimbursement.[5]

With regards, yours,

Albert.

I have sent a telegraph in case the letter arrives much too late.

34.   To Paul Ehrenfest

[Berlin,] 19 August [1914]

Dear Ehrenfest,

I wonder how your postcard came into my hands so quickly. Europe in its madness has now embarked on something incredibly preposterous.[1] At such times one sees to what deplorable breed of brutes we belong. I am musing serenely along in my peaceful meditations and feel only a mixture of pity and disgust.

My family also is in Switzerland and will stay there.[2] I am residing all by myself in my large apartment[3] in undiminished tranquillity.

My good old astronomer Freundlich will experience captivity instead of the solar eclipse in Russia.[4] I am concerned about him.

Cordial greetings, yours,

A.E.

Best regards to your family.

35.   To Hans Albert Einstein

Dahlem [10 September 1914]

Dear Albert,

I am just packing everything for all of you.[1] Affectionate greetings from me, and do write every 2 weeks. I shall answer you regularly.

Kisses to you and Tete, yours,

Papa.

36.   To Mileva Einstein-Marić

[Berlin,] 15 September 1914

D[ear] M[ileva],

Your letter of the 7th of September in which you complain about a lack of money is, after what I have done, incomprehensible to me: after sending you Fr. 150 through my uncle;[1] thereafter, a telegram with the instruction that you withdraw money from the savings passbook at the Kantonalbank. In addition, a few days ago I sent you 100 M to the Augustinerhof hotel.[2] Besides, I have provided for the move[3] and have kept only very little for myself, namely, the blue sofa, the rustic table, two beds (originating from my mother's household), the desk, the small chest of drawers from my grandparents' household, unfortunately also the electrical lamp you want, not knowing that you are attached to it. Otherwise I have kept nothing of any importance here. The furnishings can't be dispatched yet because the railway does not accept anything to Switzerland. As soon as it's possible, however, I'll telegraph you, then you can take up lodgings and will get everything delivered free of charge.

I would have transferred even more money to you, but *I have no more myself*[4] to be able to manage at all without help. I gave you 600, I paid 200 for your tickets, I sent you the above-mentioned. Then there is the move, my mother's operation,[5] etc.   That's how it came about that I have nothing left. On the 1st of October when I receive my salary I shall send you Fr. 400 immediately and the same amount at the beginning of each month.[6] I may be in the position to send more, but I consider it preferable to save up as much as possible; I myself lead the simplest, almost meager existence imaginable. In this way we can lay quite a lot aside for the children.

Do write me whether you still have not been able to withdraw anything from the Kantonalbank. I have duly noted your threat "to seek the assistance of other persons"; I know very well anyway from your previous behavior what I have to expect from you. Nothing will surprise me, no matter what you do. You have taken my children away from me and are ensuring that their attitude toward their father is vitiated. You will also take away from me other people who are close to me and will try to spoil in every way whatever has remained of my joy of living. This is the proper punishment for my weakness in allowing myself to shackle my life to yours. But I repeat: You will not surprise me, no matter what you do.

I shall return the keys to you sometime.

Best regards to the Wohlwends[7] & kisses to the children.

Albert.

Next month I am exceptionally sending Fr. 500 so that you all are very sure to get by.

I hope that you have informed the Augustinerhof of your new address;[8] I have addressed a number of telegrams and money to there.

37.   To Adolf Schmidt

[Berlin,] 30 October 1914

Highly esteemed Colleague,[1]

Thank you very much for your valuable clarification at the last meeting[2] and for sending me the description of your so superbly functioning device.[3] Meanwhile, our colleague Berliner has been so kind as to send me your article on the correlation factor.[4] I see that the essence of my reflections is not new and do not consider it justified to publish the same. Nonetheless I am sending you my manuscript[5] so that you can judge as a well-versed specialist whether there is anything new of any importance in it. I am so bold as to demand this of you only because my manuscript is just 3-$\frac{1}{2}$ pages long, thus only requiring a little time.

Your integrating apparatus appeals to me very much. Although, as it is described, it only delivers

$$\int y\,dx;$$

but you intended—if I have understood you correctly—to turn it into a measuring device for

$$\int y_1 y_2\,dx$$

which you activate through engaging a second adjustable friction clutch so that the rotation of the planimeter disk is not simply proportional to dx, but proportional to

$$y_2\,dx;$$

then, of course, the rotation of the little counter wheel becomes proportional to

$$y_1 y_2\,dx.$$

Another possibility with which you could manage with a friction element but which would require two integral operations instead would be the following: You have an instrument in which the distance of your friction roll to the center of your counter disk is proportional to y^2 (whether this is easily realizable, I do not know). With such a device you could measure

$$\int (y_1 + y_2)^2\,dx$$
$$\text{and} \quad \int (y_1 - y_2)^2\,dx,$$

thus also the difference[6]

$$2\int y_1 y_2\,dx.$$

The construction of such a device does not appear to me to be particularly difficult. When we meet again, we could, if you have the inclination and the time, weigh the various possibilities against one another. I ask in advance for your forbearance, knowing well that I am a low-grade dilettante in such matters.

With regards from your very devoted colleague,

A. Einstein.

38.   From Adolf Schmidt

(Potsdam) cur[rently] Berlin, 31 October 1914

Highly esteemed Colleague,

You can get over having once in passing on some occasion or other been inspired with an idea similar to one others had already had. When a long time ago *I* once discovered an interesting and useful important principle on spherical functions[1] about which not even the slightest mention was made in any textbook, it was much more disheartening for me finally to discover that Franz Neumann had already proven it almost 50 years beforehand.

Besides, your article (with some comments to be incorporated at the beginning) certainly does seem to me very much worth publication.[2]

The characteristic which, as mentioned, is closely connected to the character of the correlation is *not new*, as you know.[3]

The observation about *I* (along with the optical analogy) also is *not new*, for this is covered by the function introduced by Schuster under the term "periodogram."[4] (As far as I can remember, Schuster's paper appeared about 10 years ago in the *Proceedings* or *Transactions of the Royal Society*.)[5]

But to my knowledge what *is* new, at least in the literature, is *the correlation between the two* that you propose. Although this does not provide any advantage in general toward the practical (numerical) execution of the analysis, it is nonetheless interesting theoretically and may certainly also be of practical use at one time in special cases.

If I may make a comment[6] [By the way: Does the final formula $I(x) = \frac{4}{\pi} \int_0^\infty \Psi(\Delta) \cos(x\Delta) d\Delta$ not follow directly from $2\Psi(\Delta) = \int_0^\infty I(x) \cos(x\Delta) dx$ according to Fourier's integral law?] on the representation, I would think that the introduction of $\chi(\infty)$ could present problems to some readers. It is not quite clear to me, at least, how it results uniquely and clearly from the general definition of $\chi(\Delta)$. Would it not be simpler to indicate the constant in (3a) directly, which apparently comes to $\frac{A_0^2}{2}$, and to set $\psi(\Delta) = \chi(\Delta) - \frac{1}{4} A_0^2$?[7] A_0 is all too well-known as the mean value of F. Of course, $\chi(\infty) = \frac{1}{4} A_0^2$ could also be made plausible; but I have not succeeded in achieving this in a convincing, simple manner. However, perhaps I am overlooking the nearest at hand!

What you say about the planimeter, the description of which I sent you, thoroughly applies.[8] It just yields $\int y \, dx$. But you have also understood me quite correctly with regard to my suggestion of a mechanical calculation of $\int y_1 y_2 dx$. Unfortunately, I have no more descriptions left that I could have sent you of a second device in which the friction wave is the *driving* element (thus corresponding to y_2) and which has been constructed similarly to that first one.

What I have been only planning up to now but for lack of means have not yet carried out is the combination corresponding to your problem.

I have also considered arrangements for the evaluation of $\int y^2 dx$, but unfortunately have likewise not yet been able to have them carried out. The examination of this value frequently plays a role in the statistical analyses on magnetic interference. (Right now I am having the mean values for y^2 determined as a tentative test using a calculational procedure suggested by Prof. Bidlingmaier in Munich[9] in order to see if this procedure is suited for routine use.)

I would very much like to take up your proposition sometime and discuss with you the construction principles of devices for such types of integrations. You might come here once and examine the two mentioned instruments along with their operation? We could then proceed from there. I am almost always here at your disposal (aside from Monday and Thursday afternoons) upon prearrangement (poss. by telephone during the morning hours Potsdam 287).

Then I could also show you some calculations in which I determined for a particular purpose the characteristic $\int F(t)\Phi(t + \Delta)dt$.

With amicable greetings, yours very truly,

Ad. Schmidt.

Note at head of fourth page: "Namur Military Weather Station, to Officers Principal Schwarzschild."

39.   To Paul Ehrenfest

[Berlin, beginning December 1914][1]

Dear Ehrenfest,

You are just as kind a person as I am a vile beast. It cut me to the quick that you didn't reproach me for my dreadful sloth in writing, and even invite me so cordially as well. Presently, I cannot leave here easily, especially since I am just in the process of starting an interesting experimental investigation with de Haas.[2] Therefore, right at the moment I would prefer not to take a trip but would prefer to delay my visit to a time when we are both on vacation. It would also be awkward for me to cancel a few course lectures just on a whim (I am lecturing on relativity).[3] But I would come immediately anytime if in that way I could be of service to a person or a cause.

The international catastrophe weighs heavily on me an international person. Living in this "great age" it is hard to understand that we belong to this mad, degenerate species which imputes free will to itself. If only there were somewhere an island for the benevolent and the prudent! Then also I would want to be an ardent patriot.

Apart from that, I am feeling very well, despite my understandably frequent longing for my children. The decision to isolate myself is proving to be a blessing. You cannot understand this in the least at any rate; but you obviously have no need to.

In recent months I reworked extremely carefully the basis of the general theory of rel.[4] The covariance proof of last spring was not yet completely right.[5] Otherwise, I have also been able to penetrate a few things more clearly. Now I am entirely satisfied with that matter. You will soon receive the paper; read it, you will find it very enjoyable.

Cordial greetings to your wife, your little dears, and to Lorentz. So once again: I am not going to come purely for pleasure, but if I can be of some kind of service, I certainly shall. Yours,

Einstein

The statement signed by Lorentz and you has found general acknowledgment and approval. It is recognized that those wild rumors were unfounded.[6] I was delighted with the photographs from your idyllic sanctum.

40.   To Mileva Einstein-Marić

[Berlin,] 12 December 1914

D[ear] M[ileva],

I notice just now that I have paid fully for the move.[1] But the tips to the people ought to have been paid by you and, if applicable, the storage fee for Zurich during the waiting period, and the customs charge. I request that you examine the mover's bill. Should anything more have been paid, I shall claim it back.

I declare to you herewith that I will send you 5,600 M annually in quarterly installments to support you and the children,[2] at least as long as my income does not sink substantially below the present level.

Best regards to Albert and Tete. As long as Albert does not answer my letter I must assume that it has not been given to him. Otherwise I would write to him again.

A. Einstein

Greetings from me to the children do not seem to be relayed, otherwise they would have sent their regards to me at least once in this long period. Thus it is actually useless to renew the same every time.

I am going to send you the lace crocheted by Zora.[3] By the way, I don't want to be bothered with trifles anymore. Apart from furnishing one bedroom and one office with the bare essentials, I have kept nothing.

41. To Michael Polányi

Berlin. 13 Wittelsbacher St., 13 December 1914

Highly esteemed Colleague,[1]

You use the zigzag process in the accompanying illustration in order to make absolute zero accessible to thermodynamic examination.[2]

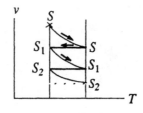

The following is immediately clear and granted:

Were the indicated reversible process carried on *ad infinitum*, or until absolute zero is reached, and if the finally attained entropy value is denoted as S_l, then you have

$$S - S_l = (S - S_1) + (S_1 - S_2) \cdot \cdot \cdot \cdot \cdot \cdot + (S_{l-1} - S_l).$$

To conclude more now, you need two assumptions:

1) $S - S_l$ is finite.
2) The right-hand side has infinitely many components $l = \infty$.

For then it really follows—as has to be proven—$\lim_{l=\infty}(S_{l-1} - S_l) = 0$. Condition (1) will undoubtedly be conceded from the current state of experience. Condition (2), however, is not evident. If, for ex., the law of osmotic pressure remains valid in principle down to absolute zero—which according to various studies is not at all improbable—then system parameters v exist for which the execution of your process would yield a finite l.[3] But then your proof fails.

Your tacitly made assumption (2) thus amounts to a *"petitio principii."*

With all due respect, yours truly,

A. Einstein.

42.   To Michael Polányi

Berlin, 30 December 1914

Dear Colleague,

From your letter I see that we do not quite understand each other yet. The tacit assumption in which I perceive a *petitio principii* is that to attain absolute zero your process needs *infinitely many* stages.[1] However, if absolute zero can be reached through a *finite* number of stages in accordance with your diagram:

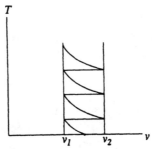

then the proof fails, even though every positive value of parameter v at each temperature may correspond to a realizable state. As long as you cannot exclude that your staged process progresses in the depicted manner, nothing can be inferred about the validity of Nernst's theorem.

Very respectfully and humbly yours,

A. Einstein.

P.S. I do not return now to my comment on osmotic pressure, because this is not necessary for an assessment of the principle issue.

43.   From Hendrik A. Lorentz

[Haarlem, between 1 and 23 January 1915][1]

You show in §12 that it is impossible to introduce a coordinate system K' other than the one used initially, K, which differs only within a finite region Σ in such a way that, like the $g^{\mu\nu}$'s, the $g'^{\mu\nu}$'s connected to K', together with their derivatives, are everywhere constant.[2]

When you nonetheless find in the following paragraphs that you can introduce in region Σ coordinate systems other than K (namely those which just like K are adapted to the gravitational field),[3] this ought to come from the fact that in introducing such a K' not all derivatives of $g^{\mu\nu}$ remain constant at the boundary of Σ. Indeed, you subject the coordinate systems adapted to the field only to the condition that at the transition from one system to another one that differs from it to an infinitely small degree, the Δx_μ and $\dfrac{\partial \Delta x_\mu}{\partial x_\alpha}$ terms vanish at the boundary.[4] ⟨If it is assumed that the second differential gradients of Δx with respect to x differ from zero at the boundary, then, as formula (63a) shows, this also applies to $\Delta g_\sigma^{\mu\nu}$. The latter do not vanish on the inner side of the boundary, and since they are zero at the outer side where nothing has been changed, an inconsistency thus arises for these values. Therefore, if to begin with you relate a given gravitational field to a coordinate system, and thereafter to a second system adapted to the field and differing from the first only in the interior by Σ, you are then introducing inconsistencies at the boundary. A description of the phenomena involving this can hardly be called satisfactory, however.

The difficulty remains⟩

Thus it is not impossible that at that place higher differential quotients of Δx with respect to x are other than zero. If this is the case, then as can be derived from formula (63), the variation Δ, with respect to the coordinates at the boundary, for certain differential quotients of the $g_{\mu\nu}$'s will also differ from zero. Now, since nothing has been changed on the outer side of the boundary, an inconsistency would necessarily occur in the values of those differential quotients, if you introduce a coordinate system that differs only in the interior of the delimited area from the one first used. The introduction of such an inconsistency in the description of the gravitational field is scarcely gratifying, though.

Were the second differential quotients of the Δx's with respect to x other than zero at the boundary, then an inconsistency would result already in the first differential quotients of the $g_{\mu\nu}$'s. Such an inconsistency can exist only when a finite amount of the attractive "agent" (thus here the energy, etc.) is distributed throughout a surface. It is clear, however, that when such a surface distribution does not exist in one description of the gravitational field, it also cannot exist in the new description.

Similar observations also necessarily follow when you imagine that material processes take place in the interior of Σ. Only, in this case attention must be directed to all the equations, therefore, not only to the gravitation eqs. but also to those that determine the material processes. By the way, it seems to me that the difficulty always exists even when, for ex., the only restriction consists in that the processes are also analyzed from a particular moment onward, so that region Σ is denoted, shall we say, by the inequality $t > F(x, y, z)$. As I see it, it is permissible to require that the description of the phenomena after the moment $t_0 = F(x, y, z)$ conform with the description of the processes preceding this time, and that therefore no discontinuities re introduced at the boundary t_0.

Can we find a way out by visualizing the four-dimensional space Σ you focus on as unlimited in all directions? Not likely, as it appears to me. For, the coordinate systs. adjusted to the gravitational field can just be found when the field in the examined region is known, and we know nothing about the physical phenomena for $t = -\infty$. Systems adapted to the field could perhaps best be found if purely periodic processes (such as in the old theory where the motion of two mutually attractive mass-points) were involved. If such processes existed also in the new theory of gravitation, $t = F(x, y, z)$ and $t = F(x, y, z) + T$ (T period) could be chosen as the boundary of region Σ and the Δx_μ's could be chosen in such a way that for $x, y, z, t + T$ the Δx_μ's would have the same value as for x, y, z, t.

As concerns the linear substitutions, these do not pose a problem, since they are introduced for *all* values of x, y, z, t. They incidentally do not belong in the class of transformations presently being examined by you, since they are incompatible with the boundary conditions adopted for the Δx_μ's.

With these remarks I do not want to deny in any way that the introduction of adapted coordinate systems is very appealing and that it has proven very useful in the derivation of the differential equations for the gravitational field.[5] If this derivation alone is involved, then we can limit ourselves to the interior of a finite region and do not need to worry about the discontinuities at the boundary. I just wanted to say that it will be difficult to uphold actually finding suitable adapted systems and that therefore in the end the need for nonlinear transformations against which the equation of the physics should be covariant is satisfied only to a limited degree.

I would like to say a few words about this "need" as well. Apparently it was much less keen in my case than in yours. I have nothing against the existence of "outstanding" coordinate systems for the physical phenomena to be "preferred" in a certain sense over all others. With this I do not mean that when basing the description of the phenomena on such an outstanding system something "absolute" is being described in the processes; only that the description is formed more simply or attractively than through another coordinate system selection, so that it offers us more satisfaction.[6] In many cases it will be easy to agree on the question of what is more and what is less simple. Incidentally the equal status of both systems is already obstructed when upon implementation only the description of the one or the other looks different.

Take, e.g., the case of a rotating system, which Newton and Mach discuss and which you also touch upon briefly.[7] Experience instructs that coordinate system I can be selected in such a way that the motion of a body in the proximity of the Earth can be described, at least to a close approximation, with the equations

$$\frac{d^2x}{dt^2} = -\alpha \frac{x}{r^3}, \ \frac{d^2y}{dt^2} = -\alpha \frac{y}{r^3}, \ \frac{d^2z}{dt^2} = -\alpha \frac{z}{r^3}. \tag{1}$$

In this coordinate system the Earth rotates at constant angular velocity ω, let us say, around the z-axis. If we now introduce a coordinate system of axes II which this rotation accepts, then we obtain the equations

$$\left.\begin{aligned}
\frac{d^2x'}{dt^2} &= -\alpha\frac{x'}{r^3} + 2\omega\frac{dy'}{dt} + \omega^2 x' \\[2mm]
\frac{d^2y'}{dt^2} &= -\alpha\frac{y'}{r^3} - 2\omega\frac{dx'}{dt} + \omega^2 y' \\[2mm]
\frac{d^2z'}{dt^2} &= -\alpha\frac{z'}{r^3}
\end{aligned}\right\} . \qquad (2)$$

The simpler form of (1) suffices to prefer I over II, and hence to say: the description becomes the simplest if we use as a basis a coordinate system in which the Earth rotates. *This* would thus be the meaning of the statement: "the Earth rotates."

However, something else can be added. Our experience has taught us that often when the motion of a body takes place according to the equations

$$\frac{d^2x}{dt^2} = X, \quad \frac{d^2y}{dt^2} = Y, \quad \frac{d^2z}{dt^2} = Z,$$

the values for X, Y, Z clearly are linked with the presence of other bodies and with the distance, size, etc., of these bodies. Thus we can establish in a rational way a connection between the terms $-\alpha\dfrac{x}{r^3}$, etc., and the existence of the Earth's body and refer to a gravitation emanating from the latter. If to begin with we had learned to describe the motion of a mass-point using equations (2), and had the idea not occurred to us to change it into form (1) through modification of the coordinate system, then the endeavor to establish a connection also between the terms $2\omega\dfrac{dy'}{dt}$, $\omega^2 x'$, etc., and the existence of some body or other would have suggested itself. Well, this has not happened, though; at least nothing clear or precise has come out of it.[8] I am thinking here of Mach's "masses of the universe" and the "average rotational motion of ponderable distant masses in the vicinity" which you are talking about.

We can imagine that for a time equations (2) were the only ones available and a "meaning" was agonizingly being sought for the terms $2\omega\dfrac{dy'}{dt}$, $\omega^2 x'$, etc. If someone then came along who by introducing coordinate system I derives (1) from equations (2), then each one of us would welcome it as a real deliverance and everyone would prefer system I.

Since it seems that general covariance in the physics equations is out of the question for arbitrary transformations, you will also naturally admit the existence of preferred coordinate systems. Your adapted systems are just such ones, and the point of departure for your theory of gravitation was precisely that the form of the equations *changes* in a specific way upon substitution of an accelerated system against the original one. ⟨I do not have to say any more about that,⟩ and I just want to add the following comment.

Supposing we could conduct our experiments in a portion of space that is very far away from all celestial bodies, and we had learned that with a suitable choice of coordinate system a mass-point approaching others not too closely moves along a straight course at a constant velocity; we also find that Maxwell's equations are valid in this coordinate system. The older theory of relativity would then show us that a whole group of coordinate systems exists for which all this equally applies, and we must hence acknowledge all of them as on a par with one another. Having arrived thus far we would certainly prefer these coordinate systems over all others, since we would not want to spoil the simpler description through the introduction of a coordinate system that does not belong with the others mentioned. If we did so, "gravitation terms" would then appear in the equations and we would not be too pleased with them, because we would not be able to indicate any bodies whose "influence" or "effect" could be attributed to such terms. The case would be similar to just now, when the meaning of the terms $2\omega\dfrac{dy'}{dt}$, $\omega^2 x'$, etc., in equations (2) was the subject of discussion.

Now that I am bothering you with such a long letter, you will surely allow me to continue on for a while. You say, p. 1031, "We are looking in vain for a sufficient reason for why one of these systems ought to be more suitable than another to serve as a reference system in the formulation of the natural laws; on the contrary, we feel compelled to postulate the equality of status of both systems." Are you not going a bit too far here by presenting a personal view as self-evident? As a matter of fact, prior physicists have thought it possible to find the "sufficient reason" you speak of in that both systems move in a different way in reference to the ether. You are correct in your observation only because you are not at all interested in the ether. This conception may ultimately be preferable to the former, but it is, of course, not the only possible one.

In your article in *Kultur der Gegenwart* I find in the discussion of the "contraction hypothesis" (Michelson's experiment) the remark: "This manner of thinking up *ad hoc* hypotheses to cope with experiments yielding negative results is very unsatisfactory."[9] Poincaré has also said this,[10] and I myself also agreed with *that*; I felt the need for a *more general* theory, as I later attempted to develop[11] and as has actually been advanced by you (and to a lesser extent by Poincaré). Nevertheless, surely my approach was not really so terribly unsatisfactory. In want of a general theory, pleasure can also be found in the explanation of a single fact, provided this explanation is not forced. This the interpretation offered by me and FitzGerald was not; it was, on the contrary, the only possible one, and I added the observation that the hypothesis is obtained when you extend to other forces what was already known about the influence of translation on electrostatic forces.[12] Had I stressed this more[13] [I must admit, though, that I made this observation only *after* I had found the hypothesis], then the hypothesis would have left less of an *ad hoc*[14] contrived impression. Whether it involves the explanation of a negative or a positive result seems to me hardly to make a difference.

I would like to attach another remark of a more didactic nature. If the "contraction" is derived from the formulas of the theory of relativity (which is naturally justified in and of itself) and nothing more is added in commentary, then there is a danger of raising the impression that only "fictitious" things were involved here and not a real physical phenomenon; at least I have occasionally encountered statements by proponents of relativity theory which appear to attest to such a view. In consideration of this, a comment can be made that when we observe a "change" by constantly comparing it to one and the same coordinate system K, this "change" according to common usage (and why can we not keep to that?) represents a physical phenomenon. The contraction of a rod moving with reference to K is just as real as expansion at raised temperatures,[15] and in molecular theory the one phenomenon is explained completely similarly to the other from the point of view of molecular forces. It can be said that these forces and their effect are modified by the translation, an interpretation which can be given slightly more vivid color by thinking of a transmission of the effects through an ether.[16] It goes without saying that without tracing these changes in detail, the statement would suffice that all actions of force must be such that they comply with the principle of relativity.

We earlier spoke repeatedly about "time" and "simultaneity." For this reason I take the liberty to examine the interpretation that most appeals to me. First of all I note that it has to do with the images we form for ourselves of the phenomena which, according to our individual characters, acquire more or less superfluous ingredients and are more or less vividly colored. Whatever they may be, however, the framework in which the image is sketched, as it were, is determined by our conceptions of space and time. I do not venture to examine the origin of these conceptions and their deeper significance; it seems indisputable to me that each one of us simply has them and that, as far as we can know, they agree among different persons. We see quite clearly not only the concepts of "next to," "behind," and "above" one another, but also of "after." In this connection, it seems to me, an unmistakable difference exists between spatial and temporal concepts, a difference which you also certainly cannot remove entirely. You cannot view the time coordinates as totally equal in status with the space coordinates. This is already evident from the circumstance that the same mass-point can very well belong on one occasion to the value system x, y, z, t and then to the values x, y, z, t', but not first to x, z, y, t and then x', y, z, t.[17] Comp. your reflections on p[age . . .].

Concerning time, we have, I think, a completely clear notion of consecutive moments and also of "simultaneity."[18] We may only have a direct notion of the length of time intervals to the extent that for three successive moments in time t_1, t_2, t_3 the interval t_1-t_3 seems "larger" to us than the subintervals t_1-t_2 and t_2-t_3.

We now imagine that we are studying the phenomena observed by us (or by someone else) taking place within a physical system. I want to assume that two groups of clocks U and U' form a part of the instruments used in the observations. Clocks U are at rest relative to one another, and likewise clocks U'. These latter ones, however, are undergoing a common, uniform translation relative to U (determined with the aid of information from U and yardsticks that are at rest with reference to system U). By the way, clocks U were set according to one another in the familiar manner (light signals), and likewise clocks U'.

Now, *two* pictures can be drafted of the phenomena taking place within the system. In one it is clocks U which simultaneously reach the same hand position, in the other one it is clocks U'. In general, both pictures are such that we have no cause to prefer one over the other. This equivalency is precisely the assertion of the principle of relativity.[19] These observations are regardless of whether we have made the observations ourselves with our senses, whether another observer has informed us of his results, or finally whether everything has been registered and photographed automatically.

Should it afford me satisfaction to adorn, or shall I say to disfigure, one or the other image with an "ether," then I am naturally free to do so[20] [The issue cannot be whether the ether "exists"; just whether it is permissible and useful to include it in our picture where simultaneity is indicated by the clocks U.] (as long as I do not dream up too much about it), and obviously (in order to keep everything as simple as possible) in one picture I am going to have the ether at rest with reference to U, in the other picture, however, with reference to U'.

When speaking about the equivalency of both pictures, this should obviously mean as far as our experience allows. I enclose the following in brackets because I am stepping beyond the bounds of physics.[21] [A "universal spirit" which, without being tied to a specific place, permeates the entire system under consideration, or "of which" this system is composed, and which could "feel" all events directly, would naturally immediately distinguish one of the systems U, U', etc., over the others. Although we are no such universal spirits, if we retain the common notion of "spirit" and "body," surely we are not so vastly different to it. For according to this view, we must feel material processes occurring in the brain; and since we can say only with uncertainty that the intellect has its seat at a specific point in the brain, it looks as if it really could perceive what occurs at different locations of the brain and (with sufficient powers of discernment) examine it directly for "simultaneity."]

I come herewith to a close. It will be of great value to me to hear sometime what your opinion is on the issues discussed.

44.   To Edgar Meyer

[Berlin,] 2 January 1915

Dear Mr. Meyer,[1]

Your letter had burrowed itself so deeply among the – tax papers (which I had naturally timorously let lie) that only today do I come by your letter's contents. I would view it as good fortune to the Zurich students if you were to be called there, and I consider this likely as well. But I am somewhat at a loss as to how I should react toward achieving this. The Zurich people are a peculiarly obstinate lot; when a suggestion is made, the opposite can easily result. Can you believe it that neither I nor Laue[2] have the least influence in the selection of the successor? At the university it is still the worst now. There is an implicit agreement among the faculty there to act as mutual yes-men or not to nose into one another's business, so that headstrong Kl.[3] is bound to be the one to exert his authority. Werner,[4] who is perhaps still developing an independence of his own, is not at all well-disposed toward me. The only one I could possibly write to is Schinz.[5] But even there the shot could backfire far too easily, namely, were Kleiner to notice that I had attempted to interfere "behind his back." Added to this, a certain stigma also weighs on me because of my flight from the country.[6]

It would be best if the people could be prompted to turn to me with their inquiries; in this way my voice would carry the greatest weight, and I would not fail to recommend you most earnestly to first place. To begin with, I am now going to do the following. I shall write to a friend who has influence among academic circles[7] that he should endeavor to have me consulted. If this succeeds, it is far preferable to my thrusting myself upon them. However, if you know of a more promising way, then write to me about it.

I am working tranquilly in my booth in spite of the distressing, abhorrent war. The general theory of relativity has now been relieved of most of its obscurities so that the clarity of the deduction has become extremely gratifying. Even Planck, who was until recently still a determined opponent, is close to accepting the new interpretation. It is just a pity that studying the theory is still quite difficult. But I console myself with the fact that at first this was also so for Maxwell's theory of electricity before Hertz provided his simple interpretation.[8] Presently I am preparing a very interesting experiment together with de Haas with which it will become possible to decide whether it is true that paramagnetism can be attributed to rotating electrons.[9]

Best regards, yours,

Einstein.

Paschen's helium paper appeals to me immensely.[10]

Why do you write me, "God chastise the English"?[11] Neither to the former nor to the latter do I have any close relations. I just see with great dismay that God punishes so many of His children for their ample folly, for which obviously only He Himself can be held responsible; I think, His nonexistence alone can excuse Him.

45.   To Paolo Straneo

Wilmersdorf, Berlin, 13 Wittelsbacher St., 7 January 1915

Highly esteemed Colleague,[1]

Simultaneously with this letter I am sending you my latest summarizing paper on general relativity theory and gravitation theory, in which the difficult problem is brought, in a sense, to a close.[2] I definitely believe that in principle the path pursued is correct and that later people will wonder why the idea of general relativity is encountering such great resistance. If motion—uniform as well as irregular—can be defined and conceived of only as *relative* motion, then this relative nature will surely be inherent to it, not only from the kinematic but also from the general physical point of view.–

I love science twice as much in these times when I feel so painfully for almost all of my fellows about their emotional misjudgments and the sad consequences. It is as if an insidious epidemic had confused their minds! The circumspect, and we scientists in particular, must foster international relations all the more and must distance ourselves from the coarse emotions of the mob; unfortunately we have had to suffer serious disappointments even among scientists in this regard![3]

It seems to me as though I had made your personal acquaintance once (ca. 19 years ago) in Italy (Pavia, Casteggio); at that time I was 16 years old.[4]

We possibly saw each other once at the home of the Marangoni family, which I frequented often.[5]

With best regards, yours,

A. Einstein.

46.   To Mileva Einstein-Marić

[Berlin,] 12 January 1915

D[ear] M[ileva],

Officially you are my wife, the same as before, and as such have together with the children a claim to my bit of monetary holdings. But I am not inclined to relinquish the Fr 10,000 which constitute the remains of a sum allotted to me personally for my achievements.[1] I find such a demand beyond discussion. If the share of your paternal inheritance had remained in your parents' hands, it would have suffered just as much from the war as in your hands.[2] As long as I live, the money in my possession will serve exclusively as security for you and the children, and when I die it will be transferred automatically to the children. The upkeep of all of you has been generously provided for, and I find your constant attempts to lay hold of everything that is in my possession absolutely disgraceful. Had I known you 12 years ago as I know you now, I would have considered my responsibilities toward you at that time quite differently.[3]

A package of returned items, which arrived through no intention of mine in my hands rather than yours, will be delivered to you sometime.

I will maintain a regular correspondence with Albert only if I can indulge in the hope that this is beneficial and pleasurable to the boy. Part of this involves that no pressures be exerted on the child aimed at giving him a distorted image of me. If it is your honest wish not to destroy the personal relations between me and the boys, you will accept the following advice. Read what I write to the children but do not discuss it with them; and above all, let little Albert write to me by himself, do not read *his* letters, do not admonish him to write me, and do not discuss with him what he ought to write me. In this way *you* could be sure that I am making no attempts to take the children away from you, and *I* could be more to the boys than their breadwinner. If I see, however, that Albert's letters are prompted, then I shall refrain from sustaining a regular correspondence out of consideration for the children.

With best wishes for 1915,

Albert Einstein.

I am going to write separately to Albert very soon.

47.   To Hendrik A. Lorentz

Berlin, 13 Wittelsbacher St., 23 January 1915

Highly esteemed and dear Colleague,

I have long been eagerly awaiting your detailed letter,[1] the existence of which de Haas revealed to me after his arrival from Holland.[2] I thank you very much for having spent so much time and effort on it. Before I answer the individual points, however, I must tell you that in the course of our collaboration I came to cherish and respect your son-in-law exceedingly. To our great delight the experiment on magnetism has turned out positively. Now de Haas has devised an even nicer investigative procedure in which the use of resonance can even be dispensed with. With it the reason for why the magnetic axis and the rotational axis of the Earth nearly coincide has now been found.[3]

If anyone had to relive exactly the same struggles here in the considerations on general relativity, I'd ardently wish that it be you. But I see from the first part of your letter that this has not yet happened completely. Had I committed the errors you rebuke me for regarding the vanishing of the Δx_ν's, then I would deserve to have ink, pen, and paper taken away from me once and for all!

In §12 the Δx_ν's refer to a coordinate transformation with Δx_ν's (& derivatives) *vanishing at the boundaries*.[4] As you know, in this § it was intended to show that acceptance of such transformations prohibits a complete mathematical formulation of the laws of gravitational fields. From §13 onward the Δx_ν's relate to the *justified* transformations for which the Δx_ν's along with their derivatives do *not* vanish at the regional boundaries.[5]

Equations 65–65b apply to arbitrary infinitesimal (constant) transformations. Now, when I say about (65b): "It vanishes if the Δ_μ and $\dfrac{\partial \Delta x_\mu}{\partial x_\nu}$ terms vanish," it was not meant to mean that this would be so in the case of "justified" transformations; for the latter, this had rather to be considered *a priori* as out of the question according to §12. When on the second half of page 1070 I examine the transformations K–K'–K'', etc., with definite border coordinates, under no condition are these transformations *justified* ones. This becomes clear, of course, from the fact that the observation serves to select from among these systems only *one* as justified, namely, the one with an extreme J. If K is this system, then K is justified, K', K'' etc., however, are *not* justified. Therefore, the transformation K–K' is *not a justified one*. In the reflections of §14, which form the crux of the whole problem, the coordinate transformation is to be conceived of in such a way that the Δx_μ and $\dfrac{\partial \Delta x_\mu}{\partial x_\nu}$ terms do *not* vanish at the boundaries; rather, the coordinate system changes (infinitesimally) over the entire expanse of the four-dimensional continuum. By contrast, the *virtual* change in the field (denoted by "δ") refers only to the interior of Σ.

I entreat you to go through the entire analysis again after correcting this construal and then inform me whether all the clouds disperse before your keen eyes. In the meantime, I have gained the conviction, after repeatedly checking most conscientiously the entire train of thought, that everything is logically correct.

From what has been said, it is also clear that linear transformations then belong among the "justified" ones. It is likewise clear that the *state of motion* of justified systems is arbitrarily selectable, since the choice of coordinates at the region's boundaries is open; this can be easily visualized from special cases.[6]

Now, to the question of the "need" for nonlinear transformations. This need has a physical and an epistemological root (1 and 2):

1) The theory must do justice to the equivalency of inertial and gravitational mass of bodies. This is achieved only when a similar relation is established between inertia and gravitation as with the orig. theory of relativity between the Lorentzian electromotive force and the effect of electrical field intensity on an electrical mass. (One or the other exists, depending on the choice of reference system.)[7] In our case this is achieved with the "Equivalency Principle," and this means mathematically that among the justified transformations nonlinear ones must exist.

2) In describing the relative motion of two coordinate systems (of any kind) K_1 and K_2, it is unimportant whether I relate K_2 to K_1, or conversely, K_1 to K_2. Nevertheless, if as a result K_1 is distinguished in that, applied to K_1, the general laws of nature were supposedly simpler than in connection with K_2, then this preference is a fact without physical cause: Among two, according to their definition, equivalent objects K_1 and K_2, one is distinguished on no physical grounds (accessible in principle to observation).—My confidence in the consistency of natural events stops most adamantly at this. A worldview that can do without such arbitrariness is preferable, in my opinion.

This second argument has already been stressed in the clearest terms by Mach. But it is best illustrated by a comparison that my friend Besso had thought up last year.[8]

Cast your mind back to the time when it was believed that the basic form of the Earth's surface was a plane. Then you could hit upon the following view:

All directions in space are of themselves equivalent; yet, two of the directions I can describe geometrically only *relatively* (through the angle they form). *Physically*, however, there is a preferred direction in the world, namely, the vertical one. There is no physical cause for the preference of this direction; it plays the part of a "*prima causa.*" Bodies fall in this direction; the basic horizontal form of the

Earth's surface is also due to this. Thus the following things can be compared:

Geometrical equivalency of directions	$\longrightarrow$	Kinematic equivalency of the variously moving reference systems
Preference for one direction without physical cause	$\longrightarrow$	Preference for a state of motion without allowing a physical cause
In reality, falling & preference for the vertical are determined by the Earth's mass	$\longrightarrow$	In reality, the inertial behavior of bodies as well as the Galilean reference system are determined by the other perceptible things of the world (their position and condition)

I believe that this comparison gets to the core of the matter. I also believe that a time will come when physicists will be very astonished that their predecessors did not balk at the *a priori* preferred (without physical cause) reference systems (Galilean).

I firmly contest the argument including coriolis and centrifugal force. The complete gravitational field always contains components that act upon the bodies like coriolis and centrifugal forces. I imagine that those components in constrained space-time regions could be made to vanish in certain cases if, for ex., the gravitational field originates from a single mass situated within the "finite realm"; but I do not believe that the equations of motion valid in this case are *general* natural laws. I think that motion equations of the kind

$$m_\nu \frac{d^2 x_\nu}{dt^2} = \sum_\mu k \frac{m_\nu m_\mu}{r_{\mu\nu}^2} \frac{x_\mu - x_\nu}{r_{\mu\nu}}$$

cannot suffice generally (quite apart from the modification produced as a consequence of the original theory of relativity). I imagine it would be impossible to cause other components (e.g., the "coriolis forces") to vanish *everywhere* through an appropriate selection of the coordinate system.

Although I also prefer certain reference systems, the fundamental difference to the Galilean preference is, however, that my coordinate selection makes no physical assumptions about the world; let this be illustrated by a geometric comparison. I have a plane of unknown description which I want to subject to geometric analysis. If I require that a coordinate system (p, q) on the plane be selected in such a way that

$$ds^2 = dp^2 + dq^2,$$

I therefore assume that then the surface can be unfolded on to a [Euclidean] plane. Were I only to demand, however, that the coordinates be chosen in such a way that

$$ds^2 = A(p, q)dp^2 + B(p, q)dq^2,$$

i.e., that the coordinates be orthogonal, then I am assuming nothing about the nature of the surface; this can be obtained on any surface.

You say that regarding coriolis and centrifugal forces as "real" field components is unsatisfactory because we cannot attribute any physical cause to their occurrence. I would like to respond to this with the supposition that we never can see the stars. According to my understanding, these force fields are determined exclusively by the boundary conditions and the field equations, if the influence of the masses belonging to the system under study can be disregarded here. It is admittedly awkward that the boundary conditions must be picked out suitably instead of being able to assume that all boundary conditions vanish into infinity. But are you so sure that you will manage with such simple boundary conditions in you view of the world? Furthermore, it must be considered that, according to my view, the multifariousness of permissible coordinate systems is immense, thus also the multifariousness of the attached boundary conditions; therefore, if these boundary conditions appear artificial in the individual case, this does not lie in the substance of the theory but in the fact that although justified, the coordinate system is not suitably chosen for the description of the case under examination. As you know, circumstances are similar even in normal mechanics. Let the world to be described be the solar system, for inst.; then it is clearly useful always to place the origin of the coordinates at its center of gravity, but the equations are obviously also valid for coordinate systems relative to which this center of gravity is moving uniformly and in a straight line. Here also, the choice of coordinates is prescribed not by the *laws of nature* but only by the need for the simplest possible description of the case at hand.

Now once again to the question of whether an unrelativistic physics violates the postulate of sufficient cause. You say sufficient cause (for preferring K over K', etc.) can be found in that both systems move in different ways relative to the ether. I understand "cause" in this connection as an *observable fact* which distinguishes K from K' not as a *merely conceptual characteristic*.

I ask you please to make allowances for my statements contained in *Kultur der Gegenwart*."[9] Although I had 3 years of time to compose it, I had completely forgotten and was reminded of my commitment by Warburg one week before the delivery deadline.[10] In this time I hastily pieced together the two articles as best I could. So please: do not punctiliously weigh every word! Regarding the erroneous view that the Lorentz contraction was "merely apparent," I am not free from guilt, without ever having myself lapsed into that error. It is real, i.e., measurable with yardsticks and clocks, and at the same time apparent to the extent that it is not present for one with moving observers.

Finally, as far as the question of *time* is concerned, we are scarcely going to be able to debate this effectivly by letter. I shall be glad to come to Holland again to discuss this and other matters, when the sorry international entanglement is finally overcome. I am satisfied if you agree with me that I did *not* set the monstrous boundary condition for the vanishing of the Δx_ν's for justified transformations (otherwise my beloved house of cards would be irretrievably lost).

With cordial greetings, yours,

Einstein

Best regards to your wife, Ehrenfest, and Kamerlingh-Onnes.[11]

48.   To Hans Albert Einstein

[Berlin,] 25 January 1915

My dear Albert,

From Hans Wohlwend[1] and your aunt in Lucerne[2] I hear that you and Tete are well and that you have a pretty apartment by the hill in Zurich.[3] I also know that you were allowed to have installed on your sled the steering fixture that you have been wanting since last year. You have found your old friends again as well, I hope, with whom you had always tussled and romped about on Zurich hill. Nowhere else is it as nice for boys as in Zurich, nor so healthy either. Boys aren't pestered too much with homework there either, nor with the requirement to be overly nicely dressed and well-mannered. Just don't forget your piano! If you can play music nicely, you can give yourself and others much pleasure with it. This evening, for example, I'm playing along in a little concert,[4] the proceeds of which are supposed to help two impoverished artists.

In a few years you will also be able to begin exercising your mind. It's splendid, once you can do it. In the last few weeks I have been doing a wonderful and important experiment on magnets with Mr. de Haas, who visited us once in Dahlem, as you know, with his wife and his little children.[5] When I spend some more time with you one day, I'll tell you a bit about it, maybe already during the next summer holidays. In the summer I hope to go on a walking tour with you in the mountains so that you also can see a bit of the world. If all goes well, the terrible war will be over by then.

I am giving the little Haber boy[6] mathematics lessons until Easter; he was quite sick and cannot go to school. I have heard nothing about your other pals here because now I am living in the city,[7] near Fehrbelliner Place. There I have a little flat in which I usually work all day. Sometimes I even cook my own lunch.

Kisses to you and Tete and write soon, yours,

Papa

Regards to Mama.

Also write me when you have any special wish.

49. To Mileva Einstein-Marić

[Berlin, 27 January 1915]

D[ear] M[ileva],

Your explanations regarding the money deposited in Zurich have convinced me.[1] You should gain control of it. Make inquiries about the way this can happen. It will surely be less easy to transfer the money deposited in Prague[2] into my hands.

Kisses to the children, y[ours],

A. Einstein.

50. To Władysław Natanson

[Berlin,] Wednesday. [27 January 1915][1]

Dear Colleague,[2]

I recently heard with great regret that you have been looking for me. The flu has caught up with me now so that I have to lie in bed in the expert care of my mother (in her apartment).[3] As soon as I am back on my feet, I shall call on you again.

With cordial greetings, yours,

A. Einstein.

Best regards to your wife and your little children.[4]

51.   To Wilhelm Waldeyer

[Berlin,] 27 January 1915

Highly esteemed Colleague,[1]

Tied to my bed with a cold, it is unfortunately impossible for me to attend Mr. Auwers's funeral today.[2] I shall probably not be able to participate in the ceremonial meeting of the Academy[3] tomorrow either.

With all due respect, yours very truly,

A. Einstein.

52.   To Hendrik A. Lorentz

[Berlin,] Wednesday, 3 February [1915][1]

Highly esteemed and dear Prof. Lorentz,

Your telegram, which I received on Saturday, moved me very much. It shows your great philanthropy and at the same time your kind sympathy for human weakness (injured author's pride!). On this, Heine's quote:

———————

But if you do not praise my verses,
I shall divorce myself from you.[2]

———————

The theoretician is led astray in two sorts of ways:

1) The devil leads him up the garden path with a false assumption.
   (For this he deserves pity.)
2) He argues inaccurately and sloppily.
   (For this he deserves a thrashing.)

According to your first letter I deserve a sound thrashing (case 2). But now from the telegram and your friendly postcard I can hope to be placed under the conciliatory and decent rubric (1).

I spend many happy hours with your children, even taking aside the gyromagnetic effect, the existence of which seems soon to be adequately established.[3]

Affectionate greetings and once again many thanks! Yours,

A. Einstein.

53.   To Erwin Freundlich

[Berlin, ca. 3 February 1915][1]

Dear Freundlich,

As a result of our conversation yesterday, the following idea occurred to me that may not be new to you. If a spectroscopic binary star exists, both components of which provide sharply observable separate spectral lines, then the mass of both components can be obtained to a certain degree of accuracy. The wavelengths λ_1 (of the 1st component) and λ_2 (of the second component) oscillate around each other. Both components possess the same average velocity within the radius of vision, however. Thus, if no influence were present other than motion, then the temporal mean values $\overline{\lambda}_1$ and $\overline{\lambda}_2$ must be equal. The variation between them may be determined by the influence of gravitation;[2] however, that is also dependent on the star's radius, which can be estimated only very roughly. What do you think of this matter?

I am writing to Planck under separate cover about the observer affair.[3]

Best regards, yours,

A. Einstein.

54.   To Erwin Freundlich

[Berlin,] Friday. [5 February 1915][1]

Dear Freundlich,

Yesterday Planck spoke with Struve about you.[2] Struve indicated that he would not oppose a habilitation petition submitted by you. Otherwise he railed about you thoroughly. He said you do not do what he asks of you, etc.[3] Planck thinks the only way for you would be to aspire for a teaching position in theoretical astronomy, and he thinks that you have good chances in this respect. He favors habilitation as soon as possible, with a thesis preferably in the field of theoretical astronomy. I think he is right to the extent that we cannot place everything on the one observer-position card.[4]

With best regards, yours,

A. Einstein.

55.   To Michael Polányi

[Berlin,] 10 February 1915

Dear Colleague,

We are now just dealing with the case that absolute zero is attainable through finite repetitions of your process.[1] We would like to arrive at A. Well, you say:

Now we enlarge V more adiabatically until we have arrived at B. After this, isothermally until C, then adiabatically to B, etc. The conclusion thus is that the isotherm $B\,C$ corresponds no more to an entropy difference than the adiabat $C\,B$ does. *q.e.d.*

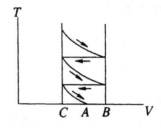

This proof applies in the case *that the adiabats $A\,B$ or $C\,B$ actually exist.* E.g., in the case of a change in volume of a solid body, it is difficult to doubt the existence of that zero-adiabat. But whether such kinds of adiabats always exist is very questionable. Then the *petitio principii* lies in the assumption of the existence of the zero-adiabat. There are cases in which this existence (in principle) is highly improbable.

Think of a fixed solution, in which the [. . .][2]

An *adiabatic* reversible change in volume of solution V at absolute zero obviously does not exist here *in principle.*[3]

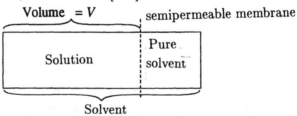

This is suspicious, anyway, in all cases where the adiabats touch the $(T = 0)$-axis at angles other than 0 [degrees].

With greetings from your colleague

A. Einstein.

I wish you luck in performing your military service.

56.   To Michele Besso

[Berlin,] (13 Wittelsbacher St.) 12 February [1915][1]

Dear Michele,

Your postcard is stirring up my guilty conscience again; I am writing now before it has subsided again. I was worn out by influenza; but now it's over.[2] Miza's address is: 30 Volta St., Zurich. I am very satisfied with the separation, despite hearing about my boys only extremely rarely. The peace and tranquillity does me tremendous good, and no less so the extremely agreeable, really fine relationship with my cousin,[3] the permanent nature of which is guaranteed by a renunciation of marriage. Presently Natanson from Lvov, theor. physicist[4] and fellow kinsman, of whom I have become very fond, is here. I corresponded with your friend of former times Straneo (now in Rome) about gravitation.[5] This theory is being received with very great interest by our Italian colleagues.[6] I am also corresponding avidly with Lorentz about it.[7]

As regards science, I have two fine matters to report to you:

1) Gravitation. Redshift of spectral lines. (Spectroscopic) binary stars have the same mean velocity within the radius of vision. The mass of the stars results from the Doppler periodic line fluctuation. The component of larger mass against the component of smaller mass ought to exhibit a *mean* redshift in the spectr. lines.[8] *This is confirmed*, since the stars' radii can also be estimated (apparently from the ⟨light intensity and⟩ spectral type); thus there is even an approximate quantitative test of the theory, with satisfactory results.

2) Experimental confirmation of Ampère's molecular current hypothesis. If paramagnetic molecules are spinning-top electrons [*Elektronenkreisel*], then each magnetic moment I corresponds to mechanical angular momentum $\mathfrak{M}$ in the same orientation of the magnitude[9]

$$\mathfrak{M} = 1.13 \cdot 10^{-7} I.$$

As I changes, a rotation moment appears $\left(-\dfrac{d\mathfrak{M}}{dt}\right)$. When the magnetization of a small hanging rod is inverted, it experiences an axial rotation moment, the existence of which I, together with Mr. de Haas (Lorentz's son-in-law), have proven with experiments performed at the Reich Institute. The experiment will be coming to an end soon.[10] With it the existence of zero point energy has also been proven in a single instance.[11] A wonderful experiment; what a pity that you can't see it.   And how traitorous nature is, when you want to deal with it experimentally! Experimenting is becoming a passion for me even in my old age.

Warm greetings also to Anna and Vero, yours,

Albert.

57. To Georg Nicolai

Berlin, 20 February 1915

Highly esteemed Sir,[1]

Although I am convinced that the voice of a handful of the informed carries little weight against the lust for power of the mighty and the fanaticism of the many, I still welcome your manifesto with pleasure and am pleased to be permitted to add my name to it. The manifesto will contribute in any case toward reestablishing the former good relations among researchers of civilized countries, which have already been clouded by some vehement and rash words.[2]

In accordance with your request I name here some men who, in my view, would want to participate in your desirable enterprise:

Prof. Diels, Berlin W. 50, 65 Nürnberger St.
Prof. Morf, Halensee, Berlin, 100 Kurfürstendamm
Prof. Planck, Grunewald, Berlin, 21 Wangenheim St.
Prof. Waldeyer, Berlin W. 62, 35 Luther St.

These four gentlemen are members of the local Academy,[3] and I am familiar with their broad-mindedness and good will.

I recommend you furthermore to the following men in my hometown, Zurich:

Prof. Stodola, Fed. Polytechnic.
Prof. H. Zangger, Zurich University.

For these two,[4] all matters concerning true social progress are always close to their hearts.

With all due respect and best wishes for the success of the good undertaking, yours very truly,

A. Einstein.

58.   To Mileva Einstein-Marić

[Berlin, 1 March 1915]

D[ear] M[ileva],

Under separate cover I am instructing the Kantonalbank to transfer the bonds as well as the savings bankbook to your name. I request that you do the same with the papers in Prague.[1]

I was *extremely* pleased with the pictures of the dear children. I thank you for that. You will receive the curtains.[2] With best regards,

A. Einstein.

59.   To Erwin Freundlich

[Berlin, between 1 and 25 March 1915][1]

Dear Mr. Freundlich,

I shall be glad to present your paper[2] to the Academy and to say at that time "that it would be very desirable if your efforts were met with support by astronomers," if you clarify for me satisfactorily the following points:

1) Please give me brief characterizations of the spectral classes of fixed stars.

2) Why do you say that the mass function

$$f = \frac{m_2^3 \sin^3 i}{(m_1 + m_2)^2}$$

for type B stars, which as you know ought to have particularly *large* mass, "takes on particularly *small* values." Surely the opposite would be expected.[3]

3) Why have you included in the examination types A and K, which are supposedly completely different spectrally?[4]

4) I do not find it justified to set the mean value

$$\overline{m_1 + m_2} = 14\odot$$

as the mean mass of simple type B stars.[5]

$$\overline{m_1} = 7\odot$$

could be taken instead; because if the mass is critical at the developmental stage, then for binary stars the mass of the components is much more critical than the overall mass.

5) You should provide me with a bit more precise information on *how* the radii of the eclipses are estimated.[6]

6) Is the absence of the effect really proven for types that yield smaller masses using the binary-star method? Why do you pass over this important question so quickly?[7]

7) Did you keep yourself open to the hypothesis that for stars with only *one* visible component, both components have masses in the same order of magnitude?

It is a shame that your descriptions are not detailed enough to be able to es-timate the uncertainty attached to your estimates. Thus a nonspecialist cannot get a notion of the reliability of your calculations. A much more in-depth pre-sentation would be desirable. The worst in this regard is the specification of the mean densities.

In the formula in the fourth column of the proofs, "$\sqrt{0.1}$" is inadvertently indicated[8] instead of "$\sqrt[3]{0.1}$."

With best regards, yours,

Einstein.

60.   To Tullio Levi-Civita

[Berlin,] 5 March 1915

Highly esteemed Colleague,[1]

By examining my paper so carefully, you are doing me a great favor.[2] You can imagine how rarely someone delves independently and critically into this subject. I also cannot help admiring the uncommon sureness with which you make use of a language that is foreign to you.

When I saw that you are directing your attack against the theory's most important proof, which I had won by the sweat of my brow, I was not a little alarmed, especially since I know that you have a much better command of these mathematical matters than I. Nevertheless, upon thorough consideration I do believe I can uphold my proof.[3]

I start with the second part of your letter in which you intend to show with an example that the result of §14, that $\dfrac{\mathfrak{E}_{\mu\nu}}{\sqrt{-g}}$ is a tensor, is not correct. For this purpose you specially choose

$$H = g_{11},$$

and show that in this case, the value calculated according to (72) and (73),

$$\frac{1}{\sqrt{-g}} \sum \delta g^{\mu\nu} \mathfrak{E}_{\mu\nu},$$

is not an invariant.

On page 1069, however, it is pointed out that H must be chosen in such a way that it is invariant for *linear* substitutions. Without this precondition, formula (65), which is fundamental to what follows, obviously is not valid.[4] As g_{11} is not an invariant for linear substitutions, your counterexample does not constitute a refutation of my stated theorem.–

As far as the first part of your letter is concerned, I do not see why the conclusion drawn from (71) ought not apply. Variation calculations are always carried out in the way in which I have done it.

I know that

$$\int d\tau \sum_{\mu\nu} (\delta g^{\mu\nu} \mathfrak{E}_{\mu\nu}) \ldots \tag{71}$$

is an invariant when the boundary conditions for the $\delta g^{\mu\nu}$'s are observed, regardless of how the $\delta g^{\mu\nu}$'s are chosen.[5]

Now, let the $\delta g^{\mu\nu}$'s differ from zero only inside an ∞ *small* area σ. The $\mathfrak{E}_{\mu\nu}$'s may be treated as constant in the integration. If you set

$$\int_{\sigma} d\tau = \tau$$

and

$$\int \delta g^{\mu\nu} d\tau = \overline{\delta g^{\mu\nu}} \tau,$$

whereby the $\overline{\delta g^{\mu\nu}}$'s signify spatial mean values of the $\delta g^{\mu\nu}$ terms, then it is possible to set instead of (71)

$$\tau \sum \overline{\delta g^{\mu\nu}} \mathfrak{E}_{\mu\nu}.$$

The theorem follows from this, taking into consideration that owing to the smallness of area σ, the $\overline{\delta g^{\mu\nu}}$'s transform at one place within σ, like the $\delta g^{\mu\nu}$'s, i.e., also like the $g^{\mu\nu}$'s.–

I urge you earnestly to inform me of your opinion of the proof upon reconsideration.

With cordial greetings, yours very truly,

A. Einstein.

61.   To Wander de Haas

[Berlin, 17 March 1915]

Dear de Haas,

According to your method, even the inducing influence of the core poles' demagnetization field can be virtually eliminated by placing the identical, oppositely wound coils right next to each other:[1]

$$\underline{H'}\ \underline{B}$$
$$I = \underline{B + H}$$
$$B = H + I$$
$$I = \underline{B - H}$$

The little rod's magnetic field produces in both coils almost the same amount of oppositely oriented e.m.f. The little coils must have the smallest diameter possible, however.

With best regards, yours,

A. Einstein.

62.   To Tullio Levi-Civita

[Berlin,] 17 March 1915

Dear Colleague,

I shall be delighted if next time you write me in Italian. I spent over half a year in Italy as a young man[1] and at that time had the pleasure of visiting the charming little town of Padova, and even now I still enjoy being able to apply my modest knowledge of the Italian language. On the other hand, I could not muster up the courage to write you in Italian, because the result would be far too clumsy and unclear.

I shall limit myself to responding to your last letter. We can go back to the earlier one anytime, of course. Your attack culminates in the argument that in reducing region Σ, in which the terms $\delta g^{\mu\nu} \neq 0$, in general, the mean values $\overline{\delta g^{\mu\nu}}$ do not approach a limit.[2] As an example you provide the function

$$\lambda = \lambda_0 \left\{ 1 - \left(\frac{x - x_0}{h} \right)^2 \right\}^2 .$$

⟨In order to arrive at a limit, one would only have to set⟩ I do not want to go into the issue of the existence or nonexistence of this limit at all, as this question is, in my view, of no importance; rather, I shall attempt to prove that the conclusion drawn from formula (71) of the paper is correct.[3] We can then refer our later observations to this proof.

Assumption:

$$\delta I = \int_{\Sigma} d\tau \sum \mathfrak{E}_{\mu\nu} \delta g^{\mu\nu} \ldots \tag{1}$$

is invariant, where the $\delta g^{\mu\nu}$'s vanish at the boundaries of Σ, but are otherwise freely selectable.

Posit:

$$\frac{\mathfrak{E}_{\mu\nu}}{\sqrt{-g}} \text{ is a covariant tensor.}$$

Proof:

As

$$g'^{\rho\sigma} = \sum \frac{\partial x'_\rho}{\partial x_\mu} \frac{\partial x'_\sigma}{\partial x_\nu} g^{\mu\nu} \ldots \tag{2}$$

thus

$$\delta g'^{\rho\sigma} = \sum \frac{\partial x'_\rho}{\partial x_\mu} \frac{\partial x'_\sigma}{\partial x_\nu} \delta g^{\mu\nu} \ldots \tag{2a}$$

I multiply (2a) by $\sqrt{-g'}d\tau' = \sqrt{-g}d\tau$ and integrate over Σ. Then I obtain

$$\int \sqrt{-g'} \delta g'^{\rho\sigma} d\tau' = \int \sqrt{-g} d\tau \sum \frac{\partial x'_\rho}{\partial x_\mu} \frac{\partial x'_\sigma}{\partial x_\nu} \delta g^{\mu\nu} \ldots \tag{3}$$

Since integration region Σ is supposed to be infinitely small, I can substitute the factors $\dfrac{\partial x'_\rho}{\partial x_\mu} \dfrac{\partial x'_\sigma}{\partial x_\nu}$ with the (constant) values that these factors obtain for any one place in the integration space. In this way I disregard only a relatively infinitely small amount on the right-hand side of (3). Therefore, if I set as a short-cut

$$A^{\mu\nu} = \int_{\Sigma} \sqrt{-g} \delta g^{\mu\nu} d\tau \ldots \tag{4}$$

it is:

$$A'^{\rho\sigma} = \sum \frac{\partial x'_\rho}{\partial x_\mu} \frac{\partial x'_\sigma}{\partial x_\nu} A^{\mu\nu} \ldots \tag{3a}$$

Thus it is proven first of all that $A^{\mu\nu}$ is a contravariant tensor and, in particular, one with components that can be selected independently from one another.–[4]

Now I am going to reformulate the expression for δI. Since $\dfrac{\mathfrak{E}_{\mu\nu}}{\sqrt{-g}}$ is a "slowly" variable function of the coordinates, I can treat the $\dfrac{\mathfrak{E}_{\mu\nu}}{\sqrt{-g}}$ quantities on the right-hand side of (1) in the integral as constants, disregarding infinitely small amounts of higher order and set

$$\delta I = \sum_{\mu\nu}\left\{ \frac{\mathfrak{E}_{\mu\nu}}{\sqrt{-g}} \int_{\Sigma} \sqrt{-g}\,\delta g^{\mu\nu}\,d\tau \right\}$$

or

$$\delta I = \sum \frac{\mathfrak{E}_{\mu\nu}}{\sqrt{-g}} A^{\mu\nu} \ldots \tag{1a}$$

Since, according to the assumption δI is an invariant and, according to the above, $A^{\mu\nu}$ is a contravariant tensor with independently selectable components, thus

$$\frac{\mathfrak{E}_{\mu\nu}}{\sqrt{-g}}$$

is a covariant tensor.–

   In the cheerful hope that you will not find any significant holes in this proof, and with best regards, I am yours,

A. Einstein.

63.   To Erwin Freundlich

[Berlin, 19 March 1915]

Dear Freundlich,
   I have thought about the question of whether matter can produce a gravitational field other than a g_{44} field. This is not the case.[1]
   The question is whether the integral covering the mass (e.g., Sun)

$$\int \mathfrak{T}_1^1\,dx\,dy\,dz$$

becomes infinitesimal. Within the accuracy applicable here[2]

$$\frac{\partial \mathfrak{T}_1^1}{\partial x} + \frac{\partial \mathfrak{T}_1^2}{\partial y} + \frac{\partial \mathfrak{T}_1^3}{\partial z} + \frac{\partial \mathfrak{T}_1^4}{\partial t} = 0.$$

$\dfrac{\partial \mathfrak{T}_1^4}{\partial t}$ vanishes in the static case. If I multiply the equation by x and integrate over the entire body, then

$$\int x \left(\frac{\partial \mathfrak{T}_1^1}{\partial x} + \frac{\partial \mathfrak{T}_1^2}{\partial y} + \frac{\partial \mathfrak{T}_1^3}{\partial z} \right) dx \, dy \, dz = 0$$

changes through partial integration to

$$\int \mathfrak{T}_1^1 dx \, dy \, dz = 0,$$

as Laue has already recognized.[3] Thus it is proven that for the planetary problem[4] a g_{11} field cannot come into consideration.[5]

     With best regards, yours,

Einstein.

64.   To Tullio Levi-Civita

[Berlin, 20 March 1915]

Dear Colleague,

     I have received your letter with the counterargument to disprove the tensorial character of $\dfrac{1}{\sqrt{-g}}\, \mathfrak{E}_{\mu\nu}$, which bases itself on the case where among the adapted coordinate systems such exist whose $g_{\mu\nu}$'s are constant.[1]

     You say under (2)

"This tensor is, to the contrary, not identical to zero for *all* adapted coordinate systems; this is especially obvious in Newton's case."[2]

     I do not consider this correct, however. It should be noted that, in general, it is not possible to alter through transformation of the coordinates any given $g_{\mu\nu}$ field into one in which the $g_{\mu\nu}$'s are constant. This will always be impossible for parts of a Newtonian field containing masses, for ex.; this is also generally true of mass-free regions, incidentally.

     I believe that the proof or disproof of the statement quoted from your letter is just as problematic as the proof or disproof of the *general* statement of the tensor character of $\dfrac{\mathfrak{E}_{\mu\nu}}{\sqrt{-g}}$.

     With cordial greetings and hoping for a prompt reply, yours,

A. Einstein.

65.   To Romain Rolland

Berlin, 15 Wittelsbacher St., 22 March 1915

Highly esteemed Sir,[1]

From the daily paper and through my connections with the highly creditable "[New] Fatherland" League I learned of how courageously you have placed your life and person at risk toward eliminating the so ominous misunderstandings between the French and German peoples.[2] I feel compelled to express my complete admiration and respect. May your magnificent example awaken other prominent men from the inexplicable delusion that has gripped like an insidious epidemic even competent men of otherwise steady reason and healthy sentiment! After three hundred years of assiduous cultural activity, must our Europe really be remembered in coming centuries for not having advanced further than from religious mania to patriotic mania? Even scholars of the various nations behave as if their cerebrums had been amputated ⟨one⟩ 8 months ago.[3]

I place my feeble powers at your disposal in case you think I can be of service to you as a tool, be it through my residence or through my connections with German and foreign individuals in the exact sciences.

With genuine admiration, yours very devotedly,

A. Einstein,
Mem. of the Prus. Acad. of Sci.

P.S. Should you be in need of the assistance of a man completely versed in the local circumstances in Switzerland, my homeland, or in need of heartwarming company, I refer you to Prof. Zangger, Berg St., Zurich, Prof. of For[ensic] Medicine at the local university.[4]

66.   To Tullio Levi-Civita

[Berlin,] 13 Wittelsbacher St., 26 March [1915][1]

Dear Colleague,

I have just received your letter of March 23rd written in the so-familiar Italian, of which I have been deprived for so long. You can hardly imagine what a pleasure it gives me to receive such a genuine Italian letter. While reading, the finest memories of my youth come alive.[2] You also do a very charming thing in your letters: First you flatter me nicely, to prevent me from making a dour face upon reading your new objections.

First of all, once again to the objection relating to the special case that with an appropriate choice of coordinates the $g_{\mu\nu}$'s are constant.[3] You state that in this case the $\mathfrak{E}_{\mu\nu}$'s do not vanish if the adapted system is selected in such a way that the $g_{\mu\nu}$'s are *not* constant. But you did not support this statement, and I am considering it incorrect as long as you have not provided an example or a general proof for this.

Now to the second objection. Although you concede that

$$\sum_{\mu\nu} \frac{1}{\sqrt{-g}} \mathfrak{E}_{\mu\nu} A^{\mu\nu} + \varepsilon = \text{invariant}, \left(A^{\mu\nu} = \int \delta g^{\mu\nu} \sqrt{-g} d\tau\right),$$

where ε is an infinitesimal amount of higher order, you contest, however, that from this it can be concluded that $\dfrac{\mathfrak{E}_{\mu\nu}}{\sqrt{-g}}$ is a tensor.

Well, I must admit that the basis of this objection is not clear to me. For, since I have proven that the $A^{\mu\nu}$'s transform contravariantly,[4] in my opinion it is quite irrelevant to the proof that these $A^{\mu\nu}$'s are infinitesimal quantities. Nevertheless, an addition must be made to the proof that instead of an infinitesimal tensor $A^{\mu\nu}$, a finite one is introduced through the formation of a limit. I shall not go into this, however, but am convinced that you will acknowledge the following derivation as sound:

Up to a relatively ∞ small quantity,

$$\sum \frac{1}{\sqrt{-g}} \mathfrak{E}_{\mu\nu} A^{\mu\nu} = \sum \frac{1}{\sqrt{-g'}} \mathfrak{E}'_{\sigma\tau} A'^{\sigma\tau}, \; \ldots \tag{1}$$

for any justified substitution, as you yourself concede.

Up to a relatively ∞ small quantity—as you likewise concede—and as i have proven in my last letter,[5]

$$A'^{\sigma\tau} = \sum \frac{\partial x'_\sigma}{\partial x_\mu} \frac{\partial x'_\tau}{\partial x_\nu} A^{\mu\nu} \; \ldots \tag{2}$$

From (1) and (2) it follows, however, that up to a relatively ∞ small quantity, the equation

$$\sum_{\mu\nu} \frac{\mathfrak{E}_{\mu\nu}}{\sqrt{-g}} A^{\mu\nu} = \sum_{\sigma\tau\mu\nu} \frac{\mathfrak{E}'_{\sigma\tau}}{\sqrt{-g'}} \frac{\partial x'_\sigma}{\partial x_\mu} \frac{\partial x'_\tau}{\partial x_\nu} A^{\mu\nu}$$

thus also applies. Since the $A^{\mu\nu}$'s can be chosen independently of one another, hence

$$\frac{\mathfrak{E}_{\mu\nu}}{\sqrt{-g}} = \sum_{\sigma\tau} \frac{\partial x'_\sigma}{\partial x_\mu} \frac{\partial x'_\tau}{\partial x_\nu} \frac{\mathfrak{E}'_{\sigma\tau}}{\sqrt{-g'}}$$

applies, i.e., $\dfrac{\mathfrak{C}_{\mu\nu}}{\sqrt{-g}}$ has a tensor character.–

It is thus quite superfluous to introduce some limit to replace the $A^{\mu\nu}$'s; the derivation is independent of it. Any consideration in which the $\delta g^{\mu\nu}$'s are subjected to more constraints than is necessary for the nature of the problem must be rejected as an unnecessary complication.

With cordial greetings, yours,

A. Einstein.

67.   From Tullio Levi-Civita

[Padova,] 28 March 1915

Dearest Colleague,

Yesterday I received your postcard of the 20th,[1] and it is with much pleasure that I reply to it immediately, as you requested.

In your view, my observation [that there are adapted (*angepasste*) coordinate systems for which tensor $\mathfrak{C}_{\mu\nu}$ does not vanish, while it is zero when the $g_{\mu\nu}$'s are constant][2] is not conclusive, because a generic gravitational field cannot be obtained with the help of a coordinate transformation starting from a Euclidean ds^2 ($g_{\mu\nu}$ constant).

This is indeed the case. Therefore it is necessary to give a concrete example in which not all $\mathfrak{C}_{\mu\nu}$'s vanish as a result of some admissible transformation from a Euclidean ds^2 (contrary to what covariance would require).

This is how we go about it.

We start from a coordinate system in which ds^2 has the canonical Euclidean form

$$ds^2 = dx_1^2 + dx_2^2 + dx_3^2 + dx_4^2$$

$$(g_{\mu\nu} = \delta_{\mu\nu}),$$

and we proceed to perform an infinitesimal transformation, putting

$$x'_\mu = x_\mu + y_\mu, \tag{1}$$

where the y_μ's designate (*a priori* any) infinitesimal functions of x.

Putting

$$\delta_{\mu\nu} + h_{\mu\nu} \tag{2}$$

for the $g_{\mu\nu}$'s relative to the new variables x', we have

$$h_{\mu\nu} = -\left(\frac{\partial y_\mu}{\partial x_\nu} + \frac{\partial y_\nu}{\partial x_\mu}\right). \tag{3}$$

Moreover, with your expression (78) for $H^{[3]}$ and, of course, always omitting terms of higher order,

$$\frac{\partial(H\sqrt{g})}{\partial g_\sigma^{(\mu\nu)}} = \frac{1}{2}\frac{\partial h_{\mu\nu}}{\partial x_\sigma}.$$

It follows (changing g to $-g$)[4]

$$\mathfrak{E}_{\mu\nu} = \frac{\partial(H\sqrt{g})}{\partial g^{(\mu\nu)}} - \sum_\sigma \frac{\partial}{\partial x_\sigma}\left(\frac{\partial H\sqrt{g}}{\partial g_\sigma^{(\mu\nu)}}\right) = -\frac{1}{2}\Delta_2 h_{\mu\nu}, \qquad (4)$$

where for brevity we designate with Δ_2 the differential operator (Laplace's)

$$\sum_1^4 \sigma\frac{\partial^2}{\partial x_\sigma^2}.$$

Having stated this general theorem, we note that infinitesimal transformation (1) will be adapted if the y's satisfy the four equations (65a)[5]

$$B_\mu = \sum_{\alpha\sigma\nu} \frac{\partial^2}{\partial x_\alpha \partial x_\sigma}\left(g^{(\nu\alpha)}\frac{\partial H\sqrt{g}}{\partial g_\sigma^{(\mu\nu)}}\right) = 0,$$

which in the present case become

$$\frac{1}{2}\sum_\nu \frac{\partial}{\partial x_\nu}\Delta_2 h_{\mu\nu} = 0. \qquad (5)$$

Because of the expressions (3) for the $h_{\mu\nu}$'s, these are linear equations, homogeneous and of the *fourth* order in y. They are therefore satisfied if for the y's we take some arbitrary polynomials of the *third* degree in x.

We assume, in particular,

$$y_\mu = \frac{1}{6}c_\mu x_\mu^3,$$

with constant c_μ (not zero).

We evidently obtain

$$h_{\mu\mu} = -c_\mu x_\mu^2,$$
$$\Delta_2 h_{\mu\mu} = -2c_\mu;$$

therefore

$$\mathfrak{E}_{\mu\mu} = +c_\mu \neq 0,$$

$$q.e.d.$$

I take the opportunity of sending you, with my compliments, a paper of which I have just received the offprints.

With warm regards, I remain yours sincerely,

T. Levi-Civita.

68.   From Romain Rolland

Beauséjour, Genève-Champel, Sunday, 28 March 1915

Dear Sir,

Your generous letter[1] touched me profoundly.

This terrible crisis has probably been a harsh lesson for all of us Europeans, writers, thinkers, and scholars alike. We ought never to have permitted it to catch us so unprepared. In the future we have to be better armed against a recurrence of such a scourge: (for we cannot delude ourselves that this folly of humanity was the last; but at least we should see to it that the intellectual élite does not participate in it anymore). Right after the announcement of this upheaval, those of us whose age excused them from military service should at the outset have delegated some from among themselves to meet in a neutral country and there together endeavored to shed light on the facts, in their own spirit of tempering the impassioned assertions by both camps and offering the voice of reason: remaining, in a word, the lucid and steadfast conscience of their nations.

Yes, we have erred. We have lived too much in the carefree or arrogant illusion that we would always be strong enough to resist the derangements of the community as a whole. The events of these last months demonstrate to us our error and impose upon us our mission. This indispensable task will be to organize ourselves later in a more European—that is to say, a truly universal—manner. And undoubtedly this will be more difficult after the war than before: because misunderstandings, grudges, and bitterness will subsist for a long time. But to start with, it suffices for a small group from all the nations to have the will to create this union. The others will follow bit by bit.—Besides, I retain the hope that after the immense sufferings and delirium of these months, a reaction will follow and the nations will reawaken ashamed, bruised, and repentant.

In the meantime, we can only maintain our composure in the storm and our belief. Little by little it will make itself felt.

You have my greatest and sincere sympathy,

Romain Rolland.

I believe that one of our most effective tasks must be to disseminate documents that can oppose the spirit of hatred.—In this sense I take the liberty of drawing your attention to the report (which will be published) by the Swiss Lieutenant Colonel de Marval, delegate of the International Red Cross, about the German prisoner camps he has just been visiting in France, Corsica, Algeria, and Tunisia.[2] Yesterday I listened to a lecture by him (including slide projections), and I would like it if many Germans could listen to it as well.[3] It would be very good if Mr. de Marval repeated the lecture in many German cities. His evidence would not be suspect, as one of his brothers is an officer in the German Army.

69.   To Tullio Levi-Civita

[Berlin,] 2 April 1915

Highly esteemed Colleague,

You letter of the 28th of March was extraordinarily interesting for me.[1] I had to ponder for one and a half days without interruption before it became clear to me how to reconcile your example with my proof. I enclose your letter so that I can refer to it without any inconvenience to you.

Your deduction is entirely correct. $\dfrac{1}{\sqrt{g}}\mathfrak{E}_{\mu\nu}$ does *not* have a tensor character with the infinitesimal transformation envisaged by you, even though the transformation follows from a justified coordinate system. Oddly enough, my proof is *not* refuted by this for the following reason: My proof fails in precisely that special case you have addressed.

In order for the tensor character of $\dfrac{1}{\sqrt{g}}\mathfrak{E}_{\mu\nu}$ to follow from the assumption[2]

$$B_\mu = 0 \qquad \Delta B_\mu = 0, \cdot \cdot \cdot \tag{1}$$

as well as from the conclusion drawn from it

$$\int d\tau \sum \delta g^{\mu\nu}\mathfrak{E}_{\mu\nu} = \text{invariant}$$

for the infinitesimal substitution under examination, it is necessary that the $\delta g^{\mu\nu}$'s be freely selectable, the fulfillment of the region's boundary conditions notwithstanding. To be more precise, this does not even have to apply to the $\delta g^{\mu\nu}$'s, but only to the integrals

$$\int \sqrt{g}\,\delta g^{\mu\nu}\,d\tau,$$

which as proven earlier, of course, have a tensor character if the chosen integration area is infinitely small.[3] Of equal importance is the requirement that for a given infinitesimal integration region, one must be able to *choose* the integrals *freely*

$$\int \delta g^{\mu\nu}\,d\tau.$$

If this is not the case, then the tensorial character of $\dfrac{\mathfrak{E}_{\mu\nu}}{\sqrt{g}}$ can*not* be concluded from (1).

In the special case examined by you, however, the $\int \delta g^{\mu\nu} d\tau$ quantities are not only not freely choosable, but they even vanish altogether.   Just as in §14, I set[4]

$$\delta g^{\mu\nu} = \delta_1 g^{\mu\nu} + \delta_2 g^{\mu\nu}.$$

The $\delta_1 g^{\mu\nu}$'s must meet the condition

$$\delta_1 B_\mu = 0,$$

which in the special case considered by you takes on the form

$$\sum_\nu \frac{\partial}{\partial x_\nu}(\Box \delta_1 g^{\mu\nu}) = 0, \text{ (comp. equation (5) of your letter)},$$

where $\Box$ is the Laplacian operator. From this follows, as is known, because of the boundary conditions,[5]

$$\sum \frac{\partial \delta_1 g^{\mu\nu}}{\partial x_\nu} = 0.$$

This is characteristic of your special case. If you multiply this equation by x_σ and integrate it over the entire region, then, upon partial integration of each index combination,

$$\int \delta_1 g^{\mu\sigma} d\tau = 0 \cdots \tag{I}$$

follows. This is a consequence of the peculiar degeneration of the equation $\delta_1 B_\mu = 0$ in the special case you have raised.

It follows furthermore from formula (63) of the paper[6] and the definition of the $\delta_2 g^{\mu\nu}$'s for an infinitesimal region in general,

$$\int \delta_2 g^{\mu\nu} d\tau = 0 \cdots \tag{II}$$

Thus from (I) and (II)

$$\int \delta g^{\mu\nu} d\tau = 0$$

follows for each index combination.

Therefore, it is inherent to the specialization you have introduced that formulas (I) of this letter are *not* adequate conditions to lend a tensor character to $\dfrac{\mathfrak{E}_{\mu\nu}}{\sqrt{g}}$ under an infinitesimal transformation.

Generally, however, equation $\delta_1 B_\mu = 0$ cannot be reduced to a first-order equation for the $\delta_1 g^{\mu\nu}$'s. Then my proof indeed holds true for all finite transformations.

This observation suggests a modification of my covariance proof in which *only* the δ_1 variations are used, since the δ_2's contribute nothing to the decisive quantities,

$$\int \sqrt{g}\delta g^{\mu\nu} d\tau.$$

With cordial greetings, yours very truly,

A. Einstein.

I have never experienced such an interesting correspondence. You should see how much I always look forward to your letters.

Many thanks for your article.

70. To Hans Albert Einstein

[Berlin, before 4 April 1915][1]

My dearest Adn,

You already wrote me two little letters that pleased me very much. Did you write them nicely by yourself as well? We can't see each other at Easter now. But in the summer I'll take a trip just with you alone for a fortnight or three weeks. This will happen every year, and Tete may also come along when he is old enough for it. I don't know yet where we'll go hiking, perhaps to Italy. Then I'll also tell you many fine and interesting things about science and much else. I am very glad that you are enjoying geometry.[2] It was my favorite pastime when I was already a bit older than you, around 12 years old. But I had no one to demonstrate anything to me, I had to learn it from books.[3] I would immensely enjoy being able to teach it to you, but that's not possible. If you write me each time what you already know, I'll give you a nice little problem to solve. I gave mathematics lessons to the little Haber boy because he was sick for a long time.[4] He is a smart boy, but ailing.

Today I'm sending off a package with some toys for you and Tete. It is as an Easter greeting. Don't neglect your piano, my Adn; you wouldn't believe how much pleasure you can give to others as well as to yourself when you are able to play music nicely![5] What are your friends like?

Another thing, dear Adn. Brush your teeth every day, and if a tooth is not quite all right go to the dentist immediately. I also do the same and am very happy now to have kept enough healthy teeth. This is *very important*, later you will realize it yourself.

Affectionate kisses to you and Tete, and regards also to Mama, yours,

Papa,

who often thinks of you.

71.   To Tullio Levi-Civita

[Berlin,] 8 April 1915

Highly esteemed and dear Colleague,

I see from your postcard of the 2nd of April that you are clinging to your limit objection.[1] I shall try to disprove it, more in order gradually to come to know the main point of your reservations.

I select the $\delta g^{\mu\nu}$'s in the following way, as shown in the figure in cross section. Within an inner zone σ, which makes up almost the entire region, let $\delta g^{\mu\nu}$ be equal to the constant $\gamma^{\mu\nu}$. A boundary zone brings about the gradual transition to zero. $\dfrac{\Sigma - \sigma}{\Sigma}$ is a genuine break,

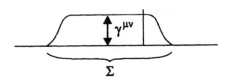

which I can make arbitrarily small without injuring the continuity conditions.

In the proof, I then proceed to $\Sigma = 0$ such that I complete the transition at constant $\gamma^{\mu\nu}$ and constant $\dfrac{\Sigma - \sigma}{\Sigma}$.

The smaller I choose the quantity $\dfrac{\Sigma - \sigma}{\Sigma}$, the more accurately the equation

$$A^{\mu\nu} = \int \sqrt{-g}\, \delta g^{\mu\nu} d\tau = \sqrt{-g}\, \gamma^{\mu\nu} \Sigma$$

is satisfied, the closer in accuracy is the validity of the following observation. If I divide the equation[2]

$$\sum_{\mu\nu} \frac{1}{\sqrt{-g}} \mathfrak{E}_{\mu\nu} A^{\mu\nu} = \sum \frac{1}{\sqrt{-g'}} \mathfrak{E}'_{\sigma\tau} \frac{\partial x'_\sigma}{\partial x_\mu} \frac{\partial x'_\tau}{\partial x_\nu} A^{\mu\nu}$$

by the region $\int d\tau = \Sigma$ and if I select the $\delta g^{\mu\nu}$'s according to the above determination, thus with the aid of the indicated expression for the $A^{\mu\nu}$'s results with greater accuracy, the smaller the selected $\dfrac{\Sigma - \sigma}{\Sigma}$ is:

$$\sum \frac{\mathfrak{E}_{\mu\nu}}{\sqrt{-g}} \gamma^{\mu\nu} = \sum \frac{\mathfrak{E}'_{\sigma\tau}}{\sqrt{-g'}} \frac{\partial x'_\sigma}{\partial x_\mu} \frac{\partial x'_\tau}{\partial x_\nu} \gamma^{\mu\nu}.$$

Hence, because the $\gamma^{\mu\nu}$'s can be chosen arbitrarily,

$$\frac{\mathfrak{E}_{\mu\nu}}{\sqrt{-g}} = \sum \frac{\partial x'_\sigma}{\partial x_\mu} \frac{\partial x'_\tau}{\partial x_\nu} \frac{\mathfrak{E}'_{\sigma\tau}}{\sqrt{-g'}}.-$$

This consideration evidently cannot capture what you mean, because it is too obvious; but I hope nonetheless that by means of the same you will be able to lead me toward understanding more easily what moves you to object.–

You have probably received my letter refuting your example.[3] I shall repeat myself.

My proof of the invariant nature of δJ fails with such infinitesimal transformations in which the $g_{\mu\nu}$'s of the original system are constant, because then the quantities $A^{\mu\nu}$ cannot be chosen freely but vanish altogether. But this does *not* apply similarly to changing $g_{\mu\nu}$'s. Thus the proof does not fail generally, but only in certain special cases.

Since the example relates to such a special case, it proves nothing about the validity of the principle in general.

I must even admit that, through the in-depth considerations to which your interesting letters have led me, I have become only more firmly convinced that the proof of the tensor character of $\dfrac{\mathfrak{E}_{\mu\nu}}{\sqrt{-g}}$ is correct in principle.

With cordial greetings, yours,

Einstein.

72.   To Geertruida de Haas

[Berlin,] 13 Wittelsbacher St. [before 10 April 1915][1]

Dear Mrs. de Haas,

I have calculated by a new method the enclosed curve, which your dear husband recorded particularly meticulously.[2] Although the linearity of the damping term is used here, the legitimacy of this procedure is proven on the basis of the curve. It is odd that the small displacements seem so systematically incorrect; I can give no explanation for it.[3]

The good agreement with the theory is by chance, of course; but it is real inasmuch as now any doubts about the theory's accuracy are necessarily silenced.

I ask you please to send back the manuscript *as soon as possible* so that I can pass it on to Mr. Scheel.[4] Do not forget also to enclose the observed curve.

With cordial greetings to you and your little children, yours,

A. Einstein.

73. To Heinrich Zangger

[Berlin, ca. 10 April 1915][1]

Dear friend Zangger,

You have the patience of an angel not to be angry about my silence. But I console myself with the fact that your memory does not extend far enough back to determine the degree of my negligence reliably. I am now starting to feel comfortable in the present-day mad turmoil, in conscious detachment from all things that occupy the deranged public at large. Why is it not acceptable to live enjoyably as a member of the madhouse staff? All madmen are respected as those for whom the building that they are inhabiting is there. The asylum can be selected freely, to a certain degree—but the difference between them is less than you expect as a young man. Romain Rolland, who currently lives in Geneva, recently sent me a suggestion, which—to continue the metaphor—amounts to the organization of the sane staff at all asylums for the purpose that they not become deranged as well.[2] Besides, he has hopes that such an organization would even cure the madmen, more or less. The optimist! If you have the opportunity, look after him; he is being persecuted for his international mentality.[3]

Concerning science, this semester I worked on a wonderful experimental problem together with Lorentz's son-in-law at the Reich Institute.[4] We have supplied firm experimental proof of the existence of Ampère's molecular currents (explanation for para and ferromagnetism). Namely, if the molecule's magnetic moment is provided by rotating electrons in this manner

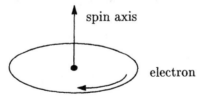

spin axis

electron

then, mechanically, the molecule is a gyroscope whose axis coincides with the magnetic one.

From this, through purely theoretical and mechanical means, we can draw the conclusion:

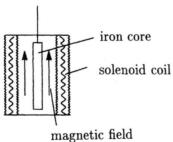

iron core

solenoid coil

magnetic field

If you commutate the magnetic poles of a little rotatable iron rod suspended within a coil by inverting the current, the rod receives a measurable and precisely predictable angular momentum.

The experiment provided detailed verification of the theory within the margin of error (about 10%). The paper will appear in a few weeks;[5] then I shall send you an offprint, of course.

The theory of gravitation will not find its way into my colleagues' heads for a long while yet, no doubt. Only *one*, Levi-Civita in Padova, has probably grasped the main point completely, because he is familiar with the mathematics used; but he is seeking to tamper with one of the most important proofs in an incessant exchange of correspondence. Corresponding with him is unusually interesting; it is currently my favorite pastime.[6]

Life without my wife is a veritable rebirth for me personally. It feels as if I had ten years of prison behind me. In matters of the emotions humans are so strange. And all this even though I love my boys more than I can say, who have now been taken away from me after all;[7] I carried them around countless times at night, took them out on walks in their pram, carried them on my shoulders, joked with them, explained things to them that their little minds were beginning to absorb, guided their attitude toward things—and now they are gone, my influence on them shrunk down to superficialities. My human and professional contacts are few but very harmonious and rewarding, my public life withdrawn and simple. I must say that to me I seem one of the happiest of persons.

I hope confidently that soon there will be peace; the madmen will soon turn their efforts to a more harmless field again. But they will remain just what they are. From where do you get your optimism to devote yourself with such resilience to the affairs of the general public? I admire it, but do not understand it. For rarely has anyone been treated so abominably out of conscious meanness as you; only educated beasts are capable of the like.[8] When I receive the promised copy of your essay,[9] I shall be glad to make notes on it, which naturally cannot yield anything *factual* for you, because I am a child in these matters; at best I am somewhat knowledgeable in logical composition.

How is your friend Huguenin?[10]

Affectionate greetings from your

Einstein.

74.   To Tullio Levi-Civita

[Berlin,] Sunday, [11 April 1915]

Dear Colleague,

You get a nice special case for the statement that $\dfrac{\mathfrak{E}_{\mu\nu}}{\sqrt{-g}}$ is a tensor[1] when you set $H = \text{const.}$ The condition $B_\mu = 0$ is then satisfied identically, and H

is invariant to arbitrary substs., thus also to linear ones. Therefore, according to the theorem in this case of arbitrary substitutions, $\dfrac{\mathfrak{E}_{\mu\nu}}{\sqrt{-g}}$ must have a tensor character. Indeed,

$$\frac{\mathfrak{E}_{\mu\nu}}{\sqrt{-g}} = \text{const} \cdot g_{\mu\nu}$$

results, which is thus in fact a covariant tensor.

With best regards, yours,

A. Einstein.

75.   To Tullio Levi-Civita

[Berlin,] 14 April 1915

Dear Colleague,

You pleased me greatly with your assentive postcard. When the occasion arises to repeat the $\mathfrak{E}_{\mu\nu}$ proof, I shall gladly include the corrections I have learned from our memorable correspondence.[1] Interest in the subject, however, is exceedingly modest for the time being. It is odd how few colleagues feel the intrinsic need for a *precise* theory of relativity. Unfortunately, the attitude of our nonprofessional fellow mortals is incomparably more peculiar. So it is doubly gratifying to have come to know more intimately a man such as you. I shall take pleasure in striving to allow our acquaintance by letter turn into a personal one: one more reason at last to cross the Alps again one day. It is to be hoped that our fatherlands will not rebel against each other as well![2]

With cordial greetings, yours,

A. Einstein.

76.   To Fritz Weishut

[Berlin, 18 April 1915]

Dear Mr. Weishut,[1]

I do not yet have the offprints of the paper but shall send them to you shortly.[2] This is the gist:

According to the theory of electrons, mechanically, every para and ferromagn. molecule is a circular current whose axis coincides with the magnetic axis. From this follows, taking into account the theorem of momentum [conservation], that upon magnetic commutation of a magnetizable body the molecules translate into one rotation moment or other, according to the formula

$$\text{rotation moment} = \frac{2 \text{ electron masses}}{\text{electron charge}} \cdot \frac{d}{dt}(\text{magnetization})$$

$$D = \frac{2\mu}{[e]} \frac{dM}{dt}.$$

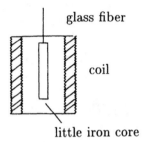

glass fiber

coil

little iron core

The experiments were done by observing the oscillation motions of a little iron rod that was hanging on a glass fiber within a coil through which an alternating current was flowing (use of resonance.)

77.   To Tullio Levi-Civita

[Berlin,] 20 April [1915]

Highly esteemed and dear Colleague,

Just now I was studying your very interesting letter of April 15th. Admittedly I do not concur—as is explained in more detail in the following—with your proof that the selection of the $\delta g^{\mu\nu}$'s as "quasi-constants" was unworkable; but I gladly acknowledge that you have put your finger on the weakest point of the proof, namely, on the independence of the $A^{(\mu\nu)}$'s. Here the proof lacks precision;[1] the statement on page 1072 of the paper "it will thus be equivalent to an *arbitrary* variation of the $\delta g^{\mu\nu}$'s" dispenses with rigorous substantiation; in the special case of a constant $g^{\mu\nu}$ it is even incorrect.[2] Yet I do cherish the firm belief that in general it applies, because the number of freely selectable variables that determine the 10 $\delta g^{\mu\nu}$'s is 10, and because both variations δ_1 and δ_2 are of fundamentally different kinds, in that a δ_2 variation generally is not a δ_1 variation.

Now to your proof. Right at the outset you say that for small regions Σ

$$A^{(\mu\nu)} = \int \delta g^{\mu\nu} d\tau = \int \delta_1 g^{\mu\nu} d\tau,$$

thus that the

$$\int \delta_2 g^{\mu\nu} d\tau$$

terms vanish. This already I contest,[3] namely for exactly the same reason that I regard your deduction of the vanishing of the

$$\int \delta_1 g^{\mu\nu} d\tau\text{'s}$$

as not sound.

In accordance with my proof, let

$$\delta g^{\mu\nu} = \delta_1 g^{\mu\nu} + \delta_2 g^{\mu\nu}$$

with the addition that $\delta g^{\mu\nu}$ should be a "quasi constant" whose value $(\gamma^{\mu\nu})$ remains unchanged by the act of forming the limit of the reduction of Σ.[4] *This is by no means to say that then the values for the $\delta_1 g^{\mu\nu}$ and $\delta_2 g^{\mu\nu}$ terms can increase more, the smaller Σ becomes.* Your method of proof implicitly assumes, however, that the quantities you have designated for $h^{\mu\nu}$ remain finite at vanishing x_v. For were this not the case, your quantities $\int K_\mu x_v^3 d\tau$ and $\int Q_\mu x_v^3 d\tau$ would not vanish at the limit.

One can say somewhat illogically: The smaller that Σ is, the more you have to struggle with the $\delta_1 g^{\mu\nu}$'s and $\delta_2 g^{\mu\nu}$'s in order to squeeze the arbitrarily given $\delta g^{\mu\nu}$'s out of the small differences in the $g^{\mu\nu}$'s within the region. Only when the $g^{\mu\nu}$'s are completely constant does this become impossible for the reason already discussed earlier.

Awaiting your answer as always with great interest, I am with cordial greetings, yours,

Einstein.

78.   To Tullio Levi-Civita

Berlin, 21 April [1915]

Dear Colleague,

I presented to you the example $H = $ const.[1] not to refute your new argument (re. the *independence* of the $A^{\mu\nu}$'s)[2] but to refute the original objection.[3] I wanted to show (with this example), that if

$$\int T^{\mu\nu} \sqrt{-g}\, \delta g_{\mu\nu} d\tau$$

is an invariant with freely selectable $\delta g_{\mu\nu}$'s vanishing at the boundary, that then $T^{\mu\nu}$ is a tensor (actually $\int T_{\mu\nu} \sqrt{-g}\, \delta g^{\mu\nu} d\tau$, which is the same, though).

From your postcard I see that you attach great weight to your *new* argument which culminates in the statement that the $A^{\mu\nu}$'s vanish, by which you once again find the theorem inapplicable. But I hope that the letter I sent off yesterday[4] will convince you that an impermissible limit exists there.

With cordial greetings, yours,

Einstein.

79.   To Hendrik A. Lorentz

[Berlin, 28 April 1915]

Highly esteemed Colleague,

I understand your telegram. Upon reviewing our sign determination, I see that there is an error in my considerations.[1] We have produced as closely as possible the case of perfect resonance. There the torque runs ahead of the swing in phase by $\frac{\pi}{2}$, which we did not take into account. It would have been correct to arrange the solenoid and the incandescent lamp in series. Then both current vectors would coincide, and the swing produced by the trembling of the filament would differ at perfect resonance in phase by π from the swing caused by the gyro effect according to the theory.[2]

Despite this oversight, however, the proof remains intact.[3] More out of luck than wit!! The corrected diagram looks like this:

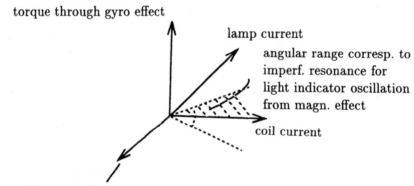

Approximate compensation for both light indicator oscillations is only possible with the correct sign of the effect.

With affectionate greetings, yours,

A. Einstein.

80.   To Tullio Levi-Civita

[Berlin,] 5 May 1915

Esteemed and dear Colleague,

I also believe that we have given our subject as full a treatment as our present state of knowledge of the same allows.[1] My proof is incomplete to the extent that it is not proven that the $A_{\mu\nu}$'s can be chosen arbitrarily.

Your formulation

$$\mathfrak{E}_{\mu\nu} = \frac{1}{2} g_{\mu\nu} \Delta\varphi - \frac{\partial\varphi}{\partial x_\mu} \frac{\partial\varphi}{\partial x_\nu}$$

cannot solve the problem in the sense of the relativity of motion. For if φ is constant for *one* coordinate system, then it is constant for all the others as well. Consequently none of the arbitrarily moving coordinate systems would have a gravitational field, which is impossible.–

Now I thank you heartily for your great patience and add the wish that we be able to make each other's personal acquaintance soon and in better times.[2]

With cordial greetings, very devotedly yours,

A. Einstein.

81.   To Michael Polányi

Berlin, 8 May 1915

Dear Colleague,

Your letter to me of February 25th indeed did not arrive. I am pleased that in this "great era" you have kept enough leisure and inclination to deliberate on scientific affairs.

Your letter shows me that we have arrived at the crux of the matter. You say: "The existence of zero-adiabats (and also isotherms) follows from the fact that the system must be determined from the dimension figures of the surroundings." However, this does not seem to me to apply to absolute zero in particular.[1]

Example:

solution (volume V)   semipermeable membrane

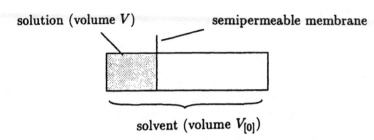

solvent (volume $V_{[0]}$)

If there is no diffusion at absolute zero, which is what I believe, then there can be no adiabatic enlargement of the solvent's volume at $T = 0$[2] (naturally no more than an isothermal one). A shifting of the semipermeable plane would not result in any diffusion (no matter how slowly it is carried out). In the case where the envisaged process is a chemical reaction, no more can I see the existence in principle of a "zero point reaction." It is my conviction, therefore, that nothing further can be achieved here from a purely thermodynamic approach.

On the other hand, there are even strong grounds for Nernst's theorem not being valid for mixed processes. Consider that with the previously envisaged diffusion process the dissolved substance and the solvent differ only slightly from each other (dissolving of one isotope metal type into the other) and observe the following reversible process (A–B–C–D):

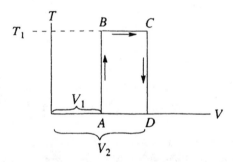

V is the volume throughout which, by virtue of the semipermeable membrane, the dissolved substance is distributed. T_1 is a temperature high enough that the law of osmotic pressure unquestionably applies.

In the special case examined here, it ought hardly be relevant to the specific heat per molecule whether the neighboring molecules are molecules of the same sort or of the isotopic sort. That is why at constant V the heat capacity is independent very close to V. Under this scarcely circumventible assumption, if the system's entropy is referred to as S, the relation will be

$$S_B - S_A = S_C - S_D$$

On the other hand, however, $S_C - S_B$ is not zero in the case of a diluted solution, e.g., equal to $R \lg \frac{V_2}{V_1}$. Thus

$$S_D - S_A = (S_B - S_A) - \overbrace{(S_C - S_D) + (S_C - S_B)}$$

$$= S_C - S_B,$$

likewise other than 0.

Through examination of the specific heat of isotopic mixtures, this consequence should, incidentally, be exactly testable. In my view, the success of such experiments can be expected quite safely.

Please let me know your view on this. With greetings and best wishes on your continued well-being, yours very truly,

A. Einstein.

82.   To Wander and Geertruida de Haas

[Berlin, ca. 10 May 1915][1]

Dear Friends,

Mrs. de Haas firstly, I have received your letter. The matter regarding your involuntary plagiarism at my hand is not so serious. For I wrote "together with *Mr.* de Haas-Lorentz."[2] As no one will have the audacity to intimate the least doubt about the gender of either of you, this specification is clear.[3] Now I just have to defend myself about why I wrote de Haas-*Lorentz* at all. I did it instinctively, without thinking. The reason is this. In Switzerland it is often customary to add the wife's maiden name to the husband's name, probably because many surnames recur very frequently. On the other hand, I made your first acquaintance, dear de Haas, as the husband of Lorentz's daughter, and it is quite natural that I should have repeatedly mentioned "Lorentz's son-in-law" in conversation as a more precise description of your person. Now I am racking my brains in vain about what I ought to correct there. Is it a disgrace to be described as Lorentz's son-in-law? I assure you first, de Haas, that I would submit to it gladly and with pride at all times, if it were I. But in any case, I regret it very much if I have produced any painful feelings in you both and am very willing to do anything and write anything you want. But I do not think it wise; put yourself in the place of the obliging reader who always has a toothy grin at the ready. Let us not give the rascal nourishment!

The business regarding the sign is definite after all; I have submitted a small correction.[4] The correct diagram now looks like this:

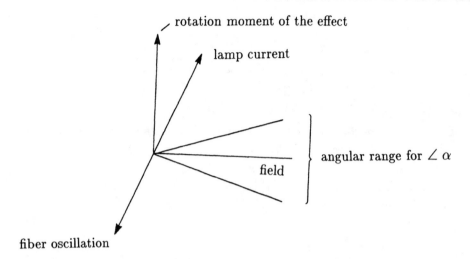

rotation moment of the effect

lamp current

angular range for $\angle \alpha$

field

fiber oscillation

An angle deviating only slightly from π between the fiber oscillation and the swing α is possible only through the selected rotation momentum sign. The angular constraint must be introduced because the phase angle changes very abruptly near the resonance. Our earlier conversation dealt with the oscillation of the loose incandescent lamp filament. There the swing phase is opposite to that of the force according to the adequately precise valid formula

$$\text{force} = \text{mass} \cdot \frac{d^2}{dt^2}(\text{swing}).$$

An addition to our experiment of the one suggested by you remains desirable in any case. I wish you the best of success. Do write to me occasionally about how it is getting on.[5]

I wish you both a happy life in rural serenity;[6] I regret very much that my like-minded friends are no longer here. I gratefully appreciated the little compliment on the lecture, likewise the friendly consignment of sweets.

With heartfelt greetings to you and your little ones.

My best regards to the estimable innocent instigator of my ineptitude.[7]

83.   To Mileva Einstein-Marić

[Berlin,] 15 May 1915

D[ear] M[ileva],

I must insist that the money in Prague be transferred to my name.[1] I will have it credited to the children and am convinced that *this* is the safest way for the amount to go to the benefit of the children. Until you have done your obligation in this matter I will not send you anything more.

Mrs. Haber shot herself 2 weeks ago.[2]

I expect to be hearing from you about when I can take a short trip with little Albert.[3] I can't go before the end of July because of my lecture course.[4] The summer and fall vacations come into consideration. Didn't the children receive the Easter package?[5]

Albert

Kisses to the boys.

84.   To Heinrich Zangger

[Berlin,] 17 May 1915

Dear friend Zangger,

I have received your detailed letter but again was unfortunately unable to read quite a lot of it. You cannot imagine how hard it is for a non-medical man to decipher your writing. This summer I am coming quite definitely to Zurich and shall spend a few days with you (if you happen to be there), so that we can make good use of the time. Presumably I shall come at the end of July,[1] but it depends upon when I can have my Albert accompany me on a short hiking trip. I also am consumed by the desire to be able to discuss thoroughly with you all the shocking things that have happened meanwhile. Even more ardently do I wish to be able to take my dear boys into my arms again. Hard though the separation is for me, it was a matter of life and death for me. I must tell you about this also, so that you do not get a more tarnished image of me from the appearances than I deserve.

In these times it becomes obvious that the only thing really worth striving for in this world is the friendship of exceptional and independent persons who cannot be convinced by any published rot to assume the perversest of mental stances. In Zurich you surely see, as well as I do here, how few such people there are.

I fear that perhaps you may well be regarding your colleagues' unfavorable attitude toward you in Zurich[2] too pessimistically. People are not so consistent as it may seem, in anything, not even in hatred and unfriendliness. A bit of impartiality might help you get over the rest easily as well. I always get by best with my naivité, which is 20% deliberate, you know. This is simply done if you are indifferent enough to your dear fellow men; all the same, you can never be as indifferent as they deserve.

Apart from the general misery we know mainly from newspapers, I am leading a happy, remarkably peaceful life, more withdrawn than in Zurich. The scarcity of persons of genuinely like mind is perhaps not more acute than what you experience in Zurich. Such as we just have to resign ourselves to loneliness in this regard, when circumstances do not provide an unusual stroke of luck. The nicer it will be when we are together in Zurich. With pleasure shall I seek to make Prof. Levin's[3] acquaintance.

It is very grave for Switzerland that the Italians are starting now as well.[4] Maintaining the food supply will be difficult. I daresay Ticino will also be in danger.–[5] You probably have received the notice on the magnetic research.[6] As a precaution I am sending you another copy. I am very much looking forward to your announced publication.[7]

With warmhearted greetings, yours,

Einstein.

Best regards to Dr. Heller. It is to be hoped that you will succeed in preserving his position.[8]

85.   From Max von Laue

Feldafing, 27 May 1915

Dear Einstein,

Cordial thanks for your letter of May 22nd. Our views on zero point energy apparently still diverge so much that it seems continued discussions about it are not very worthwhile.[1] Nonetheless, I understand your view thoroughly and gladly concede that the zero-point energy hypothesis has led to a very fine result, namely to your proof of magnetic molecular currents.[2]

I am very grateful to you for your objections to my "order" paper, albeit without being able to recognize a single one of them as irrefutable.[3] I did not take into consideration the absorption of radiation because in my opinion the order issue is independent of intensity. However, I am going to try to take this into consideration as well shortly. In *that* form, my adding another arbitrary function in p and t_n to the oscillation originating from individual resonators is very simple and does not change much.[4] In the addendum to section II of my manuscript, in which probability density ρ is no longer regarded as a constant but as a function of t_n, you already see related calculations. I want to think this through more carefully, though, in the near future.

Your second objection that in my rendition the radiation at the intersection point "successively stems constantly from different layers," I really cannot acknowledge as correct, in contrast to the first. Within the time range $[T] - \tau$, to which I have restricted the validity of my expansion, it stems from *all* layers; what happens beyond this time range is not supposed to be contained in my Fourier series at all. Thus your subsequent comment is dealt with: "If I choose the density as a function of a freely estimated location, which is surely permissible for real radiation paths, then I obtain a radiation that has a variable mean intensity against time; such radiation is not unordered." Apart from this I cannot concede to you that temporally variable radiation is not unordered. For since virtually all radiation changes intensity in time, accordingly hardly any unordered radiation would exist at all.

But now to your main objection. You say: In a Fourier series that is supposed to be zero at one section (or more sections) of its development range, the coefficients are not independent. You may gather how much this expresses my own sentiments from the fact that I myself have stressed the degree of freedom of a beam in oral discussions with Lenz and Sommerfeld.[5] It just seems to me that these relations must have the form of *equations*, not the much looser form of statistical relations as I have set them up.

Moreover, this statement of my considerations does not apply. For I do *not* require that my Fourier series vanish in the section of the development range not within the valid range; I do not require anything at all from them. They are rather entirely irrelevant to me there, as they have nothing to do with reality.

Such Fourier series are used frequently in physics. When you develop, e.g., the electrical field strength at a particular point in space from $-T$ to $+T$ according to Fourier, and then derive in the usual manner of resonance theory the forced swing of an oscillator, then you arrive at a series that naturally has a *development* range of $2T$, but is *valid* only from $-T+\vartheta$ to $+T$, where ϑ is the time in which the resonator fades away. For from $-T$ to $-T+\vartheta$ the resonator oscillation depends in part on the electrical oscillation before time $-T$, about which nothing is revealed by the Fourier series. Despite this difference between development and applicable ranges, no one has misused the coefficients of the resonator series more than those of the field strength series, and in my opinion justifiably so.

In order to knock the bottom completely out of your main objection, I can—something I had overlooked up to now—omit entirely in the last section of my

paper the distinction between τ and T.[6] For since I attribute completely different oscillations there to the resonators and merely regard the probability law as
common to the a_p's and ϑ_p's,[7] absolutely no basis remains for the assumption
$\tau < T$.[8] Besides, I can select the validity ranges for the individual oscillation
series in such a way that all these oscillations arrive simultaneously at the intersection point; thus I also obtain for the resulting oscillation a series that is valid
over the entire development range. At the same time, in taking into account
the variable distribution density of the resonators and in taking into account the
absorption, the fraction $\dfrac{\sin \pi \frac{p'-p}{T}\tau}{\pi \frac{p'-p}{T}\tau}$ is replaced by a similar fading function of
$p' - p$, *without the zero values of this fraction.*[9] Therefore, the linkage continues
to stand.

Now in closing, another comment that may not be quite necessary. You, like
many others, use for the light generation process the image of a single impact
with subsequent free, damped oscillation. Obviously, I know the reasons behind
it. But is it correct? I consider much more formal a theory with characteristics
that take into account the order in a single resonator's oscillation. Whether I can
cope with such a theory I do not yet know today, though.

I am naturally very eagerly looking forward to your reply to this letter. I
realize now already that I still have to rearrange and develop the paper very
much. This will undoubtedly occupy me up to the summer holidays. If you
presently still have reservations about it, as I am submitting it to the *Annalen*,
you will receive a carbon copy from me and can, of course, submit your opinion
directly to the *Ann*.[10] Otherwise, I remain, with cord[ial] greetings, yours,

M. Laue.

86. To Heinrich Zangger

[Berlin,] 28 May 1915

Dear friend Zangger,

Your news pleased me exceedingly. I am glad that you write me so much
and in such detail, and that my dear children are doing well, particularly that
Albert is in such good company. I can't say how much I am looking forward to
visiting you in the middle of July. Although I ought to be teaching to the end of
July,[1] I shall excuse myself from these few lectures. If Levin[2] also comes, and
your Frenchmen, that will certainly do no harm. You know me. I kindly request
that the magneton people[3] send me their objections in writing, as I am very
interested in them. Besides, I am quite willing to demonstrate the experiments

to them.[4] Initially my local colleagues were also suspicious but were convinced by the soundness of the experiments. I read with great pleasure your essay on medical confidentiality, whereas I have not yet found time to study the other paper you sent.[5] Read the remarkable little book *J'accuse*, recently published by Payot & Co., Lausanne.[6] It delights me that you have had some success in your institute affair. I am very eager to take in some Zurich air again, even though my life here is ideally pleasant if you disregard things that actually have nothing in the least to do with me.

I have read your paper on exposure risk now and understood most of it.[7] I must confess, though, that the style does not seem appropriate to me. For such an essay to get sluggish citizens into action, it must be much more graphic and must capture the emotions as well. Right at the beginning, the person should feel that he really is obliged to take the trouble, by having him feel acutely the full extent of the ills to be combatted. I consider it unproductive to present instead an accumulation of abstract, pithy distinctions. People must react to so much these days that mere allusions will always remain unfruitful, because they will not be developed by the reader. Put yourself in the place of the person forever in a hurry in whose hands such an article falls.—Do not take offense at my candor.– If you would like a bit of harmless amusement, read the booklet published by Reclam by Macaulay on Freder. the Gr. One would be tempted to add the motto: Like father, like son.[8]

Affectionate greetings, yours,

Einstein.

87.   To Walter Dällenbach

Berlin, 31 May [1915][1]

Dear Dällenbach,[2]

Your letter delighted me, especially since I can see from it that you have remained loyal to physics despite the orientation of your responsibilities toward engineering.[3] You will find a full description of the experiment conducted with de Haas on the nature of the magnetism of molecules in one of the most recent numbers of the *Berichte der Deutsch. Phys. Gesellschaft* (issue in a yellow jacket).[4] I have not received the offprints yet. On the horrendous times only this: Be glad and proud that you are Swiss! Now to scientific matters. It is not clear to me what question you had in mind regarding the molecular currents analysis. The capillarity problem, where aside from gravitation other forces originating from an electrical field are also at play, is naturally mathematically extremely complicated, because the electrical pressure forces acting on the surface depend *indirectly* on the shape of the surface; physically the approach is correct, of course.

Your observation on the potential jump between electrode and electrolyte is not new to me, because I also have already toiled along these lines.[5] But you must not forget the following.

In the electrode–electrolyte arrangement, three potential jumps occur on the observed line L, namely, at P_1, P_2, and P_3. From the field, which is produced in the headspace, only the sum of these three potential jumps can be assessed, but *never anything about the potential jump at P_3 alone*. That is why the experiment

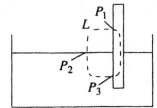

also can yield nothing of interest resulting from the measurement of just such an air-filled field. In the simpler case of the Volta effect, in which two potential jumps are at play, you have recognized this correctly yourself.

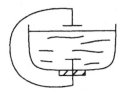

The only possibility known to me of measuring *one* boundary-layer potential difference would be the "evaporation" or "condensation" heat of the electrons, i.e., the amount of heat that an electron develops in a metal upon entering the metal from a vacuum. But an unproven assumption forms the basis here as well, namely, that *only electrical* forces are at work on this electron in the boundary layer. Thus you see that the whole undertaking is quite precarious. Scientific endeavors are quite extraordinary; often nothing is of more importance than seeing where it is not advisable to expend time and effort. On the other hand, one should not pursue goals that are too easily reached either. An instinct must be developed for what is just barely attainable upon exertion of the utmost effort.

This magnetic experiment, for ex., could have been done by any old lout.   But general relativity is of another kind. Having actually arrived at this goal gives me the greatest satisfaction of my life, even though up to now not a single colleague in the field has recognized the depth and necessity of this path. One of the two important experimental consequences has incidentally already been verified splendidly, namely, the shift of spectral lines through a gravitation potential.[6] In July I am coming to Zurich and shall tell you about it.

   With cordial greetings, yours,

                                                                A. Einstein.

88.   From Helene Katz[1]

[Berlin,] 46 Bamberger St., 11 June 1915

Dear Professor,

   Some vague feeling dictates a few words to flow from my pen to you.—Without any definable intention perhaps, also without the expectation of a reply on your part.–

   Some confounded devil is breathing down my neck and compelling me now constantly to be at the ready for you. Entirely without listening to my ego's better judgment, which feels, quite to the contrary, that things have not been rosey for you, especially recently, and you are rather in need (sorry, I know of course, you are in need of nothing—from me)—and that you are, *rather*, in need of having people be a bit nice, I should say, soothing, to you. Without danger, dear professor, for although it is not without its attractions for us to be considered slightly dangerous—I would not like to appear so to you—for reasons that you can admit without undue vanity.

   If upon receipt of this letter you should not quite be in a generous mood, I immediately beg your pardon for the disturbance;– otherwise say to yourself: Miss Katz would like to lull me a bit with a tune through hours that are peppered with gray—it is just for this that she writes.

   Maybe we can go on a walk with Elsa again sometime and talk about the clouds and trees. That would be nice.

   In the meantime, once again my regards, yours,

                                                                Helene Katz.

I want to tell you one of my mottos:

–for this is guilt, if guilt is anything:
not to multiply a loved one's freedom
by all the freedom you find in yourself–

---

unfortunately not by me: by R. M. Rilke.[2]

89.   To Michael Polányi

[Berlin,] 18 June 1915

Dear Colleague,

Regarding your thermodynamic observation on Nernst's theorem,[1] I cannot see what is *essentially* new in it to Nernst's paper published by the Pruss. Academy;[2] admittedly, it is more suggestive. However, we have already discussed this point enough,[3] so I hope I do not have to take a stance against your paper.

Regarding the isotopes, I think the following:[4] It seems to me beyond doubt that the mixture at normal temperatures is connected with a change in entropy, because it seems impossible to question osmotic pressure at normal temperatures. Nonetheless, this could still be verified specifically by the method you have suggested. At most, the only question appearing to remain is whether this entropy difference can be eliminated at absolute zero through particularly unusual variability in the specific heat of the isotope mixtures. I personally am firmly convinced that this is not the case. But I do not intend to occupy myself further with this matter, because I believe I said all that is necessary earlier already.

With best regards, yours very truly,

A. Einstein.

90.   From Hans Reissner[1]

Charlottenburg 9, 17 Tannenberg Avenue, 22 June 1915

Esteemed Mr. Einstein,

I am attempting to penetrate bit by bit into the magnificent edifice of your gen. theory of relativity, but still find myself in the anteroom. Today I would like to hear your opinion on an example that I have attempted to set up according to your theory, namely, that of the charge distribution at equilibrium (structure of unit charge).[2] However, I have succeeded only under the assumption $g_{11}\, g_{22}\, g_{33}$ constant, g_{44} variable. I cite your Academy article ("Formal Basis, etc.")[3]

(42a)
$$\sum \frac{\partial \mathfrak{T}_\sigma^\nu}{\partial x_\nu} = \frac{1}{2} \sum_{\mu\tau\nu} g^{\tau\mu} \frac{\partial g_{\mu\nu}}{\partial x_\sigma} \mathfrak{T}_\tau^\nu$$

i.e., volume force $\mathfrak{K}_\sigma = 0$
yields for polar symmetry: ($\mathfrak{E}$ electr. force)[4]

$$-\mathfrak{E} \cdot \rho = \frac{\mathfrak{E}^2}{4} \frac{\partial \ln g_{44}}{\partial r}$$

$$\mathfrak{E}\left(\frac{\partial \mathfrak{E}}{\partial r} + \frac{2\mathfrak{E}}{r}\right) =$$

(I) . . . . .
$$g_{44} = A^2 \mathfrak{E}^4 r^8$$

Potential equation:[5]

$t_4^4 = 0$ at equilibrium.

(81)
$$\sum_{\alpha\alpha} \frac{\partial}{\partial x_\alpha}\left(\frac{-1}{2\sqrt{g_{44}}} \frac{\partial g_{44}}{\partial x_\alpha}\right) = -\kappa(\mathfrak{T}_4^4 + \cancel{\mathfrak{W}}_4)$$

$$\frac{\partial}{\partial x_4} = 0$$

$$\Delta\sqrt{g_{44}} = \kappa\frac{\mathfrak{E}^2}{2} \qquad (\kappa = 1.87 \cdot 10^{-27} \text{ for } c_\infty = 1 \text{ light vel. at } \infty)$$

For polar symmetry

(II) . . . . .
$$\frac{\partial^2 \mathfrak{E}^2 r^5}{\partial r^2} = \pm\frac{\kappa}{2A}\mathfrak{E}^2 r \qquad \frac{\kappa}{2A} = k.$$

2 solutions

(a) $\mathfrak{E}^2 r^4 = c_1 e^{\frac{k}{r}} + c_2 e^{-\frac{k}{r}}$ (b) $\mathfrak{E}^2 r^4 = c_1 \sin\frac{k}{r} + c_2 \cos\frac{k}{r}$

$$\sqrt{g_{44}} = c = A\mathfrak{E}^2 r^4$$

Boundary conditions: (1) For $r = \infty$ $c = c_\infty$ $3 \cdot 10^{10}$ m/sec or 1

$c_\infty = A(c_1 + c_2)$ $c_\infty = Ac_2$

At infinity $\mathfrak{E} = \dfrac{\sqrt{c_1 + c_2}}{r^2}$ or $\dfrac{\sqrt{c_2}}{r^2}$ as usual.

If you name the charge ε, then also results

$$\sqrt{c_1 + c_2} = \frac{\varepsilon}{4\pi} \qquad\qquad \sqrt{c_2} = \frac{\varepsilon}{4\pi}$$

On the other hand, the total charge must be determined as follows:

$$\rho = -\frac{1}{r^2}\frac{\partial \mathfrak{E} r^2}{\partial r} \qquad\qquad \varepsilon = \int_{r_i}^{\infty} 4\pi r^2 dr\, \rho = -4\pi\,\mathfrak{E}\, r^2\Big|_{r_i}^{\infty}$$

$$\varepsilon = -4\pi \sqrt{c_1 e^{\frac{k}{r}} + c_2 e^{-\frac{k}{r}}}\;\Big|_{r_i}^{\infty} \qquad\qquad \varepsilon = -4\pi \sqrt{c_1 \sin\frac{k}{r} + c_2 \cos\frac{k}{r}}\;\Big|_{r_i}^{\infty}$$

If charge is to remain finite, then is required

$$c_1 = 0 \quad r_i = 0 \qquad\qquad\qquad r = r_i \quad c_1 \sin\frac{k}{r_i} + c_2 \cos\frac{k}{r_i} = 0$$

hence $\varepsilon = -4\pi \sqrt{c_2}$
                                      and hence also $\mathfrak{E} = 0$ within the cavity's boundary condition

$$\rho = -\frac{1}{r^2}\frac{\partial \sqrt{c_2}\,e^{-\frac{k}{2r}}}{\partial r} \qquad\qquad \rho = \frac{1}{r^2}\frac{\partial \sqrt{c_1 \sin\frac{k}{r} + c_2 \cos\frac{k}{r}}}{\partial r}$$

$$= \frac{k\sqrt{c_2}}{2r^4}e^{-\frac{k}{2r}} = 8k^3\sqrt{c_2}\left(\frac{k}{2r}\right)^4 e^{-\frac{k}{2r}}$$

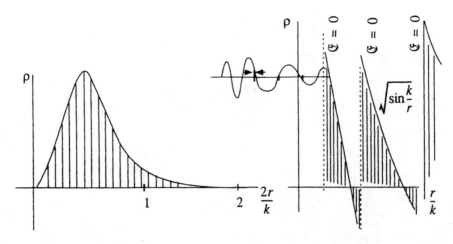

Thus there would be accordingly 2 types of polar symmetric charge at equilibrium:

(1) Charge of density one (negative elementary charge)
(2) Charge with variable densities, therefore larger mass and rings of vanishing force in which areas of embedded charges at equilibrium may exist.)

Before I continue working on this, I would like to hear your opinion, esp. on the following points

(a) Would the as yet not used eq. (42c) deliver another new condition?[6]
(b) How do (81) (42c) and (81b) read with polar symmetry, as t_σ^ν, of course, covariant only against linear transformations?[7]
(c) May it be regarded perhaps as a first-order approximation when you set $g_{11}\, g_{22}\, g_{33} = $ const.?
(d) Surely for polar coordinates eq. (54) must read differently. Is it permissible then also to call (54) invariant against arbitrary transformations?[8]

You may find these questions very idiotic, but perhaps these are precisely the hitches keeping me from moving forward.

With humblest regards, yours,

H. Reissner.

91. To David Hilbert

[Berlin, 24 June 1915]

Dear Colleague,[1]

Many thanks for both of the cheerful postcards. I am going to stay at the Gebhart Hotel and shall call on you Monday morning.[2]

With best regards, yours very truly,

A. Einstein.

92.   To Wander and Geertruida de Haas

[Berlin,] 6 July [1915]

My dear Friends,

Yesterday night upon returning from Göttingen[1] I discovered your letter. I have already written to a dependable furniture mover for a cost estimate.[2] The removal is undoubtedly possible. I shall look up the landlord[3] in order to discuss any return of rent with him. I am not going to be here the second half of July (I am going to Rügen).[4] But by no means will time be lost as a result, since I would authorize someone to represent me in the matter. You shall hear from me again soon. I am pleased for you all that you are remaining there, mainly for your children, for whom it is infinitely better. Good luck on the experiment![5]

Affectionate greetings, yours,

Einstein.

93.   To Michael Polányi

[Berlin,] 6 July [1915]

Highly esteemed Colleague,

I am in agreement with your representation to the extent that you denote as your view everything that is dubious to me. The unprepared reader will remain in the dark on many things, though. Perhaps you would also like to make use of the following in some way:[1]

1) If by your process absolute zero is reached only after ∞ many steps, then it is sufficient proof without any special assumption.
2) If by your process absolute zero is already reached through a finite number of steps, then it is sufficient proof only if you assume that at abs. zero it can be continued on in the manner

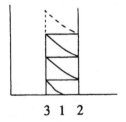

3 1 2

adiabat 1–2
isotherm 2–3
adiabat 3–1

In my view the existence in principle of such zero-point processes through boundary transitions can be made evident only, e.g., with changes in volume, el[ectrical] conduction, whereas virtually inconceivably with diffusion.[2] It remains doubtful with a chemical process.

If I spoke of a *petitio principii*,[3] I wanted to say that your proof assumes, instead of the statement $S_2 - S_1 = 0$, one that is generally by no means more plausible of the existence of reversible zero-point processes (provided (1) is not presupposed).

(By contrast, it apparently seems evident to Nernst that a process like yours would never reach absolute zero.[4] Then case (1) is sufficient.)

It would be good if you emphasized this a bit more. With best regards, yours,

A. Einstein.

94.   To Heinrich Zangger

[Berlin,] 7 July [1915][1]

Dear friend Zangger,

Now I must delay my trip to Switzerland once again. For my son has written me a very curt postcard in which he decidedly rejects going on a tour with me.[2] I would not even be able to see my children, because my wife is going on a trip somewhere or has already gone! And for this I had arranged my lecture course especially, so that I could end at the beginning of July.[3] And yet—would to God that the obstinacy of the human soul were guilty of nothing worse! The longer this dreadful state of war continues, the more grimly people hang on to insensible hatred founded on nothing. As long as one is young, one admires vibrant emotion and disdains cold calculation. But now I think that the blunders originating from blind feeling cause much more bitter unhappiness in the world than the most heartless of calculating minds.

But in these times you appreciate doubly the few people who rise well above the situation and do not let themselves be carried along in the turbid current of the times. One such person is Hilbert, the Göttingen mathematician.[4] I was in Göttingen for a' week, where I met him and became quite fond of him. I held six two-hour lectures there on gravitation theory, which is now clarified very much already, and had the pleasurable experience of convincing the mathematicians there thoroughly.[5] Berlin is no match for Göttingen, as far as the liveliness of academic interest is concerned, in this field at least. It is not without reason that they procured the highly talented Debye, who has again done magnificent things.[6] I shall tell you more about that when I come to Zurich. I would still like to come this year in the summer when I can see my children; traveling to Zurich without seeing my boys would be far too painful for me.

Abraham was in Zurich for a long time, maybe he is still there. Do take a look at him sometime. I am still of the opinion that he would be the right man for Zurich.[7]

Now I am going to Rügen with my relatives for a few weeks.[8] I am immensely looking forward to becoming acquainted with the sea.

In personal respects I have never been so at peace and happy as now. I am living a very secluded and yet not lonely life, thanks to the loving care of my cousin, who had drawn me to Berlin in the first place, of course.[9] Her having herself cooked lunch for a group of poor women every day this winter gives you an idea of her kindheartedness. Taken altogether, I cannot avoid envying myself when others fall short of doing so.

Among my truly profound colleagues, like Planck and Hilbert, the wish to uphold cordial relations with like-minded scholars abroad is very much alive. Hilbert now regrets doubly, as he told me, having neglected to maintain international relations better.[10] Planck is doing everything to bridle the chauvinistic majority at the A[cademy].[11] By the way, I must say that in this regard the hostile nations deserve one another!

I had wanted to bring your manuscript[12] with me myself. But now that I do not know when I'm going to be able to travel to Switzerland, I am sending it along to you. I wish you happy and relaxing holidays.

Best regards, yours,

Einstein

also to Heller; what's the news on his examination?[13]

95. To Wander and Geertruida de Haas

[Berlin,] 9 July 1915

Dear Friends,

Please inform me *where* the furniture should be delivered;[1] the mover is asking so that he can make a precise cost estimate. I am now going to be at Rügen (Sellin) on vacation until around August 5th. The removal of the furniture can occur right afterwards, though. I have written to the landlord[2] but have not yet received a reply. I did it in writing so that I have something in hand. So please do write me in Sellin.

Affectionate greetings, yours,

Albert Einstein.

96.   To Arnold Sommerfeld

Sellin (Rügen), 15 July [1915]

Dear Sommerfeld,

I am in favor of having the little volume published unchanged without having the general theory of relativity included, because none of the current expositions on the latter is complete.[1] Easiest would be to use the first of the *Annalen* papers and the Academy paper.[2] However, I have the intention of writing a special little book as an introduction to relativity theory, wherein the discussion leads to the general theory of rel. from the very beginning.[3]

In Göttingen I had the great pleasure of seeing that everything was understood down to the details.[4] I am quite enchanted with Hilbert. That's an important man for you! I am very curious to know your opinion.

I am glad that you have thought of Freundlich's paper;[5] it is surely fundamental. Grossmann will never lay claim to being co-discoverer. He only helped in guiding me through the mathematical literature but contributed nothing of substance to the results.–[6]

If you should be set on having the gen. theory of rel. represented in the new edition after all, that is also all right with me.

Cordial greetings, yours,

Einstein.

I have sent the magn. paper to Lenz.[7]

97.   From Heinrich Mousson[1]

Zurich, 19 July 1915

To Prof. A. Einstein, Berlin.

Highly esteemed Sir,

Allow me to turn to you with a request.

The filling of Kleiner's professorship in experimental physics is posing great difficulties for us, after the attempt to win over Prof. Debye at Göttingen for it unfortunately failed.[2]

Calling a competent employee from a German university is out of the question at the moment, as the gentlemen conc[erned] are themselves at the front or cannot leave their present positions for understandable reasons.

The next generation of young Swiss physicists seems to be very small in number; at least the faculty claims not to find among their own any suitable individual who might come under consideration for selection. The choice of Dr. Greinacher, who has admittedly acted as substitute for 2 semesters now, is probably ruled out.[3] On a visit to Lausanne we also gained the impression that Prof. Perrier would not be suitable for us.[4]

The faculty has, furthermore, also excluded Dr. Piccard,[5] assistant at the Fed. Polytechnic's phys. institute, above all because he reportedly has not yet provided enough evidence of his scientific qualifications.

Now, however, Piccard is being recommended to me from other quarters all the more highly as very intelligent, original, and enthusiastic about his subject; and I am being told that he has done some very competent work besides, which has been judged favorably by you as well.

Since I do not deem it appropriate for the interim arrangement to persist until German physicists can once again enter the contest, and I would prefer to fill the chair soon so long as there is enough guarantee that the man to be appointed is equal to the task, I take the liberty of turning to you with the polite request that you kindly inform us how you assess Piccard's scientific achievements. He is still a young man, of course, whose best work is still ahead of him in the future. Thus it seems worthwhile to learn from most knowledgeable quarters whether his existing work reveals an independent scientific approach from which one may expect fruitful development. I have been assured that Piccard can experiment ably, and his lecturing capabilities could be judged from a lecture course he holds for Prof. Weiss.[6]

You would do us a great service with a short appraisal of Piccard's scientific achievements, for which we would be obliged to you with our sincere thanks. As the handling of this business has unfortunately been dragging on considerably and the date for completion of our deliberations is close, may I presume to request additionally the favor of a quick response.

Allow me, highly esteemed Sir, to assure you of my deep respect,

Mousson
Off[icial] Adviser.

98. To Hendrik A. Lorentz

Sellin (Rügen), 21 July 1915

Dear and highly esteemed Colleague,

Recently I spoke with Planck, and we both gloomily recalled the bitter division that has arisen between us and our highly esteemed foreign colleagues as a result of the unfortunate war.[1] Whatever errors may have been committed on either side in deplorable political agitation, it is never too late to change. It is certain that we academics are all innocent of the war and that the present miserable circumstances ought to induce us even more to solidarity; whatever occurred before has to be regarded simply as not having happened.

What to do? If I did not live in Berlin, I would write personally to our closest colleagues in France and England with the request that they salvage themselves from the generally desperate condition, to the extent that the earlier friendliness in our community can be restored.[2] I would ask them to assemble, completely voluntarily and unofficially, at an appropriate location (Holland or Switzerland) already during these holidays, primarily to nurture personal contacts.

But I live in Berlin, have few contacts, and also have little skill in communicating with people. That is why I am turning confidently to you in the hope that you will be able to transform all this, which I can only dream of, into reality. Would you not enjoy devoting some time to this fine mission? Planck encouraged me very much to do everything that was in my power; he also would do everything to restore the good relations. This is all the more important as there are signs that the official relations among the societies, just as the academies, could get broken off; for there is a great surge of nationalistic blindness.[3] But I note here that especially the most highly regarded are fighting against it with all their might, and this will surely be the same in other countries as well.–

In the first days of August, I am going to be in Berlin and shall immediately take care of the move for your children.[4] I am very sorry that this matter is now being delayed a few weeks. If I receive a good offer, I will deal with the matter right away without first requesting approval of your children regarding the price.

Best regards to you, your wife and children, as well as to Ehrenfest, yours,

A. Einstein
until 1st August, Sellin (Rügen).

99.   To Wander and Geertruida de Haas

Sellin, 24 July 1915

Dear Friends,

To my great dismay I found this letter yesterday in my apartment in Berlin,[1] which complicates the moving business. Did you not terminate with Mr. Schrobsdorff *orally* either?[2] How then did he happen to draw your attention to the apartment's rentability a few weeks ago already?

I fear now Mr. S. will resist the removal of the furniture. Please inform me exactly what the situation with your termination actually is so that I can inquire about what can be done. Write me the details very soon to Berlin.

With affectionate greetings, yours,

A. Einstein.

100.   To Heinrich Mousson

Sellin, 24 July 1915

To the Office of the Director of the Department of Education for the Canton of Zurich.

Highly esteemed Sir,

Since yesterday I have been feeling most keenly the difficulty of the problem that you have to solve, i.e., since receipt of your letter.[1] I am convinced I cannot tell you anything that you have not already heard from my colleague Mr. Kleiner.

First, Mr. Greinacher.[2] By no means does this person lack talent for scientific research. He has a command of the scientific methods within a specific, albeit not very large, field and also knows how to set problems to a certain degree. However, neither his prior achievements nor that which I have come to know from personal contacts about the level of his talent makes it seem proper to entrust him with *the leading post* at the university. On the other hand, however, I would welcome it if a way were found to enable him to function as part-time academic teacher. In this way the university could perhaps gain something from him, and he would not suffer such a very severe setback.[3]

About Mr. Perrier I can be brief.[4] His character as a scientist is related to that of Piccard, but Piccard appears to me superior, so that I permit myself to turn immediately to the latter.

Piccard[5] is a man of extraordinary technical skill. He is inventive with measurement methods when a problem is already set before him. He is, moreover, extremely precise, conscientious, and critical of his own work, so that the results of his research can always be looked upon as reliable and deserving of serious consideration. Furthermore, Piccard is a man of exactitude and integrity who does not tolerate obscure and vague statements. Also as a *person* I consider him highly cultured, fine, and disciplined. He is a man *capable* and *worthy* of the post you have to fill.

Now, with all due respect and sympathy, the obverse. Piccard has not yet demonstrated that he has an *inquiring mind*. He has as yet assigned himself no new physical problems, has thought up no new approach to research. His accomplishment always lay in the technical perfection of already existing research methods. The acute mind that inspired Piccard's research is easily traced to P. Weiss.[6] With this I do not want to contend that Piccard lacks the ability to find new paths independently; only, until now he has not provided the proof, and he is still a young man. Piccard certainly is a *competent* researcher and teacher, but I am unable to foretell whether he can be a *stimulating* and invigorating influence on physical research. In any case, none of the other young Swiss physicists can come under consideration next to Piccard.

When I now consider the younger physicists of Switzerland and abroad, I come to the conclusion, completely disregarding the unfavorable conditions brought on by the deplorable war: If a Piccard is among one's number, he must be given a professorship.–

I am coming shortly to Zurich and shall inquire by telephone whether you want to speak with me.

Very respectfully,

A. Einstein.

101.   To Heinrich Zangger

[Sellin, between 24 July and 7 August 1915][1]

Dear friend Zangger,

You can hardly imagine how comforting it is to me that you are attending to my children so kindly. I am also very happy to hear that you have not forgotten me completely. Now I leave it entirely to you to fix the time for my trip to Zurich since, as you know, I am totally free. My youngest has unfortunately turned out a bit weak; it is to be hoped he will gradually develop for the better. But nature has endowed him instead with a kindhearted and cheerful disposition, more than my Albert, who cannot adapt himself so easily to everything.

I am very sorry for Kaufler[2] concerning his loss. He is not alone with it. Such hard fates have a quite different effect on people than one should have thought. Rather than aspiring to the realization that they should not let themselves be led blindly and should discard all unjustifiably acquired trust so as not to be abused in a shameful way, they hurl their hatred senselessly wherever the bellwethers want it. No education and intellectual cultivation seems to be able to protect against this wretched madness. The upright few surely have an inborn ability to remain honest. Read the moving little book *Christianity and Patriotism* by Tolstoy.[3] The obvious is said there simply and plainly, so movingly in his resignation. Tolstoy knows that what he says will be of no help, even though it ends on a more optimistic note with a hymn to the development and cleansing of public opinion.[4] He has not like me traveled with soldiers and relished their stupidity and crudeness.

Do not delay my trip to Zurich out of consideration, because you think the matter might turn out to be partly unpleasant. In this regard I am prepared for anything. Only be guided by when you are available. The education director's office asked me about Piccard. I have recommended him.[5]

It is wonderful here. Never have I been able to rest so well since adulthood. Surely I have already written you that I held 6 lectures at Göttingen in which I was able to convince Hilbert of the general theory of relativity. I am quite enchanted with the latter, a man of astonishing energy and independence in all things.[6] Sommerfeld is also beginning to agree;[7] Planck and Laue are remaining aloof.

I do not yet understand your suggestion regarding applying statistics.

Again sincere thanks & affectionate greetings, yours,

Einstein.

102.   To Wander and Geertruida de Haas

[Sellin,] Monday [2 August 1915]

My Dears,

I am very cheered by your letter, from which I gather that you have terminated [your lease] properly.[1] Now certainly no obstacle remains for the dispatch of the furniture. Thursday evening I am coming to Berlin and shall provide then for a prompt settlement.

Affectionate greetings, yours,

Einstein.

103.   To Hendrik A. Lorentz

Sellin, 2 August 1915

Highly esteemed and dear Colleague,

Your negative reply did not come as a surprise, as I already had indications of the mood of our colleagues abroad.[1] It is curious in Berlin. Professionally, scientists and mathematicians are strictly internationally minded and guard carefully against any unfriendly measures taken against their colleagues living in hostile foreign countries. Historians and philologists, on the other hand, are mostly chauvinistic hotheads. The well-known and notorious "Manifesto to the Civilized World" is being deplored by all level-headed people here.[2] The signatures had been given irresponsibly, some without prior reading of the text. That is how it was for Planck and Fischer, for ex., who have supported upholding international ties in a very resolute manner.[3] I am going to talk to Planck about your suggestion. But I believe that these persons cannot be prompted to retract their words.[4]

I must admit that the narrow nationalistic sentiment even of people of high standing is bitterly disappointing to me. Moreover, I must say that my respect for the politically more advanced states has diminished significantly on perceiving that they are all in the hands of oligarchies that own the press and wield the power and can do what they like. A malicious person has altered a fine proverb thus

"vox populi, vox ox."

Add to this that the perceptive and powerful have no heart for the many; there you have the sad picture of what is revered as the "fatherland" by those who belong to it. This does not change with the boundary posts but is everywhere essentially the same. And before this threadbare ideal, relations between persons who have come to respect one another privately and professionally must pale? It is beyond belief and refuses to enter my head. It seems that people constantly need a chimera for the sake of which they can hate one another; earlier it was religious faith, now it is the state.

Now to your second letter. It is disconcerting to see oneself get carried away by social pressure to say things for which one can only offer feeble support.[5] The second comment may certainly be printed with your addition; a correction would just be in order:

If we [want to] introduce zero-point energy to the Planckian resonator (instead of "in Planck's sense") . . .[6]

The editing of the first comment is unfortunate. If you consider it permissible, the following essentially congruent version could perhaps be chosen:[7]

There are serious doubts about the assumption of zero-point energy existing in elastic oscillations. For if at falling temperatures the (thermal) elastic vibrational energy does not drop to zero but only *drops to a finite positive value*, then an analogous behavior must be expected of all temperature-dependent properties of solids, i.e., *the approach toward constant finite values* at low temperatures. But this contradicts Kamerlingh Onnes's important discovery, according to which pure metals become "superconductors" on approaching absolute zero.[8]

It is very good of you to take such trouble with the conference proceedings. If I can relieve you of some of it, I shall do so gladly (e.g., correcting the discussion contributions by the German speakers).

With cordial greetings, yours,

A. Einstein.

[6]Recipient's appended note: "If we want to introduce zero-point energy in Planck's sense, in my view we may not deny that it consists in motions, i.e., here specifically in elastic oscillations in Debye's sense. Thus it is furthermore inevitable that we attribute to that energy an influence on the intensity of Laue's interference points."

104.   To Wander de Haas

[Berlin,] 7 August 1915

Dear de Haas,

I just received the offer from the mover (Schur).[1] He asserts that the price of removals has risen greatly as a result of the war; he asks for a total of 590 M. I am going to have a second cost estimate made by another mover and shall take Schur if the second is not cheaper.–

Lots of luck on your work.[2] I am now trying resonance-free alternating current through a refinement of the optical method, but I fear effects at double the frequency.

Affectionate greetings to you & yours, from your

Einstein.

105.   From Knud A. Nissen

Copenhagen W., Denmark, Gl. Kongevej No 101 I, 9 August 1915

To Prof. Albert Einstein.

Highly esteemed Professor,

I ask you most humbly for permission to add, with reference to my letters to you of Aug. 2nd, 5th, & 7th, that the motion equation of relevance with regard to the equivalency condition to be attached to Bohr's stability condition

$$\mu \cdot \omega_0 \cdot r_0^2 = M \cdot \omega_0 \cdot e + E \cdot e$$

has been provided by Prof. Störmer in Christiania, Norway, and was mentioned by Dr. Stanley Allen in *Phil. Mag.*, Jan. 1915.[1] M is the momentum of a magnet in whose equatorial plane an electron describes a circular path[2] (number of revolutions $n_0 = \dfrac{\omega_0}{2\pi}$); I believe one can insert $E_1 = 8 - \dfrac{e}{4} \displaystyle\int_0^7 \dfrac{1}{\sin\frac{s\pi}{8}} = (8 - 2'81e)$ instead of E for Ne, A, Kr, & Xe (compare Riecke's report on Bohr's theory).[3]– Is it permissible to adhere to Planck's dispersion theory and regard σ *as constant* for Ne, A, Kr & Xe?[4] Is there no contradiction?

I humbly hope I am not unworthy of receiving a discussion of the difficulties that undoubtedly must be clearly apparent to you with regard to this formulation and its applicability. I cannot thank you enough for having urged me to think over the weakness of my supporting point, the "viscosity radii," and I just hope that I have not tired you with my letters now, when I *believe* to be somewhat better supported in that I rely on radii proportionate to $\sqrt[3]{T_e/\varphi_e}$. Please kindly forgive the strong impulse to write you, which made me forget that you probably have very much to do. Most humbly and respectfully,

Knud Aage Nissen.

106.   To Wander and Geertruida de Haas

[Berlin,] Tuesday. [10 August 1915]

Dear Friends,

Today the packing began.[1] The move will now be done without a furniture van by a very rel[iable] mover (Franzkowiak).[2] My cousin negotiated with him. Careful packing will be supervised by me. Omitting packers, tips & customs, it costs 270 M. The landlord[3] is not causing any trouble. But there is a lot to be done otherwise. I hope that the things leave in 3 days.

107.   To Wander and Geertruida de Haas

[Berlin,] Saturday, 14 August [1915]

Dear Friends,

The things are already loaded.[1] The landlord caused no problems at all. Only the export license from the Imperial Office of the Interior is still missing. The entire move will come to about 350 M, at most 400 M.[2] Shortly after the 20th of this month the things ought to arrive in Deventer.[3] Then you can pay the freight immediately. I shall inform you of my disbursements when everything is finished. Should I ever get into the position of declaring my love to someone, I would say to the person: "For you I would arrange a move abroad in wartime." Warm greetings to you both, yours,

A. Einstein.

I am looking forward to the results of your experiments.[4]

108.   To Paul Hertz

Berlin, Saturday [between 14 August and 9 October 1915][1]

Dear Mr. Hertz,[2]

If I understand your letter correctly, you have a completely inappropriate notion of what I call an "adapted coordinate system."[3] How do you come to requiring that such a pair of coordinate systems—and added to that, with congruency at the region's boundary—should exist;

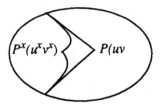

that for $u^x = u$
$$v^x = v$$
also $\quad E^x = E$
$$F^x = F$$
$$\langle G^x = G \rangle \; \varphi^x = \varphi?^{[4]}$$

I am rather convinced that (aside from very special cases perh[aps]) *this could never be possible.* Nowhere have I assumed the existence of systems equivalent in this sense.

Independently of this, I understand how you define a special coordinate system within a two-dimensional manifold by the curves of constant curvature and those of maximal curvature gradient. It is suspicious, though, that on structures with a constant degree of curvature, the curves (or planes) of constant curvature move infinitely far away from one another.   The fundamental dissimilarity between both introduced coordinates is also suspicious. You could try out, nonetheless, whether such a thing can be done in a four-dimensional manifold.

I did not understand the suggestion for setting up a gravitation law, because I cannot read your writing on page 5. As you know, in my paper I showed that a useful gravitation law cannot be generally covariant.[5] Do you not agree with this consideration?

So once again: I have not the slightest intention of requiring "that the world be able to unwind upon itself" and do not grasp why you would expect me to do such an atrocious thing. In my meaning there surely is an enormous variety of adapted systems that, however, do *not* conform at the boundary.

With best regards to you, your wife, and your already astonishingly affable and avid writer of a son,[6] and eagerly awaiting more reports, I remain yours,

A. Einstein.

109.   To Pieter Zeeman

Berlin, 15 August 1915

Highly esteemed Colleague,[1]

Your two articles on the Fizeau experiment delighted me.[2] They fill a previously unpleasantly perceptible gap. I can imagine how difficult it was to attain such a high degree of precision! Now the validity of Lorentz's formula has surely finally been secured.[3]

With many thanks, yours,

A. Einstein.

110.   To Wander and Geertruida de Haas

[Berlin,] Monday. [16 August 1915][1]

Dear Friends,

The furniture has now left (luckily and thank God).[2] I ask that you meet the same in Deventer,[3] so that no storage fees accrue for you. The entire business will cost you a bit more than 400 M, which you must pay immediately in Deventer. The war makes the matter much dearer. The bill will be issued to me as well, and I am going to check whether everything is in order. The mover is Franzkowiak, 83 Uhland St.[4] No furniture vans were used, but everything was packed in crates instead, because a van would have incurred *great* expense owing to the war. I have just paid the taxes for you (19.95 M); I am enclosing the receipt. The booksellers' bill you mentioned has also arrived, but has not yet been paid by me.

I am curious about the experiments.[5] Describe them to me please in a letter, also mentioning the detours and difficulties. I am feeling quite fine; at Rügen I recuperated well,[6] particularly after making the tough decision to give up smoking entirely.

At the end of June–beginning of July I held six detailed lectures at Göttingen on general relativity theory. To my great delight I succeeded in convincing Hilbert and Klein completely.[7] I have not found out anything new of consequence recently; I have even put aside the attempt to verify our effect in a new way because of the substantial optical hurdles.[8] At the end of the month I am going to Switzerland to see my children & friends, no small undertaking in these times.

With best wishes for your new position in life, and affectionate greetings, yours,

A. Einstein.

D[ear] de Haas. I am very glad for you that you have preferred a settled teacher's career over a restless existence in industry and city life. It is more pleasant for you all and healthier for your children.

111.   To Paul Hertz

[Berlin,] 22 August [1915][1]

Dear Mr. Hertz,

He who has wandered aimlessly for so long in the chaos of possibilities understands your trials very well. You do not have the faintest idea what I had to go through as a mathematical ignoramus before coming into this harbor. Incidentally, your idea is very natural and would by all means be worth following up, if it could be carried through at all, which, based upon my experiences gathered during my wayward wanderings, I doubt very much.

Given an arbitrary manifold of 4 dimensions (given $g_{\mu\nu}(x_\sigma)$). How can one distinguish a coordinate system or a group of such?[2] This appears not to be possible in any way simpler than the one chosen by me.[3] I have groped around and tried all sorts of possibilities, e.g., required: The system must be chosen such that the equations

$$\sum_\nu \frac{\partial g^{\mu\nu}}{\partial x_\nu} = 0 \ (\mu = 1 - 4)$$

are satisfied throughout.[4]

At least it seemed definite to me *a priori* that a transformation group exceeding the Lorentz group must exist, because those observations summed up in the words "relativity principle" and "equivalency principle" point to it.

The coordinate limitation that was finally introduced deserves particular trust because it establishes a link between it and the postulate of the event's complete determination.

A theoretical differential geometric interpretation of preferred systems would be of great value. The weakest point of the theory as it stands today consists precisely in this, that the group of justified transformations are by no means closely assessable.[5] There is not even any *exact* proof that arbitrary motions can be transformed to motionlessness. This is because the difficulties connected with the dissimilarity of elliptic & hyperbolic types of differential equations stand in the way of a general observation. The equation

$$\frac{\partial^2 \varphi}{\partial x^2} + \frac{\partial^2 \varphi}{\partial y^2} = 0,$$

can be solved with arbitrarily given boundary values for φ

φ given

The equation

$$\frac{\partial^2 \varphi}{\partial x^2} - \frac{\partial^2 \varphi}{\partial y^2} = 0$$

by contrast, cannot. Now, how does it look for the complicated transformation conditions of the gen. theory of rel.? I am stuck there like a bewildered ox. Maybe we could gain an overview of the question if the geometric interpretation you are looking for is found.

Cordial greetings and best wishes for progress in your efforts! Yours,

A. Einstein.

I am leaving for Switzerland for about 3 weeks (26 Aug. until about 15 September. Address there for any letters: Prof. H. Zangger, Berg St., Zurich).

112. To Paul Ehrenfest

[Berlin, 23 August 1915]

Dear Ehrenfest,

Many thanks for your articles. The one on osmotic pressure is very fine and amazingly simple.[1] The suggestion I made to Lorentz was naive; I realize that now. The wish was greater than the thought. I would so much like to do something to keep our colleagues from the various fatherlands united.[2] Is this little clutch of busy intellectuals not our only "fatherland" to which such as we have any serious attachment? Should even *these* people have mentalities that are solely a function of residence?

I am sending two offprints for Nordström.[3] Do tell him that the "Tolman principle" is no more than the requirement of covariance under similarity transformations, i.e., under a very special type of *linear* transformation,[4] and that he unjustifiably contends that my theory is not in keeping with it;[5] as you know, the former maintains the covariance of *arbitrary* linear transformations. Also inform him that he has been recommended as second-place candidate to the extraordinary professorship at Berlin University; this was an indication of Planck's sense of justice.[6]

At the end of this week I am traveling to Switzerland for about 3 weeks. Then I shall be done with traveling for the time being. But I shall gladly descent on you another time, if possible with my Albert.[7]

Cordial greetings to you all, yours,

Einstein.

113. To Władysław Natanson

[Berlin,] Tuesday. [24 August 1915]

Dear Colleague,

I accept your friendly invitation with great pleasure and shall arrive punctually.[1]

Cordial greetings, yours,

Einstein.

I also extend my sincere thanks to Mrs. Natanson.

114.   To Elsa Einstein

[Heilbronn,] Monday. [30 August 1915][1]

Dear Else,

Yesterday (Sunday) afternoon 4 o'clock I arrived here safe and sound and was collected from the train by Mama.[2] She was tolerably satisfied with my appearance, & we very much enjoyed being together. Eating and mode of life strictly according to regulation. No visits are being paid. This morning I was at the police station and am sending you the passport herewith. Summon up all of your skills so that it goes quickly now; it will not be easy.[3] The photographs are enclosed.

I haven't experienced anything yet. Mama has still not given up hope of traveling to Switzerland.[4] I am supposed to help by sending a telegram from Switzerland.

Kisses to you and the children, yours,

Albert.

115.   To Elsa Einstein

[Heilbronn,] Friday. Evening [3 September 1915][1]

Dear Else,

Scarcely had my telegram left when your liberating postcard arrived. I hurried straight to the police station and head office, and an hour later had my passport ready in my pocket.[2] I know what hurdles and pains you must have endured for me and that you are my trusty lady and companion upon whom I can safely rely. When I return again, we'll meet and spend some time peacefully in the countryside. Don't let it be spoiled; we are going to be stuck in Berlin for long enough. I intend to arrive in Berlin Sunday evening.[3] There was little to complain about my health. I never had any pain. I did not move more than was necessary for my well-being and followed the diet strictly. It is even better here than in Berlin. Milk and honey may not be flowing any more, but there is still a decent trickle. You cannot get any fruit here, but there are vegetables.

The emotions surrounding my trip to Switzerland are somewhat mixed. Maja wrote me she would like to speak with me before I appear in Zurich.[4] This doesn't bode well.– I have read through almost all of Spinoza's *Ethics* here, much of it with great admiration.[5] Kraft[6] was very right in drawing my attention to this profound work. I believe it will leave a lasting impression on me.

Here people are much more confidently and aggressively inclined vis-à-vis politics than in Berlin, no doubt because the food situation is better. I have restrained myself much in conversations about such matters, if only because it is boring to say the same thing over and over again. Also, observation has also shown me that it is not words and ideas but direct, tough experience which guides the world.

When you write me, please send the letters & postcards to Maja. I don't know yet where I'll be living in Zurich, how long I'm going to stay there, or whether I'll go to Arosa first.[7] But I'll keep you nicely up-to-date about everything so that you don't get the feeling again that you've been abandoned by thankless me.

Kisses to you and the little minxes, yours,

Albert.

Kind regards to Uncle & Aunt.

116.   To Elsa Einstein

[Lucerne,] Saturday. [11 September 1915]

Dear Elsa,

My letter-writing sloth is indulging itself in veritable orgies. But I hope that your indignation will have dissipated by the time we see each other again. Twice I was together with the children.[1] After that, standstill. Cause: Mother's fear of the little ones becoming too dependent on me. Yesterday I arrived here, Monday morning I am returning to Zurich. Thursday or so I am going to return home to Berlin.[2] I am looking forward to it very much and shall not concede a single day. The outcome of the trip was moderate. Zangger has at least seen beyond doubt which side is the more decent. He will kindly ensure that I have the children for a while every year.[3] I am not visiting Romain Rolland;[4] there is no point. I really like being here; the two lead a very peaceful, pleasant life.[5] Today Edith Einstein will probably also be coming.[6] She wants to study mathematics. Yesterday we went sailing but unfortunately with no wind. In Zurich I was together with Besso very often;[7] my stay in Zurich was very much improved by it, but thus I neglected my duties to others.

Affectionate greetings, yours,

Albert.

117. To Elsa Einstein

Zurich, Monday [13 September 1915]

Dear Else,

I just received your express postcard from Heilbronn, late unfortunately, because I was at Maja's in Lucerne until this morning. Edith was also there.[1] Yesterday we drove up the Stanserhorn.[2] It was breathtakingly beautiful (sea of mist). In a few days I'm also going home. Why didn't you have me come for the homeward journey? I was so much looking forward to it. But *even so* I am looking forward to our reunion more than I can say. I saw the children only twice; *she*[3] appears to have become distrustful. Zangger will kindly try to smooth over the affair with the children (for the future).[4] I'm keen to know whether your coup has succeeded; I hope so!

Don't scold me for having written so little. You know me! I ought to go to Geneva for the sci[entists] meeting,[5] to Vevey to visit Romain Rolland,[6] and to see Uncle Caesar[7] in Geneva. But I am dispensing with all of this in order to be back in Berlin soon.

Heartfelt greetings, yours,

Albert.

118. To Romain Rolland

Zurich, 15 September [1915][1]

Highly esteemed Sir,

Your hearty invitation to visit you in Vevey has strengthened my inclination to drive there to make the acquaintance of one of the rare conciliatory Europeans. Well, my various obligations do not leave enough time for this trip;[2] so I am going to use my time in Switzerland to send you an uncensored letter at least. The "New Fatherland" association is going through quite difficult times; it is being harassed by the inspection authorities and being condemned (on the whole) by the press.[3] It looks as if the pro-military party and the pan-Germans have gained influence on the government through the successes in Russia.[4] But on the other hand, those among my acquaintances who are most discerning on economic matters do not have particularly great confidence; this seems to be connected to a shortage in certain raw materials.[5] Strangely enough, in Germany one finds, besides a curious self-conceit, a love for France and its population, whereas great animosity against England is quite universal.[6] Among the uncritical masses there is a quite general confidence in victory and an equally prevalent annexation greed. It is peculiar how, for his heavy sacrifices, the man on the street can feel recompensed by stolen territory, from which he certainly benefits little personally.—I hope it does not come to that! A decisive victory for Germany would be a misfortune to the whole of Europe, but especially to this country itself.[7]

One of the most disheartening phenomena of this terrible time is that in many cases intellectuals have completely lost their composure. I regret to have to say that the unfortunate and ridiculous war of words has already started among Berlin writers. You have certainly wondered that so many men, who in times of peace had justifiably been regarded as sensible, had signed the notorious "Manifesto to the Civilized World."[8] Condemnation of this step is now quite universal in Berlin as well. Incidentally, the best names had in fact lent their signatures *by telephone*, without having read the appeal![9] Recently there was a great furor at the Academy, because someone had responded to Violle's Academy speech by submitting a petition to sever ties with the French Academies. A curious thing happened then: almost all the historians, philologists, etc., supported the petition, while almost all the scientists and mathematicians worked avidly to maintain the international ties. Thank God, the latter won, if only by a small majority; Planck (a physicist) and Fischer (a chemist) deserve particular merit for their great resolve and firmness.[10]

I wish you good success in your philanthropic endeavors and send you my cordial greetings, your truly devoted servant,

A. Einstein.

Document description: "Confidential."

119.   To Heinrich Mousson

Zurich, 17 September 1915

Highly esteemed Education Director,

The question you raised[1] has been going through my mind very much still. I have the impression that I was not clear enough in one respect.[2] Under the prevailing circumstances the view suggests itself that *ceteris paribus*, selecting a Swiss would be preferable, and I must confess that I became fully aware of the importance of this motive only on speaking with men of high regard.

In consideration of this, I take the liberty of repeating to you that in my opinion, Mr. Piccard should be regarded as thoroughly on a par with Mr. Meyer, both as an academic member of the staff and as a person.[3]

With all due respect, yours very truly,

A. Einstein.

120.   To Heinrich Zangger

[Kreuzlingen,] Sunday, 3 o'clock [19 September 1915][1]

Dear friend Zangger,

He who goes on a journey has something to tell, but I particularly so. In Constance they found fault with my passport because *a visa from the German consulate in Zurich was missing. Ergo,* I was conveyed by honorary military escort (a "gray boy") to Kreuzlingen,[2] where I first looked around for a means of communication with Constance. Then I sent an initiated cyclist—naturally nothing in writing—over to the object of my desires in the Land of –Promise. Said object had already changed location, however, so that I would anyway have only found a vacated nest had my "brave field-grays" not saved me from disappointment. Now for the moment I am stuck in the Lions Hotel in *Kreuz*lingen* to rest on my momentary laurels. I am enclosing my passport with the request to have it viséed and thereupon sent back to me. Should the "powers-that-be" refuse to attach their blessing to the passport, however, please summon me back to Zurich by telegraph.

I shall always remember with gratitude the many interesting conversations and hours enjoyably spent otherwise together, in Zurich and on the road, no less for the kind solicitude with which you have brought my boys closer to me again. No less do I thank your wife,[3] for her friendly and at the same time salubrious hospitality.

Best regards from your stranded

Einstein

My farewells to Heller.[4] Greetings also to my boys, in case you see them. The young Pole appealed to me very much;[5] he is definitely a decent fellow, critical and yet not unhealthily skeptical.

*'Kreuz' alone translates as 'cross, crucifix'.

[1]Einstein's note to Hans Tanner: "I am returning home now."

121.   To Heinrich Zangger

[Kreuzlingen,] Tuesday. [21 September 1915][1]

Dear friend Zangger,

This printed leaflet is very sweet and there is some truth to it. But why is Delcassé made minister?[2] Why is revenge taught in schoolbooks? Why are parts of Turkey consumed while hissing vigorously at the Algeciras Treaty?[3] Why did Jaurès have to be murdered?–[4]

And yet! How nice France looks when I think of the others! Such a leaflet, even though its content is not quite accurate, proves nevertheless that people have a sense for decent and noble ways of thinking. It is this that I miss so sorely on the other side. A craving to dominate and a thirst for power have poisoned the minds of the upper class everywhere; but in France at least there was a lack of uniformity and consistency in the evil. Moreover, I am convinced that the gospel that brute armed force is the basis of all existence finds its main support in Germany. Even the disgraceful regiment in Russia finds powerful support in the German ruling class.

I am very pleased that in Switzerland awareness of the seat of the real danger is making headway, and I hope confidently that this simple recognition will prevail despite the press. The press! The article in *Volksrecht* on the submission by intellectuals regarding the annexation question[5] has been silently ignored by the bourgeois press. Is it fear? Have scholars of distinction lent their names elsewhere to such a thing?

I am sorry to bother you with my passport.[6] But I could hardly come back, as the duties of the various farewells I had escaped so lightly seemed so menacing to me. I quite like being stuck here for a few days in rural tranquillity; I would be inclined to wish the same to you.

Many thanks and cordial greetings, yours,

Einstein.

Just now your letter and the passport arrived. Heartfelt thanks once again. I really liked the Pole.[7] He has potential.

Today out of boredom I attended the local teachers' seminar to see what they are up to, and there I meet a teacher who had formerly attended my lectures.[8] Isn't that amusing?

122.   To Hendrik A. Lorentz

[Berlin,] 23 September 1915

Dear Colleague,

Your letter came into my hands only yesterday evening, as I was in Switzerland for some weeks to visit my children.[1] I agree completely with the deletion you have suggested and acknowledge fully the correctness of your arguments in this regard. I should think I may presume Grüneisen's consent with the deletion and willingly assume responsibility for it.[2]

I am pleased that you and de Haas have met with success in analyzing the magnetic effect.[3] That method of attaching the coil to the iron core in order to render harmless the obscuring interactions between coil and core had been conceived of by de Haas alone, by the way, which he in his modest way does not seem to have told you.[4] Now I also have thought up a simple method to eliminate that source of error; but as long as I have not tried it, I do not want to be too trusting. It involves commutating the core by a single *brief* electrical impulse in the (stationary) coil. I shall see if I can manage to do it. It is a very simple experiment, should it work; I shall write you and de Haas about it.[5]

I am very glad that de Haas now has the move successfully behind him and is living and working in a healthy and enjoyable environment.[6] Warmest greetings to them both and to their children. My thanks for now for Mrs. de Haas's friendly letter; I am going to write to them when I have something to report on the planned experiments. Today I had great pleasure seeing many of the local scholars of high regard declare their strong opposition to the annexations in a submission to the Chancellor. You are right after all with your fine optimism and your faith in advancement. It is to be hoped that the good and reasonable gain the upper hand.[7]

I did not understand your conclusion regarding the damping of sodium light in the atmosphere. I do not see why g_2 should not be able to be arbitrarily larger than g_1, because of the validity of Rayleigh's formula; for the weakening of light intensity related to g_2 would clearly contribute nothing to the *scattering*, but would relate to a real absorption of light (conversion into heat).[8] You say nothing, however, about how we know that such real absorption does not take place.—I have probably misunderstood you.

Your article delighted me.[9] I have also found a proof for the validity of the [relativistic] energy-momentum conservation principle for the electromagnet. field taking gravitation into consideration, as well as a simplified covariant theoretical representation of the vacuum equations, in which the "dual" six tensor [*Sechservektor*] concept proves unessential.[10] At the moment I am occupied with studying your paper. Besides, I am reading with admiration the clever pamphlet

No. 3 by Nico Van Suchtelens.[11] I shall probably scarcely be able to bring about a formal revocation of the notorious manifesto,[12] even though the realization is forcing its way through that it was a quite unfortunate and ill-considered step.[13] I think it more important that persons of good will unite in things of *genuine* importance for the future. It is not for revocation that the remorseful are primarily known.–

Warmest greetings, yours,

A. Einstein

Please kindly reciprocate for me your wife's greetings.

123.   To Erwin Freundlich

[Berlin,] 30 September [1915][1]

Dear Freundlich,

I shall be glad to write to Naumann, as soon as in the next day or so. Tomorrow morning I am going to see Planck, with whom I shall also discuss it.[2] I am writing you now about a scientific matter that electrifies me enormously. For I have come upon a logical contradiction of a quantitative nature in the theory of gravitation, which proves to me that there must be a calculational error somewhere within my framework.

Imagine an infinitely slowly rotating coordinate system (rotation velocity ω). In this one, as can easily be shown by simple transformation, the gravitational field is given by the $g_{\mu\nu}$ system

$$
\begin{array}{cccc}
-1 & 0 & 0 & \omega y \\
0 & -1 & 0 & -\omega x \\
0 & 0 & -1 & 0 \\
\omega y & -\omega x & 0 & 1
\end{array}
$$

Using the formulas, I can now calculate the closest approximation (terms proportional to ω^2) and find from the last gravitational field equation

$$
g_{44} = 1 - \frac{3}{4}\omega^2(x^2 + y^2),
$$

while the direct transformation from the Galilean case yields

$$
g_{44} = 1 - \omega^2(x^2 + y^2).
$$

This is a blatant contradiction.[3] I do not doubt, therefore, that the theory covering perihelion motion is suffering from the same fault.[4] Either the equations are already numerically incorrect (numerical coefficients), or I am applying the equations in a principally incorrect way.   I do not believe that I myself am in the position to find the error, because my mind follows the same old rut too much in this matter. Rather, I must depend on a fellow human being with unspoiled brain matter to find the error. If you have time, do not fail to study the topic.

With best regards, yours,

A. Einstein.

124.   To Otto Naumann

[Berlin, after 1 October 1915][1]

Letter to Naumann.

A day or so ago Dr. Freundlich from the N[2] Observatory called on me. He told me that on the occasion of an official meeting you had indicated to him allusively the possibility that he be released from his duties as assistant at the Observatory for some years,[3] without his salary as assistant being cut off. I was extremely pleased about this information and likewise my colleague Planck, who recently encouraged me to ask you by letter to be sure not to abandon this liberating idea.

This man is crushingly burdened with the responsibilities of an extremely tiring and encumbering job that is preventing him from pursuing the questions currently so important to us. This work (measurement of the sky coordinates of thousands of fixed stars) can be performed just as well by any trained individual equipped with a healthy mind as by a person who has his own ideas and the initiative necessary to implement them! The smaller independent researches Dr. Freundlich had accomplished previously[4] under the burdensome pressure on him are already of great value and justify expecting worthwhile results if he were given the freedom and opportunity to conduct astronomical observations.

It is correct that by relieving Dr. Freundlich of his present responsibilities, a temporary increase in expenditures for the Observatory could arise. ⟨But if one considers the minimal practical efficiency that research-oriented budgeted institutes generally can show for themselves, in my view, one should not think of abandoning the goal for reasons of economy in a case where there are such good prospects of success, and the funds to be applied, relatively low.⟩ It would possibly involve employing provisionally an additional worker who would continue that cataloguing work from which Mr. Freundlich [would] be released. Maybe this would not be required either. In any case, the prospects of scientifically valuable results are very great against any funds to be expended. I therefore sincerely hope that the matter will not fall through because of the extreme cost-saving constraints brought on by these hard times.[5]

125.   To Paul Hertz

[Berlin, before 8 October 1915][1]

Dear Colleague,

Your carefulness can in any case do no harm, although it does seem to me to be a bit exaggerated. That article impresses me little; the paper is forbearant.[2] An associated member is only of value to the association in that it is allowed to send its printed material to such members, while it is not permitted to do so to completely private individuals. For you, membership only has the value that you receive such written material.[3] If you prefer, you could also do it via a straw man, if you do not like personally being an assoc. member. I must say, though, that this type of cautiousness, i.e., not standing up for one's rights, is the cause of the entire wretched pol[itical] situation.[4] We can discuss this when we see each other again. Best regards, also to your wife, yours,

A. Einstein.

126.   To Paul Hertz

[Berlin, before 8 October 1915][1]

Dear Mr. Hertz,

Each of your letters has given me a quarter of an hour of amusement. Resign instantly from the association;[2] for you have that type of valiant mentality the ruling powers love so much in Germans. People like you from the most highly educated class offer the best guarantee for the preservation of the political quagmire (to life everlasting, amen). Do not take offense at my coarseness, but rest assured that I am fond of your mentality, even though I regret the weakness of your spine. The latter I do not consider an inborn trait but a fruit of one's upbringing . . . .

Cordial greetings from your sweet-tempered

Einstein.

127.   From Paul Hertz

Göttingen, 8 October [1915][1]

Highly esteemed Professor,

Had you devoted as much care to understanding people as to understanding nature, you would not have written me any insulting letter.[2] I am firmly convinced that this was not legitimized by my letters, which I certainly have no cause to justify to you, leaving out of consideration that I find a continuation of our correspondence impossible.

Respectfully,

P. Hertz.

128.   To Paul Hertz

[Berlin,] 9 October 1915

Dear Mr. Hertz,

I cannot stand knowing that I have offended you.[1] You *must* forgive me, particularly in consideration of the fact that I—as you yourself say with justification—have *not* bestowed the same care to understanding people as to understanding nature. (How would one in the contrary case get around to dealing with nature?) I am willing to do anything to make amends to you and beg you only to inform me soon that you again want to hold out your hand in friendship. My emotions ran away with me. Even in case your indignation prompts you to be permanently angry with me, I shall maintain a friendly attitude toward you.

With cordial greetings, yours,

A. Einstein.

129.   To Hendrik A. Lorentz

[Berlin,] 12 October 1915

Highly esteemed, dear Colleague,

Subsequent reflections on the last letter I sent you have revealed that I made erroneous assertions in that letter. In actual fact the invariant theory method does not yield more than Hamilton's principle when determining your function $Q(= H\sqrt{-g})$.[1] The reason why I did not notice this last year is that on page 1069 of my article[2] I had frivolously introduced the condition that H was invariant against *linear* transformations. If this condition is relinquished, then the following result is obtained.

However Q is selected, if the coordinate system is chosen so that with a given gravitational field J becomes an extremum[3] through the choice of coordinates, or that

$$C_\mu = B_\mu - \sum_\lambda \frac{\partial S_\mu^\lambda}{\partial x_\lambda} = 0$$

where $S_\mu^\lambda = \sum_{\sigma\nu} \left(g^{\nu\lambda} \frac{\partial Q}{\partial g^{\mu\nu}} + g_\sigma^{\nu\lambda} \frac{\partial Q}{\partial g_\sigma^{\mu\nu}} - \frac{1}{2} g_\mu^{\sigma\nu} \frac{\partial Q}{\partial g_\lambda^{\sigma\nu}} + \frac{1}{2} Q \delta_\mu^\lambda \right)$,

then

$$\frac{\partial Q}{\partial g^{\mu\nu}} - \sum \frac{\partial}{\partial x_\sigma} \left(-\frac{\partial Q}{\partial g_\sigma^{\mu\nu}} \right)$$

is always a tensor related to such coordinate systems.[4] The postulate of covariance or relativity thus cannot serve to determine function Q.

This determination is best based upon the following physical postulate.[5]

The field equations read in mixed form as

$$-\sum \frac{\partial}{\partial x_\sigma} \left(g^{\nu\lambda} \frac{\partial Q}{\partial g_\sigma^{\mu\nu}} \right) = \kappa \mathfrak{T}_\mu^\lambda + \left(-\sum_\nu g^{\nu\lambda} \frac{\partial Q}{\partial g^{\mu\nu}} - \sum_{\nu\sigma} g_\sigma^{\nu\lambda} \frac{\partial Q}{\partial g_\sigma^{\mu\nu}} \right),$$

the conservation equations,[6]

$$\left. \begin{aligned} &\sum_\lambda \frac{\partial}{\partial x_\lambda} (\mathfrak{T}_\mu^\lambda + t_\mu^\lambda) = 0, \\[2mm] &\text{where } t_\mu^\lambda = \frac{1}{2\kappa} \sum_{\sigma\nu} \left(-g_\mu^{\nu\sigma} \frac{\partial Q}{\partial g_\lambda^{\nu\sigma}} + Q \delta_\mu^\lambda \right) \end{aligned} \right\} .$$

The gravitational field's divergence must be determined according to the field equations from the sum of the gravitational masses (energies) *of both the matter and the gravitational field*. This applies in our field equation only when the second term on the right-hand side is equated with the gravitational field's energy tensor t_μ^λ multiplied by κ.

Thus we arrive at the condition

$$S_\mu^\lambda \equiv 0.$$

This is simultaneously the condition for QdV being an invariant in relation to *linear* substitutions.[7] The latter situation would be trivial on its own. But it facilitates seeking out Q. For it follows directly from this invariance that $\dfrac{Q}{\sqrt{-g}}$ must be a linear homog. function of the five expressions on page 1075 below of my article. That $\dfrac{Q}{\sqrt{-g}}$ had been set by me as equal to the fourth expression given there can be justified by the fact that only with this choice does the theory contain Newton's in approximation. That I believed it possible to support this selection on the S_μ^λ equation was based on error.[8]

Cordial greetings from your

A. Einstein.

130. To Heinrich Zangger

[Berlin,] Friday. [15 October 1915][1]

Dear friend Zangger,

I very much like the printer's proof you sent me; not exaggeratedly abstract and yet sufficiently general.[2] Since coming here[3] I have been working very solidly in my cubbyhole. It has unfortunately become clear to me now that the "new stars" have nothing to do with the "lens effect," moreover that, taking into account the stellar densities existing in the sky, the latter must be such an incredibly rare phenomenon that it would probably be futile to expect one of the like.[4] I wrote a supplementary paper to my last year's analysis on general relativity.[5] Presently I am researching a bit in heat theory.[6] However, I have not yet begun the experiment. Warburg requested that I wait a few days.[7] I haven't seen any of my colleagues, as the Academy has not started yet[8]—I still think back with bliss on the Swiss times of free speech [*Schnabelfreiheit*]. . . .

The serious accident in the Bieler Lake region, about which you wrote me, upset me very much. You complain about the thoughtless manner in which people allow disasters to happen.[9] But it is much more dire when things become a calamity precisely because they are in the focus of human attention!

I am very glad that my boy has come to see you. Do ask him whether he has received my long letter; I almost fear he hasn't, because I have not received a reply. Today I am going to try once again. But if I see that the woman is thwarting all of my efforts to remain in contact with the boy, I shall use legal means after all to enforce that the boy spend a holiday month with me every year. Only in this way will he lose the feeling that he has lost his father. I can't understand at all that she doesn't consider it her responsibility to leave the boy with me occasionally. I noticed distinctly that in front of others Albert and his mother were uncomfortable that I was in Zurich and did not stay with them.[10] I understand that. But precisely for this it would be right for the boy to be together with me somewhere else, where regard for other people's gossip does not interfere obtrusively.

I am very much looking forward to being able to receive you at my home. You had led me to expect you to come quite definitely. Now, however, you didn't mention anything more about it in any of your letters.

Affectionate greetings, yours,

Einstein.

131. To Walther Schücking

Berlin, 13 Wittelsbacher St., 22 October 1915

Highly esteemed Colleague,[1]

From Miss Jannasch I hear that you would like to have me included in the Anti-Oorlog Raad's Gr. Council.[2] I have absolutely no experience and am not an efficient person in political affairs. Nonetheless, I am quite willing to support this splendid cause.[3] So do include me, if you consider it fruitful. I must inform you, however, that I am *Swiss* and consequently cannot figure as a German.

With all due respect, yours very truly,

A. Einstein
A brother of the unfortunate "Fatherland."

132. To Berliner Goethebund

Berlin, [after 23 October 1915][1]

To the Berliner Goethebund.[2]

Highly esteemed Sirs,

On the enclosed sheet I have—pursuant to your kind letter of October 23rd— noted down briefly my opinion on war.[3] I was encouraged to do so by your closing remark in your letter, according to which complete freedom was given with regard to form and inclination. Of course, I shall not be at all surprised or even annoyed if you do not make use of my comments. In this case, however, I request that you send the same back to me.

I beg you not to be angry with me for my words; I assure you that what I said is my conviction.[4]

With all due respect, yours very truly,

A. Einstein.

133.   From Michele Besso

[Zurich, ca. 30 October 1915]

We acknowledge the legitimacy of your wish to communicate with your children without disturbance, but also the legitimacy of your wife's reservations that this not take place close to your Berlin relatives.[1] The discord resulting *from this* can only eventually place the child's emotional harmony at risk. If you enforced it, then in both of our views[2] the time would inevitably come when the suffering caused by it, also for you, would far outweigh the present conveniences. For *the children*, no contact would without a doubt be better than contact *in Berlin*, therefore the desirability of a stay, initially of the older boy, together with you at a good Swiss inn or boarding house or of a trip through Switzerland for the duration of time you can devote to your boy.

We also acknowledge that you both, your wife and you, are entitled to mutual good faith and respect, which you are to show toward each other; in particular, there is no basis (to the extent that lawyer's tricks are disapproved of, as is *your* way of thinking) for denying the woman a written confirmation of the conditions under which you are willing to meet the (specified) financial obligations, so as to exclude the possibility that a means of financial pressure remain that could be used toward achieving some unforeseen purpose.

Independent, in principle, of the above is the question of your divorce and remarriage. We both certainly dearly hope that your contacts with the children are arranged favorably, also because it would make it easier for you to maintain your present single state, which undoubtedly ought to be the *only* desirable and tolerable state for you during your holidays, knowing how people unfortunately are, as well as for your mental strength, which is being taxed by powerful "otherworldly" interests. You told us, as you know, how an almost churchlike atmosphere is pervading your desolate house now. And justifiably so, for unusual divine powers are at work in there.

––––––––––––

Your Albert will visit us (Besso) this winter during the Christmas holidays simultaneously with Vero; perhaps you would like to spend that time in part or entirely at the same place.[3] You know, of course, how much this would please me personally as well.

⟨As concerns your money matters, I point out that you are not dependent on the friendly efforts of anyone, but that every large bank, e.g., the Schweizerische Kreditanstalt, which has a special fiduciary division, will handle everything for you at a reasonable price.⟩ [Should the matter of Albert E.'s sister, Lucerne, also be brought up?][4]

134.   To Hans Albert Einstein

[Berlin,] 4 November [1915][1]

My dear Albert,

Yesterday I received your dear little letter and was delighted with it. I was already afraid you didn't want to write to me at all anymore. While I was in Zurich you told me that you find it unpleasant when I come to Zurich.[2] That is why I thought it better if we were together somewhere else where we can relax cozily with no one disturbing us.[3] In any case, I shall press for our being together every year for a month so that you see that you have a father who is attached to you and loves you. You can learn a lot of fine and good things from me as well that no one else can offer you so easily. What I have gained from so much strenuous labor should not be there only for strangers but especially for my own boys. In the last few days I completed one of the finest papers of my life;[4] when you are older I'll tell you about it.

I am very glad that you enjoy playing piano. This and carpentry are, I think, favorite pastimes at your age, even more so than school. These are things that fit a young person like you very well. On the piano, play mainly the things that you enjoy, even if your teacher doesn't assign them to you. You learn the most from things you enjoy doing so much that you don't even notice time passing. I am often so engrossed in my work that I forget to eat lunch. Also play ring toss with Tete. You gain dexterity from that. Go to my friend Zangger once in a while as well. He is a nice man.

Kisses to you and Tete from your

Papa.

Regards to Mama.

135. From Mileva Einstein-Marić

Zurich, 5 November 1915

Dear Albert,

I have received the life insurance money and have paid it out. I would also like to turn your attention to a few things: Wouldn't you prefer to write your communications to little Albert on postcards, at least some of them, since it seems to me that longer letters are subject to longer delays along the way than postcards. Some of my acquaintances have also made this observation. I am writing this because I find that long intervals like this first one before your first letter arrived are more liable to aggravate bitterness in the child, and I would like to spare him this and hope that this is your wish too. In addition, I'd also like to request that matters relating to the children first be arranged *with me.* If you were also to inform me *in good time* of your particular wishes and opinions, then a way could surely always be found to address them, even if they couldn't be carried out in the form presented by you. Also believe me that if Albert had the feeling that what is being demanded of him is done with the consent of both parents, he would much sooner succeed in calmly appreciating you than if he has the feeling that you were working as an enemy of this little world I have built up here for the children in which they are living and which they love. I remember so well how much you used to love this little Albert and consider it impossible that you would not want to help me remove bitterness from his life, and not increase it.

Miza.

136. To David Hilbert

[Berlin,] Sunday. [7 November 1915][1]

Highly esteemed Colleague,

With return post I am sending you the correction to a paper in which I changed the gravitation equations,[2] after having myself noticed about 4 weeks ago that my method of proof was a fallacious one.[3] My colleague Sommerfeld wrote that you also have found a hair in my soup that has spoiled it entirely for you.[4] I am curious whether you will take kindly to this new solution.

With cordial greetings, yours,

A. Einstein.

When may I expect the mechanics and history week to take place in Göttingen? I am looking forward to it very much.

137.   From Max Planck

Grunewald, 7 November 1915

Dear Colleague,

I have now performed the comparison of my formulas to those of Tetrode and proceeded in the following manner.[1]

If formulas (17) and (16) in Tetrode's paper are subtracted from each other, the *entropy originating from the rotations of the diatomic molecules* results:[2]

$$S_{(17)} - S_{(16)} = kN\{\ln(kT) + \ln(2\pi J) - 2\ln h + \ln(4\pi) + 1\}$$

The same quantity results from the theory developed by me based on the formula for the thermodynamic function I recently set down in writing:[3]

$$\Psi = -\frac{F}{T} = Nk \ln \sum_{0}^{\infty} (2n+1) \cdot e^{-(n^2+n+\frac{1}{2})\sigma},$$

where F is the free energy, $\sigma = \dfrac{h^2}{8\pi^2 JkT}$, when you assume T there very large.[4]
Then the sum may be written as an integral, namely:

$$F = -TNk \ln \int_{0}^{\infty} dn \cdot 2 \cdot \left(n + \frac{1}{2}\right) \cdot e^{-[(n+\frac{1}{2})^2 + \frac{1}{4}]\sigma}$$

$$= -TNk \ln \left(e^{-\frac{\sigma}{4}} \cdot \frac{1}{\sigma}\right) = TNk \left(\frac{\sigma}{4} + \ln \sigma\right).$$

If you consider now that

$$S = -\frac{\partial F}{\partial T},$$

then, through substitution of the value for σ, we get:

$$S = Nk \ln \frac{8\pi^2 JkTe}{h^2}.$$

This is *precisely* Tetrode's value.[5]

I now applied the theory for an arbitrary (also "incoherent," or however else you want to express it) degree of freedom as well.[6] It all works very well and reliably,[7] [Some of the sidetracks were horrendous, though. But you probably understand that.] so that even the specific heat of polyatomic (rigid) molecules as well as the energy of spatial oscillators are easily calculated. I shall tell you about that in person.

With cordial greetings, yours,

Planck.

138.   To Berliner Goethebund

[Berlin,] 11 November 1915

To the Berliner Goethebund.

Highly esteemed Sirs,

If I am to express my opinion on war in a truthful manner,[1] I must say in some form that the glorification of war in peace as well as the fostering of these idea-emotion complexes which prepare for war in peace, must be counteracted energetically by all genuine friends of human progress. This includes in my view *everything* that falls under the term "patriotism." I must agree unconditionally with Tolstoy in this respect.[2] If I cannot express this idea resolutely, then my contribution to the work undertaken by you awakens a wrong impression about my attitude in this matter, which I would definitely like to avoid.

I ask you please to inform me upon assessment of this point whether you consider the publication of a contribution by me desirable. Should this be the case, I would make an effort to express my opinion in another way.[3] But perhaps it is better to abandon the contribution entirely, so that it is not placed, quite rightly, under the heading of a memorable article written by a political commentator recently: "The Expert as a Layman"!

With all due respect, yours very truly,

A. Einstein.

139.   To David Hilbert

[Berlin,] Friday. [12 November 1915][1]

Highly esteemed Colleague,

I just thank you cordially for the time being for your kind letter. The problem has meanwhile made new progress. Namely, it is possible to exact *general* covariance from the postulate $\sqrt{-g} = 1$; Riemann's tensor then delivers the gravitation equations directly. If my present modification (which does not change the equations) is legitimate, then gravitation must play a fundamental role in the composition of matter.[2] My own curiosity is interfering with my work! I am sending you two copies of last year's paper.[3] I have only two other intact copies myself. If someone else needs the paper, he can easily purchase one, of course, for 2 M (as an Academy offprint).

Cordial greetings, yours,

Einstein.

140.   From David Hilbert

[Göttingen, 13 November 1915][1]

Dear Colleague,

Actually, I first wanted to think of a very palpable application for physicists, namely reliable relations between the physical constants, before obliging with my axiomatic solution to your great problem. But since you are so interested, I would like to lay out my th[eory] in very complete detail on the coming Tuesday, that is, the day after the day after tomorrow (the 16th of this mo.).[2] I find it ideally handsome math[ematically] and absolutely compelling according to ax-iom[atic] meth[od], even to the extent that not quite transparent calculations do not occur at all and therefore rely on its factuality.[3] As a result of a gen. math. law, the (generalized Maxwellian) electrody. eqs. appear as a math. consequence of the gravitation eqs., such that gravitation and electrodynamics are actually nothing different at all.[4] Furthermore, my energy concept forms the basis:[5] $E = \sum(e_s t^s + e_{ih} t^{ih})$, which is likewise a general invariant, and from this then also follow from a very simple axiom the 4 missing "space-time equations" $e_s = 0$. I derived most pleasure in the discovery already discussed with Sommerfeld that normal electrical energy results when a specific absolute invariant is differentiated from the gravitation potentials and then g is set $= 0.1.-$[6] My request is thus to come for Tuesday. You can arrive at 3 or $\frac{1}{2}$ past 5. The Math. Soc. meets at 6 o'clock in the auditorium building. My wife[7] and I would be very pleased if you stayed with us. It would be better still if you came already on Monday, since we have the phys. colloquium on Monday, 6 o'clock, at the phys. institute. With all good wishes and in the hope of soon meeting again, yours,

Hilbert.

As far as I understand your new pap[er], the solution giv[en] by you is entirely different from mine, especially since my e_s's must also necessarily contain the electrical potential.[8]

Document description: "Continuation on Sheet I with the invitation to come for Tues-day, 6 o'clock. Best regards, H."

141.   To Wander and Geertruida de Haas

[Berlin, before 15 November 1915][1]

My dear de Haases,

I was very pleased with your exquisite little letter, from which one can immediately feel how happily and contentedly you are living in your little nest. I definitely believe that you have done the right thing with your decision. 23 hours are a lot, though. But teaching is very amusing, especially when facing sturdy, unspoiled country children with healthy nerves.[2] I believe that I also could accustom myself with relish to such a position even now.[3]

As far as I can see, not much can be done in the laboratory with gravitation. The negative result on the independence of gravitational acceleration from matter is probably adequately secure. It would suffice for me, even without any experiments with radioactive substances whatsoever.[4] In the end, because we can observe their transformations, these substances will not exhibit any *unusual* inertial behavior.

I am very eager to see your paper on the effect.[5] I have also conducted experiments in which I reversed the remanent magnetism by means of a condenser's discharge current.[6] But the thing has not worked until now because, despite the short duration of the field ($10^{-3''}$), an intense vibration of the rod ensued, obliterating the effect.[7] This will naturally be avoided with your method. I hardly believe that your 10% discrepancy with the theory is real. If it were so, however, it would be very important. As a matter of fact, I still do not know what Maxwell has done on this subject.[8] At any rate it is evident here as well what a good nose this man had! Barnett's comment actually gave me little confidence.[9]

Working analogously with metal electrons is very difficult. But you are at the best location for this enterprise, of course.

I congratulate you particularly on your lathe and on its attached mechanic;[10] the thing has style.

Cordial greetings to you, your wife, and the children and a happy New Year, yours,

A. Einstein.

[10]Recipient's note: "who could work only with the bad lathe and not with the good one."

142.   To Hans Albert Einstein

[Berlin, 15 November 1915]

My dear Albert,

I would like to come to Switzerland around New Year's, in order to spend a couple of days with you. Prof. Zangger has surely already made the suggestion to you.[1] Where would you like to go with me most? Naturally, not very far away from Zurich. Prof. Zangger thought we ought to go to Mr. Besso's.[2] I am not sure whether you'd like that, though. Maybe it would be better if we were alone somewhere. What do you think?

Kisses to you and Tete from your

Papa.

143.   To Mileva Einstein-Marić

[Berlin, 15 November 1915]

D[ear] Mileva,

Your letter[1] honestly pleased me, because I can draw from it that you don't want to undermine my relations with the boys. I tell you, for my part, that these relations make up the most important *personal* element in my life. In the future I'll settle everything that involves them with *you*,[2] if I find your letter confirmed. I intend to travel to Switzerland at the turn of the year, in order to see at least Albert outside of Zurich and to spend a few days with him. Do you have any particular wish regarding the choice of location?

Best regards, yours,

Albert Einstein.

144.   To David Hilbert

[Berlin,] Monday. [15 November 1915][1]

Highly esteemed Colleague,

Your analysis interests me tremendously, especially since I often racked my brains to construct a bridge between gravitation and electromagnetics. The hints you give in your postcards[2] awaken the greatest of expectations. Nevertheless, I must refrain from traveling to Göttingen for the moment and rather must wait patiently until I can study your system from the printed article;[3] for I am tired out and plagued with stomach pains besides.   If possible, please send me a correction proof of your study to mitigate my impatience.

With best regards and cordial thanks, also to Mrs. Hilbert, yours,

A. Einstein.

145.   From Max Planck

[Berlin, 15 November 1915]

Dear Colleague,

Priceless—I nearly split my sides, I really did. I am not quite satisfied with one portrait, though, that is your own. May I suggest a few corrections to choose from? I do it at the risk of being misunderstood: either: genius, or: woodworm. Now, take your pick.

Yours,

Planck.

Unfortunately, I must postpone a closer study of Ritz.[1]

146.   To Berliner Goethebund

[Berlin,] 16 November 1915

Highly esteemed Sirs,

I have tried to fulfill your request without becoming in any way untruthful, and without having to say anything directly about patriotism.[1] Naturally, I do not delude myself in the least that I could say anything new or even original about these general matters. But I certainly do believe that in matters concerning the general public, an effect can only be achieved through relentless repetition; the success of advertising certainly proves this in a comical way.

Should you find that the thus modified content also jars against the finer senses that ally you with the local citizenry, please send me another note.

With all due respect,

A. Einstein.

147.   To Michele Besso

[Berlin, 17 November 1915]

Dear Michele,

Just now I received your letter.[1] I am coming, and am very much looking forward to it. I am enclosing the magnetism paper.[2] In these last months I had great success in my work. *Generally covariant* gravitation equations. *Perihelion motions explained quantitatively.* The role of gravitation in the structure of matter.[3] You will be astonished. I worked horrendously intensely; it is strange that it is sustainable. I am very much looking forward to our reunion. Affectionate greetings to you, Papa,[4] Anna, Vero, yours,

Albert.

148.   To David Hilbert

[Berlin, 18 November 1915]

Dear Colleague,

The system you furnish agrees—as far as I can see—exactly with what I found in the last few weeks and have presented to the Academy.[1] The difficulty was not in finding generally covariant equations for the $g_{\mu\nu}$'s; for this is easily achieved with the aid of Riemann's tensor. Rather, it was hard to recognize that these equations are a generalization, that is, a simple and natural generalization of Newton's law. It has just been in the last few weeks that I succeeded in this (I sent you my first communication[2]), whereas 3 years ago with my friend Grossmann I had already taken into consideration the only possible generally covariant equations, which have now been shown to be the correct ones. We had only heavy-heartedly distanced ourselves from it, because it seemed to me that the physical discussion yielded an incongruency with Newton's law.—[3] The important thing is that the difficulties have now been overcome. Today I am presenting to the Academy a paper in which I derive quantitatively out of general relativity, without any guiding hypothesis, the perihelion motion of Mercury discovered by Le Verrier.[4] No gravitation theory had achieved this until now.

Best regards, yours,

Einstein.

149.   From David Hilbert

[Göttingen, 19 November 1915]

Dear Colleague,

Many thanks for your postcard and cordial congratulations on conquering perihelion motion.[1] If I could calculate as rapidly as you,[2] in my equations the electron would correspondingly have to capitulate, and simultaneously the hydrogen atom would have to produce its note of apology about why it does not radiate.[3]

I would be grateful if you were to continue to keep me up-to-date on your latest advances.

Greetings, yours,

Hilbert.

150.   To Hans Albert Einstein

[Berlin, 23 November 1915]

My dear Albert,

So, we would like to meet in Winterthur and drive together to Krummenau where Mr. Besso lives.[1] We shall both stay at the inn there, though. I'm looking forward to it very much, especially since I've been working terribly hard recently. Write to me very soon about which day we should meet, since preparations for such a trip take up a lot of time now during the war. Do write me sometime soon! You are almost as lazy as your old father at writing.

Kisses to you and Tete from your

Papa.

Regards to Mama.

151.   To Erwin Freundlich

Berlin, Wednesday [24 November 1915][1]

Dear Freundlich,

I just came back from Naumann,[2] who has answered me now after all. I poured him a heavy dose of the truth, commended you highly, and took Struve's protectionism soundly to task.[3] I asked him sincerely to give you the observer position.[4] The man is imposing and not unlikable. He listened to everything I said but promised nothing, so that I do not know whether my errand has had any direct effect. I did have that impression, though. I prepared him for the pretexts S. would respond with, to which no importance ought to be given, and asked him also to consult with Planck.[5] On the other hand, I noticed that he is reluctant to communicate directly with the people from the observatory. Thus I do not believe that he will have you summoned.[6] Undoubtedly I did arouse in him a powerful distrust for all time in Struve's suggestions.

Best regards, yours,

Einstein.

152. To Heinrich Zangger

[Berlin,] Friday. [26 November 1915][1]

Dear friend Zangger,

With dismay I see that you must rest and that you are in pain;[2] I did not understand the Latin name of the cause. Please tell me in German or, preferably, tell me that you are completely healthy again.

The affair involving Heller makes me sorry but does not surprise me.[3] As a worker he is decidedly second-rate. There is no room for sentimentality. You will probably have to decide on a long-term dismissal. In the end you do owe it to the others and to yourself.

To R. R.[4] I have nothing to say that could interest him and that could be told to him along this route. It is of no relevance anyway. I am going to try to procure for myself the booklet by Klein.[5]

The general relativity problem is now finally dealt with.[6] The perihelion motion of Mercury is explained wonderfully by the theory.[7]

Rotation of the entire orbit around the Sun in the amount of ca. 45 arc sec. in 100 years in the sense indic[ated]

Astronomers have found from observations[8]

$$45'' \pm 5''.$$

I have found with the general theory of rel.

$$43''.$$

Add to this the line shift for fixed stars which, as you know, has also been established securely,[9] thus this is already considerable confirmation of the theory. For the deflection of light by stars, the theory now provides an amount twice as large as before.[10] I shall tell you verbally how this comes about.

The theory is beautiful beyond comparison. However, only *one* colleague has really understood it, and he is seeking to "partake" [*nostrifizieren*] in it (Abraham's expression) in a clever way.[11] In my personal experience I have hardly come to know the wretchedness of mankind better than as a result of this theory and everything connected to it. But it does not bother me.

My boy still has not responded to my inquiry about the meeting in Krummenau.[12] This must surely be traced back to the woman's influence. You will see more and more on which side to look for good will and honesty. There are reasons why I could not endure being with this woman any longer, despite the tender love that ties me to the children. At the time we were separating from each other, the thought of the children stabbed me like a dagger every morning when I awoke; I have never regretted the step in spite of it.

I have become a member of the Gr[eat] Council of the Dutch Anti-Oorlog Raad.[13] The times show that everyone must do his bit toward the organization of the whole. Unfortunately so many intelligent people are acting to the contrary.

Your allusion to avoiding distractions surely regards my relationship with my cousin.[14] With all due respect for her decent character and kindheartedness, and also taking into consideration that she has an adult little daughter of 18 years of age,[15] I cannot resolve on a second marriage, not even one with two households. Snobbism here has developed to such an extent that these women do not lose but, on the contrary, increase their standing through me. The attempts to force me into marriage come from my cousin's parents[16] and is mainly attributable to vanity, although moral prejudice, which is still very much alive in the old generation, also plays a part. If I let myself be trapped, my life would become complicated, and above all, it would probably be a heavy blow for my boys. Therefore, I believe I must not allow myself to be moved either by my inclination or by tears, but must rather remain as I am.

Affectionate greetings, yours,

Einstein.

153.   To Arnold Sommerfeld

Berlin, 28 November [1915][1]

Dear Sommerfeld,

You must not be cross with me that I am answering your kind and interesting letter only today. But in the last month I had one of the most stimulating, exhausting times of my life, indeed also one of the most successful. I could not think of writing.

For I realized that my existing gravitational field equations were entirely untenable! The following indications led to this:[2]

1) I proved that the gravitational field on a uniformly rotating system does not satisfy the field equations.[3]

2) The motion of Mercury's perihelion came to 18″ rather than 45″ per century.[4]

3) The covariance considerations in my paper of last year do not yield the Hamiltonian function H. When it is properly generalized, it permits an arbitrary H.[5] From this it was demonstrated that covariance with respect to "adapted" coordinate systems was a flop.[6]

Once every last bit of confidence in result and method of the earlier theories had given way, I saw clearly that it was only through a link with general covariance theory, i.e., with Riemann's covariant, that a satisfactory solution could be found. Unfortunately, I have immortalized the final errors in this struggle in the Academy contributions, which I can send to you directly.[7] The final result is as follows.[8]

The gravitational field equations are generally covariant. If

$$(ik, \ lm)$$

is Christoffel's tensor of the fourth order, then $G_{im} = \sum_{kl} g^{kl}(ik, \ lm)$ is a symmetrical tensor of the second order.[9] The equations read

$$G_{im} = -\kappa \left(T_{im} - \frac{1}{2} g_{im} \underbrace{\sum_{\alpha\beta}(g^{\alpha\beta}T_{\alpha\beta})}_{} \right).$$

Scalar derived from the energy tensor of "matter," for which I write "T" in the following.

It is naturally easy to set these generally covariant equations down; however, it is difficult to recognize that they are generalizations of Poisson's equations, and not easy to recognize that they fulfill the conservation laws.

Now the entire theory can be simplified conspicuously by selecting the frame of reference in such a way that $\sqrt{-g} = 1$. Then the equations take on the form

$$-\sum_l \frac{\partial \begin{Bmatrix} im \\ l \end{Bmatrix}}{\partial x_l} + \sum_{\alpha\beta} \begin{Bmatrix} i\alpha \\ \beta \end{Bmatrix} \begin{Bmatrix} m\beta \\ \alpha \end{Bmatrix} = -\kappa(T_{im} - \frac{1}{2} g_{im} T)$$

I had considered these equations with Grossmann already 3 years ago, with the exception of the second term on the right-hand side,[10] but at that time had come to the conclusion that it did not fulfill Newton's approximation, which was erroneous.[11] The clue leading to this solution was my realization that not

$$\sum g^{l\alpha} \frac{\partial g_{\alpha i}}{\partial x_m}$$

but the related Christoffel's symbols $\begin{Bmatrix} im \\ l \end{Bmatrix}$ are to be regarded as the natural expression of the gravitational field "component."[12] Once one sees this, the above equation is the simplest conceivable, because there is no temptation to reformulate it for the purpose of a general interpretation through computing the symbols.

The splendid discovery I then made was that not only Newton's theory resulted in first-order approximation, but also Mercury's perihelion motion (43″ per century) in second-order approximation. Solar light deflection came out twice the previous value.[13]

Freundlich has a method of measuring light deflection by Jupiter.[14] Only the intrigues of pitiful persons prevent this last important test of the theory from being carried out.[15] But this is not so painful for me anyway, because the theory seems to me to be adequately secure, especially also with regard to the qualitative verification of the spectral-line shift.[16]

Now I am going to study both of your articles[17] and then send them back to you. Cordial greetings from your raving

Einstein.

I shall send you the Academy papers all at the same time.

154.   From Michele Besso

Krummenau, 29 November 1915

Dear Albert,

I duly received your nice postcard of the 17th relatively quickly as well as the magnetism paper.[1] Many thanks!

We shall not be granted the joy of being together in Krummenau.[2] It seems, your boy had not meant so seriously the consent he had given me and is glad to be alone with you somewhere; I believe one of his wishes is Zugerberg mountain.[3] I naturally would not insist upon my illusion now anyway, against his wishes; but besides, there have been a few cases of diphtheria here again, so that we ourselves would have suggested a change in plan. We, my wife and I, were at Mileva's in Zurich on Friday in order to reiterate the invitation *officially* when we learned of the above-mentioned, as far as little Albert's wishes are concerned—we heard about the diphtheria only upon our return here in Krummenau.

But we will meet anyway! And then you'll tell me what could be changed in the earlier efforts regarding the planetary motions—whether a term of second order corresponding to the centrifugal force perhaps plays a role after all,– whether assumptions are to be made on the density distribution in the Sun.[4]

155.   To Michele Besso

[Berlin, 30 November 1915][1]

Dear Michele,

I'm not coming to Switzerland until Easter after all, because my Albert has written very nastily to me, so that I have lost all inclination. I have to come to Berne at Easter anyway, where a meeting of the Anti-Oorlog Raad is taking place, in whose Great Council they have elected me.[2] You will be receiving my papers soon.[3] The problem has now finally been solved (general covariance). My colleagues are acting hideously in this affair.[4] You will have a good laugh when I tell you about it. *Very definitely visit Zangger when you come to Zurich.* He asked me explicitly to report this to you. You really will find him a matchless friend; such as he are few and far between.

Affectionate greetings, yours,

Albert.

156.   To Hans Albert Einstein

[Berlin, 30 November 1915]

Dear Albert,

I received your letter just now, the unkind tone of which dismays me very much.[1] I see from your long delay and from the unfriendliness of your letter that my visit would bring you little joy. Therefore I consider it not right that I sit in the train 2 · 20 hours without the result of making someone happy. I'll come to visit you again only when you ask me to do so yourself. At Easter I'm going to be in Switzerland anyway, because I must attend a meeting in Berne.[2] I'll send you the Christmas present in cash as you wish. But I do think *that a luxury gift costing Fr. 70 does not match our modest circumstances.*

Regards to you and Tete, yours,

Papa.

157.   To Erwin Freundlich

Tuesday, [30 November 1915][1]

Dear Mr. Freundlich,

Many thanks for your kind notes. Something is still missing, though, namely

1) The length of the ascending node for the planets, relative to the spring equinox.

2) The length of the Earth on December 6th, which according to your information equals the length of the ascending node for the solar equator (I could actually calculate that from your data; but you probably have this in tables there).

3) The common notation in astronomy for the orbit elements of planets.[2]

Please add this information to the slip and then send the same back to me.–

On Friday, Rubens will give a talk at the Phys. Soc.[3] I am enclosing herewith Naumann's letter. Upon reconsidering the man, he actually appealed to me. He is one of those not used to relying on the authority of others but on his own judgment. As his own reasoning became rusty, he consequently had to reject everything he was no longer able to grasp entirely.[4]

Best regards, yours,

A. Einstein.

158.   From Michele Besso

[Berlin, after 30 November 1915][1]

My pastoral manner permits me to make you aware of things you obviously already know—that you will not seriously take any offense at the boy,[2] nor allow yourself to be disturbed by it, which naturally has, aside from him, an additional source hostile to you—that this additional source also is made not only of meanness and troublesomeness but of goodness tipped off balance by the mix of social circles, which had to become even more unstable by your stance on religion, but also on duty and truth. It is more difficult for her to understand you than you her—and it always has been more difficult for her, if only because the role as the wife of a genius *never* is easy.– All this just set against decisions arising out of the emotions. Now you will probably have written to the effect that *now* you cannot come;[3] but through communication by letter, which is your undeniable right and a growing responsibility, you can *prove* to your son what he has in you *this way* as well and avoid having him reply to you inappropriately again (the next time).

159.   To Mileva Einstein-Marić

[Berlin, 1 December 1915][1]

D[ear] Miza,

A string of misunderstandings has prevailed. I was also a bit disappointed that I'd not get Albert to myself but only under Besso's protection, as it were. Since the thought did not occur to me at all that this had been on Besso's sole initiative, but assumed as obvious that this had been arranged between Besso and you all, I consented, for God's sake.[2] So I was altogether disappointed yesterday by Albert's letter, which was really very unfriendly. For this reason I immediately wrote him to cancel.[3] But your letter leads me to think, on the other hand, that the unfriendliness was more childish clumsiness than nasty intent.

Well, this is how the matter stands. In any case, I cannot make the long journey only to spend two or three days with Albert. The meeting is only worthwhile if it lasts at least one week. I must go to Berne anyway at Easter, now, in order to be present at an important meeting.[4] So it would be preferable to me if I could be together with Albert around this time, so that I don't have to take the long and presently somewhat inconvenient trip twice within a few months. Should his vacation not be such that he is on holiday just then, we'll take him out of school for a couple of days.

Albert is now gradually entering the age at which I can mean very much to him. You can safely entrust him to me from time to time. Your relationship with him will not suffer from it in the least, as you will soon notice. For my influence is limited to the intellectual and the aesthetic. I want to teach him mainly to think, judge, and appreciate objectively. A number of weeks a year are necessary; I hope they will not only be enjoyable for him but also beneficial. A meeting of only a few days, however, is merely a thrill without any deeper value.

Greetings to you and kisses for the boys from their

Papa.

160.   To Otto Naumann

[Berlin,] 7 December 1915

Your Excellency,

I have just been informed by my colleague Planck by telephone that he spoke with you about the Freundlich affair.[1] It affords me great pleasure to see that you have not lost sight of the matter, on account of which I had recently taken the liberty to call on you.[2] My colleague Planck recommended I now provide some more factual comments to orientate you on the state of the problems and on the related analyses to be taken up.

The issue involves testing the so-called general theory of relativity. This theory bases itself on the assumption that no physical reality can be lent to time and space; it leads to a very specific theory of gravitation, according to which Newton's classical theory is valid only in admittedly superb first-order approximation. Verification of the results of this theory can be done only with methods used in astronomy. Three results could be found up to now for which a comparison against experiment is feasible at present.

1. The theory provides that the spectral lines in the light of fixed stars must be shifted slightly toward the red end of the spectrum against the corresponding spectral lines from terrestrial light sources, namely, the larger the mass of the light-emitting fixed star is, the more pronounced it is.

This result was confirmed qualitatively by Mr. Freundlich using the already available observational data gathered mostly by American observatories.[3] Such a redshift was shown to be present on average, especially for fixed-star classes in which significant stellar masses had been determined through astronomical means. Establishing an average value through the measurement of numerous stars was necessary because the unknown individual motions of the separate stars also produce line shifts.

A more stringent test of this prediction of the theory could be performed on a binary star composed of two stars of substantially different sizes. We would then no longer be dependent on establishing a mean value, but could test the theory directly from the observational data gained from a single binary star of that type. This is one of the projects Mr. Freundlich aims to carry out.[4]

2. The theory provides that (owing to the imprecise validity of Newton's theory) the elliptic orbits of the planets rotate slowly around the Sun (in the orbital direction) instead of being fixed in space. This consequence of the theory has been verified brilliantly with the planet Mercury.[5]

3. The theory provides that a light ray passing by a celestial body is deflected by it.

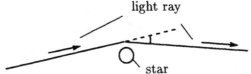

light ray

star

Now, this consequence is the most interesting and astonishing of all, and probably also undoubtedly the theory's most characteristic one; and precisely this consequence has not yet been subjected to any test. It is true that our Academy has sent an expedition to Russia, led and organized by Mr. Freundlich for the observation of the latest solar eclipse in order to study the effect produced by the Sun. But unfortunately war (and weather) deprived this expedition of success.[6] Through close examination of the available observational material, Mr. Freundlich now arrives at the result that the light-deflection effect could be demonstrated also with the planet Jupiter, although only with the subtlest of photographic measurements and through an increased number of observations. The most important specialists in the pertinent field of precision technology confirm that the observational method conceived of by Mr. Freundlich ought to lead to the objective.[7]

This is the most important project for Dr. Freundlich. It could be carried out at the observatory in Neubabelsberg without any substantial cost. It would just be necessary that Mr. Freundlich be released for a few years from the routine duties of measuring star positions so that he could devote himself in peace to the projects indicated here.[8]

If Your Excellency wishes more detailed information on this or that point, I am naturally gladly prepared to provide them for you personally. I preferred to remain a bit unspecific here so as not to tax your patience excessively!

(sig.) A. Einstein.
Telephone Pfalzburg 7273.

161.   To Arnold Sommerfeld

[Berlin,] 9 December 1915

Dear Sommerfeld,

Here you have both manuscripts back, which I have looked through with interest.[1] Planck is also working on a similar problem as you (quantization of the phase space of molecular systems).[2] He is also laboring on spectral questions. General relativity is unlikely to be able to assist you, because it practically coincides with the more restricted theory of relativity for these problems.[3] From what I know of Hilbert's theory, it makes use of an approach for electrodynamic events that—apart from the treatment of the gravitational field—is closely connected to Mie's.[4] Such a specialized approach is not in accordance with the point of view of general relativity. The latter actually only delivers the law for the gravitational field, and quite unambiguously so when general covariance is required. You will see this in the papers I sent out to you this morning.[5] (Be sure to have a look at them; it is the most valuable finding I have made in my life). Conversely, any other theory that corresponds with relativity in the more restricted sense can be taken over in the general theory of relativity through simple transformation, without the latter delivering any new criteria. Thus you see that I cannot help you in the least.

The result of the perihelion motion of Mercury gives me great satisfaction.[6] How helpful to us here is astronomy's pedantic accuracy, which I often used to ridicule secretly!

Do not let yourself be dissuaded from looking at the papers more closely by the fact that as you read it the final stage in the battle over the field equations is being fought out before your eyes!

Best regards, yours,

Einstein.

Tell your colleague Seeliger that he has a frightful temper. I relished it recently in a reply that he directed to the astronomer Freundlich.[7]

162.   To Michele Besso

[Berlin, 10 December 1915]

Dear Michele,

Many thanks for the letter.[1] I am coming to Switzerland now after all, because my wife has assured me by letter that my Albert is eager to be with me.[2]

Now I am going up the Zugerberg with him.[3] Of course we'll see each other on that occasion. I sent you the papers today.[4] The boldest dreams have now been fulfilled. *General* covariance. Mercury's perihelion motion wonderfully precise. The latter is completely secure from the astronomical standpoint, since Newcomb's mass determinations for the inner planets are from *periodic* perturbations (not from secular ones).[5] This time the nearest at hand was right; but Grossmann & I believed that the conservation laws were not satisfied and Newton's law did not result in first-order approximation.[6] You will be surprised by the appearance of the $g_{11} \ldots g_{33}$'s.[7] Cordial greetings also to Anna & Prof. Winteler from your contented but quite worn-out

Albert.

163. To Mileva Einstein-Marić

[Berlin, 10 December 1915]

D[ear] M[ileva],

Your letter, which just arrived, prompts me to travel to Switzerland now after all. For there is a faint chance that I'll please Albert by coming.[1] Tell him this and see to it that he receives me fairly cheerfully. I'm quite tired and overworked, you see, and not capable of enduring new agitations and disappointments.[2]

With my greetings to you and kisses for the boys,

Albert.

164. From Michele Besso

Krummenau, 11 December 1915

Dear Albert,

A week ago yesterday I received your postcard in which you announce your intention not to come to Switzerland for Christmas[1]—on Monday at Prof. Zangger's I then saw your letter addressed to him[2] and little Alb[ert's] letter to you[3]—then [looked] at Prof. Z.'s letter to you. He was furious with your wife; I for my part must admit that I had not expected much else of her. It is now obviously not a matter of establishing harmony, for which fundamental preconditions are simply lacking[4]—I would like to lend you my glasses, which scale things down to size and provide some objectivity, only because I think that you could defend yourself much better without rage. That the boy would naturally be completely different as soon as you had spent a couple of days with him[5] is as clear to you as to me.   His abrasive, almost impudent manner is not entirely suggested to him, but is probably to a large part that self-defense of his heart against painful impressions which you yourself know only too well. [. . .][6] I am therefore *entirely* in agreement with good old Z. when he reiterates to you that you ought to come; he would take care that you still get the boy on the 26th or 27th. Your wife instinctively or consciously includes in her calculations your more refined sensitivity, which propels you to a kind of flight from emotional up-heavals. I should think that the proximity of dear, good friend Z., can assure you the necessary composure, whose natural manner offers you a kind of guarantee that you don't go too far by impulse either against yourself or against others.

165.   To Moritz Schlick

[Berlin,] 13 Wittelsbacher St., Tuesday. [14 December 1915][1]

Highly esteemed Colleague,[2]

Yesterday I received your paper and have studied it thoroughly already.[3] It is among the best that has been written on relativity to date. From the philosophical perspective, nothing nearly as clear seems to have been written on the topic. At the same time, you have a complete command of the subject material.[4] I have nothing to criticize about your representations.

The relationship of the theory of relativity to Lorentz's theory is presented superbly, your relation to the teachings of Kant and his successors is truly masterful. Confidence in the "apodictic certainty" of "synthetic judgments *a priori*" is severely shaken by the acknowledgment of the invalidity of even only a single one of these judgments.[5] Your representations that the theory of rel. suggests itself in positivism, yet without requiring it, are also very right. In this also you saw correctly that this line of thought had a great influence on my efforts, and more specifically, E. Mach, and even more so Hume, whose *Treatise of Human Nature* I had studied avidly and with admiration shortly before discovering the theory of relativity.[6] It is very possible that without these philosophical studies I would not have arrived at the solution.

Your comments on the general theory of relativity are also entirely correct, as far as this theory was right by then.[7] The new finding is the result that a theory exists that agrees with all previous experience whose equations are covariant with arbitrary transformations in the space-time variables. Thus time & space lose the last vestiges of physical reality. There is no alternative to conceiving of the world as a four-dimensional (hyperbolic) continuum of 4 dimensions. The fact that the theory's equations can be simplified *a posteriori* by selecting the frame of reference in such a way that the determinant equation

$$|g_{\mu\nu}| = -1$$

is satisfied is of no epistemological significance.[8] The feasibility of checking the theory empirically is not as dismal as you indicate.[9] The theory explains quantitatively the perihelion motion of Mercury discovered by Le Verrier.[10] The influence of gravitational potential on the color of emitted light required by the theory has already been confirmed qualitatively in astronomy (Freundlich).[11] There are also good prospects of testing the result reg[arding] the deflection of light rays by a gravitational field.

In requesting that you visit me when your path leads you to Berlin, I rem[ain] with best regards, yours very truly,

A. Einstein.

166.   To Hans Albert Einstein

[Berlin, 18 December 1915]

My dear Albert,

It's so difficult to come across the border that many acquaintances here had to return without being able to travel to Switzerland. That's why I cannot come to see you now.[1] At Easter, however, I definitely will visit you; then I'll wait at the border until I'm allowed across.

With affectionate greetings for the holidays, yours,

Papa.

P.S. The Christmas gift in the form of money is being sent off to you all today.

167.   To David Hilbert

[Berlin,] Monday, 20 December 1915

Highly esteemed Colleague,

I thank you for your kind communication regarding my election as corr. member.[1] On this occasion I feel compelled to say something else to you that is of much more importance to me.

There has been a certain ill-feeling between us, the cause of which I do not
want to analyze. I have struggled against the feeling of bitterness attached to it,
and this with complete success.

I think of you again with unmixed geniality and ask you to try to do the same
with me. Objectively it is a shame when two real fellows who have extricated
themselves somewhat from this shabby world do not afford each other mutual
pleasure.[2]

Best regards, yours,

A. Einstein.

168.   To Michele Besso

[Berlin,] 21 December [1915]

Dear Michele,

I cannot come now because the border is almost constantly closed. A number
of my acquaintances had to return in spite of passport, etc. So I prefer to postpone
the trip until Easter. *I will wait then until I can enter.* I am very much looking
forward to the Swiss air and the associated muzzle release! Read the articles![1]
They provide final deliverance from misery. Most gratifying is the agreement
with perihelion motion and the general covariance; strangest, however, is the
circumstance that Newton's theory *of the field* is incorrect already in the 1st order
eq. (appearance of the $g_{11}-g_{33}$).[2] It is just the circumstance that the $g_{11}-g_{33}$'s do
not appear in first-order approximations of the mo[tion] eqs. which determines
the simplicity of Newton's theory. Now Planck also is beginning to take the
matter more seriously; he is still resisting a bit, though.[3] But he is a splendid
person. My other experiences with colleagues reveal an alarming predominance
of all-too-human attributes![4] I'll tell you about everything then. *If only it were
now already!*

Affectionate greetings to you, Prof. W.,[5] Anna, and Vero, yours,

Albert.

169.   From Karl Schwarzschild[1]

[at the Russian front,] 22 December 1915

Esteemed Mr. Einstein,

In order to become versed in your gravitation theory, I have been occupying myself more closely with the problem you posed in the paper on Mercury's

perihelion[2] and solved to 1st-order approximation.[3] Initially, one factor made me very confused. I found for the coefficients $g_{n\nu}$ in first-order approximation, in addition to their solution, also the following second one:[4]

$$g_{\rho\sigma} = -\frac{\beta x_\rho x_\sigma}{r^5} + \delta_{\rho\sigma} \left[\frac{\beta}{3r^3} \right] \qquad g_{44=1}.$$

According to this there would be a second α in addition to yours and the problem would be physically ambiguous. Thereupon, I took my chances and made an attempt at a complete solution.[5] A not overly lengthy calculation yielded the following result: There is only one line element that satisifies your conditions (1) to (4),[6] aside from the field and determinant eqs.,[7] and is singular at the origin and only at the origin.

Let $\qquad x_1 = r \cos \varphi \cos \vartheta \qquad x_2 = r \sin \varphi \cos \vartheta \qquad x_3 = r \sin \vartheta$

$$R = (r^3 + \alpha^3)^{1/3} = r \left(1 + \frac{1}{3} \frac{\alpha^3}{r^0} \cdots \right),$$

then the line element reads:[8]

$$ds^2 = \left(1 - \frac{\gamma}{R} \right) dt^2 - \frac{dR^2}{1 - \frac{\gamma}{R}} - R^2 (d\vartheta^2 + \sin^2 \vartheta d\varphi^2).$$

R, ϑ, φ are not "admissible" coordinates with which the field equations could be formed, because they do not have the determinant 1, but the line element is written most neatly in them.

The equation for the orbit remains exactly the one obtained by you in first-order approximation (11),[9] except under x not $\frac{1}{r}$, but $\frac{1}{R}$ must be understood, which is a difference of the order of 10^{-12}, thus practically absolutely irrelevant.

The problem of the two arbitrary constants α and β, which the first-order approximation had yielded, is solved in that β must have a specific value of the order of α^4,[10] the way α is given, otherwise in continuing the approximations the solution would be divergent.

Thus the uniqueness of your problem is also in the best of order.[11]

It is a wonderful thing that the explanation for the Mercury anomaly emerges so convincingly from such an abstract idea.

As you see, the war is kindly disposed toward me, allowing me, despite fierce gunfire at a decidedly terrestrial distance, to take this walk into this your land of ideas.

[5]Draft version: "In order to become versed in your gravitation theory, I set myself the task of solving completely, if possible, the problem you posed in the paper on

Mercury's perihelion and approximately solved. It is easy to specify the most general line element which has the necessary symmetry properties. Proceeding from this, I initially determined the first-order approximation in your way and found:

$$g_{\rho\sigma} = -\alpha\frac{x_\rho x_\sigma}{r^3} - \beta\frac{x_\rho x_\sigma}{r^5} = \delta_{\rho\sigma}\left[\frac{\beta}{3r^3}\right] \qquad g_{44} = 1,$$

therefore, *two* arbitrary quantities α and β, which was peculiar. Then I moved to the solution as a whole."

170.   To Hans Albert Einstein

Berlin, 23 December 1915

My dear Albert,

I have been working so hard in the last few months that I urgently need a rest during the Christmas holidays. Aside from this, coming across the border is very uncertain at present, since it has been almost constantly closed recently.[1] That is why I must unfortunately deprive myself of visiting you now. But instead I'm definitely coming in the first days of April. Then I have more time and can wait at the border until I am allowed into Switzerland. I have sent my Christmas gift to you both in the form of money, according to your wish. I'm glad that you romp around a lot out of doors. I hope there will continue to be a decent amount of snow as well so that you can do your skiing. Ask Mama to send me a photo of you both again. You have grown again and have changed since the last one.

Write me about the things you are up to besides school, and who your friends are. Is there a subject in or outside of school that interests you particularly? Then I would also like to hear from you what schooling you have ahead of you, because I am not exactly familiar with the Zurich school system.[2] How is Tete's health? Does he ever have complaints? Does he look well?

Now to something else of importance. I have heard that there is an excellent remedy for tooth and bone development especially for growing children, which you both should take. You prepare a saturated solution of calcium chloride ($CaCl_2$) and store it in a corked medicine bottle. After every meal you put one little teaspoonful of it in half a glass of water (or milk) and drink it. It would be a true blessing for Tete if he were to take it. You see, calcium (Ca) is a metal that plays a major role in teeth and bones. It is present in a digestible form in natural food, but during cooking it precipitates out in an indigestible form so the body cannot benefit from it any more. Talk also to Mama about this; she will immediately see the sense in it.

Write to me again soon, my dear Albert, and on replying have my letter before you and answer everything I have asked you. In this way it's also easier for you to reply, and you don't have to struggle as much as when you have to think up everything yourself.

Kisses to you both from your

Papa

Kind regards to Mama.

171. To Royal Society of Sciences in Göttingen

[Berlin,] 23 December 1915

To the Roy. Society of the Sciences in Göttingen.

Highly esteemed Sirs,

I thank you cordially for the kind acknowledgment, which your election of me demonstrated.[1] It is my pleasure to belong to a society to which science is indebted for so much.

Most respectfully,

A. Einstein.

172. To Hans Albert Einstein

[Berlin,] 25 December [1915]

My dear Albert,

Nothing has delighted me as much for a long time as your postcard with the well-drawn pictures by both of you and the little Zürcher boy.[1] Time and again I take it out of my pocketbook and look at it. But now you must patiently wait out the couple of months as well, my son. I am too tired out for all that traveling, have only little time, and also dread the great expense. For it, I'll devote all the more attention to you in April. In the meantime you can frolic about on Zurich Hill on your new sliding boards;[2] go outside very often by all means, so that you become strong. Be sure you and Tete take the calcium chloride I wrote you about yesterday.[3]

Kisses to you and Tete from your

Papa.

173.   To Paul Ehrenfest

[Berlin,] 26 December [1915][1]

Dear Ehrenfest,

I just received your postcard with the kind invitation. As a matter of fact, every muscle in my body is itching to set off. But my mother is visiting[2] and I cannot act on it. I am even booked for Easter, as then I am going to my Albert. So since I am already tied up and cannot drive out to meet you, at least I shall prove in writing my desire to visit you with unusual talkativeness. Einstein has it easy. Every year he retracts what he wrote in the preceding year; now the sorry business falls to me of justifying my latest retraction.

In §12 of my paper of last year, everything is correct (in the first 3 paragraphs) except for what is printed at the end of the third paragraph in spaced type.[3] A contradiction to the uniqueness of the event does not follow at all from the fact that both systems $G(x)$ and $G'(x)$, related to the same frame of reference, satisfy the conditions of the grav. field. The seemingly compelling part of this reflection founders immediately when you consider that

1) the reference system has no real meaning

2) that the (simultaneous) materialization of two different g systems (more aptly put, two different grav. fields) within the same area of the continuum is, according to the nature of the theory, impossible.

In place of §12 the following consideration must appear. Whatever is physically real in events in the universe (as opposed to that which is dependent on the choice of a reference system) consist *in spatio-temporal coincidences* and in nothing else![4] For ex., the intersection points of two world lines are real, or the statement that they do *not* intersect each other. Therefore, those statements relating to the physically real do not lose validity from the absence of a (unique) coordinate transformation. When two systems in the $g_{\mu\nu}$'s (or gen., the variables used to describe the world) are constituted in such a way that the second can be obtained from the first by mere space-time transformation, then they are entirely equivalent. This is because they have in common all the spatio-temporal point coincidences, that is, all the observables.

This consideration shows simultaneously how natural the requirement of general covariance is.

Now to the second point. The main equations

$$\sum \frac{\partial t_\sigma^\lambda}{\partial x_\lambda}$$

(I) and (II) of your postcard are written as they already appear for the reference system specialization $g = -1$.[5] First imagine them written in the general covariant form, which is also known, of course. Certainly you can eliminate the T_σ^λ's from these equations and you obtain equations that are only satisfied by the $g_{\mu\nu}$'s. But these equations are then, just as (I) and (II), *generally covariant; they therefore do not require any specialization of the reference system, but are valid in any reference system* (if they are valid in *one*). Specialization to ($g = -1$) systems changes nothing in the consideration).

Cordial greetings to your wife, your children, Lorentz, both de Haases, and Fokker. I am going to write to de Haas soon. My congratulations and gravitation paper for Fokker both came back as undeliverable; I had addressed them to Gravenhage.[6] Best regards to you; write again soon, yours,

Einstein.

174. To Paul Ehrenfest

[Berlin, 29 December 1915]

Dear Ehrenfest,
  Your relation

$$0 = \sum_\kappa \frac{\partial \Omega_i^\kappa}{\partial x_\kappa} - \cdots \tag{IV}$$

may not be an identity, but (with the restr[iction] $\sqrt{-g} = 1$) it certainly is a general covariant equation.[1] Thus it cannot serve to fix a preferred system.–

I probably disproved my "philosophical" consideration[2] adequately enough for you. The field equations naturally do not provide any unique determination of the $g_{\mu\nu}$'s; randomness remains, which the "philosoph.' consideration yields and which results from the randomness of the reference system's continuation.–

Your relation (IV), or its generalization,[3] remains satisfied at every transformation. Thus it can pose no obstacle here for the specialization of the system according to $g = -1$.

Third question.[4] The passage ought to have read: Gen. relat. *does not restrict the possibilities with regard to the other processes more than* special relativity does; but it provides the influence of the gravitational field on those processes.[5]

Fourth question. With "recently" my addendum "On the gen. theory of rel."[6] was meant, where I believed that general covariance demands the hypothesis $\sum T_\mu^\mu = 0$, which then turned out to be unnecessary.[7]

Now I believe I have answered, admittedly briefly, everything. It would be much nicer orally.

Write again soon, yours,

Einstein.

A little letter to "the charmer [*den Holder*]" had accidentally slipped into your letter to me.

175.   To Władysław Natanson

[Berlin,] 29 December 1915

Dear Colleague,

I am very ashamed of myself that I still have not answered your kind little letter. I was most acutely ashamed yesterday when in addition I received your friendly congratulations. I also wish you health and happiness at work for the coming year. As long as you were here, you were my favorite Berliner; now I miss our relaxed relations very much.[1]

I am sending you some papers.[2] You see that I have upturned my house of cards once again and have built a new one; the middle section at least is new. The explanation for the perihelion motion of Mercury, which is empirically completely secure, gives me great pleasure, no less the fact that now general covariance of the gravitation law could be carried out after all.

Cordial greetings to you, your family & our coll. Smoluchowski,[3] yours,

A. Einstein.

176.   To Karl Schwarzschild

[Berlin,] 29 December 1915

Highly esteemed Colleague,

Your calculation providing the uniqueness proof for the problem[1] is extremely interesting. I hope you publish it soon![2] I would not have thought that the strict treatment of the [mass-]point problem was so simple.

That your particular solution is of the 3rd order is immediately apparent for reasons of dimension. For $\dfrac{\kappa m}{r}$ is a dimensionless number. As your β must depend on m, your expression

$$g_{\rho\sigma} = -\beta \frac{x_\rho x_\sigma}{r^5} + \delta_{\rho\sigma}\frac{\beta}{3r^3},$$

in which $\dfrac{\beta}{r^3}$ is a (dimensionless) number, thus requires that β be equal, apart from a numerical factor, to $\left(\dfrac{\kappa m}{r}\right)^3$.[3]

I am very satisfied with the theory. It is not self-evident that it already yields Newton's approximation; it is all the more gratifying that it also provides the perihelion motion and line shift, although it is not yet sufficiently secure.[4] Now the question of light deflection is of most importance.

With my best regards and wishes for the New Year, yours,

Einstein.

177.   To Hendrik A. Lorentz

[Berlin,] 1 January 1916

Highly esteemed Colleague,

Your enticing invitations make it hard for me to stay put here.[1] I can visualize a visit in your midst, can depict for myself the very interesting conversations, can imagine that for a few days I was allowed to walk around without a muzzle, so to speak, and can see myself sitting in Ehrenfest's cozy little home.[2] But I must forgo all this because I cannot easily get away for a number of reasons. With all of your permissions, however, I am going to extend the date set for me to a time when I really can travel; I shall certainly not postpone it for longer than is necessary.

Trying times awaited me last fall as the inaccuracy of the older gravitational field equations gradually dawned on me. I had already discovered earlier that Mercury's perihelion motion had come out too small. In addition, I found that the equations were not covariant for substitutions corresponding to a uniform rotation of the (new) reference system. Finally, I found that the consideration I made last year on the determination of Lagrange's H function for the gravitational field was thoroughly illusory, in that it could easily be modified such that no restricting conditions had to be attached to H, thus making it possible to choose it completely freely.[3] In this way I came to the conviction that introducing adapted systems was on the wrong track and that a more broad-reaching covariance, preferably a *general* covariance, must be required. Now general covariance has been achieved, whereby nothing is changed in the subsequent specialization of the frame of reference.[4] I had considered the current equations in essence already 3 years ago together with Grossmann, who had brought my attention to the Riemann tensor. But because I had not recognized the formal importance of the

{ } terms, I could not obtain an overview and prove the conservation laws.[5] I was equally unable to recognize that Newton's theory was contained in it in first-order approximation; I even believed to have understood the contrary. That is how I got caught in the jungle! Now I am more pleased than ever about the arduously won lucidity and about the agreement with Mercury's perihelion motion. I am conducting a discussion with Ehrenfest at present essentially on whether the theory really does fulfill the general covariance requirement.[6] He also indicated to me that you had encountered problems or objections to it as well; you would do me a great favor if you were to inform me of them briefly. I have broken in my hobbyhorse so thoroughly that with a short hint I certainly also would notice where the crux of the problem lies.

I wish you and yours a happy New Year, and Europe fair and permanent peace. With affectionate greetings, yours,

A. Einstein.

178.   To Michele Besso

Berlin, 3 January 1916

Dear Michele,

I was delighted with your little letter and find your idea of corresponding with Albert very good.[1] I also tried to do so often and believe that it will yet become very beneficial. At the moment the boy is still not skilled enough to inject the necessary animation into the written word; I can remember only too well the time when I also was like that. Nevertheless, I'm going to continue to try it from time to time. Once he has acquired a taste for it, much will have been gained. In any case, I am coming to Switzerland at Easter and am staying there for a few weeks, even if I have to wait long at the border.[2] I am very much looking forward to seeing my boys and you. Going now would have been nonsense. I am tired out from a lot of work, had to reckon with an unavoidably long wait at the border, had little time, and was also hesitant about the quite considerable cost. Personally I am fine. My relatives especially continue to welcome me with open arms, despite my having stated in no uncertain terms that I will not enter into the projected marriage.[3]

The great success in gravitation pleases me immensely. I am seriously contemplating writing a book in the near future on special and general relativity theory, although, as with all things that are not supported by a fervent wish, I am having difficulty getting started.[4] But if I do not do so, the theory will not be understood, as simple though it basically is.[5]

Studying Minkowski would not help you further. His papers are needlessly complicated.[6] Based on the judgment of local astronomers here, only Mercury's perihelion motion is secure, which drops very rapidly $\left(\text{like } \dfrac{1}{R^{\frac{5}{2}}}\right)$ with the orbital radius. If the orbits of Venus and the Earth were more eccentric, the effect could also have been demonstrated with these. The great magnification of the effect against our calculation[7] stems from that, according to the new theory, the g_{11}–g_{33}'s also appear in the first order[8] and hence contribute to the perihelion motion. The amount is determined very reliably by means of the Sun's transits. The other effects hardly come into consideration (solar rotation virtually not, perturbation deviations from Newton's law, not at all).[9]

In the hole argument, everything was correct up to the last conclusion.[10] There is no physical substance to two different solutions $G(x)$ and $G(x')$ existing with reference to *the same* coordinate system K. Imagining two simultaneous solutions within the same manifold makes no sense, and system K obviously is not a physical reality. Taking the place of the hole argument is the following consideration. Nothing is *real* physically except for the entirety of the spatio-temporal point coincidences.[11] If, for ex., physical events were to be constructed out of the motions of mass-points alone, then the meeting of the points, i.e., the intersection points of their world lines, would be the only real, that is, principally observable, things. These intersection points naturally remain intact in all trans-formations (and no new ones are added), only if certain uniqueness conditions are maintained. It is thus most natural to demand that the laws not determine *more* than the spatio-temporal coincidences as a whole. This is accordingly already achieved with generally covariant equations.

The first paper along with the addendum still suffers from want of the term $\frac{1}{2}\kappa g_{\mu\nu}T$ on the right-hand side; therefore the postulate $T = 0$.[12] The matter must naturally be executed as in the last paper, whereby no conditions result on the structure of matter.[13] The dimensional observation, according to which electrons and quanta require a special h hypothesis independent of gravitation, thus still stands, and rightly so.[14]

I join you ardently in your wishes for 1916. The prospects are miserable, though.[15]

Affectionate greetings, yours,

Albert.

179.   To Paul Ehrenfest

[Berlin,] 3 January 1916

Dear Ehrenfest,

An addition is still needed to my replies to your comments.[1] Namely, it is conceivable that your four relations (IV) contradict the field equations, that is, do not agree with them. That is why it is of interest to see that your four equations be satisfied *identically*. This can be most easily understood with reference to the calculations developed in my papers as follows. (In the citations, the paper "On the gen. theory of rel." is denoted as A, and the pap[er] "The field eqs. of gravitation" as B.)[2]

If instead of (7)B, the equation[3]

$$\sum \frac{\partial T_\sigma^\lambda}{\partial x_\lambda} = -\sum \Gamma_{\sigma\beta}^\alpha T_\alpha^\beta + U_\sigma \quad (U_\sigma \neq 0) \cdots \tag{7a}$$

is taken as a basis, then (9)B is valid regardless of this.[4] On the other hand, from the method indicated in the papers,[5]

$$\frac{\partial}{\partial x_\mu} \left[\sum \frac{\partial^2 g^{\alpha\beta}}{\partial x_\alpha \partial x_\beta} - \kappa(T+t) \right] = -2\kappa U_\sigma$$

results in contradiction to (9)B.

Therefore (7a)B is a necessary consequence of (6)B.[6] That is why nothing can be inferred from the combination of both these systems that is not already contained in (6)B. From this I conclude that your equation (IV) must be satisfied identically; for in any other way it is not conceivable that (IV) contains no new conditions to (6)B.[7]

With best regards, yours,

Einstein.

180.   To Paul Ehrenfest

[Berlin,] Wednesday [5 January 1916][1]

Dear Ehrenfest,

The great interest you[2] are showing in my theory delights me. Were the trip not so difficult now, I would come immediately, despite lecture course[3] and Academy. But it seems that the border is closed a lot, causing people to turn back at the Dutch border, mission unaccomplished, in spite of regulation passports.[4] So I am going to try to answer your question in writing. I shall make an effort not to leave a word of your letter out of consideration but shall address everything.

First of all I note that your equation (IV) is satisfied identically, as you will have gathered from my last postcard.[5] Should the proof not have convinced you, I shall provide you with another one that is less bumpy mathematically but which cannot be linked so conveniently to the papers.

I cannot hold it against you that you have not yet understood the admissibility of generally covariant equations, because I myself needed so long to arrive at total clarity on this point.[6] The root of your difficulty lies in that you instinctively treat the reference system as something "real." Your somewhat simplified example: You examine two solutions with the same boundary conditions at infinity, in which the coordinates for the star, the material point at the aperture, and the plate are the same. You ask whether "the direction of the wave normal" at the aperture always comes out the same. As soon as you speak of "the direction of the wave normal *at the aperture*," you are treating this space with regard to the $g_{\mu\nu}$ functions as an infinitesimal space. *This and the definiteness of the coordinates for the aperture points have as a consequence that for all solutions the direction of the average waves at the aperture* is the same.

This is my contention. For a finer illustration, the following: In the above special case you obtain all the solutions that are a consequence of general covariance in the following way. Trace the above little diagram on to a completely flexible piece of tracing paper. Then deform the tracing paper randomly along the paper plane. Then make another tracing on the letter paper. You then obtain, e.g., the diagram

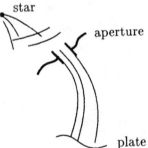

If you now again relate the diagram to orthogonal letter-paper coordinates, the solution is mathematically a different one to beforehand, naturally also with regard to the $g_{\mu\nu}$'s. But physically it is exactly the same, simply because the

letter-paper coordinate system is only a contrivance. Always the same points are illuminated on the plate. If you perform the distortion of the tracing paper only within the finite realm and in such a way that the image of the star, the aperture, and the plate remain unshifted without damaging the constancy, then you obtain the special case your question refers to.

The essence is: As long as the drawing paper, i.e., "the space," is unreal, both diagrams do not differ at all. "Coincidences" are what count, e.g., whether the plate points are hit by the light or not. Thus the distinction between your solutions *A* and *B* is merely a difference in presentation with physical congruency. This will surely become evident to you upon closer contemplation.

If the equations of the physics were not generally covariant, you would also be able to make the above consideration; however, relative to the letter-paper system, the same laws would not be valid in the *second* diagram as in the first. To this extent then, both of them would not be equivalent. This distinction falls away, however, with general covariance.

I also note that I have purposefully left out the fourth (time) coordinate, which is insignificant in principle, however.

Cordial greetings to you and everyone, yours,

Einstein.

181.   To Karl Schwarzschild

[Berlin,] 9 January 1916

Highly esteemed Colleague,

I examined your paper with great interest.[1] I would not have expected that the exact solution to the problem could be formulated so simply. The mathematical treatment of the subject appeals to me exceedingly. Next Thursday I am going to deliver the paper before the Academy with a few words of explanation.[2]

Meanwhile, I received another letter from you yesterday evening, which I would also like to answer right away.

1) The theory is fully developed, as far as the fundamental formulas are concerned, so no other difficulties remain in the treatment of the individual problems aside from the computational ones, which are, however, inordinately large. But you will gather from the following reflection that no notable modification is made for the perturbation problem.

The modifications that the theory yields are of a relative order of magnitude determined by $\dfrac{\kappa M}{r}$. If *M* is taken as the solar mass, then this quantity is barely sufficient to produce *secular* effects accessible to observation. If M is taken for planetary masses, then this relative quantity is diminished significantly. However, secular variations that are produced by the interaction of the planets only amount to at most 1,000″ in 100 years. According to the theory, they would be modified by the tiny fraction indicated. Thus a more exact development of the perturbation calculation to modify orbital motion theory cannot provide anything within the reach of observation.

2) The statement that "the fixed-star system" is rotation-free undoubtedly is meant in a relative sense, which is described by a comparison.

The Earth's surface is irregular as long as I envisage very small sections of it. But it approaches the flat basic form when I envisage larger sections of it, whose dimensions are still small against the length of the meridian. This basic form becomes a curved surface when I envisage even larger sections.

Likewise for the gravitational field. On a small scale the individual masses produce gravitational fields that even with the most simplifying choice of reference system reflect the character of a quite irregular small-scale distribution of matter. If I regard larger regions, as those available to us in astronomy, the Galilean reference system provides me with the analogue to the flat basic form of the Earth's surface in the previous comparison. But if I consider even larger regions, a continuation of the Galilean system providing the description of the universe in the same dimensions as on a smaller scale probably does not exist, that is, where throughout, a mass-point sufficiently removed from other masses moves uniformly in a straight line. Ultimately, according to my theory, inertia is simply an interaction between masses, not an effect in which "space" of itself were involved, separate from the observed mass. The essence of my theory is precisely that no independent properties are attributed to space on its own.

It can be put jokingly this way. If I allow all things to vanish from the world, then following Newton, the Galilean inertial space remains; following my interpretation, however, *nothing* remains.

3) As concerns Jupiter, I understand that it is a difficult proposition for astronomers.[3] However, in my view the importance of the matter supports only *one* standpoint, and that is: It *has* to work! Jupiter's moons could serve in studying closely the systematic errors of which you speak; for the apparent displacement of Jupiter's moons through light deflection is entirely negligible owing to the smallness of the moon–Jupiter distance. The angle to be confirmed amounts to $2 \cdot 0.02''$ and is thus within the order of currently attainable precision.

4) It never occurred to me to think of a clique against Freundlich. It is generally far from my mind to think of such things. Struve's attitude is understandable.[4]

He is an old man and no longer has the flexibility needed to delve into new issues. That is why he assumes a negative stance on technical matters, and this negative attitude also extends to Freundlich, whom he envisions somewhat as an incarnation of these things. I readily believe that Freundlich on his side has little tact and lacks social skills, and in general has little psychological understanding for his fellows, which only makes the circumstances more disagreeable.

I do not take Freundlich for a very great talent, but for a person with a burning interest and a remarkable tenacity. He was the first astronomer to understand the significance of the general theory of relativity and to address enthusiastically the astronomical issues attached to it.[5] *That is why I would regret it deeply if he were deprived the possibility of working in this field.* I know from my own experience that the necessary technical skills can be acquired, if the requisite understanding and a great interest are combined. Should such deficiencies nevertheless hamper the enterprise, help from a well-meaning expert could lead to valuable results.

Best regards, yours truly,

A. Einstein.

182.   To Paul Ehrenfest

[Berlin,] 17 January 1916

Dear Ehrenfest,

I am enormously pleased about your and Lorentz's concurrence, which has such a cheerful and bright ring to it![1] You form a brilliant nook on this barren planet. Cleverness there is aplenty, but goodness and generosity are pitifully scarce.[2]

Your truly Ehrenfestian description of the telescope affair must be accepted without protest.[3] Now to some of your questions.

(1) The equation[4]

$$\sum_{\lambda} \frac{\partial}{\partial x_{\lambda}}(T_{\sigma}^{\lambda} + t_{\sigma}^{\lambda}) = 0$$

is a covariant equation, whereby t_{σ}^{λ} is not a tensor and the operation performed on a tensor $\sum_{\lambda} \frac{\partial}{\partial x_{\lambda}}$ does not lead to any covariant. The equation is covariant on subst. of det[erminant] 1, because both of the main equations from which it is derived are as well.

(2) Nevertheless, there is something superfluous in both equation systems. Namely,

the conservation equations

$$\sum \frac{\partial T_\sigma^\lambda}{\partial x_\lambda} + \sum \Gamma_{\sigma\nu}^\mu T_\mu^\nu = 0$$

are already a consequence of the field equations. Proof roughly as on the postcard you did not understand.[5] I repeat[6]

$$\sum \frac{\partial \Gamma_{im}^l}{\partial x_l} - \sum \Gamma_{i\rho}^l \Gamma_{ml}^\rho = -\kappa(T_{im} - \frac{1}{2} g_{im} T). \dots \tag{1}$$

$$\sum \frac{\partial T_\sigma^\lambda}{\partial x_\lambda} + \sum \Gamma_{\sigma\nu}^\mu T_\mu^\nu = A_\sigma. \dots \tag{2}$$

$$\left(+\frac{1}{2} \sum \frac{\partial g^{\mu\nu}}{\partial x_\sigma} T_{\mu\nu} \right).$$

The contention is that from (1) it follows that the four spatial functions A_σ vanish. Proof: If you multiply (1) by g^{im} and perform a summation over i and m, you obtain through computation

$$\sum \frac{\partial^2 g^{\alpha\beta}}{\partial x_\alpha \partial x_\beta} - \kappa(T + t) = 0. \dots \tag{3}$$

If you also multiply (1) by $g^{m\mu}$, sum over m, and have the result undergo the operation $\sum_\lambda \frac{\partial}{\partial x_\lambda}$, then you obtain with the aid of the equation drawn from (1) and (2)[7]

$$\sum \frac{\partial}{\partial \lambda}(T_i^\lambda + t_i^\lambda) - A_\lambda = 0$$

the equation[8]

$$\frac{\partial}{\partial x_\mu} \left[\sum_{\alpha\beta} \frac{\partial^2 g_\alpha^{\alpha\beta}}{\partial x_\alpha \partial x_\beta} - \kappa(T + t) \right] + 2\kappa A_\mu = 0. \dots \tag{4}$$

From (3) and (4), $A_\mu = 0$ results. Thus the conservation law of matter is a consequence of (1). You will find the necessary hints for performing the math in both of the brief papers.[9] (There must naturally be a less bumpy method for this proof, but I do not know it.)[10] In this dependence of both fund. equa. systs. also lies your "inevitability" guarantee requirement for the additional term[11] $-\frac{1}{2} g_{im} T$. The tensor character of the right-hand side requires generally[12]

$$-\kappa(T_{im} + \lambda T), \qquad (\lambda = \text{number})$$

obviously since the energy tensor should surely enter *linearly*. So that the outlined consideration yields the conservation law of matter ((2) with $A_\sigma = 0$), $\lambda = -\dfrac{1}{2}$ must necessarily be chosen. Otherwise a contradiction arises. It was precisely this that I had not noticed in my first communication.[13]

De Sitter has asked me for a copy of my last year's paper.[14] Unfortunately I have none left. Please lend it to him and relay my best regards.

Imagine my delight at realizing that general covariance was feasible and at finding out that the equations yield Mercury's perihelion motion correctly. I was beside myself with joy and excitement for days.

Is *Tetrode* at Leyden? His papers on the entropy constants are superb.[15] Congratulate him for me, if you see him. Recently I gave a presentation on it at the German Phys. Society.[16]

Accept, with your wife and the little ones, my warm greetings, yours,

Einstein.

Greetings to Fokker.[17]

183.   To Hendrik A. Lorentz

[Berlin,] 17 January 1916

Dear and highly esteemed Colleague,

I am in possession of your three letters and very happy about your concurrence, especially since I see that you have considered the principal parts of the theory thoroughly and have taken to the idea that all of our experience in physics relates to coincidences.[1] This point of view quite consequently requires the formulation of generally covariant equations. I had taken this view together with Grossmann already three years ago but had then come to the false notion that it was in contradiction to the requirement of unique causal dependence. I had hit upon this notion, which corresponds to the standpoint held by you in the first of your letters,[2] after all my efforts at that time to find a link between covariant gravitation equations and Newton's theory had failed. My series of gravitation papers are a chain of wrong tracks, which nevertheless did gradually lead closer to the objective. That is why now finally the basic formulas are good, but the derivations abominable; this deficiency must still be eliminated.

In both your letters you presented the sense of the general covariance requirement in an exemplarily clear fashion. It would certainly be of tremendous benefit to the issue if you made your considerations available to other physicists as well, by writing an article on the foundations of the theory, as you kindly proposed in your third letter.[3] I could do this myself, of course, to the extent that it is all clear to me. However, nature has unfortunately deprived me of the talent of being able to communicate in writing, so that what I write may well be correct but is quite indigestible.

Judging from your second letter, it seems to have escaped you that I already indicated the generally covariant form of the field equations. It is provided in equations (2a) and (1) of the paper, "The Field Equations of Gravitation."[4]

I am of the conviction that the depiction of the theory would gain much clarity if Hamilton's formulation were taken as a point of departure, as you have done in your fine paper published by the Amsterd[am] Acad.[5] The path toward this seems to be the following. The V-scalar[6]

$$G = \sqrt{-g} \sum_{iklm} (ik, lm) g^{kl} g^{im}$$

is a function of the quantities $g^{\mu\nu}$, $\dfrac{\partial g^{\mu\nu}}{\partial x_\sigma}$, $\dfrac{\partial^2 g^{\mu\nu}}{\partial x_\sigma \partial x_\tau}$. The integral

$$\int G d\tau$$

is hence invariant. Through partial integration (in only *one* way) the same can be presented in the form

$$\int L d\tau + \text{surface integral},$$

whereby L is now just dependent on $g^{\mu\nu}$ and $\dfrac{\partial g^{\mu\nu}}{\partial x_\sigma}$. In this way I find (calculating only *once!*)

$$L = \sqrt{-g} \left[\sum g^{\sigma\tau} \Gamma_{\sigma\alpha}^\beta \Gamma_{\tau\beta}^\alpha + \sum \frac{\partial \lg \sqrt{-g}}{\partial x_\sigma} \left(-\frac{\partial g^{\sigma\alpha}}{\partial x_\alpha} - g^{\sigma\beta} \frac{\partial \lg \sqrt{-g}}{\partial x_\beta} \right) \right].$$

If, therefore, the generally covariant field equations can be presented in the form of Hamilton's principle, of which I am quite certain, then this must be the Hamiltonian function to use. The second term in brackets can also be written in the form

$$-\sum g^{\alpha\beta} \left\{ \begin{matrix} \alpha\beta \\ \sigma \end{matrix} \right\} \left\{ \begin{matrix} \sigma\rho \\ \rho \end{matrix} \right\} = -\sum g^{\alpha\beta} \Gamma_{\alpha\beta}^\sigma \Gamma_{\sigma\rho}^\rho.$$

The computation of $\dfrac{\partial L}{\partial g^{\mu\nu}}$ and $\dfrac{\partial L}{\partial g_\sigma^{\mu\nu}}$ is quite arduous, though, at least with my limited proficiency in calculation.[7]

It is important that the equations initially be given in a generally covariant form because only in this setup for the equations is all arbitrariness avoided. For if you confine yourself from the start to the case $\sqrt{-g} = 1$, then a factor $(\sqrt{-g})^n$ could be added to scalar G without disturbing the thus restricted covariance. The same is valid analogously to the equations covering matter.

It would undoubtedly be a step forward if later the frame of reference could be specialized even further in a natural way. My efforts in this direction have been unsuccessful up to now, though. In any case, the problem must obviously be arranged so that dx_1, dx_2, dx_3 is of a spatial nature throughout and dx_4 of a temporal one. But this is a specialization solely by means of inequalities, not equalities.

Your remark on extinction is very convincing.[8] If only light were finally shed on the absorption process! But the proof of the existence of zero-point energy shows us how far away we are from a genuine understanding here.[9]

In thanking you again for your burning interest and even more for your intention to place your efforts in the service of this problem, I remain with heartiest greetings to you and yours, in friendship, yours truly,

A. Einstein.

Cordial thanks for sending your lectures on statistical mechanics.[10] Feb. 23. I admire very much Tetrode's splendid analyses on the entropy constant.[11] I recently gave a report on it.[12]

184.   To Hendrik A. Lorentz

[Berlin,] 19 January 1916

Dear and highly esteemed Colleague,

Only too well do I understand your attempt also to derive gravitation from the field equations in the manner of Hamilton's principle. I myself am compelled to derive the Hamiltonian function retroactively, in order to derive the expression for the conservation laws conveniently.[1] Nevertheless, I must admit that I actually do not see in Hamilt. princip. anything more than a means toward reducing a system of tensor equations to a scalar equation for which the conservation laws are always satisfied and easily derived. I definitely believe that it is possible to find a Hamiltonian form also for the *generally covariant* form of the equations, as I already indicated in yesterday's letter.[2]

I must emphasize again that my field equations (2a) given in my contribution, "The Field Equations of Gravitation"[3]

$$G_{im} = \kappa \left(T_{im} - \frac{1}{2} g_{im} T \right)$$

are *generally* covariant. I also contend that these are the only equations that fulfill the following conditions:

1) general covariance.

2) first order compons. T_μ^ν of matter and der[ivations] of the g^{im}'s higher than the second, do not appear in the $\dfrac{\partial^2 g^{im}}{\partial x_\alpha \partial x_\beta}$'s and the energy tensor.

3) consistent with the "conservation law" of matter without any other restrictions for the $T_{\mu\nu}$'s.

This statement is based above all on the knowledge that aside from the tensors

$$G_{im}$$

$$\text{and } g_{im} \sum_{\alpha\beta} g^{\alpha\beta} G_{\alpha\beta}$$

there are no ⟨*general*⟩ (arbitrary substitutions for covariant) tensors (that satisfy condition (2)). *Your function G vanishes identically,* because—as you can easily calculate—the extension already vanishes, thus all the more so the divergence of the fundamental tensor $g_{\mu\nu}(g^{\mu\nu})$.[4]

Hence it is clear that a consideration according to Hamilton's principle would have to be connected to the V-scalar

$$Q = \sqrt{-g} \sum_{\alpha\beta} g^{\alpha\beta} G_{\alpha\beta},$$

as I already indicated in yesterday's letter. I avoided the somewhat involved computation of the $\dfrac{\partial Q}{\partial g^{\mu\nu}}$'s and $\dfrac{\partial Q}{\partial g_\sigma^{\mu\nu}}$'s by setting up the tensor equations directly. But the other way is certainly also workable and even, more elegant mathematically.

With cordial greetings, yours,

A. Einstein.

185. To Paul Ehrenfest

Berlin, Monday, [24 January 1916 or later][1]

Dear Ehrenfest,

Today you should finally be content with me. I am delighted about the great interest you are devoting to this problem. I am not going to support myself at all on the papers but shall calculate everything out for you.[2] Then, if anything should remain incomprehensible, the gap can be easily filled.–

1) Lagrangian form of the equations.

Statement: Let $\sqrt{-g} = 1$. In addition, let $L = g^{\sigma\tau}\left\{\begin{matrix}\alpha\beta\\\alpha\end{matrix}\right\}\left\{\begin{matrix}\tau\alpha\\\beta\end{matrix}\right\}$. Then

follows, if $\mathfrak{L}$ is conceived of as a function of the $g^{\sigma\tau}$'s and $g_\alpha^{\sigma\tau} = \dfrac{\partial g^{\sigma\tau}}{\partial x_\alpha}$:[3]

$$\left.\begin{matrix}\dfrac{\partial L}{\partial g^{\sigma\tau}} = \left\{\begin{matrix}\sigma\beta\\\alpha\end{matrix}\right\}\left\{\begin{matrix}\tau\alpha\\\beta\end{matrix}\right\}\\[3mm]\dfrac{\partial L}{\partial g_\alpha^{\sigma\tau}} = -\left\{\begin{matrix}\sigma\beta\\\alpha\end{matrix}\right\}\end{matrix}\right\} \tag{1}$$

Comment: I always omit the summation symbol. An index must always be summed when it appears twice.[4]

Proof: From differentiating $\mathfrak{L}$—always considered as a funct. of $g^{\sigma\tau}$ & $g_\alpha^{\sigma\tau}$—follows

$$dL = \left\{\begin{matrix}\sigma\beta\\\alpha\end{matrix}\right\}\left\{\begin{matrix}\tau\alpha\\\beta\end{matrix}\right\}dg^{\sigma\tau} + 2g^{\sigma\tau}\left\{\begin{matrix}\sigma\beta\\\alpha\end{matrix}\right\}d\left\{\begin{matrix}\tau\alpha\\\beta\end{matrix}\right\}.$$

(Two terms differing only in index name are summarized).

From this follows furthermore from $g^{\sigma\tau}d\left\{\begin{matrix}\tau\alpha\\\beta\end{matrix}\right\} = d\left(g^{\sigma\tau}\left\{\begin{matrix}\tau\alpha\\\beta\end{matrix}\right\}\right) - \left\{\begin{matrix}\tau\alpha\\\beta\end{matrix}\right\}dg^{\sigma\tau}$

$$\begin{aligned}dL &= -dg^{\sigma\tau}\cdot\left\{\begin{matrix}\sigma\beta\\\alpha\end{matrix}\right\}\left\{\begin{matrix}\tau\alpha\\\beta\end{matrix}\right\} + 2\left\{\begin{matrix}\sigma\beta\\\alpha\end{matrix}\right\}d\left(g^{\sigma\tau}\left\{\begin{matrix}\tau\alpha\\\beta\end{matrix}\right\}\right)\\&= -dg^{\sigma\tau}\{\}\{\} + 2\left\{\begin{matrix}\sigma\beta\\\alpha\end{matrix}\right\}d\left(g^{\sigma\tau}g^{\beta\lambda}\left[\begin{matrix}\tau\alpha\\\lambda\end{matrix}\right]\right)\end{aligned}$$

In addition

$$\left\{\begin{matrix}\sigma\beta\\\alpha\end{matrix}\right\} = \left\{\begin{matrix}\beta\sigma\\\alpha\end{matrix}\right\}\cdots \tag{α}$$

Taking into consideration that the second term does not change when the sum of the indices α & β are exchanged simultaneously with λ and τ, the second term is then also equal to

$$\left\{\begin{matrix}\sigma\beta\\\alpha\end{matrix}\right\}d\left(g^{\sigma\tau}g^{\beta\lambda}\left(\left[\begin{matrix}\tau\alpha\\\lambda\end{matrix}\right]+\left[\begin{matrix}\lambda\alpha\\\tau\end{matrix}\right]\right)\right)$$

or equal to

$$\left\{ \begin{matrix} \sigma\beta \\ \alpha \end{matrix} \right\} d \left(g^{\sigma\tau} g^{\beta\lambda} \frac{\partial g_{\lambda\tau}}{\partial x_\alpha} \right)$$

or equal to[5]

$$-\left\{ \begin{matrix} \sigma\beta \\ \alpha \end{matrix} \right\} dg_\alpha^{\sigma\tau}$$

For from $g_{\rho\sigma} g^{\sigma\tau} = \delta_\rho^\tau = 1$ or 0 follows

$$\frac{\partial g_{\rho\sigma}}{\partial x_\alpha} g^{\sigma\tau} = -g_{\rho\sigma} \frac{\partial g^{\sigma\tau}}{\partial x_\alpha}, \tag{β}$$

& from this through multiplication by $g^{\rho\lambda}$

$$g^{\rho\lambda} g^{\sigma\tau} \frac{\partial g_{\rho\sigma}}{\partial x_\alpha} = -\frac{\partial g^{\lambda\tau}}{\partial x_\alpha}$$

analogously also

$$g_{\rho\lambda} g_{\sigma\tau} \frac{\partial g^{\sigma\tau}}{\partial x_\alpha} = -\frac{\partial g_{\lambda\tau}}{\partial x_\alpha} \tag{β'}$$

It thus follows[6]

$$dL = \left\{ \begin{matrix} \sigma\beta \\ \alpha \end{matrix} \right\} \left\{ \begin{matrix} \tau\alpha \\ \beta \end{matrix} \right\} dg^{\sigma\tau} - \left\{ \begin{matrix} \sigma\beta \\ \alpha \end{matrix} \right\} dg_\alpha^{\sigma\tau},$$

from which statement (1) follows. From this it follows that the gravitation equations can be written in the form[7]

$$\frac{\partial}{\partial x_\alpha} \left(\frac{\partial L}{\partial g_\alpha^{\sigma\tau}} \right) - \frac{\partial L}{\partial g^{\sigma\tau}} = -\kappa \left(T_{\sigma\tau} - \frac{1}{2} g_{\sigma\tau} T \right) \dots \tag{2}$$

2) Conservation laws.

If you multiply (2) by $g_\beta^{\sigma\tau}$, then you obtain upon partial differentiation reformulation of the first term,[8] because of interchangeability of α and β,

$$\frac{\partial}{\partial x_\alpha} \left(g_\beta^{\sigma\tau} \frac{\partial \mathfrak{L}}{\partial g_\alpha^{\sigma\tau}} \right) - \underbrace{\left(\frac{\partial L}{\partial g^{\sigma\tau}} g_\beta^{\sigma\tau} + \frac{\partial L}{\partial g_\alpha^{\sigma\tau}} \frac{\partial g_\alpha^{\sigma\tau}}{\partial x_\beta} \right)}_{\frac{\partial L}{\partial x_\beta}} = -\kappa T_{\sigma\tau} g_\beta^{\sigma\tau}.$$

The second term on the right-hand side disappears because

$$g^{\sigma\tau} \frac{\partial g^{\sigma\tau}}{\partial x_\beta} = -g^{\sigma\tau} \frac{\partial g_{\sigma\tau}}{\partial x_\beta} = -\frac{\partial \lg g}{\partial x_\beta} = 0.$$

Now I write the conservation law of matter by introducing it formally without assuming its validity[9] [[A_μ's are] unknown space functions]

$$\frac{\partial T_\mu^\sigma}{\partial x_\sigma} + \frac{1}{2}g_\mu^{\alpha\tau}T_{\alpha\tau} = A_\mu \ldots \tag{3}$$

The second term on the left-hand side can be converted because of (β) into the form $-\frac{1}{2}\frac{\partial g_{\alpha\tau}}{\partial x_\mu}T^{\alpha\tau}$. The { } can likewise be introduced, which I do not need here, however. With the aid of this, the right-hand side of the last equation can be substituted by

$$2\kappa\frac{\partial T_\beta^\alpha}{\partial x_\alpha} - 2\kappa A_\beta,$$

and you obtain:

$$\frac{\partial}{\partial x_\alpha}\underbrace{\left(\frac{1}{2\kappa}\left[L\delta_\beta^\alpha - g_\beta^{\sigma\tau}\frac{\partial L}{\partial g_\alpha^{\sigma\tau}}\right] + T_\beta^\alpha\right)}_{t_\beta^\alpha} = A_\beta \ldots \tag{4}$$

If the A_β's were to vanish, this would be the conservation equation both for matter and for gravitation. For the moment we shall work with (4). Taking into account (1)[10] it follows from the definition of the t_β^α's

$$t_\alpha^\alpha = t = \frac{1}{\kappa}L \ldots \tag{5}$$

3) *Mixed form of the gravitation equations.* We write the latter in the form

$$-\frac{\partial\left\{\begin{array}{c}\sigma\tau\\\alpha\end{array}\right\}}{\partial x_\alpha} + \left\{\begin{array}{c}\sigma\beta\\\alpha\end{array}\right\}\left\{\begin{array}{c}\tau\alpha\\\beta\end{array}\right\} = -\kappa(T_{\sigma\tau} - \frac{1}{2}g_{\sigma\tau}T) \cdots \tag{2a}$$

and multiply by $g^{\tau\nu}$. The first term on the left yields through partial diff. reformulation

$$-\frac{\partial}{\partial x_\alpha}\left(g^{\tau\nu}\left\{\begin{array}{c}\sigma\tau\\\alpha\end{array}\right\}\right) + \left\{\begin{array}{c}\sigma\tau\\\alpha\end{array}\right\}\frac{\partial g^{\tau\nu}}{\partial x_\alpha},$$

of which the second term can be modified with the aid of the formula (γ), deduced here to the side,[11] so that you obtain

$$-\frac{\partial}{\partial x_\alpha}\left(g^{\tau\nu}\left\{\begin{array}{c}\sigma\tau\\\alpha\end{array}\right\}\right) - g^{\tau\varepsilon}\left\{\begin{array}{c}\varepsilon\alpha\\\nu\end{array}\right\}\left\{\begin{array}{c}\tau\sigma\\\alpha\end{array}\right\} - g^{\nu\varepsilon}\left\{\begin{array}{c}\varepsilon\alpha\\\tau\end{array}\right\}\left\{\begin{array}{c}\sigma\tau\\\alpha\end{array}\right\}.$$

Calculation aid [*Hilfsrechnung*]:

$$\frac{\partial g^{\tau\nu}}{\partial x_\alpha} = -g^{\tau\varepsilon}g^{\nu\xi}\frac{\partial g_{\varepsilon\xi}}{\partial x_\alpha} \text{ (because of } \beta')$$

$$= -g^{\tau\varepsilon}g^{\nu\xi}\left(\left[\begin{array}{c}\varepsilon\alpha\\\xi\end{array}\right] + \left[\begin{array}{c}\xi\alpha\\\varepsilon\end{array}\right]\right)$$

thus

$$\frac{\partial g^{\tau\nu}}{\partial x_\alpha} = -g^{\tau\varepsilon}\left\{\begin{array}{c}\varepsilon\alpha\\\nu\end{array}\right\} - g^{\tau\varepsilon}\left\{\begin{array}{c}\varepsilon\alpha\\\tau\end{array}\right\}\dots \qquad (\gamma)$$

The third of these terms[12] cancels out with the one formed from the second of (2a). Hence you obtain initially

$$\frac{\partial}{\partial x_\alpha}\left(g^{\tau\nu}\left\{\begin{array}{c}\sigma\tau\\\alpha\end{array}\right\}\right) + g^{\varepsilon\tau}\left\{\begin{array}{c}\varepsilon\alpha\\\nu\end{array}\right\}\left\{\begin{array}{c}\tau\sigma\\\alpha\end{array}\right\} = \kappa(T^\nu_\sigma - \tfrac{1}{2}\delta^\nu_\sigma T)\dots \qquad (6)$$

From the definition of t^α_β and the equations (1) & (5) you obtain

$$t^\alpha_\beta = \frac{1}{2}t\delta^\alpha_\beta + \frac{1}{2\kappa}\left\{\begin{array}{c}\sigma\tau\\\alpha\end{array}\right\}\frac{\partial g^{\sigma\tau}}{\partial x_\beta}.$$

On transforming the second term according to (γ) and combining both the thus formed terms

$$t^\alpha_\beta - \frac{1}{2}\delta^\alpha_\beta t = -\frac{1}{\kappa}g^{\sigma\varepsilon}\left\{\begin{array}{c}\sigma\tau\\\alpha\end{array}\right\}\left\{\begin{array}{c}\varepsilon\beta\\\tau\end{array}\right\}\dots \qquad (7)$$

Disregarding the factor $-\dfrac{1}{\kappa}$ and the index designation, the right-hand side of (7) corresponds to the second term in (6), so that you can write

$$\frac{\partial}{\partial x_\alpha}\left(g^{\tau\nu}\left\{\begin{array}{c}\sigma\tau\\\alpha\end{array}\right\}\right) = \kappa\left((T^\nu_\sigma + t^\nu_\sigma) - \tfrac{1}{2}\delta^\nu_\sigma(T + t)\right)\dots \qquad (8)$$

This equation is interesting because it shows that the source of the gravitation lines is determined solely by the sum $T^\nu_\sigma + t^\nu_\sigma$, as must obviously be expected.[13]

(Second sheet)

4) *Proof that the A_μ's vanish.* Now comes the main issue.
     a) If (8) is multiplied by δ^σ_ν, then your obtain the scalar equation

$$\frac{\partial}{\partial x_\alpha}\left(g^{\tau\nu}\left\{\begin{array}{c}\nu\tau\\\alpha\end{array}\right\}\right) = -\kappa(T + t)\dots \qquad (9)$$

The left-hand side reads fully as

$$\frac{1}{2}\frac{\partial}{\partial x_\alpha}\left(g^{\nu\tau}g^{\alpha\beta}\left(\frac{\partial g_{\beta\nu}}{\partial x_\tau}+\frac{\partial g_{\beta\tau}}{\partial x_\nu}-\frac{\partial g_{\nu\tau}}{\partial x_\beta}\right)\right).$$

The third term yields nothing, because $g^{\nu\tau}\dfrac{\partial g_{\nu\tau}}{\partial x_\beta}=\dfrac{\partial \lg g}{\partial x_\beta}=0$. The first two

coincide through an exchange of ν and τ, such that they can be combined. By applying (β') you obtain finally

$$-\frac{\partial^2 g^{\alpha\tau}}{\partial x_\alpha \partial x_\tau}.$$

From (9) thus results

$$\frac{\partial^2 g^{\alpha\beta}}{\partial x_\alpha \partial x_\beta}-\kappa(T+t)=0\cdots \tag{9a}$$

We execute the operation $\dfrac{\partial}{\partial x_\nu}$ on (8) and allowing for (4) obtain[14]

$$\frac{\partial}{\partial x_\alpha}\frac{\partial}{\partial x_\nu}\left(g^{\tau\nu}\left\{\begin{matrix}\sigma\tau\\ \alpha\end{matrix}\right\}\right)=-\frac{1}{2}\frac{\partial(T+t)}{\partial x_\sigma}+\kappa A_\sigma\cdots \tag{10}$$

The left-hand side is more completely

$$\frac{1}{2}\frac{\partial}{\partial x_\alpha}\frac{\partial}{\partial x_\nu}\left(g^{\tau\nu}g^{\alpha\beta}\left(\frac{\partial g_{\sigma\beta}}{\partial x_\tau}+\frac{\partial g_{\tau\beta}}{\partial x_\sigma}-\frac{\partial g_{\sigma\tau}}{\partial x_\beta}\right)\right).$$

If you exchange in the first term α and ν as well as β and τ, then disregarding the index, it coincides with the third. Only the second remains, which allowing for (β') changes into

$$-\frac{1}{2}\frac{\partial^3 g^{\alpha\nu}}{\partial x_\alpha \partial x_\nu \partial x_\sigma}.$$

That is why you obtain instead of (10)[15]

$$\frac{\partial}{\partial x_\sigma}\left(\frac{\partial^2 g^{\alpha\nu}}{\partial x_\alpha \partial x_\nu}-(T+t)\right)=-2\kappa A_\sigma\cdots \tag{10a}$$

From (9a) and (10a) $A_\sigma=0$ results, that is, according to (3) of the conservation law of matter, as a consequence of the field equations (2).–

You will certainly not encounter any more problems now. Show this thing to Lorentz as well, who also does not yet perceive the need for the structure on the right-hand side of the field equations. I would appreciate it if you would then give these pages back to me, because nowhere else do I have these things so nicely in one place.

With best regards, yours,

Einstein.

186.   To Arnold Sommerfeld

[Berlin,] 2 February 1916

Dear Sommerfeld,

I've been cracking my brains about your letter, especially since I must acknowledge quite much of what you say as correct. Fr.[1] is of the "greyhound" breed, more or less, as defined by a good acquaintance of mine. His way of bolting is also not particularly distinguished. I have known this person's weaknesses for a long time[2]—and have also been more or less irritated by him. It is undoubtedly justified to raise the question: Is Einstein right when he attempts to remove all obstacles in this person's career?[3] You answer in the negative. I have thought about the matter thoroughly and also have discussed it with an intelligent and well-intentioned person to whom I had presented the "evidence" and whose objectivity in the matter is entirely beyond doubt.[4]

First to the personal characteristics. I would *not* choose Fr. as an intimate friend of mine but would always maintain a healthy distance from him either way. And yet I come to the conclusion: If the devil were to unseat all of those among our professorial colleagues whose self-criticism and decency are not above that of Freundlich's, then the trusty ranks would be considerably thinned. I would even—*horribile dictu*—fear for your informant S![5] On the other hand, Freundl. offers something worth its weight in gold—an enthusiastic dedication to the problem; that is a rare trait he does not share with very many.

Now to the professional qualities. Freundl. is not really creatively talented, but he is intelligent and resourceful. The greyhound nature of his, mentioned above, comes to a large part from his pounding heart when he investigates a scientifically important issue. We must not forget that Fr. had devised the statistical method that makes possible using fixed stars in addressing the line-shift question. Although the nasty calculation error did slip by him and some other things there are greyhound-like as well (density definition), the overall value of the matter ought not to be forgotten because of it.[6] Errors can be corrected and in time are always corrected. The task involves discovering a way and smoothing it until it becomes passable.

Seen from my point of view, the affair looks like this. Freundl. was the only colleague in that profession until now to support my efforts effectively in the area of general relativity.[7] He has devoted years of thought, and of work as well, to this problem, as far as was possible beside the exhausting and tedious duties at the observatory. What a pitiful rascal I would be if, now that the idea has become accepted, I dropped the man on the thought that I am now no longer dependent on him. Just put yourself into my shoes. Then you will stop applying the watchword "landgrave, steel yourself."[8]

Fr. has a second attribute to his merit as well. I do not want to talk about the refutation of von Seeliger's theory of Mercury's perihelion motion, since this deed may possibly be described as battering down open doors.[9] But Fr. has shown that modern astronomical tools are good enough to demonstrate light deflection by Jupiter,[10] which *I* would not have thought possible despite having considered the case years ago.[11] I simply lack contacts in astronomy.

Now, I gladly admit that Fr.'s weaknesses do not make it seem at all desirable that the execution of this important matter be placed solely in his hands. But up to now no one has made any effort toward participating in the undertaking, so that I am *de facto nolens volens* dependent on Freundlich alone in this endeavor of promoting the resolution of this eminently important question.

Recapitulation: (1) Fr. deserves my attempts at making it possible for him to collaborate in the undertaking here in question.

(2) It would be extremely desirable if others also were to take on the problem, be it together with Fr. or independently, without any collaboration.

Cordial greetings, yours,

Einstein.

187. To Mileva Einstein-Marić

Berlin, 13 Wittelsbacher St., 6 February 1916

D[ear] M[ileva],

I propose herewith to turn our now tested separation into a divorce,[1] whereby we could essentially keep to our friend Haber's draft.[2] I believe that it is in the interest of both of us that in this way the duties and rights of each of us be laid down clearly, so that we each can arrange the rest of our lives independently as far as the situation allows.

Once you have informed me that you agree in principle, it will surely be possible to have the details settled to your satisfaction. This clarification of our relationship does not change in any way that I regard it as my first and foremost duty to accumulate savings for my boys; in these $1\text{-}^1/_2$ years[3] I have already put aside 8,000 M for them.

Also write me something about Albert's school affairs soon.[4] He said a bit about this in his last, very dear letter.

With regards to you and kisses for the boys, yours,

A. E.

P.S. Give the boys calcium chloride[5] after the midday meal (1 teaspoon of concentrated solution in $\frac{1}{2}$ a glass of water). This way the calcium deficiency in cooked food is compensated.

188.   From Karl Schwarzschild

[at the Russian front,] 6 February 1916

Esteemed Mr. Einstein,

Many thanks for your letter of January 9th.[1] I wrote about Jupiter to Hertzsprung,[2] who immediately brought my attention to the fact that in the next few years Jupiter will be orbiting too much to the south of us. This extreme precision is only attainable between the zenith and perhaps a declination of 50°. The problem must therefore be left to observatories more to the south. Mr. Freundlich could be of service if he were to pick out stellar transits and occultations. (It appears to me, though, that Mr. Banachiewicz[3] has already been engaged in this for other purposes.) In other respects, we shall not be able to agree too easily on Freundlich, and I only want to add: Debating this way and that about him is pointless. I just think that he has already fallen out with Struve to such a degree that it would be best if you exerted your influence toward obtaining another occupation for him.[4]

With regard to the inertial system, we are in agreement. You say that beyond the Milky Way conditions could exist under which the Galilean system is no longer the simplest. I only contend that within the Milky Way such conditions do not exist. As far as very large spaces are concerned, your theory takes an entirely similar position to Riemann's geometry, and you are certainly not unaware that elliptic geometry is derivable from your theory, if one has the entire universe under uniform pressure (energy tensor $-p$, $-p$, $-p$, 0).[5]

I cannot deny that you have put the freedom extending beyond it to most fortunate use.

Meanwhile, in order to acquaint myself with your energy tensor, I have been working on the incompressible liquid sphere problem (energy tensor $-p$, $-p$, $-p$, $+\rho_0$). I would not have done so, had I known that it would cause me so much trouble.[6]

The line element in the interior reads:

$$ds^2 = \left(\frac{3\cos\sigma_0 - \cos\sigma}{2}\right)^2 dt^2 - \frac{3}{\kappa\rho_0}[d\sigma^2 + \sin^2\sigma(d\vartheta^2 + \sin^2\vartheta\, d\varphi^2)]$$

ϑ, φ are the normal polar coordinates. σ would be related to the radius vector according to:

$$\left(\frac{\kappa\rho_0}{3}\right)^{\frac{3}{2}} r^3 = \frac{9}{4}\cos\sigma_0\left(\sigma - \frac{1}{2}\sin 2\sigma\right) - \frac{1}{2}\sin^3\sigma,$$

whereby σ_0 is determined by the sphere's radius r_0 according to

$$\left(\frac{\kappa\rho_0}{3}\right)^{\frac{3}{2}} r_0^3 = \frac{9}{4}\cos\sigma_0\left(\sigma_0 - \frac{1}{2}\sin 2\sigma_0\right) - \frac{1}{2}\sin^3\sigma_0.$$

For pressure p,

$$(\rho_0 + p)\left(\frac{3\cos\sigma_0 - \cos\sigma}{2}\right) = \text{const.}$$

applies. The line element outside is the old one, where only in

$$R^3 = r^3 + \rho$$

is $\rho = \alpha^3$ not valid, but is instead:

$$\alpha = m\frac{2}{3Z - 1} \qquad \rho = \frac{3}{2}\frac{\alpha^3}{\sin^6\sigma_0}(1 - Z) \qquad m = \frac{\kappa\rho_0}{3}r_0^3$$

$$Z = \frac{3}{2}\frac{\cos\sigma_0}{\sin^3\sigma_0}\left(\sigma_0 - \frac{1}{2}\sin 2\sigma_0\right).$$

The strange thing is that, for a finite globe of $\cos\sigma_0 = \frac{1}{3}$, the pressure at the center ($\sigma = 0$) becomes infinite; smaller spheres of a given mass are not possible. The transition to an infinitesimal radius, where $\rho = \alpha^3$ must then result, also results in physically meaningless solutions—I always thought I had miscalculated before realizing this. Do you see concretely where the limit $\cos\sigma_0 = \frac{1}{3}$ comes from?

The square bracket is the line element in spherical geometry! Thus, spherical geometry governs within the sphere. The interior of the sphere is not the entire spherical space but only a sphere of radius σ_0 within this space. It is just from the increase in the velocity of light that it becomes noticeable whether you are nearer to the center or nearer to the surface.

Here it is even more obvious than in the case of the outer field, that r has absolutely no physical meaning, but is just a variable making the $|g_{\mu\nu}|$'s equal to 1. This looks as if it must be possible to find a condition other than $|g_{\mu\nu}| = 1$, making the field equations even simpler.

Interlineations at end of second paragraph: "that observations within the Milky Way are significant for the entire system."

189.   To Arnold Sommerfeld

[Berlin, 8 February 1916]

Dear Sommerfeld,

Your letter pleased me very much, your report on the theory of spectral lines *enchanted* me. A revelation! Just one thing I still did not grasp. If in

$$\nu = N \left(\frac{1}{n^2} - \frac{1}{m^2} \right)$$

$$/$$
$$n = 2$$

$n = 2$ causes a division into two,[1] then surely $m = m$ should also cause a division into m parts, so that in general a division of $n \cdot m$ would have to be expected. This is not supposed to be an objection, though. I am much rather convinced that you are right.– I have found a lecture experiment to demonstrate conveniently Ampere's currents (already tried out).[2]

You will be convinced of the general theory of rel. when you have studied it. That is why I am not mentioning a word in its defense.

Hearty congratulations on your fine discovery and best regards, yours,

Einstein.

190.   To Hermann Struve

[Berlin,] 13 February 1916

Highly esteemed Colleague,

Many thanks for the *Astron. Nachrichten* issue, which I am returning herewith. Seeliger's article told me nothing new.[1] Of great interest was Ludendorff's, from which I learned how incomplete the available observational data still are.[2] But the good news is that one gets the impression that data leading to a definite decision are bound to come in bit by bit.

The following can be asked in connection with Ludendorff's article:

Do weaker stars of the same spectral class have on average a smaller mass as well? This could be answered in the case of binary stars whose mass quantities can be obtained through averaging. Should there turn out to be a significant correlation, then Ludendorff's approach could prove very valuable in the future.

As long as the average error exceeds the determined redshift (p. 79, first column), the result remains completely uncertain, at least if "w[ith] e[rror]" for the result indicates the dreaded error, not those attached to the individual determinations of stellar velocity, which seems to be the case.

In any case, I see that Freundlich's result is by no means secure, not even qualitatively.[3] However, credit should be given to Freundlich for having been the first to point out a practicable way to test the question.–

I thank you very much for your amiability during my visit with you. I am very glad to see that our good personal relations do not suffer from our differences of opinion in certain scientific questions.[4] I am also glad that now there appears to be a possibility to solve the unpleasant Freundlich situation in a satisfactory manner.[5]

With best regards, yours very truly,

A. Einstein.

191.   To Otto Stern

[Berlin, 15 February 1916]

Dear Mr. Stern,

This time you have shot wide of the mark, as you will recognize from the following comparison.

On a solid horizontal base lies a pencil (1st state). It can also stand (2nd state). Energy difference E for both states (owing to gravity) very substantial.

$e^{-\frac{E}{kT}}$ for $T = 0$ dizzyingly small. Hence the statement: The pencil cannot stand at absolute zero!

Corresponding assertion. Solid carbon and solid oxygen cannot be contiguous to one another at $T = 0$. Only CO_2 exists. Proof $e^{-\frac{E}{kT}}$ is frightfully small! You stuntman![1]

General relativity is now, almost exactly since we last saw each other, *finally* resolved. *General* covariance of the field equations. Perihelion motion of Mercury explained exactly. Theory very transparent and fine. Lorentz, Ehrenfest, Planck, and Born are convinced adherents, likewise Hilbert.

Reflect on the objection; you will surely come to the conviction that it is justified.

Cordial greetings, yours,

A. Einstein.

192.   To Otto Stern

[Berlin, after 15 February 1916][1]

Dear Mr. Stern,

I remain with the example of the pencil.[2] I assert: The fact that *we may apply thermodynamics to chemical systems at all, ultimately rests on that we do treat a change, happening so quickly that Brownian motion does not have time to knock down the standing pencil, as an infinitely slowly reversible change, in the thermodynamic sense.*

On the basis of your reasoning you can propose at most one of two statements:

a) The mixed crystal (in the sense of your correctly given definition)[3] does not exist at $T = 0$.

b) Although it exists, you cannot speak about its entropy, because the state cannot be produced by reversing the process.

Now, (a) could be asserted with the same justification (*mutatis mutandis*) of any chemical system that is not at the absolute entropy maximum (system C | O_2).

With (b), if it were correct, you could slaughter almost all of physical chemistry.

But for the statement:

c) the mixed crystal existing at $T = 0$, unordered by definition, can be produced (sufficiently exactly) by the reverse process *and has entropy* 0, you have not provided the least argum. in support.[4]

Right from your detailed correspondence I gather that you do not separate the cases (a) (b) and (c) distinctly, whence in my view your error originates. You hold the following impossible position:

"*Thermodynamically*, only *ordered* crystals exist at absolute zero. Consequently, the entropy of mixed crystals, even when real, i.e., *unordered*, must be calculated as if they were ordered."

Write me your answer briefly and to the point (such as a postcard). It is easier then to answer concisely.

With best regards, yours,

Einstein.

193.   To David Hilbert

[Berlin, 18 February 1916][1]

Highly esteemed Colleague,

Many thanks for taking my wishes into consideration.[2] I will not inform you of the exact time of my arrival so that you do not inconvenience yourself. At all events, it will only be on Thursday. I gladly accept your and your wife's kind invitation to stay with you. That the affair with H. promises to turn out favorably pleases me greatly, especially since I am not going to receive the call to Vienna because Smoluchowski will receive the appointment there.[3] Important matters can be carried out at the observatory even with relatively modest means, in particular, beside the Jupiter matter,[4] the question of redshift for fixed stars.[5] It will be a joy when everything has straightened itself out!

Best regards, yours,

A. Einstein.

194.   To Karl Schwarzschild

[Berlin, 19 February 1916]

Esteemed Colleague,

Unfortunately because of a lot of work I was not yet able to answer your earlier letter.[1] Also, the special cases discussed there awoke my interest to a lesser degree. But I find your new communication very interesting. I have found your calculation confirmed. My comment in this regard in the paper of November 4 no longer applies according to the new determination of $\sqrt{-g} = 1$, as I was already aware.[2] The choice of coordinate system according to the condition $\sum \dfrac{\partial g^{\mu\nu}}{\partial x_\nu} = 0$ is not consistent with $\sqrt{-g} = 1$. Since then, I have handled Newton's case differently, of course, according to the final theory.[3]—Thus there are no gravitational waves analogous to light waves. This probably is also related to the one-sidedness of the sign of scalar T, incidentally. (Nonexistence of the "dipole".)[4]

Cordial greetings and many thanks for the interesting communication. Yours,

A. Einstein.

195.   To Max Born

[Berlin,] Sunday [27 February 1916][1]

Dear Mr. Born,

This morning I received the correction to your paper for the *Physikalische Zeitschrift*, which I read, not without embarrassment, but with the happy feeling of having been understood thoroughly and acknowledged by one of my most highly qualified colleagues.[2] But aside from the objective content, I was also filled with the happy sensation of cheerful goodwill that emanates from the paper and that otherwise so rarely lingers undiluted in the pale light of the study lamp.[3] I thank you from my heart for granting me the privilege of this fine joy.

With best regards, yours,

A. Einstein.

196.   To Wilhelm Wien

Berlin. 13 Wittelsbacher St., Monday, 28 February 1916

Highly esteemed Colleague,

I am engaged in developing fully the general theory of relativity. In about two weeks the manuscript will be finished.[1] I would like to publish this detailed paper in the *Annalen*, if I were granted the right to separate publication of the article by a publisher of my free choice. I ask you please to inform me whether such a procedure is permitted on the part of the journal and the publisher.

With regards from your very devoted colleague,

A. Einstein.

197.   To Hans Albert Einstein

Göttingen. [3 March 1916]

My dear Adn,

I send you greetings from here.[1] In a month I'm going to be with you in order to go on a walking tour with you. It'll be very nice. Write to me soon when you have holidays. If it's not convenient, I'm going to take you out of school for one week. What's the news on your examination?[2]

Kisses to you and Tete from your

Papa.

198.   To Otto Stern

[Berlin,] 10 March 1916

Dear Mr. Stern,

Your example is very instructive.[1]

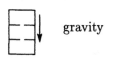 gravity

At $T = 0$ the molecule is at the bottom. That means the mixed crystal does not exist, only the chemical compound does.[2] In this case entropy actually does vanish

Counter-example (ideal case)

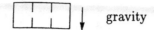 gravity

Here the mixed crystal certainly exists: Its entropy does not disappear.[3]

Now you will say that this case does not exist in reality, since the potential energies of conceivable cases never are exactly identical. Therefore, the following case as a true analogy to the real one:

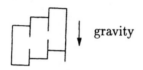 gravity

At high temperatures the distribution is practically uniform despite gravity. We cool it down now to absolute zero. As a result two cases are possible, namely,

1) Transition from one chamber to another is so rare that the case where transition from one chamber to another during cooling can take place occurs very rarely.

2) Transition from one chamber to another is so frequent that it occurs very frequently during the cooling period.

In case (1) the mixed crystal still exists at $T = 0$ and does not satisfy Nernst's theorem.

In case (2) the particle is certainly at the bottom in the first chamber at $T = 0$. The mixed crystal as such does not exist at $T = 0$. Nernst's theorem applies.

In any case, it cannot be asserted that Nernst's theorem applies *to a mixed crystal* at $T = 0$. For in your opinion it simply does not exist at $T = 0$. According to your view, case (1) is not present in nature.

But I hold another view. As I do not see why mixed crystals should not behave like chemical mixtures, about which thermodynamically realizable cooling is known to be possible without the creation of a chemical or statistical equilibrium affecting the *structure*.

Do you not agree with this? With best regards, yours,

Einstein.

199. To Hans Albert Einstein

[Berlin,] Saturday. [11 March 1916][1]

My dear son,

The letter I received from you a moment ago pleased me greatly. If you wrote it by yourself, you have developed very well indeed.[2] Keep it up. You don't have to worry about secondary school [*Gymnasium*]. If you don't manage to get in at the moment,[3] you can always catch up easily. I just don't want you to lose a year. You have to be at school far too long as it is. When Tete is better again, take the calcium chloride with him; it is very healthy.[4] I am taking it too.

I am very glad that you enjoy the piano so much.[5] I have one as well in my little apartment[6] and play it daily. I also play the violin a lot.[7] Perhaps you can practice something to accompany violin; then we can play it at Easter when we are together. It would have been silly for me to try to come for New Years. I would have had to wait almost the entire time at the border.[8] But at Easter (in the first days of April) I will definitely try it.

It is a fine achievement that Tete has learned to read without school.[9] He must have a good memory for him to have mastered it so easily. Earlier he had absolutely no aptitude for singing; my father was also completely unmusical.

In the last few months I accomplished very fine things again.[10] How much I am looking forward to the day when I can tell you about such things! But it will have to wait a couple of more years.

Recently I had an odd experience. I was at a small social gathering. An artist was also there who could read a person's character off the pattern of lines on the palm of one's hand. She looked at my hand as well and, without knowing me at all, said some very applicable things. That's very strange, isn't it?

Tell me also something about your friends and what you do in your free time for entertainment. Do you play a lot with Tete? Do you come out much into the fresh air? I would like to know all that.

When do you have your Easter holidays?[11] I would like to know this so that I can plan ahead a bit.

Kisses to you and Tete from your

Albert.

Greetings to Mama!

200.   To Mileva Einstein-Marić

[Berlin,] Sunday, [12 March 1916][1]

D[ear] M[ileva],

I had not answered the letter before your last one yet because I am quite overwhelmed with work on various manuscripts. You don't need to worry at all. For you it involves a mere formality, for me, however, an imperative duty.[2] Try to imagine yourself in my position for once. Elsa has two daughters, the elder of whom is 18 years old, that is, of marriageable age.[3] This child, who is anyway seriously disadvantaged by the loss of one eye, must suffer from the rumors that are circulating with regard to my relationship with her mother. This weighs on me and ought to be redressed by a formal marriage.[4]

You gain from this formal change as well, to the extent that in this way your rights are clearly established. I wish to do even more than I had obligated myself to before:[5]

1) 5,600 M yearly money for your disbursement.
2) Deposit of my Prague money[6] as well as 6,000 M savings made here for the benefit of our children at a place agreed on by both of us.
3) Deposit of at least *3,000 M* annually to create the previously planned reserve fund.[7]

By making myself such a frugal bed of straw, I am proving to you that my boys' well-being is closest to my heart, above all else in the world. I am also there personally for *them* in the first place. Our divorce has nothing to do with my relationship to the boys. That is a very peculiar interpretation of yours. However, I also want to make one condition on my part that no one can take as unjustified. I want the right, when peace has returned, to have my children not only on trips but also in my apartment during the short period I am granted to be together with them; there they are alone with me and not exposed to outside influences.[8] For I shall never give up the state of living alone, which has manifested itself as an indescribable blessing.

Something else I had forgotten above.

4) In case of my death the pension will be guaranteed to you in such a way that you would be guaranteed to receive it even in case of Else's death from her pension capital.
And finally
5) The new marriage will be arranged to keep the properties completely separate, thus amounting to no material loss for you.

So you really can be content and have some peace of mind in looking to the future. When you have informed me of your opinion, I shall entrust the matter to an attorney-at-law who must see to it that everything is put nicely in order.

In 14 days I'll set off and try to travel to Zurich. I am going to stay at an inn there. But if mortal providence does not allow me to cross, I shall stay in Singen[9] and request that Albert travel the couple of stations across to me. He will not encounter any difficulties. I'll write to him very soon.

Greetings to you and kisses for the boys, y[ours],

Albert.

201. From Otto Stern

Lomscha [Poland], 13 March 1916

Dear Prof. Einstein,

I measure the specif. heat (and entropy) of the model:

 Gaseous molecules in gravity, where I cool it down to $T = 0$.[1]

Question.

What do I measure in case (1) and in case (2)?

Case (1)

Transition from one chamber to another very rare (in my view also, always realized at sufficiently low temperatures).

Answer

Initially I measure no quantity at all with which I can do anything as a thermodynamics expert. Only from that moment on when the molecule is located in the chamber (e.g., 2) from which it can no longer escape any more do I measure the spec. heat of the system in this state (2), that is, the spec. heat of a specific chemical bond.

Case (2)

Transition frequent (*practically, never* realizable at sufficiently low temperatures; *in principle, always* [realizable] through sufficiently gradual cooling).

Answer.

I measure what is referred to as the solution's spec. heat in all thermodynamic calculations.

––––––––––––––

I thus contend:

Strictly speaking, only the spec. heat of the solution is measurable in a manner of use to thermodynamics. The fact that I can also measure that of a particular chemical compound is because I can choose the reaction rate at will within broad limits. In the present case, for ex., I can reduce it optionally by making the bond openings between two chambers smaller and in this way keep the molecule in a particular chamber throughout the entire duration of cooling. If I do not succeed in this, then my measurement is useless; s[ee] Nernst, iodine's spec. heat distorted by latent heat.[2] The situation is exactly the opposite when I want to measure the solution's spec. heat. Then I must make the bond openings very large (add a diffusion catalyst) and cool very slowly. It is clear that as the temperature drops, it soon comes to a limit at which measurement becomes unfeasible practically, but this does not disturb the thermodynamics theorist in the least, since he has an infinite amount of time.

That these questions of rate play no role can be seen very nicely in the model when the energy differences between the chambers are made very large. The solution's entropy is then very close to equaling zero already at temperatures high enough for the diffusion rate still to be very great.[3]

Unfortunately the letter has again become very long; orally I would certainly have come to concurring with you within a couple of minutes.[4]

With cordial greetings, yours,

Otto Stern.

[3]Draft version: "That these questions of rate are inconsequential for statistical thermodynamic considerations can be seen very nicely in the model, in the following way. We make the energy differences between the chambers very large ⟨and these chambers themselves very small⟩, then ⟨the molecule generates⟩ there are quite high temperatures at which, on the one hand, the molecule's velocity determined by T is so large that during any time period needed for practical measurement, however short it may be, the molecule frequently is present in the chamber with the highest energy, and the probabilities (determined by $e^{-E/T}$) of the individual chambers varies so much that the molecule spends the largest part of the time in the chambers with the lowest energy. As the temperature rises, the probabilities of the individual chambers will then gradually and steadily become increasingly similar to one another. Then clearly

we call the system a solution, at lower temperatures, a chemical compound, which is completely unjustifiable as absolutely no qualitative differences exist. Of course, this impression is unimportant practically, because at low temperatures the entropy of the solution differs only very little from the most probable representative compound."

[4]Draft final paragraph: "Your distinction between case 1 and 2 also seems to me legitimate. For whether 1 or 2 occurs depends on entirely secondary conditions: the size and position of the open connections between the chambers. Such conditions surely cannot possibly have an influence on the system's entropy value. But—and this is the main point—one can always transform case 1 into case 2, provided that a long enough cooling time is chosen."

202. To Hans Albert Einstein

[Berlin, 16 March 1916]

My dear Albert,

The explanation for the curious signature on my last letter is that, in my absent-mindedness, instead of signing my own name, I frequently sign for the person to whom the letter is directed.[1] So, in the beginning of April I'm going to try to come to Zurich. If I'm not allowed out, we'll meet at the border (Gottmadingen near Schaffhausen).[2] At all events I'll be staying at an inn so that we'll be completely by ourselves without any strangers along. Congratulations on passing your examination. Which school are you going to now?[3] You still make so many writing errors. You must take care in this regard; it gives an odd impression when words are written incorrectly.[4]

I would like most to be with you outside of Zurich, because in Zurich there are too many people I know. The main question is whether I can come across.

Kisses to Tete and greetings to Mama from your

Papa.

Give Mr. Besso my regards when you see him.

203. To Wilhelm Wien

Berlin, 18 March 1916

Highly esteemed Colleague,

Enclosed herewith is my complete revision of the general theory of relativity for the *Annalen*. As I have consigned publication of the offprint to Ambr[osius]

Barth,[1] no more obstacles ought to be standing in the way of publication; therefore, I take the liberty of submitting the paper to you without first waiting for your reply to my earlier inquiry.

With regards from your very devoted colleague,

A. Einstein.

204.   From Wilhelm Foerster[1]

Bornim near Potsdam, 25 March 1916

Esteemed colleague Einstein,

You would do a great service in bringing peace of mind to many qualified people if in the near future you were able to find a way of addressing the German public to counteract the *anxiety* of large segments respecting doubts about previously held basic tenets of our knowledge of the world and to allay excessive skepticism.

It has perhaps been through certain figures of speech in some recent popular explanations by our colleague Weinstein or in newspaper reports about his descriptions of your "theories" that this unease was amplified and disseminated.[2] You would not believe how much one is being pestered now by inquiries and complaints about this from the most disparate social circles.

This agitation is probably connected to the almost psychopathic state of the current, widely spread sentiments among the populace. For ex., some are happy that you had now put an end to the global confusion caused by the Englishman Newton, etc.[3]

Surely you will find words free of scholarly jargon to introduce the German public to a sound and sober-minded interpretation of your so extremely important ideas and problems; but there really is a need for this now.

Most respectfully,

Prof. Wilhelm Foerster.

205.   To Otto Stern

[Berlin,] 27 March 1916

Dear Mr. Stern,

Your last letter indicates to me that we are not going to be able to agree now in writing.[1] As I have already relayed to you my reasons and I actually have nothing else new to put forward, I shall close our discussion for the time being. Recently I broached this subject with Planck; he sees the matter as I do. When you come to Berlin again, let us straighten it all out.

With best regards, yours,

A. Einstein.

206.   To Hans Albert Einstein

[Berlin,] 30 March 1916

My dear Albert,

My passport, which I need for the trip to see you, is causing me problems.[1] I have already been to the Swiss embassy for that reason,[2] which I hope will soon put the matter in order. I'll travel as soon as I can. I am looking forward to you all very much and have in mind to go on a small trip or hike with you.

Farewell for now, and kisses to you and Tete from your

Papa.

207.   To David Hilbert

[Berlin,] 30 March 1916

Highly esteemed Colleague,

With much pleasure do I recall the extremely interesting and agreeable days spent at your hospitable home.[1] To you and Mrs. Hilbert many thanks for all your friendliness to me (including the Freundlich–ness* in the special sense the word has for us).[2]

This affair, which is somewhat euphemistically attached to the word "friendly," has elicited in me a peculiar kind of reaction. The plan, formed out of necessity, that Freundlich take up the important experimental questions in Göttingen under my responsibility, has awakened serious doubts in me after all. For such a half-marriage, not only is a certain respect for the other person necessary, but also some of that personal congeniality which makes frequently repeated meetings pleasant and which sweetens shared disappointments. But in the present case this is decidedly lacking; that is why I fear disappointment would result from such a step. Therefore, I told Freundlich nothing about our conversations in this regard but only about our foiled assault on the astronomical fortress defended by an invincible phlegm.[3] I also refrained from writing to Herglotz.[4] However, I shall still try to change Struve's mind, bit by bit, through gentle persuasion.[5] Unfortunately the latter comes only rarely to the Academy meetings, the only opportunities to get hold of him if you do not dare to venture into his pompous den. Who knows whether his hardened heart cannot be softened into submission?

The error you found in my paper of 1914 has now become completely clear to me.[6] If I perform an infinitesimal transformation which converts the

$$g^{\mu\nu} \text{ terms into } g^{\mu\nu} + \Delta g^{\mu\nu}$$

and the

$$g_\sigma^{\mu\nu} \text{ terms into } g_\sigma^{\mu\nu} + \Delta g_\sigma^{\mu\nu},$$

then the relation

$$\Delta g_\sigma^{\mu\nu} = \frac{\partial}{\partial x_\sigma}(\Delta g^{\mu\nu})$$

is not valid. That is why there is no variation, in the sense of variational calculus, corresponding to the change Δ.[7] I came upon this because I caught myself making the same error in another problem.

Cordial greetings, yours very truly,

A. Einstein.

*Literally translated, the surname Freundlich means "friendly."

208. To Mileva Einstein-Marić

[Berlin,] 1 April 1916

D[ear] M[ileva],

I have prodded myself into action now and, with your consent in principle, have discussed the matter of the divorce[1] with a lawyer. The court proceedings are to be held in Berlin and should neither cause you inconvenience nor incur you any costs. Your rights will be protected in the broadest sense. Now, be so kind and write to the lawyer assigned to you,

Dr. Albert Pinner,[2] 46 Tauben St., Berlin,

that you are inclined in principle to instituting divorce proceedings against me and hiring him as your lawyer. In addition, outline to him what terms you are setting—upon agreement with me. That you will not allow the boys over here even later on is an unjustified chicanery, despite my promise not to bring them into contact with anyone in the family.[3] Nevertheless, I'm not going to insist on this demand, so that you see that I will oblige you in every way possible.

As it stands, I am coming to Zurich to go on a short hike of approx. 10 days duration with Albert, if possible also with Tete. I am going to stay in Zurich at a hotel or a guest-house, as Albert had wished the last time.[4] I hope confidently that this time you will not withhold the boys from me almost entirely again.[5] If you repeat what you did the last time, I shall not return to Zurich so soon. I don't want to force anything, but also don't want to have to beg for my right to be together with my boys. I'll be arriving in Zurich in about a week if I manage to come across the border right away.

I emphasize once again that the circumstance that I could not live with you personally has no bearing on my relations with the boys. I would be happy if I could have the children with me and am attached to ⟨him⟩ them no less than before. My leaving them to you alone means a severe sacrifice on my part, not a lack of attachment and paternal feeling.

With kind regards,

A. E.

Kisses to the boys.

209.   To Michele Besso

[Zurich, 6 April 1916][1]

Dear Michele,

I, Albert, have run down your empty snail shell of a hide-away[2] with great effort, after having arrived here just this afternoon. Telephone me tomorrow morning at the Gotthard Hotel (train station), where I have taken a room. Or even better, look me up there.

I am very eager to see you. Yours,

Albert.

210.   To Hans Albert and Eduard Einstein

[Zurich,] Thursday. [6 April 1916]

Dear Adn and Tete,

I arrived here just now. I am staying at the Gotthard Hotel, Bahnhof St., right next to the train station, room no. 50, *and am enormously happy to be seeing you.* I expect you tomorrow (Friday) morning between 9 and 12 o'clock at the hotel.

Kisses from your

Papa.

Regards to Mama.

211.   To Mileva Einstein-Marić

[Zurich,] 8 April [1916][1]

D[ear] M[ileva],

I just received your letter. First of all, my compliments on the good condition of our boys. They are in such excellent physical and mental shape that I could not wish for more. And I know that this is for the most part thanks to the proper upbringing you provide them. I am likewise thankful that you have not alienated me from the children. They came to meet me spontaneously and sweetly. I am very much looking forward to being able to spend a few days alone with Albert.

There would be no point in a conversation between us and it could only serve to reopen old wounds.[2] I am sorry that I misinterpreted your last letter. A divorce between us can—as far as I know—only take place on the basis of a complaint coming from you.[3] For, since I must figure as the guilty party and I cannot sue myself, this seems to be the only possibility. The first question is now: Are you inclined in principle to filing a divorce claim against me? If not, then the following questions are inapplicable. It seems to me that you do not risk anything by doing so, as you can yourself set the conditions under which you would agree to a divorce, of course.

If you are in principle inclined to submitting a divorce petition, once we have agreed on the terms, the question arises whether the proceedings are to take place here or in Berlin. As far as I know, it is unclear whether this can be done here. In Berlin it is certainly possible and less protracted. Since nothing can happen except under conditions to which you agree, you are not sacrificing anything by having the petition filed in Berlin. Everything will be done openly and honestly.

Therefore, I expect an answer to the following questions.

1) Are you prepared, provided that we agree on the conditions, to submit the claim?

2) In the affirmative case, what might your conditions be?

3) Are you prepared, in the case of agreement regarding (1) and (2), to entrust the Berlin lawyer[4] already advised with performing the formalities?–

Send the reply to Michele's address, with whom I'm going to be staying after the excursion.[5]

The statement regarding my traveling here was not a threat. I was quite serious about wanting to be together with the children only properly or not at all. Now I'm very glad that you have not caused me any difficulties with this.

Regards, y[ours,]

A.E.

212.   To Elsa Einstein

[Seelisberg, 12 April, 1916]

My dear Else,

Today we are snowed in at Seelisberg[1] but are enjoying ourselves immensely together. The boy delights me, especially with his clever questions and his undemanding way. No discord exists between us.

In a good week (bet[ween] the 20th & the 25th) I'll be on my way back to you and am already looking forward to it very much. I am staying with Besso, who has a place in Zurich, the last few days.[2] We shall not utter a word about creatures of the opposite sex!

Greetings and kisses from your

Albert

Greetings to the little daughters & Uncle & Aunt.

213.   To Elsa Einstein

[Zurich, 15 April 1916]

Dear Else,

We had to break off our mountain trip a bit early because of the weather.[1] We were good companions, the boy and I. I have many obligations here, some of them enjoyable ones. Yesterday I was with Abraham.[2] Tomorrow Maja is coming over here.

Warm greetings also to Ilse & Margot from your

Albert.

Best regards to Uncle & Aunt. The N. case seems to be repairing itself, thank God.[3]

214.   To Hans Albert Einstein

[Zurich, 15 April 1916]

Dear Adn,

I am expecting you Monday, 10 o'clock, at the Phys. Institute.[1]

Regards from your

Papa.

215.   To Michele Besso

[Zurich, 21 April 1916]

Dear Michele,

Please pick up Reichinstein's book at the hotel.[1] It is brilliantly written. You all could fill a few charming hours with it.

Our time together was very nice.[2] I hope it can be repeated soon. Zangger was good-humored and looks fine.

Affectionate greetings also to Anna and Vero, yours,

Albert.

Please give the landlady[3] Fr. 5 from me and remind me of it later.

216.   To Elsa Einstein

[Zurich,] Friday. [21 April 1916][1]

My dear Else,

The reproach for my silence weighs heavily on me again. I don't know, will you get accustomed to my laziness sooner, or I to writing; the former almost threatens to become true. I enjoyed myself very much with Besso, who is very comfortable with his new duties.[2] Personal family matters were as good as not at all discussed but we rode on ahead on our little hobbyhorses. During the entire week of my stay here, I was not together with my children, even though relations with my Albert had been so affectionate during the trip.[3] This is how it came about: Already on the second day he was at the Institute with me (Monday morning) to watch me prepare the experiment. As we were leaving he urged me to call on his mother. When I resolutely declined,[4] he became stubborn and refused to return in the afternoon. That is how it remained; and I saw neither of the children since and also did not arrange any more meetings. I should see the children only when they are not also under their mother's influence; only in this way are serious conflicts avoided.

Today I'm traveling to Lucerne and am staying until Tuesday morning.[5] This long delay in my homeward trip is an involuntary one; the visa endorsement of my passport required time-consuming formalities,[6] which prevented me from finishing yesterday. Tuesday morning I'll still be occupied with it; then I'll depart immediately. I want to stay in Dettingen for a day;[7] but then I am coming to your house immediately. I am, to be honest, famished, because I've been deprived of your and the little minxes' nice company for so long. This pleasant feeling is already motivation enough to travel. One appreciates more what one has at home! I gathered the news for you very diligently, but never had it in front of me when I was writing to you. This habit also has an unfavorable influence on my correspondence . . . But in return, you may make fun of me when we are together again.

Kisses also to the little minxes from your

Albert.

Cord[ial] greetings to Uncle & Aunt.

217.   To Michele Besso

[Lucerne, 22 April 1916]

Dear Michele,

I am arriving at 10h 28 in Zurich, and am then going straight to the Consulate[1] and shall then look you up, if at all possible, at your apartment. We're having a very good time here. Paul looks well.[2] That lady was presumably the mother of Geiger,[3] who used to work with Rutherford. I calculated during the trip that the amplitude must grow by about 6 cm at each full swing, using the experimental arrangement at the Phys. Institute.[4] Show the thing to Vero, as I wasn't able to get to it. Don't forget to poke your noses into the book by Mann; it's well worth it.

Affectionate greetings to you all, yours,

Albert.

218.   To Paul Ehrenfest

[Berlin, 29 April 1916]

Dear Ehrenfest,

Upon returning from Switzerland I found your postcard. I'll gladly send you what you request, quite musty stuff, God forbid. I could not find the Solvay publication.[1] But these things are developed impeccably in Lorentz's Parisian lectures,[2] so you lose nothing. I'll also send you an uncorrected correction of my summarizing paper on gravitation;[3] write me a frank critique of this portrayal of the subject sometime! I wanted to write a comprehensible introduction but don't know whether this has succeeded. Show it to Nordström[4] as well when he comes.

I was very pleased about Planck's open letter and no less about Lorentz's fine cooperation.[5] The world would be a true paradise if such persons were leaders of the general public.

I constantly feel like popping in to see you all. Maybe this will happen sometime unexpectedly.

With warm regards to you, your family, Lorentz, and de Haas, yours,

A. Einstein.

219.   To Michele Besso

[Berlin,] 14 May, [1916][1]

Dear Michele,

Everything else went well (on the trip and later).[2] My laziness in writing has not gotten worse, you know, it's just that you've improved, strangely and happily enough. Our real-life Sterne novel is at least as fine as the original and calls for continuation.[3]

That thing about Brownian motion is Stodola's[4] barroom idea, which I have already tried to talk him out of once, in vain. The   curve is not an extremum problem. The allowed lift is given by the stability conditions of the fluid flow;[5] when the lift is great, the airflow does not follow the surface, resulting in vortices.

airfoil

I hope you're enjoying the lecture course.[6] I still recall very well that a hefty push is needed to overcome an initial aversion and that one always thinks everything one has to say is obvious. But this is an optical illusion. Do you remember how in Berne you always used to attend mine so nicely?[7] And now I can't reciprocate. I have another quite amusing expert opinion to give for a patent case.[8] When we see each other again I'll tell you about it.

At the moment I'm working quite moderately, so I'm feeling nicely well and am living peacefully along without any discord. In gravitation I'm now looking for the boundary conditions at infinity; it certainly is interesting to consider to what extent a *finite* world exists, that is, a world of naturally measured finite extension in which all inertia is truly relative.[9] The funeral for Schwarzschild, director of the observatory in Potsdam, was held today.[10] Surely I have already told you about him; he is a real loss. He would have been a gem, had he been as decent as he was clever.[11] Of the photographs, keep one, give one to Maja, one to Zangger, and save the remainder. You may also give them away, though, if you believe you could make someone happy with it. I discovered a neat simplification of the thermodynamic derivation of the photochemical $h\nu$ law, somewhat in the manner of Van't Hoff.[12] I'm glad that my boys are in good spirits and that you are concerning yourself with them. Now I can soon send you the detailed paper on gravitation in which everything is calculated explicitly.[13]

Affectionate greetings also to Anna and Vero, yours,

Albert.

220.   To Paul Ehrenfest

[Berlin, 24 May 1916]

Dear Ehrenfest,

I have such an urge to see you all again that I can scarcely wait for peace without having at least made an attempt at coming to visit you. I'll try to do so during summer vacation. My specialization of the frame of reference is not based *solely* on laziness.[1] I might present the matter once also without the specialization, somewhat like Lorentz in his paper.[2] Hilbert's description doesn't appeal to me.[3] It is unnecessarily specialized regarding "matter,"[4] is unnecessarily complicated, and not straightforward (= Gauss-like) in set-up (feigning the super-human through concealment of the methods). I am receiving only 20 offprints of the paper.[5] In return, Ambr[osius] Barth was willing to have the paper appear as a pamphlet in the book trade.[6] It will probably be appearing in the next few days. It is better this way than if I had received 100 offprints which would then not have been enough either. I have already sent a copy to Brouwer.[7]

Cordial greetings, yours,

Einstein.

221.   To David Hilbert

[Berlin,] (13 Wittelsbacher St.) Friday. [25 May 1916]

Highly esteemed Colleague,

I am sitting over your relativity paper,[1] which I must review in Rubens's colloquium, and am honestly toiling over it.[2] I do admire your method, as far as I have understood it. But at certain points I cannot progress and therefore ask that you assist me with brief instructions.

I have been able to derive your fine formula (6), albeit only with the aid of the first form of the same on page 398. By contrast, I have not succeeded in seeing directly the invariance of the expression $P(J)$, p. 399.[3] How is it derived? It undoubtedly is connected with the following. On page 400, slightly below the middle you say: "Now $\dfrac{\partial H}{[\partial g]_{[k]l}^{\mu\nu}}$ is a mixed tensor . . ."[4] But with the best of my efforts I see no basis for this and for this reason do not understand the following. I very much like the proof on page 403.[5] I still do not grasp the energy principle at all, not even as a statement. In the boundary case of the vanishing of the $g^{\mu\nu}$ derivatives, for ex., the energy principle is, according to (18) and the gen. form[6] $\sum \dfrac{\partial \sqrt{g}e^l}{\partial w_l} = 0$, in the form[7]

$$\sum_{ls} \frac{\partial(T_s^l p^s)}{\partial w_l} = 0$$

for an *arbitrary* selection of p^s. From this, however, the vanishing of not only the $\sum \dfrac{\partial T_s^l}{\partial w_l}$'s but also of the T_s^l's themselves would follow. How is this resolved?

Requesting kind enlightenment and with best regards, yours,

A. Einstein.

222.   From David Hilbert

Göttingen, 27 May 1916

Dear Colleague,

First of all, many thanks for your friendly letter of last March,[1] to which I had been wanting to respond for so long now. "Friendly" also in the literal sense; although, I hope we are going to discuss this "Freundlich" affair soon in person.–[2] At Easter I was in Lugano for 4 weeks and gorged myself there on meat and whipped cream, etc.– Schwarzschild's death certainly is terrible[3] and is being painfully felt even by people who everywhere have become so heartless.– Now as regards this semester's Göttingen physics lectures, von Smoluchowski—as has now been confirmed—will lecture on the days 19–21 June on diffusion and coagulation theory of colloid particles,[4] and Szigmondy will perform experiments on it at the same time.[5] Furthermore, Mie has agreed to come here to present a talk on his electrodynamics. The time for that is not yet settled, however; I am thinking of the last week in the semester.[6] My wife is already very much looking forward to receiving you in our home again on these occasions;[7] she is still very struck by your modesty but will nevertheless attempt not to put you to too severe a test: If owing to grocery assortment we have nothing to eat, we shall all go together to the hotel or on excursions to neighboring villages.–[8] Finally, I shall immediately respond to your postcard of today,[9] which also pleased me very much:

P is nothing more than the always generally invariant polar process.[10] Proof is thus.[11] If $J(g^{\mu\nu} g^{\mu\nu} \ldots q_s, \ldots)$ is an invariant, then obviously $J(g^{\mu\nu}+\kappa p^{\mu\nu}, g^{\mu\nu}_\sigma + \kappa p^{\mu\nu}_\sigma, \ldots q_s + \kappa p_s)$ is also, where κ is a constant and $p^{\mu\nu}$ to $g^{\mu\nu}$ a co-gradient, and p_s to q_s.[12] Consequently every coefficient in the power expansion in κ is indeed an invariant. However, the coefficient of κ^1 is $P = P_g + P_q$.[13]

But this is not connected, as you assume, with the theorem that $\dfrac{\partial H}{\partial g^{\mu\nu}_{kl}}$ is a tensor. Rather the latter follows directly from the manner in which the derivatives of the $g^{\mu\nu}$'s transform themselves; this always happens in such a way that upon transformation the nth $g^{\mu\nu}$ derivatives just combine with derivatives to the $n - 1^{\text{th}}$—thus with those of *lower* order.[14] And from this you can see, of course, that an invariant's differential quotients to the *highest* $g^{\mu\nu}$ derivatives occurring in them must always be tensors.[15]

My energy law is probably related to yours; I have already assigned this question to Miss Noether.[16] As concerns your objection, however, you must consider that in the boundary case $g^{\mu\nu} = 0, 1$ the vectors a^l, b^l do not vanish by any means,[17] as K is linear in the $g^{\mu\nu}_{\sigma\kappa}$'s and is differentiated according to these quantities.[18] For brevity's sake I send you the enclosed slip by Miss Noether.

We are expecting you, therefore, for the 19th of June and hope that you stay the entire week so that we can go on an excursion with you. There is a very pleasant and also politically very inspired and interested circle here.

With most cordial regards, yours,

D. Hilbert.

223.   To David Hilbert

[Berlin,] 30 May 1916

Dear Colleague,

Your invitation to the Smoluchowski event[1] tempts me very much, especially considering that the fine days spent with you have stayed with me as a dear memory.[2] I cannot promise yet to come, because I am tied down quite inextricably here by various ongoing commitments. But in any case, this time I would stay at an inn, if only not to appear to myself as a nuisance.

I also find Schwarzschild's death distressing.[3] Among the living there are probably only a few who know how to apply mathematics with such a virtuosity as his.

Your explanation of equation (6) of your paper[4] was charming. Why do you make it so hard for poor mortals by withholding the technique behind your ideas? It surely does not suffice for the thoughtful reader if, although able to verify the correctness of the equations, he cannot have a clear view of the overall plan of the analysis.

In the Freundlich affair a turn for the better seems to be in the making, thank heavens.[5] Fr. went to see *Küstner* in Bonn.[6] The latter received him very amicably and is prepared to support him in every way. If Küstner came to Potsdam as Schwarzschild's successor,[7] a conscientious examination of the experimental gravitation questions would then be safeguarded, and simultaneously Freundlich would be saved from a situation that has become positively heartrending.[8] I have written a letter to Naumann, in order to place my bit of weight on the scale toward the above.[9] For, if Hartmann were to come to Potsdam, it would be disastrous.[10] (Anyway, how could you possibly find the strength to console yourself about the separation?)

In your paper everything is understandable to me now except for the energy theorem. Please do not be angry with me that I ask you about this again.[11] My difficulty consists in the following:

In your equation[12]

$$\sum \frac{\partial \sqrt{g}\, e^l}{\partial x_l} = 0$$

$\sqrt{g}e^l$ can be ordered according to p^α and derivatives p_β^α.
Then results

$$\sqrt{g}e^l = \sum_\alpha A_\alpha^l p^\alpha + \sum_{\alpha\beta} B_\alpha^{l\beta} p_\beta^\alpha,$$

where the A_α^λ's and $B_\alpha^{\lambda\beta}$'s are no longer dependent on vector p. If I substitute this in the above equation, I can then write out

$$\sum_{\alpha l} \frac{\partial A_\alpha^l}{\partial w_l} p^\alpha + \sum_{\alpha\beta l} \left(A_\alpha^\beta + \frac{\partial B_\alpha^{l\beta}}{\partial w_l} \right) p_\beta^\alpha + \sum_{\alpha\beta l} B_\alpha^{l\beta} p_{\beta l}^\alpha = 0.$$

If this equation should be valid for any selection of p^α, the $B_\alpha^{l\beta}$'s must first vanish owing to the independence of the $p_{\beta l}^\alpha$'s. Because of the independence of the p_β^α's, the individual A_α^l's must then also vanish, however, which would rob the theorem of its sense. How is this cleared up? It would suffice, of course, if you would charge Miss Noether with explaining this to me.[13]

Cordial greetings to you and Mrs. Hilbert, yours,

A. Einstein.

224.   To David Hilbert

[Berlin,] 2 June 1916

Dear Colleague,

Your new explanation has dispelled my doubts on this point as well.[1] I really had failed to notice that $p_{\beta l}^\alpha = p_{l\beta}^\alpha$. Thus an equation in the form

$$\sum \frac{\partial A_\sigma^\nu}{\partial x_\nu} = 0$$

remains, which describes precisely the conservation law in common form, while the other two relations just define the $B_\alpha^{l\beta}$'s, which are uninteresting of themselves, as it seems. Now your entire fine analysis is clear to me, also with respect to the heuristics. Our results are in complete agreement.

Unfortunately I cannot come to Göttingen now. However, I shall try to break myself loose at the end of July. What I appreciate long after every discussion with you is the far-reaching concurrence of our views and convictions.

Thanking you heartily for the kind explanations, I am with best regards, yours,

A. Einstein.

225.   From Hendrik A. Lorentz

Haarlem, 6 June 1916

Dear Colleague,

In recent months I have been occupying myself much with your gravitation theory and general theory of relativity and have also lectured on it, which was very useful to me.[1] Now I believe I understand the theory in its full glory; every difficulty I encountered I was able to overcome upon closer consideration. I have also succeeded in deriving your field equations

$$G_{im} = -\kappa(T_{im} - \frac{1}{2}g_{im}T)$$

out of the variation principle; at least, but a minor detail is still lacking in this derivation, which required what I consider long calculations.[2]

But now I have come across a consideration I would like to present to you and which is based upon the examination of a fictitious experiment. We can imagine that Lecher's experiment is done with two completely conductive wires which are stretched around the Earth along the equator and each of which is self-contained.[3] In order to avoid the danger of a "derailing" by the electromagnetic waves (owing to the Earth's curvature), we can also use instead of the two wires a single wire enveloped concentrically by an entirely conductive sheath. At a particular point A on this self-contained "cable"[4] [space between the conductors void of air] let there be a device that permits the excitation of waves and a detector with which we can observe at A the returning waves after their having gone through the circuit. The cable as well as point A are firmly affixed to the Earth.

From all that we know, we can probably say with certainty what we would observe with adequately refined means. Waves produced at the same moment at A and that go through the circuit in opposite directions will *not* return to A at the same instant.

Among the various ways in which we can describe this result, only two exist that are particularly simple.

a. We can choose coordinate system I, $OX\,OY$ (OZ coincides with the Earth's axis), such that in this system the propagation velocity of the waves is the same for both directions around the cycle. We then find that the Earth rotates in this coordinate system.

b. We introduce coordinate system II, which is firmly attached to the Earth. In this one, unequal propagation velocities c_1 and c_2 exist for both directions around the cycle.

It goes without saying that just this necessary difference in the propagation velocities results from your general formulas, when moving from I to II; and to the extent that an equation of the form

$$c_1 - c_2 = a$$

is valid both in systems I and II, and likewise also in many other systems (each time with another a), it can be said, it expresses the experiment's result in a *covariant* form. But this does not have to prevent us from regarding the equation $c_1 - c_2 = 0$ as *different* from $c_1 - c_2 \neq 0$. In this sense we shall conclude: The phenomena in the cable do not take place in the same way with reference to coordinate systems I and II.

If you now try to make this comprehensible in some way or to visualize it graphically, you can scarcely confine yourself to speaking *only* of the Earth, the cable, and this "space" or "vacuum" contained in the latter; you will be inclined, of course, to imagine that there is nothing in and of itself in the space or vacuum that would behave differently toward systems I and II.

The idea certainly suggests itself,[5] [I am going to speak about another one further down.] and formerly would surely have seemed very natural to all physicists, that in the cable there is a medium (the ether) *in which* the waves propagate in such a way that the propagation velocity is always the same relative to the medium but that this medium can be at rest with reference to one axial system, and moving with reference to the other. If we assume this point of view, we can thus say the experiment had shown us the relative motion of the Earth against the ether. If then in this way we have acknowledged the possibility of establishing a relative *rotation*, we therefore may not deny from the outset the possibility of obtaining changes in such a *translation* as well, i.e., we may not hold the basic principle of relativity theory up as a *postulate*. We must rather (and this was also the actual course of development) seek the answer to the question in observation. Once these observations have instructed us that no influence owing to translation can be found, we may through generalization (and quite extensively at that) express the law as a basic *hypothesis*, whereby we still grant the possibility, however (no matter how improbable we may consider it to be), that future observations will force us to abandon the hypothesis.

These observations can also be clothed in another way. Namely, we can produce *standing* waves in the self-contained cable and observe the position of the nodes at every moment. It will then be shown that they rotate in a circle relative to the Earth. You could, of course, limit yourself to establishing the relative motion of the nodes against the Earth (or *vice versa*). But if you consider that the same rotation occurs in standing waves of various lengths and various intensities, it is then quite obvious (shall we say, as a graphic summary of what is common to all these phenomena) to think of an ether in which the standing waves are located.

Mach, whose interpretation you have espoused,[6] in discussing similar experiments also felt the need for assuming something situated externally to the Earth that determined the phenomena. Following his chain of reasoning a defining moment would be sought in the influence of "distant bodies in the universe," let's say, of fixed stars. Thus one would say, it is the fixed stars which determine the cyclical course (or the static state) of the nodes in the circular cable. Now, although this interpretation seems to me very much less manifest than the hypothesis of an ether, I could accept it nonetheless, if in comparison to this hypothesis it offered some advantage. But such an advantage I am unable to see. For if we must assume that the Earth's *rotation*, with reference to the fixed stars, has an observable influence on electromagnetic phenomena, then we may not deny from the start the possibility of a similar influence of a *translation* by the Earth or the solar system relative to the fixed stars. Then we are exactly as far as with the ether hypothesis and we must examine through experiment whether perhaps any effect of a translation exists. Even now a relativity *postulate* would be out of the question.

By the way, both interpretations, influence of fixed stars and ether hypothesis, are basically not even very different from each other, it seems to me. Supposing I assumed the nodes' motion or rest in our annular cable were determined by the fixed stars' influence. Then, in order to establish to some extent the nature of this influence, I can simultaneously assume a system of points rigidly connected to one another within the cable as a connecting link between the fixed stars and electromagnetic waves. I shall say, the said influence manifests itself in that with reference to this point system, which is itself connected to the fixed stars, the nodes have fixed positions. The step from this point system to an ether is not far.

Obviously, other experiments also, e.g., the ones discussed by you and Mach,[7] give occasion for very similar observations, and the above considerations will by no means be new to you. The main point in them is actually that departures from relativity theory to the "fixed-star hypothesis" would also be very well conceivable. I do not need to mention that, also in my opinion, not only the theory of relativity but also your gravitation theory can remain valid in their entirety. They will just not impose themselves on us so much as the only possible ones.

I hope very much that you will stay well. As regards us, unfortunately, since many years now already, my wife often suffers under changes in the weather, which is connected with a deficiency in the thyroid gland; but otherwise we are nice and healthy. De Haas also, likewise his wife and children (the third is expected next month).[8] He has now been named *Conservator* at the local Teyler Laboratory, which makes me very happy for him and for myself.[9] He will now be able to devote himself exclusively to scientific work. That I, that is, in agreement with my colleagues, could suggest that he be entrusted with the position, which had become available is, above all, thanks to the research you conducted with him[10] and thus to the interest and kindness you have shown toward him, and for which I am very grateful to you.

I am sending you under separate cover the two first parts of an analysis of your theory of gravitation. Unfortunately the translation is not ready yet.[11] I made an effort to clothe the fundamentals geometrically, so that the use of coordinates is avoided as far as possible.[12]

With warm regards, faithfully yours,

H. A. Lorentz.

226.   To Hendrik A. Lorentz

Berlin, 17 June 1916

Dear and highly esteemed Colleague,

Your lengthy letter[1] pleased me very much, mostly because of the extremely welcome news that de Haas has become *Conservator* of the Teyler Laboratory.[2] You should not thank me repeatedly for doing the Ampère investigation with him.[3] For I chose him out of egotism, because I happened to like him best, and my impression proved to be right. Our colleagues here have preserved a lukewarm memory of him because of his attitudes about impersonal matters. But it is notable that two people were enchanted by the de Haases: the institute mechanic[4] and the landlady of their private apartment. Relay my warm greetings to both from me.[5] I hope the new little grandchild turns out very well. In the summer I want to make an attempt at coming to Holland; but it is difficult to obtain the permit.

I am very pleased that you are engaging yourself so productively with the theory of gravitation. I liked particularly well the direct interpretation of tensor K_{ab}, which was new to me.[6] I myself worked on the integration of the field equations in first-order approximation and examined gravitational waves.[7] The results are in part astonishing. There are three kinds of waves, though only *one* type transports energy. I am not quite finished yet with the theory of emission in material systems. But this much is clear to me: that the quanta difficulties affect the new theory of gravitation just as much as Maxwell's theory.[8] I was delighted that in your Parisian lectures you gave the fluctuation properties of radiation deservedly thorough discussion;[9] there the theories' inaccuracies become most clearly evident.

Now to your interference consideration! It amused me that you hit upon exactly the same example that I also had often turned over in my mind in the last few years. I grant you that the general theory of relativity is closer to the ether hypothesis than the special theory of relativity. But this new ether theory would not violate the relativity principle any more. For the state of this $g_{\mu\nu} =$ ether would not be that of a rigid body in an independent state of motion. Rather, a state of motion would be a function of position, determined by the material processes. Example:

If the Earth were not there or if it were not to rotate, the interference nodes of rings I and II would remain at rest relative to the "fixed stars," thus also relative to each other. But if the Earth rotates, both nodal systems indeed rotate with it by a tiny proportion, specifically, that of I more than that of II because of the lesser distances. Nodal systems I and II thus rotate at a minuscule velocity against each other in proportion to the Earth's rotation and the distances. Foucault's pendulum plane also rotates a little with the Earth, by about 0.01″ per year.[10] What a pity that it does not amount to *more*! I must admit, though, that I prefer the $g_{\mu\nu}$ system to an incomplete comparison with anything material. For the preference of uniform motion is not expressed in this modified ether hypothesis, but it certainly is in the abstract system. Namely, if a region of the world with constant $g_{\mu\nu}$'s is assumed, a *linear* substitution for the x_ν's thus changes nothing in the constancy of the $g_{\mu\nu}$'s, but a *nonlinear* substitution for the x_ν's certainly does. From this it follows that uniform relative motion does not "generate" a gravitational field, that is, it is imperceptible as opposed to irregular motion. However, this fundamental difference between uniform and irregular [motion] does not find direct expression in the ether notion; it rather attempts to *prove* constantly *uniform* motion.

With affectionate greetings and best wishes for your wife's health,[11] yours,

A. Einstein.

227.   To Willem de Sitter

[Berlin,] 22 June 1916

Highly esteemed Colleague,

Your letter pleased me very much and inspired me tremendously.[1] For I found that the gravitation equations in first-order approximation can be solved exactly by means of retarded potentials, if the condition for the coordinate choice $\sqrt{-g} = 1$ is abandoned.[2] Your solution for the mass-point is then the result upon specialization to this case.[3] Obviously your solution differs from my old one[4] merely in the choice of coordinate system, but not intrinsically.

As it is, one could think that the coordinate choice $\sqrt{-g} = 1$ was not at all natural. However, I have found a very interesting physical justification for the latter. I denote the $\sqrt{-g}$ system as K, the generalized de Sitter system[5] as K'. We now inquire about plane gravitational waves.[6]

In system K' I find 3 types of waves, of which only one is connected to energy transportation, however. In system K, by contrast, only this energy-carrying type is present. What does this mean? This means that the first two types of waves obtained with K' do not exist in reality but are simulated by the coordinate system's wavelike motions against Galilean space.[7] The $(\sqrt{-g} = 1)$ = system thus excludes undulatorily moving reference systems, which simulate energyless gravitational waves. System K' is nonetheless useful for the integration of the field equations in first-order approximation. I am curious about your Moon paper[8] and am anyway delighted that you like the general theory of relativity so much.

Cordial regards, yours very truly,

A. Einstein.

228.   From Théophile de Donder[1]

[Brussels,] 27 June 1916

Sir and highly esteemed Colleague, (Einstein)

I just noted with pleasure that your system of 10 diff. eqs. for the gravitational field is equivalent to the sys. of 11 diff. eqs. that I had found, as a result of Hamilton's principle extended to the generalized Maxwell-Lorentz electromagnetic field.

The same method also provided me simultaneously with the gravitational tensor $t_{\lambda\mu}$ $(\lambda, \mu = 1 \ldots 4)$ in its full generality. These results were summarized in a note appearing, I believe, on the 27th of May in the *Verslag[en]* of the Acad. of Amsterdam.[2] The develop[ment] of this note, which I have the honor of offering to you respectfully with the enclosed, will furnish chapter VII of a report that will be appearing soon, I hope, in the *Archives du Musée Teyler*.[3]

I compared this tensor $t_{\lambda\mu}$ to your functions[4] t_λ^μ in the particular case where $(-g)^{\frac{1}{2}} \equiv 1$. Well, the result of my calculations is that there is no identity; it is in this regard that I turned to your insights.

The method adopted, the generality of the result, and the simple covariance of $t_{\lambda\mu}$ nevertheless instill in me great confidence in the values I attributed to $t_{\lambda\mu}$.

I have also calculated your function t_1^1 and my function t_{11}, setting out from the quadratic formula related to your admirable application to the mo[tion] of Mercury around the Sun.[5] I find your $t_1^1 = R^{-2}$,[6] and my $t_{11} = 0$. It likewise appears that your $t_3^3 \neq 0$, while my $t_{33} = 0$. *All my $t_{\lambda\mu}$'s are zero* in the problem considered except for t_{22} which is equivalent to $-R^{-2}$.

The result of this is that everywhere and always (except at the origin) the K_λ's are zero $(\lambda = 1 \ldots 4)$.[7]

I am also sending you a note that appeared on 6 July 1914 in the *Comptes Rendus*.[8] You will see that at that time I already had the *method* for calculating T_{ij}; some calculation errors led me to inexact values for this tensor; since September 1914 I have been in possession of the *exact* values for T_{ij}. You will notice that my method also applies to systems in which $DI^* \neq 0$, systems that you have already considered.[9]

Please, Sir and highly — Colle[ague], allow me to assure you of my greatest respect.

229.   From Michele Besso

Zurich, 28 June 1916

Dear Albert,

I'm a lazy loaf: first, I never write you; and second, when I do write you it is in order to save myself work.

First: I heard your boys' voices Friday evening. They were bathing and are well. Little Albert is a bit earnest.

Second: I am fighting an uneven battle with my lecture material;[1] I simply can't succeed in gaining and retaining an overview that would make the work productive. And I'm too unorganized for a more external order. I therefore ought to concentrate my little bit of diligence on this self-imposed responsibility. But the devil has taken my friends at the Physical Society and they want a talk from me on your newest papers: even though there are at least three, Abraham, Grossmann, and Weyl,[2] who are a hundred[3] times more familiar with the subject than I. I feel like someone to whom Beethoven had whistled his symphony and who must now whistle back a part of it—albeit with the score in front of his eyes, but being able to read it only as well as I can read sheet music . . .

Well, I did enjoy myself when I had a look at the score again. But I must prove to you right away how insecure I am in the details[4] despite the demonstration whistling and despite the score.

First: $\kappa = 1.87 \cdot 10^{-27}$gr cm^{-1} is an absolute natural constant, is it not?[5] I do not even know about that *definitely*!

Second: Although I have your Viennese lecture as a paradigm, equations (1d) and (7e'), page 19,[6] I cannot manage to develop the corresponding relations according to the new gravitation equations. The result will probably differ in the numerical factors.[7]

⟨Third (and if misunderstandings by me exist, then my case is severe indeed!): ds^2 is an invariant. I.e.: The world distances between every two space-time points yield the same numerical result, whatever might be the coordinate system selected for their categorization. The transformation $x'_i = ax_i$ is not a permissible transformation?⟩

Third: In a finite portion of a mass-point's vicinity, coordinates can be introduced in such a way that with reference to the same the centripetal acceleration vanishes in the entire finite portion. How do the $g_{\mu\nu}$'s look for this?[8]

230.   To Théophile de Donder

Berlin, 13 Wittelsbacher St., 30 June 1916

Esteemed Colleague,

Many thanks for sending me your interesting analysis.[1] Our lack of agreement regarding the energy components for the gravitational field surely arises from your determination

$$C = 0,$$

which is always inconsistent with the gravitation field equations when the scalar of the mass energy tensor does not vanish.[2] This can be seen by forming the appropriate scalar equation out of the field equations through inner multiplication by $g^{\mu\nu}$.

It is not correct that in my formula (2a), p. 845 of my pap[er] *Sitz. Ber.* XLVIII 1915,[3] a $\sqrt{-g}$ factor is missing. On both sides of the equation a covariant tensor of second order must appear; a $\sqrt{-g}$ factor on the left-hand side would destroy the equations' invariant character.[4]

I have presented the entire theory in summary in a paper appearing in the most recent *Annalen*.[5] In this paper you can also find a simplified description of vacuum electromagnetics, in which it is shown that there is no need for the "dual" six-tensor [*Sechservektor*] concept.[6]

Very respectfully yours,

A Einstein.

231.   From Théophile de Donder

[Brussels,] 4 July 1916

S[ir] and esteemed Coll[eague] (Einstein)

I thank you — postcard [of] 30 June 1916.[1] — exact meaning of your symbols.

I.

May I ask you to consider first the case of the elect[rical] vacuum permeated by electr[icity] but *devoid of actual matter*; it is exclusiv[ely] to such a field that my note relates. In appl[ying] Hamilton's gener. principle I find the 11 eqs.; (330, 352, 353, ch. VII);[2] by means of the *identity*[3]

$$(1 + \varepsilon_{im})\Diamond^{im}\overline{l} \equiv k(-g)^{\frac{1}{2}} \sum_k \sum_l g^{kl}(ik, lm) - k(-g)^{\frac{1}{2}} C g_{im},$$

these *eleven* eqs. can be written:

$$\begin{cases} k(-g)^{\frac{1}{2}}\sum_k\sum_l g^{kl}(ik,\,lm) - k(-g)^{\frac{1}{2}}Cg_{im} = \sum_k g_{km}T_{ik} & (\alpha) \\[2em] C \equiv \dfrac{1}{2}\sum_i\sum_m\sum_k\sum_l g^{kl}g^{im}(ik,\,lm) = 0 & (\beta) \end{cases}$$

These eleven eqs. are *equivalent* to your 10 eqs.[4]

$$K(-g)^{\frac{1}{2}}\sum_k\sum_l g^{kl}(ik,\,lm) = \sum_k g_{km}T_{ik},$$

since, here $\sum_k T_{kk} = 0$.

$\rightarrow$ I would like to know whether your[5] $[T_{im}]$ is identical, in the case considered, to my $(-g)^{\frac{1}{2}}\sum_k g_{km}T_{ik}$, where[6]

$$T_{ik} \equiv \frac{\varepsilon_{ik}}{4}\sum_\alpha\sum_\beta (-1)^{\alpha+\beta+1}M_{\alpha\beta}M^{*\alpha\beta} + \sum_\alpha (-1)^{\alpha+\beta}M_{i\alpha}M^{*k\alpha}.$$

In drawing inspiration from your earlier works (1914),[7] I found some $t_{\lambda\mu}$'s (341)[8] like the one (347, ch. VII)[9]

$$(-g)^{\frac{1}{2}}F_\lambda = \sum_\mu \frac{d(T_{\lambda\mu} + t_{\lambda\mu})}{dx_\mu}. \tag{347}$$

$\rightarrow$ I would like to know whether your $[t_\lambda^\mu]$'s, which figure in the relations (8), page (846), (1915),[10] are assigned to play the same physical role as $t_{\lambda\mu}$ (gravitational tensor). *I do not understand why the second members of these relations (8) are zero.*

II.

([. . .]) I would be very grateful if you would please give me your opinion on these second cases.

My method is also applicable in the case where the field considered *contains some actual matter*; in this case your 10 eqs. (2a), p. 845, (1915)[11] are equivalent to the eleven eqs. obtained in expressing that[12]

$$\delta(A - B + kC)(-g)^{\frac{1}{2}}dx\,dy\,dz\,dt = 0\ldots \tag{\gamma}$$

with the complementary eq.

$$kC(-g)^{\frac{1}{2}} = -\frac{1}{2}\sum_k \mathcal{T}_{kk}. \tag{δ}$$

The symbols A, B, $\mathcal{T}_{ik}$ have meanings analogous to those for A, B, T_{ik}, which appear in my eqs. (ch. VII);[13] thus, for ex., $\mathcal{T}_{ik} \equiv T_{ik} + T_{ik}^{\mu}$, where T_{ik} relates to the electrical f[ield] and where T_{ik}^{μ} is due to actual matter.

The 10 eqs. arising from (γ) will be able to read (352, 353, Ch. VII):

$$(1+\varepsilon_{im})\Diamond^{im}\overline{l} \equiv k(-g)^{\frac{1}{2}}\sum_k \sum_l g^{kl}(ik,\,lm) - k(-g)^{\frac{1}{2}}Cg_{im}$$

$$= \sum_k g_{km}\mathcal{T}_{ik}. \ldots \tag{α'}$$

If to these 10 eqs. (α') the compl[ementary] eq. (δ) is added, the 11 eqs. are obtained.

$$\left\{\begin{array}{l} k(-g)^{\frac{1}{2}}\sum_k \sum_l g^{kl}(ik,\,lm) = \sum_k (g_{km}\mathcal{T}_{ik} - \dfrac{1}{2}g_{im}\mathcal{T}_{kk} \ldots \qquad (\alpha'') \\[4mm] kC(-g)^{\frac{1}{2}} = -\dfrac{1}{2}\sum_k \mathcal{T}_{kk} \ldots \qquad (\delta'') \end{array}\right.$$

But from the 10 eqs. (α'') the 11th (δ'') can be easily deduced. Thus, it is demonstrated that (γ) and (δ) provide the physical meaning of your ten eqs.

The result of this is that $t_{\lambda\mu}$ (341, Ch. VII) agrees without modification with this case.

Please — gr[eatest] re[spect].

232.   To Théophile de Donder

[Berlin,] 8 July 1916

Esteemed Colleague,

I hasten to answer your questions.[1] Beforehand, however, a general remark. It is, in my opinion, of disadvantage not to look at the indices of a tensor, whether they have a covariant or contravariant transformation character; you obscure the general idea unnecessarily through this shortcoming in the designation. When you therefore write, e.g., my field equations with vanishing energy scalar for matter, you must not write

$$k\sqrt{-g}\sum_{kl} g^{kl}(ik,\,lm) = \sum_k g_{km}T_{ik}$$

but

$$k\sqrt{-g}\sum g^{kl}(ik,\,lm) = \sum g_{km}\mathfrak{T}_i^k$$

or

$$-\;\;-\;\;-\;\;-\;\;= \sqrt{-g}\sum_k g_{km}T_i^k.$$

There T_i^k is a (mixed) tensor, $\mathfrak{T}_i^k$ a tensor multiplied by $\sqrt{-g}$. The right-hand side can also be written more briefly $\sqrt{-g}T_{im}$ by inserting the covariant instead of the mixed tensor. Hence I can answer your first question thus:

$$(T_{ik})_{\text{Donder}} = (\mathfrak{T}_i^k)_{\text{Einstein}} = \sqrt{-g}(T_i^k)_{\text{Einstein}}$$

$$= \sqrt{-g}\sum_\sigma g_{i\sigma}(T^{\sigma\kappa})_{\text{Einst.}}$$

$$= \sqrt{-g}\sum_\sigma g^{k\sigma}(T_{i\sigma})_{\text{Einst.}}$$

In my paper XLVIII 1915[2] I chose (eq. (3a) p. 845) the coordinate system so that $\sqrt{-g} = 1$. In this case

$$(t_{\lambda\mu})_{\text{Donder}} = (t_\lambda^\mu)_{\text{Einstein}}.$$

With this I have answered your questions I.

It is not necessary to add to

$$\delta\{\int (A - B + kC)(\sqrt{-g})dx_1\ldots dx_4 d\tau\} = 0 \ldots \tag{γ}$$

the equation

$$kC\sqrt{-g} = -\frac{1}{2}\sum_k T_{kk} \ldots, \tag{δ}$$

because (δ) is a consequence of (γ), as you also note, of course. The (γ) equations *alone* are already equivalent to my field equations. This is also valid in the case where the "material" process is other than a purely electromagnetic one.

In the purely electromagnetic case your earlier determination

$$C = 0$$

does not lead to contradictions. But these contradictions occur if the formulas arrived at in this way are applied to cases in which the energy scalar of matter does not vanish everywhere[3] (mass-point).

Best regards, yours very truly,

A. Einstein.

233.   To Michele Besso

[Berlin,] 14 July 1916

Dear Michele,

First of all I thank you from the bottom of my heart for being such a loyal helper to my children and my wife. From your letter it appears that my wife really does seem to be seriously ill.[1] At first I regarded it as obvious that I must travel there but after careful reflection am now of another opinion. If I go to Zurich, my wife will demand to see me. This I would have to refuse, partly on an inalterable resolve, partly also to spare her the agitation. The children on their part would perceive it as an unbearable unkindness if I did not fulfill this wish. Besides, you know that the personal relations between the children and me deteriorated so much during my stay at Easter (after a very promising start)[2] that I doubt very much whether my presence would be reassuring for them. So if my wife can stay at home and continue to manage the household, then I won't come at all. If my wife must go to the hospital, however, I'm very willing to meet the children at a neutral place, e.g., in Schaffhausen, and spend the holidays with them. In this case, though, after the bad experiences at Easter, I would avoid coming to Zurich.

Dear Michele, I understand your remark about the effect of "emotional tension" as the cause of the present circumstances; this is a fair hypothesis.[3] But I personally have the suspicion that the woman is leading both of you kindhearted men down the garden path.[4] For she is not afraid to use all means when she wants to achieve something. Unfortunately it can't be determined clearly in retrospect whose impression is more true to the mark. But as the burned child, I really ought to know the fire better. It is true that I obviously can't deny the possibility that the woman really is ill. But last year it was similar with the little one, at Easter with Albert. I always had to listen to the most dubious excuses, always at the beginning of the vacation period, that is, ... Dear Michele, you have no idea of the natural craftiness of such a woman. I would have been physically and mentally broken if I had not finally found the strength to keep her at arm's length and out of sight and earshot. Didn't you feel at our last meeting that you had an emotionally more stable fellow before you than formerly, one who again has a harmless joy in sheer existence? Perhaps I shall see the day, after all, when your eyes are opened to these things.

Now please just keep me up-to-date on the state of affairs by postcard, while ignoring the continuity of the substance of your remarks. In the age of quantum theory this is all the more permissible.

Affectionate greetings to you and your family, yours,

Albert.

Cordial greetings and many thanks for now to Zangger, likewise to the Zürchers.[5]

234.   From Théophile de Donder

[Brussels,] 14 July 1916

S[ir] and esteemed Colleague,

I acknowledge receipt of your letter of the 8th instant;[1]

I. I was happy to note that the eqs. (371) of my chap. VII are *identical* to yours (2a, p. 845, 1915) when you set:[2]

$$\sqrt{-g}(T_i^k)_{\text{Einstein}} \equiv (T_{ik})_{\text{DeDonder}}.$$

However, then I note your eq. (14) p. 782 (XLIV, 1915) must be modified in the following manner:[3]

$$\sum_\nu \frac{\partial (\mathfrak{T}_\sigma^\nu)_{\text{Einstein}}}{\partial x_\nu} = \text{etc.},$$

in other words, among these 4 relations the $\mathfrak{T}_i^k$'s (i.e., my T_{ik}'s) *must* appear. I have verified this fact *very carefully* in 3 different ways.

II. You note that in the case where $\sqrt{-g} = 1$ and where $T \equiv 0$ my $(t_{\lambda\mu})_{\text{DeD}}$ *must coincide*, according to your indications, with your $(t_\lambda^\mu)_{\text{Einst}}$; hence, as I wrote you in my first letter,[4] this coincidence *does not* take place. I have verified this with the ds^2 which appears on p. 194, eq. (4) (report by the late K. Schwarzschild 1916).[5] This still leaves me perplexed?—.

III. I found $\boxed{\sum_\mu \dfrac{d(t_{\lambda\mu} - t^{\lambda\mu})}{dx_\mu} = 0}$ The $t^{\lambda\mu}$'s diff. from the $t_{\lambda\mu}$'s *and* $(t_\lambda^\mu)_{\text{Einst}} \equiv$

general case??

IV. ⟨It is accurate to say that *of* your ten equations my *eleven* eqs. can be deduced, i.e., those that result from[6]

$$\delta \int \int \int \int D(dx_1 \ldots dx_4) = 0$$

and from the compl. eq. $l = -\dfrac{1}{2} \sum T_{kk}.$⟩

235.   To Willem de Sitter

[Berlin,] 15 July 1916

Dear Colleague,

Many thanks for your long letter. I agree with your calculation.

You are quite right, of course, that my specialization of the coordinate system by the condition $\sqrt{-g} = 1$ is not a complete one, so boundary conditions would not suffice to make the mathematical problem of the determination of the $g_{\mu\nu}$'s unique.[1] After all, substitutions with functional determinant 1 exist that leave the boundary region untransformed. It would be very nice if the [coordinate] system could be specialized further in a natural way, if only in the interest of better comparability of found solutions. But I have not been able to find anything of the sort.

The conditions added by me in the case of the mass-point overdetermine the problem without, however, contradicting one another. The three conditions $g_{i4} = 0$ do not yet suffice.[2] For, a purely spatial transformation can still be performed with subst. determinant 1, which destroys the point symmetry.

It is not good if in my letter I called the $\sqrt{-g}$ system "Galilean."[3] Only a space in which all the $g_{\mu\nu}$'s are constant can be called that. But there it is not just the "space" that is "Galilean" but the "space" together with the frame of reference, which makes the $g_{\mu\nu}$'s into constants. Nevertheless, these are only questions of definition about which one does not have to rack one's brains much. What you say there about "true" and "apparent" is correct in principle. What I mean in my addendum is the following.[4] When I find some process or other, e.g., a wave process, as a solution to the differential equations, two possibilities exist. Either such wave processes exist, no matter how I may have chosen the reference system, or such kinds of wave processes do not exist when [I] choose the coordinate system in a particular manner. If the latter is the case, I can then denote the relevant process (because it can be "transformed away"), in a certain sense as an "unreal" process. "Unreal" then simply means "can be transformed away." But it is better, of course, to avoid such words, which give rise to unclarity. One can say: the coordinate choice according to the condition $\sqrt{-g} = 1$ is simple or advantageous to the extent that with this choice only waves of the 3rd type[5] occur. (For the calculation, however, your coordinate choice is preferable.)

Best regards, yours,

A. Einstein.

236.   To Théophile de Donder

[Berlin,] 17 July 1916

Dear Colleague,

I. My equation (14) p. 782, 1915[1] is only valid for the coordinate choice $\sqrt{-g} = 1$. Generally, instead of T_σ^ν, $\mathfrak{T}_\sigma^\nu = \sqrt{-g}T_\mu^\nu$ ought to be inserted into the same so that everything is in full agreement.

II. That your $t_{\lambda\mu}$ does not agree with my t_λ^μ has a profound reason. Since in your equation (341) you may *not* in general set $l = 0$,[2] your [. . .] yield[3] as [ours?]?

V. The constant κ has no physical meaning, because it is determined, as you know, by the unit of mass. A variation principle of the form

$$\delta \int \int \int L d\tau = -\frac{1}{2}\sum T_{kk}$$

is peculiar. Besides, the equations[4] $k\sqrt{-g}\sum g^{kl}(ik, lm) = \sum(gT_{ik} - \frac{1}{4}g_{im}T_{[kk]})$ are *not* consistent with the energy law of matter.

Best regards, yours truly,

A. Einstein.

237.   From Michele Besso

Zurich, 17 July 1916

Dear Albert,

I just received your nice letter of the 14th.[1]

There is not the least reason to thank me. Whatever could be done practically was done by Mrs. Zürcher and, as far as was necessary, directed by Prof. Zangger.[2] I had to play an all the more passive role as my presence seems to affect your wife unfavorably.

A deception ⟨absolutely cannot be involved. She did not complain about her condition⟩ is not at issue here. Prof. Zangger reported to me his very factual observations; these I passed on to you, in essence.[3] I mentioned the emotional stress as a concurring cause because, as may be determined objectively, [it] has improved again, as I reported to you yesterday; also the boys are, as I wrote you, provided for very well for the holidays.[4]

For the rest, you know what I mean: "we are sinners, one and all." If a person is held to blame for all his misdeeds, and if that which belongs on the other side of the scale is not counted accordingly, then surely we all belong in hell. (It may be that I am generalizing too much: but *I*, for one, certainly.). The woman's sufferings had left their mark clearly on her appearance for a long time already; she did not let herself go either, but overburdened herself with too much work[5]—which was only discovered by chance when an unexpected visit was paid there once.

All this does not mean to say, though, that any pair of horses could be used in a single team, and with *all* due respect for both, they might be *used* separately. That *she* does not quite understand this I can grasp better than that you cannot understand it.

And joy in sheer existence ⟨I hope you will [feel that] for long and unmixed⟩ I grant you with all my heart. May heaven preserve it for us!

238.   To Michele Besso

[Berlin,] 21 July 1916

Dear Uncle Toby,[1]

I am exceedingly pleased that the nasty matter seems to be blowing over so smoothly. I didn't suspect a *pretense* at all, but surely, conditions brought on purely by the nerves.[2] Dear Michele, we men are deplorable, dependent creatures, this I admit gladly to anyone. But compared to these women, every one of us is a king; for he stands more or less on his own two feet, not constantly waiting for something outside of himself to cling on to. They, however, always wait for someone to come along who will use them as he sees fit. If this does not happen, they simply fall to pieces.

As concerns my wife, please consider the following. She has a worry-free life, has her two fine boys with her, lives in a lovely area, can use her time freely, and basks in the aura of abandoned innocence. The only thing she is missing is someone to dominate her!

Now, is it really so terrible in your eyes that I fled from this responsibility after many attempts this way and that?[3] Who would tolerate having something which to him has an odious smell stuck in his nose his life long, for no purpose at all, with the secondary obligation of putting on a friendly face?

I gladly acknowledge that each one of us would go to hell, were the Lord (in ignoring His absolute responsibility) to want to pass severe judgment on us.[4] But far from wanting to stand before the court together with the woman, I make do with my own defense; and this defense takes place *only* before you and Zangger;

I leave all others to form their opinion of me at their own discretion. In the end, I cannot make you two change your minds either, but must take what comes.

I sincerely regret having apparently annoyed you so much in your selfless kindness that you address me formally in the postscript to your letter[5] and herewith take back everything that may have provoked your anger. I have avoided the way you had indicated, out of the frying pan into the fire,[6] without substantial hardship. The latter is even alleviated considerably by the recent events. I would like Mileva to know this, as she may be able to be calmed by it. But I can't tell her this myself.

Affectionate greetings, yours,

Albert.

Dear Michele, we have understood each other well for 20 years.[7] And now I see you developing a bitterness toward me for the sake of a woman who has nothing to do with you. Resist it! She would not be worth it, even if she were a hundred thousand times more in the right!

239.   To Michele Besso

[Berlin, 21 July 1916]

Dear Michele,

Now I see that I made a very grave mistake in that the letter's postscript is not by you but by Anna.[1] So I wrongly experienced seeing you punish me with the formal address "Sie." It is only thus that you can understand the tone of my letter.[2] And this should happen to a scientist!

Affectionate greetings to you both, yours,

Albert.

240.   To Théophile de Donder

[Berlin,] 23 July 1916

Dear Colleague,

I must admit to you that, unlike most of our colleagues, I am not at all of the opinion that every theory must be put into the form of a variation principle. I attribute this common urge simply to the fact that everyone is used to scalars but not to tensors.[1] If the Hamiltonian form is desired nonetheless, the following device is best used.

The difficulties discussed arise from the fact that scalar l contains *second* derivatives.[2] In $L = \sqrt{-g}l$, terms in the form of

$$G_{iklm} \frac{\partial^2 g_{ik}}{\partial x_l \partial x_m}$$

occur. (G_{iklm} *only* dependent on $g_{\mu\nu}$.) However, $\int L d\tau$ can be modified through partial integration so that[3]

$$\int L d\tau = \int L^* d\tau + \text{surface integral},$$

where L^* just contains the *first* derivatives of $g_{\mu\nu}$. This is because

$$\int G_{iklm} \frac{\partial g_{ik}}{\partial l \partial m} d\tau = -\int \frac{\partial G_{iklm}}{\partial x_l} \frac{\partial g_{ik}}{\partial x_m} d\tau + \underbrace{\int \frac{\partial}{\partial x_l} \left(G_{iklm} \frac{\partial g_{ik}}{\partial x_m} \right) d\tau}$$

$$/ \qquad\qquad \text{convertible into surf. integral.}$$

contains only *first* derivatives.

That is why

$$\delta\{\int L d\tau\} = \delta\{\int L^* d\tau\},$$

and in the Hamiltonian consideration, L is replaceable by L^*. Then no second derivatives appear in the energy components and

$$\delta\{\int (L^* + M) d\tau\} = 0$$

can safely be set up as a variation principle, where variations are to be made with respect to $g_{\mu\nu}$ (or $g^{\mu\nu}$) and M refers to matter. In this way you will undoubtedly arrive at my gravitational field energy components. I did not perform the somewhat tedious calculation of

$$t_\sigma^\nu = \frac{1}{2\kappa} \left(\frac{\partial L^*}{\partial g_\nu^{\alpha\beta}} g_\sigma^{\alpha\beta} - \delta_\sigma^\nu L^* \right).$$

It seems to me only of interest that an energy law thoroughly analogous to mine also exists if the reference system specialization according to the condition $\sqrt{-g} = 1$ is omitted.[4] In any case, this specialization of the coordinate system makes the formulas clearer without impairing the generality of the theory.

Best regards, yours very truly,

A. Einstein.

241.   To Hans Albert Einstein

[Berlin,] Tuesday. [25 July 1916][1]

My dear Albert,

From Zangger and Besso I hear that Mama is quite ill[2] and that now you're home alone with the maid. I myself had decided that Mama should go to a sanatorium[3] so that she is taken very good care of and does not have to get upset about anything. I am constantly thinking of you both and about how you may be doing. But I cannot get away right now, because I have a lot of work to do.[4] See my dear friend Besso often when he is back again. He will gladly do things with you, and he is so clever and so good. *Do write me again!* It is sad when one has such a big son and hears nothing from him. I'll always be happy to answer immediately. Everything you occupy yourself with interests me. What are they teaching you at school? How is Tete? Is he always healthy? Now you must look after him well, especially that he always eats well. Do you also read books yet? Are you still taking woodworking? Do you like doing Latin?

When you write me, take this little letter in hand and answer all of my questions. And Tete should also scribble something underneath.

Kisses to you both from your

Papa.

Perhaps Aunt Zora will be coming to visit you.[5]

My address is always 13 Wittelsbacher St.

242.   To Heinrich Zangger

[Berlin,] 25 July 1916

Dear friend Zangger,

Besso's last letter[1] removes all doubt about the seriousness of the situation. Although it doesn't seem to be clear what my wife is suffering from, a very grave illness is involved from which she is unlikely to be able to recover rapidly. Now the difficult problem arises of how I can provide moral support for the children in case of a long period of illness. I am ready to come to Zurich without hesitation, to keep the boys company throughout the day when they return from holiday[2] so that they do not get the feeling of being alone. I am very afraid, though, that my wife will express the wish that I visit her. Under the current grievious circumstances I would have to satisfy this wish. On this occasion I could be forced to make promises regarding the children that result in the boys being taken away from me, also in case of the woman's death. I am very afraid of that. But if you think it appropriate that I come, I will come by all means.[3] I have no requisite duties in the near future. I beg you to assuage the woman in every way possible; in particular, I ask you please to give her a good nurse, if she has nothing against it.

My boys should not get the distressing feeling that they are getting no support from their father. In case the woman falls victim to the illness, I would raise the two boys myself, without having them go to a Berlin school. They would be taught at home, as far as possible by me personally.

Without your, the Zürchers', and the Bessos' help,[4] I would lose my mind in this distressing situation. I am very sorry for the woman, and I also believe that her difficult experiences with me and through me are at least partly to blame for her serious illness. Affectionately yours,

A. Einstein.

243. From Willem de Sitter

Loenen, 27 July 1916

Dear Colleague,

Your letter reached me yesterday. I am here in the countryside,[1] far away from my bookshelves, but shall try nevertheless to answer your question, to the best of my knowledge. For the directorship of a large astrophysical observatory, obviously, an astrophysicist will be selected;[2] but in my view, the choice must be guided, in the first place, by the consideration that the director ought to be a man with a broad overview of the entire field of astronomy, who not only can do his own research but is also capable, in particular, of grasping and solving major problems. The current situation is that astrophysics has largely perfected its own methods, and the primary requirement now is a collaboration between astrophysical and classical astronomical research methods and a combination of the findings. Unfortunately many astrophysicists still fail to recognize truly the value of traditional astronomical precision methods, whereas the genus of those classical astronomers who are unable to value astrophysics, is becoming extinct or has already died out.—But in general, one can more easily make an astronomer into the director of an astrophysical observatory than an astrophysicist into the director of an astronomical observatory.–

Now, as concerns the persons coming into question: aside from Küstner I would name Hertzsprung (who, however, although now employed at Potsdam, is not a German), Ludendorff, and Hartmann.[3] The last was not chosen on the previous occasion, and the same considerations are perhaps still valid now as well. He may be the most highly skilled operator of the instruments, but he surely remains far behind Küstner in an overall command of these sciences. Among the named, Hertzsprung is the most original; he has done many very fine things, yet always within a limited area. I am inclined to doubt whether he has a sufficient command of the major problems of today to direct a *large* observatory (which is quite another matter than doing research on one's own). The same applies to Ludendorff, whose researches on spectroscopic binary stars[4] I value very highly, incidentally. Küstner has perhaps not achieved as much as the others in the area of astrophysics: The Bonn Observatory's limited equipment also forced him to specialize, but his determinations of motion along the line of vision[5] and his papers in photometry,[6] in addition to his catalog of 10,000 stars,[7] are certainly research of the highest order. That with such limited means he turned to new methods at all, besides his extensive, purely astronomical work, shows that he is a man—I would say he is the only man in Germany now—who thoroughly grasps the major modern-day problems and who can still contribute much, particularly in the field of precision astrophysics, as well as toward the fusion of astrophysical and astronomical methods—where, in my view, the future lies.

As theoreticians, none of the named are comparable to Schwarzschild, of course—as general astronomers, only Küstner is on the same level—as observers, Küstner is undoubtedly superior to Schwarzschild as well as the others, probably. Among the named, I just know Hertzsprung personally and thus cannot make a comparison of their personal qualities.

I am now occupied with writing a small paper for the *Monthly Notices of the Royal Astron. Soc.* (London) on the new theory of gravitation and its astronomical consequences.[8] Your theory still seems to be almost entirely unknown in England.[9] If you have made new discoveries recently, I would be very pleased if you could report them to me as soon as possible; and your advice on my work is anyway very welcome.

Yours very truly,

W. de Sitter.

244.   From Willem de Sitter

Loenen, 27 July 1916

Dear Colleague,

Your postcard arrived a few hours after my previous letter[1] had been put in the mail. Mr. Müller[2] certainly has done very good astrophysical work; specifically, photometry at Potsdam, which was probably carried out principally by him, is a first-class achievement.[3] I did not mention him because, frankly speaking, I consider him too old. I do not know his exact age, but his entire work actually belongs to an earlier epoch. While Küstner's research is characterized precisely by the search for new avenues, Müller is better characterized for his careful application of tested methods (previously also by himself).[4] But even if the type of method is disregarded and success of the application alone is considered instead, Küstner's researches are certainly not inferior to Müller's. In my opinion, Küstner is the man who can be expected to make Potsdam move into the forefront: Müller would *perhaps* be able to keep it at the level it is.

Yours truly,

W. de Sitter.

245.   To Michele Besso

[Berlin,] Monday. [31 July 1916]

Dear Michele,

Your postcard made me very happy, as did your prior letter, according to which my wife is feeling better again.[1] Do none of you have a clear insight into her malady? I again firmly believe in the resilience of our friendship.[2]

The "Uncle Toby" salutation is a reminder of Sterne's novel, the main characters of which we compared ourselves to many times in Zurich.[3] The rotating ring's field in the proximity of the axis is found as follows. The field in *first*-order approximation easily results from a direct integration of the field equations.[4] The second-order approximation results from the vacuum field equations to closest approximation.[5] The first-order approximation yields the coriolis forces,[6] the second, the centrifugal forces. That the latter come out correctly is self-evident from the general covariance of the equations, hence actually calculating it out is of no interest whatsoever anymore. This interest is there only if it is not known whether rotational transformations are among the "admissible" ones, that is, if the equations' transformation properties are not clear, a stage which has finally been overcome, thank heavens.[7]

Heartfelt greetings also to Anna and Vero, yours,

Albert.

[6]Recipient's appended note: "How is this supposed to work where the coriolis forces are independent of location?"

246.   To Arnold Sommerfeld

[Berlin, 3 August 1916]

Dear Sommerfeld,

I shall be glad to send Mr. Lenz a copy of my summary paper on general relativity.[1] Schwarzschild's calculations are correct. I have examined them closely for the point.[2] h is the energy constant,[3] must therefore remain arbitrary. Your spectral analyses number among my finest experiences in physics.[4] It is just through them that Bohr's idea becomes entirely convincing. If only I knew which little screws the Lord is adjusting there!

With best wishes for your holidays, yours,

A. Einstein.

247.   From Gunnar Nordström

Leyden, 3 August 1916

Dear Prof. Einstein,

There are so many here in Leyden who would like to meet you. Do make arrangements so that the city of Leyden, with all its residents, comes to you. If you can demonstrate practically that a reference system exists in which such a phenomenon occurs, all your friends in Holland would be very pleased.

I also have been working here on your theory and would like to discuss some matters.[1]

In the hope that we really shall see each other soon, I send you my most cordial greetings, yours,

G. Nordström.

248.   From Théophile de Donder

[Brussels,] 5 August 1916 (modified 6 August 1916)[1]

Sir, (Einstein)

I just received your letter of the 23rd of July.[2] The envelope of this one bore the following inscription: "delayed because the required sender's information is missing on the exterior."

I thank you heartily for kindly communicating to me the device by which you make the terms $G_{iklm}g_{ik,lm}$ of $(-g)^{\frac{1}{2}}C$ disappear by replacing them with

$$-\frac{dG_{iklm}}{dx_l}g_{ik,m}.$$

Unfortunately, the t_σ^ν's which this method provides are unacceptable; indeed, the 10 eqs. of gravitation deduced from

$$\delta \int (L^* + M)d\tau = 0$$

will not be invariant for all the chang[es] in variables x_1, x_2, x_3, x_4. Therefore, one will obtain, in general:[3]

$$\lozenge^{ab}\bar{l} \neq \lozenge^{ab}L^*.$$

Thus (chapter VII, (352))[4] it is necessary that

$$\lozenge^{\alpha\beta}\bar{l} = -\frac{dM}{dg^{\alpha\beta}}.$$

Hence,

$$\boxed{\lozenge^{\alpha\beta}L^* \neq -\frac{dM}{dg^{\alpha\beta}}.}$$

Thus, so that[5]

$$t_\sigma^\nu = -\frac{1}{2\kappa}\left(\frac{dL^*}{dg_\nu^{\alpha\beta}}g_\sigma^{\alpha\beta} - \delta_\sigma^\nu L^*\right),$$

one could deduce the "energy momentum conservation law":[6]

$$\sum_\nu \frac{dt_\sigma^\nu}{dx_\nu} = -\sum_{\alpha\beta}\frac{dM}{dg^{\alpha\beta}}g_\sigma^{\alpha\beta} = -K_\sigma \quad \text{(ch VII, 344)}.$$

It is *necessary* to have:

$$\boxed{\lozenge^{\alpha\beta}L^* = -\frac{dM}{dg^{\alpha\beta}}.}$$

Any spec[ific] case amounts to *leaving aside* a certain number of terms. The result of this is that the *only* known rigorous method of establishing the energy momentum conservation law is the one presented in my note.

249.   From Théophile de Donder

Brussels, 11 Forestière Street, 8 August 1916

Sir and esteemed Colleague,

The lengthy investigations and innumerable calculations I have devoted to your theories oblige me to take a few weeks of relaxation: I shall thus have to postpone until later the calculation in question in my preceding letter.[1]

I am happy to note that we are now in agreement concerning the energy momentum conservation law; I saw with pleasure *that you deduce your [?] t_λ^μ's from my 10 equations*, which I [*set for*] the first time as[2]

$$\lozenge^{\alpha\beta} l = -\frac{d\overline{L}}{dg^{\alpha\beta}} \quad \text{(ch. VII, 352 or *Verslag Amsterd.* of May 1916)[3]} \qquad (1)$$

and from my relations[4]

$$\frac{d\overline{L}}{dg^{\alpha\beta}} = \left(\frac{\sum \varepsilon_{\sigma\mu}}{2} - 1\right) \sum_i g_{\beta i} T_{\alpha i} \quad \text{(ch. VII, 353)} \qquad (2)$$

equations, *to which you have added the important remark*

$$\lozenge^{\alpha\beta} l \equiv \lozenge^{\alpha\beta}(L^*)_{\text{Einstein}}.$$

May I hope that it will [always remain possible] for me to be able to count on this collaboration? Your manner of calculating your — t_ν^μ's (*arbitrary g*)

The identification of my eqs. (1) and (2) with yours (10 eqs. of gravitation) require very tedious calculations.

250.   To Michele Besso

[Berlin, 11 August 1916]

Dear Michele,

Your friend Ugo Russi[1] sent me a postcard for you. He reports being well and not worried. He has no fresh news about his mother.[2] The latest was satisfactory, though. His address is Mittergrabern, Oberhollabrunn, L[ower] A[ustria], Barrack 4.–[3]

A brilliant idea dawned on me about radiation absorption and emission; it will interest you. An astonishingly simple derivation, I should say, *the* derivation of Planck's formula. A thoroughly quantized affair. I am writing the paper right now.[4] Your silence calms me down; I hope all is well on the way to recovery in Zurich.[5] Many thanks for your last letter from the Planalp.[6] I hope you were able to relax well. In a few weeks I will be traveling—if possible—to Holland for a fortnight.

Warm greetings also to Anna & Vero, yours,

Albert.

On what are you lecturing next semester? Did you have hard-working students?[7] Are you content with your teaching duties? Write me about all this at length! When there is peace, you must come soon to visit!

251. To Michele Besso

[Berlin, 24 August 1916]

Dear Michele,

Many thanks for the postcard from Basel. Unfortunately, I misplaced your friend Russi's postcard,[1] so I can't pass on your instructions to him. I don't recall anymore whether I answered him; so send me his address. I am very glad that my wife is slowly getting better.[2] Although if it is cerebral tuberculosis,[3] as seems probable, a quick end would be better than long suffering. I am not going to Holland, because traveling is terribly complicated;[4] I am thinking of waiting until peacetime. But I definitely want to go to Switzerland next spring. My Albert is not writing to me. I believe that his negative attitude toward me has fallen below the freezing point. In his place, under the given circumstances, I also would probably have reacted in the same way.[5] The papers on gravitational waves and Planck's formula have been lying around at your place for a long time now.[6] You will enjoy the latter. The derivation is purely quantized and yields Planck's formula. In connection with this, it can be demonstrated convincingly that the elementary processes of emission and absorption are directed processes. One just has to analyze the (Brownian) motion of a molecule (in the sense of that derivation) within a radiation field. In this analysis, which is being published in the Zurich Phys. Soc.'s issue in honor of Kleiner, there are no undulatory theory considerations either.[7]

Warm regards, yours,

Albert.

252.   To Heinrich Zangger

[Berlin, 24 August 1916]

Dear Zangger,

Please let me know briefly whether you received my reply to your last letter along with the papers.– Besso, who is in Berne right now, informs me that my wife is feeling a bit better. But in combination with your diagnosis, this is cold comfort.[1] In any case, I am glad that the woman is receiving the best care possible. Aside from this worry, I am living peacefully and contentedly in my quiet cell, into which no newspapers penetrate. I have the feeling as if people had all gone into hibernation; because whatever is active does not suggest human feeling. How is Huguenin doing?[2] Send him my regards.

I wish you happy holidays, yours,

A. Einstein.

Document description: *"Please forward!"*

253.   To Paul Ehrenfest

[Berlin,] Friday. [25 August 1916]

Dear Ehrenfest,

Today I was at the foreign office about the trip. Any *official* invitation, which you can easily engineer in some form or other, would facilitate the trip immensely. Once I have it, it takes about one more week to obtain permission for the trip. I shall come then, if nothing very unexpectedly arises.[1] I hope Nordström will also be there then.[2]

I also considered the energy tensor matter[3] once but have since abandoned it completely. If all positive energy of matter is compensated by the negative energy of gravitation, then the energy concept loses all meaning. Moreover, then it is not understandable why the energy law is satisfied to such a close approximation for matter alone. More on this orally.

Cordial greetings to all of you, yours,

Einstein.

254. To Michele Besso

[Berlin,] 6 September 1916

Dear Michele,

I transmitted the message to your friend Russi.[1] Your thorough report on Mileva's condition appeased me greatly.[2] From now on, I'll not trouble her any more with the divorce.[3] The corresponding battle with my relatives has been fought. I have learned to withstand tears.[4]

Planck's papers provide no correlation between h and ε. You have in mind the dimensional equivalence and virtually identical order of magnitude of $\dfrac{\varepsilon^2}{c}$ and h, which no theory has elucidated yet, though.[5] To derive Wien's displacement law, the Doppler principle and the law of radiation pressure are needed, which have been described only according to undulatory theory, however, on the whole similar to the concept of frequency, of course.[6] But what is crucial is that the *statistical* considerations leading to Planck's formula have become *uniform* and thus have become the most general imaginable, in that nothing more is assumed about the special properties of the molecules involved than the most general idea of the quantum.[7] The result (not yet contained in the paper sent to you) thus obtained is that at each elementary transfer of energy between radiation and matter, the impulse quantity $\dfrac{h\nu}{c}$ is passed on to the molecule.[8] It follows from this that any such elementary process is an *entirely directed* process. Thus light quanta are as good as established.

I'm going to attempt the trip to Lorentz.[9] To do it, I need an original of my citizenship certificate. I wrote to Zurich for it and requested that the document be sent to you, in case the Registry Office does not prepare mailings abroad. In that case, be so good as to send me the document by return mail.[10]

With best regards to you and your family, yours,

Albert.

With whom are you staying, and what is tying you down in Berne for such a long time?

255.   To Constantin Carathéodory

[Berlin,] 6 September 1916

Dear Colleague,[1]

You held out the prospect of writing an intuitive derivation for me of the Hamilton-Jacobi relation. Well, I succeeded in doing so myself and am showing you my simple considerations only to spare you the effort. For the Lagrange function L

$$\delta\{\int L dt\} = 0 \ldots \tag{1}$$

$$\text{or } \frac{\partial L}{\partial q_\nu} - \frac{d}{dt}\left(\frac{\partial L}{\partial \dot{q}_\nu}\right) = 0 \ldots \tag{1a}$$

applies. Now, we set

$$\int L dt = J(q_\nu, Q_\nu, t, T) \ldots \tag{2}$$

Here the Q_ν's are the initial coordinates to a specified initial time T.

I now consider a neighboring path (between the same times t and T) to be reached through a virtual shift in the path. Through variation of (2) you obtain, taking (1a) into account,

$$\sum \frac{\partial L}{\partial \dot{q}_\nu} \delta q_\nu - \sum \frac{\partial L}{\partial \dot{Q}_\nu} \delta Q_\nu = \sum \frac{\partial J}{\partial q_\nu} \delta q_\nu + \sum \frac{\partial J}{\partial Q_\nu} \delta Q_\nu \ldots \tag{3}$$

From this, both Jacobian equation systems result. Since, first,

$$\frac{\partial L}{\partial \dot{q}_\nu} = p_\nu = \frac{\partial J}{\partial q_\nu} \ldots \tag{3a}$$

Second,

$$\frac{\partial J}{\partial Q_\nu} = -\frac{\partial L}{\partial \dot{Q}_\nu}.$$

For one and the same path, however, $\dfrac{\partial L}{\partial \dot{Q}_\nu} = \dot{P}_\nu$ is given as an initial condition, thus is constant; therefore, the $\dfrac{\partial J}{\partial Q_\nu}$'s are also constant on a single path. If instead of Q_ν arbitrary functions α_ν of these quantities are introduced, we then naturally also have

$$\frac{\partial J}{\partial \alpha_\nu} = \beta_\nu = \text{const.} \tag{3b}$$

By differentiating (2) according to time,

$$L = \frac{dJ}{td} = \frac{\partial J}{\partial t} + \sum \frac{\partial J}{\partial q_\nu} \dot{q}_\nu$$

$$= \frac{\partial J}{\partial t} + \sum p_\nu \dot{q}_\nu$$

results, or

$$\left. \begin{aligned} \frac{\partial J}{\partial t} + H &= 0 \\[2mm] H &= \sum p_\nu \dot{q}_\nu - L \end{aligned} \right\} . \tag{4}$$

This is the Hamilton-Jacobi differential equation, where H is initially expressed as a function of the q_ν's and p_ν's and then the p_ν's are substituted by $\dfrac{\partial J}{\partial q_\nu}$ according to (3a).–

Naturally, the Jacobian transformation is by no means yet proven with this. But for this the formal, less transparent proof offered by Appell suffices for me.[2] What I was missing was a natural way to arrive at equations (3a) and (3b) from the Lagrange equations.

Wouldn't you like to reflect a bit more on the problem of closed time lines?[3] Here lies the core of the as yet unresolved portion of the space-time problem.

Best regards, yours very truly,

A. Einstein.

P.S. Naturally, I do not imagine that these trivialities are in any way original or new. These things just give me the sense of being versed in the subject.

256.   To Paul Ehrenfest

[Berlin, 6 September 1916]

Dear Ehrenfest,

You have no idea how difficult it is to venture out nowadays. Yet, after Lorentz's kind help,[1] I will do my utmost to be with you all as soon as possible. First, I must obtain an original of my certificate of citizenship, located in Zurich;[2] then a long chain of other still obscure obstacles awaits me. So don't be surprised if many more delays occur. I'm happy beyond words to be seeing you all again. Also scientifically, I have something nice to show you.[3]

Cordial greetings to you and yours, to Nordström, Lorentz, and de Sitter, yours,

Einstein.

257.   To Hedwig Born

[Berlin,] 8 September 1916

Esteemed Mrs. Born,[1]

Your poem delighted me, above all because it signals a cheerful disposition but also is a sign that you entertain the best of relations not only with the Parnassian Muse but also with the Flemish sow.[2] The latter is truly not necessary, though, to make the prospect of a couple of pleasant evening hours spent in your home appear to me in the most enticing colors!

I read the book with great interest. It is interestingly written, without a doubt by a man who knows the pitfalls of the human soul. Incidentally, I believe I have made the acquaintance of this man in Prague.[3] He apparently belongs to a small philosophically and zionistically infested circle, which was loosely associated with the university philosophers, a small troop of unrealistic people, harking back to the Middle Ages, with whom you have become familiar on reading the book.[4]

Best regards to you both, yours,

Einstein.

I am sending you the two papers you wanted at the same time. I will bring the book along myself.

[2] "Flämisch" also meaning "of the flames."

258.   To Helene Savić

Berlin, Wilmersdorf. 13 Wittelsbacher St., 8 September 1916

My dear Helene,[1]

Your letter pleased me greatly, first because you talk in detail about my dear boys,[2] then because you do not judge me superficially, as most of my acquaintances do. Separation from Mieze[3] was a matter of life and death for me. Our life together had become impossible, even depressing; *why*, I cannot say. Thus I deprive myself of my boys, whom I still love tenderly. During the two years of our separation, I saw them twice; last spring I went on a little trip with Albert, and to my great dismay, I noticed that my children, not understanding my behavior, feel secretly angry at me,[4] and I find that, although it is distressing, it is better for them if their father does not see them anymore. I would be satisfied if they became useful and respected men; everything points toward believing that this will be so; for they are quite gifted, and although in general I do not have a high opinion of the influence of education, I have great confidence in that of their mother.–[5] I was pleased to learn that Tete has recovered so well; he was always delicate and nervous, and what's more, precocious in his intellectual development. Mieze's illness worried me; fortunately, she is on the way to a complete recovery.[6] Despite this interest on my part, she *is* and always *will remain* an amputated limb to me. I am never going to be close to her again; I shall end my days far away from her, feeling that this is absolutely necessary. I believe that Mieze herself sometimes suffers from her great reserve. Her parents and her sister,[7] with whom she had always lived harmoniously, were completely unaware of this skill of hers. You, dear Helene, can be of great use to her in this respect, in helping her overcome moments of discouragement. I am profoundly grateful to you for all that you have done for Mieze and, above all, for the children.– Do not feel sorry for me; despite my external troubles, my life goes by in perfect harmony; I devote all of my thoughts to reflection. I resemble a presbyope, who is charmed by the vast horizon and who is disturbed by the foreground only when an opaque object obstructs his view.– If your path takes you northwards sometime, do me the pleasure of coming to see me. I wish you and yours better days and a prompt return. Always your old

A. Einstein.

I sent Fr 200—write me upon receipt of it. Has Miss Rougea[8] received *her brother's*[9] postcards sent out a few days ago?

259.   To Paul Ehrenfest

[Berlin,] 14 September [1916]

Dear Ehrenfest,

Everything seems to be working out for the trip, to my great joy. Also thank Lorentz again on my behalf for his effort.[1]

I shall probably be able to travel in about 10 days.

With cordial greetings, yours,

Einstein.

260.   To Michele Besso

[Berlin,] 26 September 1916

Dear Michele,

Tomorrow I am going to Leyden to Ehrenfest. I'm very much looking forward to meeting our colleagues there and to the discussions. Nordström is there as well.[1] I have been accomplishing little recently but am living peacefully and contentedly; that's not bad either. When you are back in Zurich, write me your address there. For I do not consider myself justified in assuming that you will still be living at 10 Zehnder Lane. Did you find what you wanted in Berne? Did you see Spoerri?[2] He wanted to look you up. We shall see each other again in the spring, God willing. I must admit, though, that I hesitate to imagine myself traveling to Zurich again. My last trip resulted in my Albert no longer answering me.[3]

Affectionate greetings also to Anna and Vero, yours,

Albert.

261.   To Hans Albert Einstein

[Berlin,] 26 September [1916]

Dear Albert,

I am writing you now for the third time without receiving a reply from you. Don't you remember your father anymore? Are we not even going to see each other again? I hear with great relief from my friends Zangger and Besso that Mama is feeling better again. I wrote twice to your aunt in Novy Sad (Kissacer Alley). She wanted to visit you all.[1] If she is not yet with you, do write to her very soon. She is very attached to all of you and never gets any letters.

Kisses also to Tete from your

Papa.

Regards to Mama.

262.   To Wander and Geertruida de Haas

[Leyden, 3 October 1916]

My Dears,

Very touched by your letter, I am going to follow your good advice. Tomorrow (Wednesday) evening I shall come directly from Mr. Lorentz to you.[1] Then we can chatter away pleasantly all evening, and I can still go to the laboratory Thursday morning.[2] Thursday midday I have to be in The Hague, though, for a meeting.[3]

With cordial greetings, yours,

Einstein.

263.   To Hans Albert Einstein

[Berlin,] Friday, 13 October 1916

My dear Albert,

Your letter pleased me exceedingly. I received it this morning. Yesterday evening I returned from a trip to Holland where I was staying with Ehrenfest in Leyden. He asked many questions about you,[1] and we decided to go on a trip together with you once peace has returned. Be sure to write to your aunt again soon, because she is very worried about Mama and is not allowed to visit you now.[2] Because of my trip I was not able to send you the money yet. I am sending it tomorrow to Prof. Zangger, who is going to pick it up from the bank for you all. I am very glad that Mama is gradually feeling better; Prof. Zangger promises definitely that she will get healthy again.[3] But she has to have a complete rest for a long while yet. I am particularly glad that you both are already so understanding and independent that you can get by quite well with the maid.[4] But I am sorry that you are not taking piano lessons anymore. How did this happen? Don't you still enjoy it a bit? It is at least as important to me as what you are learning at school. You don't need to worry about marks. Just make sure that you keep up and that you don't have to repeat a year. But it's not necessary to have good marks in everything.

I miss you and Tete very much and am very eager to see you both again. When is Tete starting school? One is not allowed to wait too long, you know.[5]

Don't get anxious when you are alone with Tete. Although I am over here, you do have a father who loves you more than anything else and who is constantly thinking of you and caring about you. Send Mama my kind regards, and fond kisses for both of you from your

Papa.

I would like to see for myself how our little tyke plays chess!

264.   To Werner Weisbach

[Berlin,] 14 October 1916

Highly esteemed Colleague,[1]

Your new organization's leanings seem to me entirely above criticism.[2] I am convinced the malady of our times is that moral ideals have almost lost their force. In short, Bismarck-Treitschke personify the history of the ailment.[3] And now we are witnessing the crisis:

If Bismarck-Treitschke emerge with the glory of a foreign victory, then the world will be morally contaminated for an incalculable amount of time. Humanity will then have to endure an endless series of disloyalties and abominable acts of violence.

If Bismarck-Treitschke do not meet with success, however, people will lose faith in the empty ideal of power and will extend the principles of justice willingly and fairly to the states. Then our hotly pursued goal of an organization of states eliminating war (at least for Europe and America) will get implemented in a short time.[4]

I support this conviction publicly,[5] even though I am very aware that personal advocacy of this conviction can just have a weak influence on the masses (including professors!).[6] As depressing as it is, victories or defeats abroad are decisive; let us not delude ourselves about that! Therefore include me on your list, so that I am left with the consolation: *dixi et salvavi animam meam.*[7]

With best regards, yours truly,

A. Einstein.

[4]Deleted text in draft version: "This is my conviction. And yet, I hesitate to join this new association. This is the reason: Joining an association means offering up time. One must thus ask oneself whether this time will be well spent."

265.   To Carl Kormann

Berlin, 13 Wittelsbacher St. [15 October 1916][1]

Esteemed Sir,[2]

I am contacting you, without knowing you, because the following matter impels me to do so. In the household of my relatives, which I frequent daily,[3] a young woman, Miss Margarethe Telle, has been employed for sometime. The girl has won the family's confidence through her impeccable behavior; that is how she came to reveal to them her serious troubles, after we had already noticed for a long time how deeply depressed she was. She confessed to us that she has an illegitimate child and identified you as the father. The girl's proven truthfulness, the inquiries we made about her life *at that time*, her precise and graphic description of her experience in your villa, as well as the time of birth are adequate proof to me of the accuracy of her account.

It is completely incomprehensible to me and my family why the poor girl has not received what she is entitled to up to now. We are firmly resolved to help her obtain her rights from you. We have already applied to the Charlottenburg Council for the placement of your child and I ask you herewith, before I take further steps in the matter, whether you are inclined to declare yourself the father of the child.

Very respectfully,

Prof. A. Einstein
Memb. of the Roy. Pruss. Academy of Sciences.

266.   From Carl Kormann

Lichterfelde, Berlin, 109 I Carl St., 16 October 1916

To Professor Einstein, Berlin.

I consider the letter of 15 October[1] addressed to me as intimidation. Before placing your name and—what, in judging the situation, carries particular weight—your capacity as member of the Academy of Sciences at the foot of such a letter based on utterly false facts, you really ought to have considered properly the consequences of your line of action.[2] It cannot be unknown to you—unless out of gross negligence—that final legal decisions do exist by two courts.

When you nevertheless write me that the (contradictory) allegations are adequate evidence for you, that "you are firmly resolved to help obtain her rights with regard to these (contradictory) allegations"[3] and "you ask me herewith, before you take further steps in the matter, whether I am now inclined to declare

myself the father of the child," then I must come to this conclusion ⟨You want
to coerce me into this by stressing your official capacity⟩ and react by taking the
appropriate steps. I am fed up with continuing to suffer injury.

Respectfully.

267.   To Wilhelm Wien

[Berlin,] 17 October 1916

Highly esteemed Colleague,

Enclosed you will find a reply to a critical paper appearing in the *Annalen*
by Kottler with the request that it be included in the *Annalen*.[1] On the other
hand, I am not going to respond to Gehrcke's tasteless and superficial attacks,
because any informed reader can do this himself.[2]

With kind regards and best wishes for the new semester, yours truly,

A. Einstein.

268.   To Paul and Tatiana Ehrenfest

[Berlin,] 18 October 1916

Dear (P + T),

The reinvigorating days spent with you have melted into a beautiful dream
which I relive tirelessly in my imagination.[1] Now I am sitting in my room and
am battling with letters and papers that had been piling up, lying in wait for me
upon my return. Now it has been cleared up a bit. I have already written a reply
to Kottler (*Annalen* critique).[2] I have not yet spoken with Rotten,[3] nor with
any of the players of the chorale for 93.[4] Bach's are finer; in you I see developing
a new admirer of these magnificent things.[5] T also will join, despite the alarming
straightness of her intellectual world line (exception to the law of motion?).

Affectionate greetings also to the three little ones, yours,

A. Einstein.

Tell de Haas about the magnetic experiment. He may be interested in it.[6]

269.   To Paul Ehrenfest

[Berlin,] 24 October 1916

Dear Ehrenfest,

You are lucky to be able to play Bach and not have to wait until someone plays it to you.[1] But I am also lucky, because I can travel to Leyden where not only Bach but also all the other choice pleasures are bestowed upon me. Recently, while sitting innocently together with Planck and Rubens, I planted Lorentz's ideas into their heads, at which their previously normal faces fell noticeably. We shall see whether it will be of any help. In any case, Planck did receive the letter at the time.[2] I shall speak with Waldeyer and Diels as well.[3] I'm not going to visit them specifically, if possible, but shall try to speak to them at the Academy so that they do not gain the impression that a systematic assault is being made. When matters such as these are involved, these otherwise truly highly respectable men are inhibited, roughly like someone who had been given a command in a hypnotic state. In our case, this hypnosis happens during childhood. Tell Lorentz about it; I don't want to write to him myself before I have carried out his instructions *completely*. I have presented the principal parts of the general theory of relativity in a Hamiltonian manner now, as well, in order to show the link between relativity and the energy principle. I am submitting it Thursday.[4] Soon you will receive the correction proofs. Then show them to Lorentz and the other X-brothers over there. I wrote a reply to Kottler (*Annalen* paper).[5] Hume made a really powerful impression on me. Against him, Kant seems quite weak to me, but to save time I have given up maintaining this thesis.[6] I inquired about Gomperz at the bookshop.[7] It is a two-tome unfinished work of over 1,000 pages already, which would thus definitely require more energy than I have available to read it. Or were you perhaps mistaken about the title? Do think about it again. What you said about the matter made very much sense to me. Your observation about the semipermeable membrane is correct, of course; but I do not recall the conversation to which it relates, so I do not understand with what objective you are advancing your arguments. The trip to Holland did me an indescribable amount of good, physically and mentally; I am much more refreshed and cheerful. Solitude can be tolerated only up to a certain limit, you know.

Cordial greetings to all of your family, yours,

Einstein.

Best regards to Nordström, de Sitter, and Fokker.

270.   To Michele Besso

[Berlin,] 31 October 1916

Dear Michele,

In the interim, I had a lovely time in Holland.[1] The general theory of relativity has already come very much alive there. Not only are Lorentz and the astronomer de Sitter working independently on the theory but a number of other young colleagues as well.[2] The theory has also taken root in England.[3] I spent unforgettable hours with Ehrenfest and not only stimulating but also reinvigorating ones with Lorentz especially. In general, I feel incomparably closer to these people. Nordström, whom you know, of course, was also there. Zangger has been keeping me informed about my wife's condition and my boys' welfare. I'm very glad that she is recovering now after all, if only slowly.[4] I'll take care that she doesn't get any more disturbance from me. I have abandoned for good proceeding with the divorce.[5] Now to scientific matters!

The objective significance of space and time is primarily that the four-dimensional continuum is hyperbolic; so from each point there are "temporal" and "spatial" line elements, that is, those for which $ds^2 > 0$ and those for which $ds^2 < 0$. The x_r coordinates do not have a spatial or temporal character *per se*. To preserve the habitual way of thinking, you can prefer systems for which throughout $g_{44}dx_4^2 > 0$, $g_{11}dx_1^2 + 2g_{12}dx_1dx_2 \ldots g_{33}dx_3^2 < 0$. However, there is no objective justification for such a choice. Thus the "spatial" or "temporal" nature is real. But it is not "natural" for one coordinate to be temporal and the others spatial.

On Dällenbach:[6] The reduction of the Riemann tensor (single or double) does not result in the vanishing of the former. For it should be easy to demonstrate, in the case of the field of a mass-point at rest (external to the latter), that the

$$(ik, lm)\text{'s}$$

do not become infinitesimal even though

$$\sum_{kl} g^{kl}(ik, lm)$$

all vanish.

On Grossmann:[7] He was mistaken. The case of normal relativity is the case of vanishing curvature; more precisely: all the components of (ik, lm) vanish.

Definition of the tensors: Not "objects that transform in such and such a way." Rather: "objects that may be described, with reference to an (arbitrary) frame of reference, by a number of $(A_{\mu\nu})$ quantities, with a particular transformation law applying to the *latter*.[8] The independence from the reference system arises, in general, from the fact that the transformation law is known, specifically, that as a result of this law, all $A'_{\mu\nu}$'s vanish when all $A_{\mu\nu}$'s vanish. ($f \cdot dx_\nu$ is a first-rank tensor only when f is a scalar.)

Within the framework of the special theory of rel., covariants and contravariants do not differ from one another when $x_4 = ict$ is set. This is because the $g_{\mu\nu}$ tensor degenerates into

$$
\begin{array}{cccc}
1 & 0 & 0 & 0 \\
0 & 1 & 0 & 0 \\
0 & 0 & 1 & 0 \\
0 & 0 & 0 & 1
\end{array}
$$

Therefore,

$$ A^{\mu\nu} = \sum_{\alpha\beta} g^{\mu\alpha} g^{\nu\beta} A_{\alpha\beta} = A_{\mu\nu}. $$

The equivalence (duality) of the $g^{\mu\nu}$'s and $g_{\mu\nu}$'s is not a complete one, because the *extension* is *covariant* in character.

Your remark about the equivalence of measuring rods or clocks that are physic. different (and that have different prehistories) is quite right. But this condition also figures tacitly in Galilean-Newtonian theory.

The "Coriolis field" is given in first-order approximation by

$$ g_{14} = -\omega y $$
$$ g_{34} = 0 $$
$$ g_{24} = \omega x $$

The second-order approximation then results from the second-order terms from it, thus becoming of the type $g_{44} = \omega^2 r^2$, which is indeed in the form of a potential of the centrifugal force.[9]

It would be permissible to argue as you did regarding the generalization of relativity. The consequences regarding an induction effect of the dilation are also correct, no doubt. This approach has the disadvantage, however, of requiring that we proceed from *the universe as a whole*. Although, it is more favorable to proceed from one *portion* and to give the boundary conditions, as I have done for the equivalency hypothesis.

What you said about Mr. Dolder's paper is absolutely correct.[10] Only by making use of the observational data as a whole can the necessity for the condition of the principle of the constancy of light velocity be obtained. Naturally, Lorentz's ether can serve as a synopsis. Local time is not needed for Fizeau's experiment.[11] One examines the relation between $\mathfrak{e}$, $\mathfrak{h}$, and $\mathfrak{d}$ in a moving medium, which is resolved, by means of the Lorentz force, in the following way: $\mathfrak{p} = \mathfrak{d} - \mathfrak{e} =$

$(\varepsilon - 1)\left(\mathbf{e} + \dfrac{1}{c}[\mathfrak{v}, \mathfrak{h}]\right)$. If this is incorporated into Maxwell's equations, you obtain Fizeau's result through simple calculation.[12]

I don't know the paper by Cailler;[13] at least, I do not recall having read it. If you have it, show it to me the next time I visit you in Switzerland. Do also look after my boys, if you conveniently find time for it.[14] What's Vero doing? When is he going to fly from the nest?[15]

Warm regards, yours,

Albert.

You're going to receive a small paper from me soon on the basis of the general theory of rel., in which it is demonstrated how the rel[ativity] requirement is linked with the energy principle.[16] It's very amusing.

271. To Hans Albert Einstein

[Berlin, after 31 October 1916][1]

Dear Albert,

I am very glad that all three of you are happily together again now.[2] I heard from Mrs. Besso that you supported your mother like a grown man in the difficult times now behind you all. This makes me happy and proud; it is not through joys and pleasantness that a decent fellow develops, but through suffering and injustice. Your father's path was also not always strewn with roses like now, but rather more with thorns! Just have your mother [tell] you about the early days sometime. Also tell me about Tete. How does he look? Go out with him frequently into the fresh air, and you all should take his temperature often, so that we can take him away from Zurich again right away. I am probably coming to Switzerland in July and going somewhere at high altitude to relax. There we'll be able to be together with Tete; you have holidays in July, you know.[3]

Do write me a bit about school, music, and whatever else you are doing, whether you have made a closer friendship with another boy, and with whom else you are associating. You are now at an age in which the most important impressions of life are made. Later it all runs off you like water off a crocodile's back. Write me back soon and warm regards and kisses also to Tete from your

Papa.

272.   From Willem de Sitter

Leyden, 1 November 1916

Dear Mr. Einstein,

I am sending you today a separate offprint of a little popular exposition of the general theory of relativity which I have published in an English astronomical journal.[1]

I have been thinking much about the relativity of inertia and about distant masses,[2] and the longer I think about it, the more troubling your hypothesis becomes for me.[3] I mean the hypothesis that (a) at infinity the g_{ij}'s are such that the Minkowski cones become planes (i.e., three-dimensional flat spaces), and that (b) far beyond all known material bodies, as-yet-unknown masses exist which produce the effect that, in regions of space and time we know about, the special theory of relativity is valid in the absence of mass, thus that the Minkowski cones have a finite aperture. First, a question. As I understand the hypothesis, not only does it predict that the g_{ij}'s degenerate in the way mentioned for infinite values of the space variables $x_1 \, x_2 \, x_3$, but also for infinite values of the time variables x_4. Is this correct, or do the g_{ij}'s remain Galilean—or approximately Galilean—for $x_4 = \infty$ but x_1, x_2, x_3 finite? If I am right, the hypothesis would thus make the universe finite not only in space but also in time. We know nothing about the infinitely distant past and about the infinitely distant future—therefore, no observations can tell us that there has always been a universe and that there always will be a universe. It is not the finiteness, in principle, in space and time which bothers me, but the conviction that the boundary, the "envelope," will always remain hypothetical and will never be observed. Now we can say: the sources of inertia lie beyond the Milky Way, but when our grandchildren make an invention that enlarges the known world in the same proportion that it was enlarged 300 years ago through the invention of the telescope, then the envelope will simply have to shift farther outwards again. From this I conclude that the envelope is *not* a physical reality.

If the hypothesis is accepted, one would first want to get a [crude] idea of *where* these distant masses are and of what they are composed; second, *how* the inertia comes over here from there. An artificial mechanism will be invented. The coordinate system with reference to which the envelope and this mechanism are at rest will also be defined. Although the principle of relativity will still hold *formally*, effectively, we shall have the old absolute space with the ether back.

And another thing as well. At infinity, only transformations in which t' is just a function of t are permissible,[4] hence, no Lorentz transformations, for ex. Thus in the finite realm, no *exact* Lorentz transformations may be performed either, only one that, as far as our universe extends, coincides very precisely with a Lorentz transformation but degenerates into another transformation at infinity.

Naturally, all of this is *not* in conflict with the principle of relativity, I acknowledge. But if I am to believe all of this, your theory will have lost much of its classical beauty for me. With it an "explanation" of the origin of inertia is gained, which is actually not an explanation, for it is not an explanation from known or verifiable facts but from masses invented *ad hoc*.[5] I am convinced—but this is only a *belief* that cannot be proven, of course—that these masses will go the way of the "ether wind." New efforts will continually be made actually to observe them, but it will never work until we finally come to the conviction that they do not exist. [The stars and nebulae obviously are *not* part of the "envelope," because there the g_{ij}'s are still approximately Galilean; otherwise, we could not possibly identify spectral lines.][6] Is it not possible that, in the end, the explanation for inertia must be sought in the infinitely *small* rather than in the infinitely large? I am not a physicist, and this is probably just an entirely meaningless hallucination. But it is very hard for me to believe in the distant masses. I would prefer having *no* explanation for inertia to this one.

I would not dare to write you all this if I did not know, from the very pleasant hours we spent together, that you will not take it amiss. You know that it is only my deep admiration for your theory which impels me to do so.

With all good wishes, yours very truly,

W.dS.

273.   To Willem de Sitter

[Berlin,] 4 November 1916

Dear Colleague,

Your letter,[1] which puts me back in time to the pleasurable Leyden days,[2] I read with great interest and am looking forward to your popular paper in English, which will probably also be arriving soon.[3] I am sorry for having placed too much emphasis on the boundary conditions problem in our discussions. This is purely a matter of taste which will never gain scientific significance. I must add right away, though, that I *never* did think of a *temporally* finite extension of the world; also spatially, *finite extension* is not the issue.[4] Rather, my need to generalize just led me to the following interpretation:

Let it be possible to indicate a spatial envelope (a massless geometric surface) (in four dimensions, a tube), outside of which a gram atomic weight can have as little inertia as I please.[5] Then I can say that, within the envelope, the inertia is determined by the masses present there and *only* by these masses. A *specific* inertia-generating envelope is not assumed; rather, all inertia-generating matter will consist of stars, as those in the portion of our universe accessible to our telescopes. This is compatible with the facts only when we imagine that the portion of the universe visible to us must be considered extremely small (with regard to mass) against the universe as a whole. This view played an important role for me psychologically, since it gave me the courage to continue to work at the problem when I absolutely could not find a way of obtaining covariant field equations.[6]

Now that the covariant field equations have been found, no motive remains to place such great weight on the total relativity of inertia. I can then join you in putting it this way. I always have to describe a certain portion of the universe. In this portion the $g_{\mu\nu}$'s (as well as the inertia) are determined by the masses present in the observed portion of space and by the $g_{\mu\nu}$'s at the boundary. Which part of the inertia stems from the masses and which part from the boundary conditions depends on the choice of the boundary.

In practice I *must*, and in theory I *can* make do with this, and I am not at all unhappy when you reject all questions that delve further. On the other hand, you must not scold me for being curious enough still to ask: Can I imagine a universe or the universe in such a way that inertia stems entirely from the masses and not at all from the boundary conditions? As long as I am clearly aware that this whim does not touch the core of the theory, it is innocent; by no means do I expect you to share this curiosity! Have a look at the printer's proof I sent to Ehrenfest. The link between the relat. postulate and the energy conservation law emerges particularly clearly there.[7]

Cordial greetings, yours,

Einstein.

274. To Wilhelm Ostwald

[Berlin, 6 November 1916]

Highly esteemed Colleague,[1]

I thank you cordially for the paper on color theory,[2] which I am reading with enchantment for the second time already. Science is indebted to you for a significant advance here. I want to present it to our colleagues in Rubens's seminar;[3] they also will be delighted with it.

With respectful regards, yours truly,

A. Einstein.

275.   To Paul Ehrenfest

[Berlin,] 7 November [1916]

Dear Ehrenfest,

I have spoken now with *Waldeyer* (Secretary of the Academy) about Lorentz's suggestion.[1] He had *not* received the relevant letter.[2] He welcomed the suggestion *very warmly*. However, he considers implementation only possible after the conclusion of peace. Then he intends to support it, though. He wants to write to Lorentz immediately himself. I was very pleased with the man's reaction. Tell this to Lorentz. There is no doubt that he will write.[3] Read W. Ostwald's paper in the *Phys. Zeitschrift* on chromatics.[4] I liked it very much. Did you receive a correction proof on Hamilton's principle and relativity?[5] This is, I believe, exactly what you wanted. In my earlier portrayal with $\sqrt{-g} = 1$, the identity is established directly through calculation; here, it is presented as a result of the invariance.[6] It is nice that the Hamilton function $\mathfrak{G}^*$ does not get too complicated, even when the coordinate system is left general.[7] Droste's paper is beautiful; he calculates truly gracefully.[8] We still do not know whether the complete system with *finite* mass, which de Sitter detests so much, can be construed;[9] but it will soon become apparent in a relatively general case. Nothing is growing on my tillage at the moment; I am being lazy.

Cordial regards to you and your family, yours,

Einstein.

276.   To Hendrik A. Lorentz

[Berlin,] 13 November 1916

Highly esteemed and dear Colleague,

I am still living fully under the impression of the revitalizing trip to Holland.[1] It was not just meeting with highly esteemed men with similarly oriented goals that made this experience so refreshing, but especially, the concurrence in opinion on non-scientific matters. I have now spoken with Planck, Rubens, Waldeyer, and Nernst about the matter close to your heart,[2] and I can tell you, much to my delight, that I have had better success than I had expected. But I encountered a kind of skittish rejection by Planck and Rubens which, however, must not be ascribed to ill will but to a kind of timidity toward actions with the faintest political flavor. For you know yourself that the former is an extraordinarily scrupulous and upright person. Planck believes that he did receive the letter at the time but was not quite sure.[3] Rubens could not remember. I spoke with Waldeyer recently after an Academy meeting. He assured me that the letter had not come into his hands and spontaneously promised to write to you himself about the matter.[4] He considers execution feasible only after the war but wants to give it his full and dedicated support. He considers the suggestion reasonable and fair and hopes to gain, from an eradication of all confusion, an improvement in relations between scholars on the opposing sides. His honest stance, not stained by any utilitarian considerations whatsoever, encouraged me a great deal. Nernst also welcomes the suggestion. He does not know whether he has received the letter and admonished the unpractical manner in which the affair was staged. People are swamped with printed documents originating from strangers. The letter could therefore have wandered unread into the wastepaper basket with many others, for which neither he nor other persons ought to be legitimately reproached. He would consider it proper that such a suggestion come from locally known men or corporate bodies of neutral foreign countries. Then, no one would disregard the matter.[5] Prudence, rather than the unmitigated wish to be fair, leads him to approve. But this also is better than the absence of prudence *and* a sense of justice, as is usually found. I see best how necessary an objective examination of the state of affairs would be, incidentally, from the fact that Nernst provides me a *bona fide* account of the facts that differs in essential points from your picture.—I now think I should not speak with anyone else, since Nernst's opinion that the proposal ought to be revived in a more effective manner seems to me to be correct.–

I am sending you simultaneously with this letter a short paper in which I explained how, in my view, the relation between the conservation laws and the relativity postulate should be construed.[6] I made an effort to present the matter as succinctly as possible, free of all unnecessary trimmings. Particularly, I wanted to show that the general relativity concept regarding matter does *not* limit the variety of possible choices for the Hamilton function to a higher degree than the postulate of special relativity, since the conservation laws are satisfied by any choice of $\mathfrak{M}$. The selection made by Hilbert thus appears to have no justification.[7]

Cordial greetings to you, your wife, and your children from your

A. Einstein.

277.   To Paul Ehrenfest

[Berlin, 17 November 1916]

Dear Ehrenfest,

I remember that you showed me the wonderful experiment by Hering;[1] but I had forgotten that you had called my attention to Ostwald's paper.[2] My most stupid conduct concerned Brownian motion, though. Now I can't find your instructive letter anymore (likewise my house keys, so, to make matters worse, I am under house arrest); thus I do not know what I should ask Miss R.[3] God grant me more insight (for your letter); I can use it. I have reported my conversations with academicians to Lorentz.[4] On closer look at these people, all animosity dissolves. Lack of insight, sincere good will, but narrow-minded worship of false gods who send ruin.

Cordial greetings, yours,

Einstein.

278.   To Hermann Weyl

[Berlin,] 23 November 1916

Highly esteemed Colleague,

I am extremely pleased that you have welcomed the general theory of relativity with such warmth and enthusiasm. Although the theory still has many opponents at the moment, I console myself with the following situation: The otherwise established average brain power of the advocates immensely surpasses that of the opponents! This is objective evidence, of a sort, for the naturalness and rationality of the theory.

I make the following comment on your interesting presentations.[1] I also came belatedly to the view that the theory becomes more perspicuous when Hamilton's scheme is applied and when no restrictions are put on the choice of the frame of reference.[2] It is true that the formulas then become somewhat more complicated but more suitable for applications; for it appears that free choice of the reference system is advantageous in the calculations.[3] The connection between the general covariance requirement and the conservation laws also becomes clearer.[4] It turns out, though, that the Hamilton function to be used for the gravitational field, which gives the generally covariant equations, is not

$$\frac{H}{\sqrt{-g}} = \frac{1}{2} \sum g^{\mu\nu} \Gamma^{\alpha}_{\mu\beta} \Gamma^{\beta}_{\nu\alpha}$$

but[5]

$$\frac{H}{\sqrt{-g}} = \frac{1}{2} \sum g^{\mu\nu}(\Gamma^{\alpha}_{\mu\beta}\Gamma^{\beta}_{\nu\alpha} - \Gamma^{\alpha}_{\mu\nu}\Gamma^{\beta}_{\alpha\beta}).$$

The elements given by the second term drop out only when $g = -1$.

Unfortunately, even a preliminary statement of the Hamilton function for matter is still quite involved, so I prefer to do this only in special cases. Thus, for inst., your matter (in the literal sense) is nothing but infinitely fine, electrically charged dust. The reason for this is that you did not furnish your matter with surface tension or cohesion. Neither electrons, nor atoms, nor macroscopic matter can be described in such a way. Otherwise, I fully agree with your statements. I have not yet checked whether your numerical calculation concerning point charges is affected by the incorrect choice for H.[6]

Hilbert's assumption about matter appears childish to me, in the sense of a child who does not know any of the tricks of the world outside. I am searching in vain for a physical indication that the Hamilton function for matter can be formed from the φ_{ν}'s, without differentiation.[7] At all events, mixing the solid considerations originating from the relativity postulate with such bold, unfounded hypotheses about the structure of the electron or matter cannot be sanctioned. I gladly admit that the search for a *suitable* hypothesis, or for the Hamilton function for the structural makeup of the electron, is one of the most important tasks of theory today. The "axiomatic method" can be of little use here, though.[8]

Best regards, yours very truly,

A. Einstein.

I am pleased that you are associating with Dällenbach. He was my most capable student. With his sincere and noble manner, he is also likable as a person.[9]

279. To Hans Albert Einstein

[Berlin,] 26 November 1916

My dear Albert,

Your letter, which I received yesterday evening, pleased me greatly. I am very glad that Mama is home again with you now,[1] so you two are not all on your own; not to worry, her condition will slowly get better, and everything will go back to the normal routine. I also had to grind away at Latin in my day. It's not a bad mental exercise and doesn't hurt anyone.[2] But don't be ambitious, and don't think that you must speak it better than others. You should just not lose a year, because that's boring and wastes time. I shall send you the money for the life insurance; be sure to have it taken care of immediately so that there is no loss.

I was most pleased with your ship, though.   You should know that when I was your age this was my favorite hobby as well.[3] I really would like to have a closer description of it. Does the hull consist of two parts? Is it solid or hollow? How did you taper it at the bow? Is that which is sticking out downwards at the stern a rudder or only a weight? Where do you let it float?

I so much wish to see you both again. In the coming year there will surely be peace, so it will be easier for us to meet. When does Tete start school?[4] Do go to see Mr. Besso sometimes; you can learn many fine things from him, and he likes all of you.

Kind regards to Mama and kisses for you, my dear boys. Yours,

Papa.

Which Mozart sonata did you play for Mama? Can you note down a few bars for me?

280.   To Wilhelm Röntgen

[Berlin,] 29 November 1916

Highly esteemed Colleague,[1]

As far as I can observe, dear vanity is thriving so splendidly in our land that, in my view, there is no need for a new hothouse facility for its cultivation. It is in accordance with this conviction that I ask you courteously to dispense with my participation in this affair.[2]

In begging you, esteemed colleague, not to take my stance amiss, I am yours very truly,

A. Einstein.

281.   From Gunnar Nordström

Leyden, 30 November 1916

Dear Prof. Einstein,

I have been intending to write you for a long time, but all sorts of things got in the way. Here at Leyden we are still living in the aftermath of your visit.[1] What was discussed at that time is still continuing to be debated. For example, yesterday at the colloquium, Fokker spoke about the problem of the $g_{\mu\nu}$'s at infinity,[2] prompted by the second solution by Droste to the gravitational field of a central mass.[3]

I also have been working on your theory in the last few weeks, and I found that Herglotz's mechanics of deformable bodies (*Ann. der Phys.*, 36, p. 493)[4] can easily be generalized for your theory. For rest deformations e_{ij}, Herglotz's expressions (16′) are valid also in the general theory of relativity, but the A_{ij}'s are no longer defined by equations (20) but by the more general

$$A_{ij} = -\sum_{\mu\nu} g_{\mu\nu} a_{\mu i} a_{\nu j}.^{[5]} \tag{1}$$

The function $\Phi = \Omega(e_{ij}, \varepsilon) \cdot \sqrt{-A_{44}}$ in eq. (45)[6] thus becomes a function of the quantities a_{ij}, $g_{\mu\nu}$, and ε,

$$\Phi = \Phi(a_{ij}, g_{\mu\nu}, \varepsilon), \tag{2}$$

and because the a_{ij}'s and $g_{\mu\nu}$'s occur only in connection with A_{ij}, the relation

$$\sum_n a_{jn} \frac{\partial \Phi}{\partial a_{in}} = \sum_n g_{in} \left\{ \frac{\partial \Phi}{\partial g_{jn}} + \frac{\partial \Phi}{\partial g_{nj}} \right\} = 2 \sum_n g_{in} \frac{\partial \Phi}{\partial g_{jn}} \tag{3}$$

can be proven. Furthermore, for each φ function of the $g_{\mu\nu}$'s:

$$\sum_\sigma g_{\mu\sigma} \frac{\partial \varphi}{\partial g_{\nu\sigma}} = -\sum_\sigma g^{\nu\sigma} \frac{\partial \varphi}{\partial g^{\mu\sigma}}, \tag{4}$$

hence

$$\sum_n a_{jn} \frac{\partial \Phi}{\partial a_{in}} = -2 \sum_n g^{jn} \frac{\partial \Phi}{\partial g^{in}}. \tag{5}$$

If this equation is divided by the determinant

$$D. = |a_{ij}|,$$

Herglotz's expression (68) for the strain components is obtained on the left-hand side.[7] Because Φ and $\sqrt{-g}D$ are scalars, the strain tensor $-\mathfrak{T}_i^j$ thus found is a mixed volume tensor, and

$$-\mathfrak{T}_i^j = \frac{1}{D} \sum_n a_{jn} \frac{\partial \Phi}{\partial a_{in}} = 2 \sum_n g^{jn} \frac{\partial \mathfrak{F}}{\partial g^{in}}, \tag{6}$$

where

$$\mathfrak{F} = -\frac{\Phi}{D} = \text{volume scalar.} \tag{7}$$

To obtain your formula for $\mathfrak{T}_i^j$, one must thus set

$$2\mathfrak{F} = \mathfrak{M}. \tag{8}$$

Then your formulas agree with those by Herglotz.[8] From Hamilton's principle it can be deduced that the first expression (6) for $\mathfrak{T}_i^j$ really is valid in the general theory of relativity, by designating spatial variations to the mass-points, leaving the $g_{\mu\nu}$'s unchanged, however. From

$$\delta \int \mathfrak{F} \, dx \, dy \, dz \, dt = 0,$$

the differential equation[9]

$$\sum_j \frac{\partial \mathfrak{T}_i^j}{\partial x_j} + \frac{1}{2} \sum_{\mu\nu} \mathfrak{T}_{\mu\nu} \frac{\partial g^{\mu\nu}}{\partial x_i} = 0$$

is thus obtained, which contains nothing new, of course.– I hope you can see what the issue is from this short summary. I have written a brief communication about it and am going to publish it here in Holland.[10]

From inside information, I gather that the following four persons are being considered as the experts solicited for opinions on the filling of the professorship at the polytech[nic]: two Helsingfors professors Sundell and Melander, Bjerkén at the Stockholm Polytech., and Birkeland in Norway.[11]

I hope that my cousin Carl Hirn has come to see you.

I send you most cordial greetings and hope that we shall be seeing each other again next year. Yours,

Gunnar Nordström.

282.   To Paul Ehrenfest

[Berlin,] 4 December [1916]

Dear Ehrenfest,

I wrote to Knudsen a while ago and have received his *consent* in reply.[1] Unfortunately, according to Nordström's letter, he is apparently not being asked.[2] Should I write to anyone else about the matter? Lorentz has received my letter, I hope.[3] Has Waldeyer written to him? He had definitely promised to.[4] I recently spoke with Miss R. Prisoners are not permitted to send money abroad because of the exchange rate.[5] Maybe it would be best if you wrote concerning Nordstr[öm] and requested that I also be asked. Please send me your *Annalen* paper so that I can consider it at home at my leisure.[6] It is very quiet here. Even the Phys. Society's meeting next Friday has been called off because no one is here to give a talk. The little bit of reason has retired. One speaks, moves about, and eats little, and sleeps a lot. Cordial greetings to you and yours, as well as to Nordström, yours,

Einstein.

If I must write to someone quickly regarding Nordström, send a telegraph. I am reluctant, because I do not know the people. The opposite can too easily be achieved. It would be good if the gentlemen heard that Planck had suggested N. as second choice.[7] It is disadvantageous that the foreign experts appear to be pure technologists.–[8] Carl Hirn has not visited me yet.[9]

283.   From Michele Besso

Zurich, 5 December 1916

Dear Albert,

Yesterday evening Zangger gave me another account of Mrs. Mileva. The condition has worsened a bit again since two weeks.[1] She has to lie still again and is understandably discouraged by the return of the attacks after about five weeks respite.[2] It seems that the recent deterioration coincides temporally with a letter (from you?) that little Albert had supposedly received and that he did not want to show her. I could not find out anything more specific about it, without myself causing damage.– The attending physician wants to consult the neurologist Veraguth,[3] which Zangger approves of and which should be happening in the next few days. Zangger thinks no substantial change can be expected throughout the winter and hopes to be able to take her to Lugano in the spring, to her benefit.– Since coming home, she has, as Mrs. Zürcher[4] has told me, managed the household calmly and dependably from her bed, despite physical weakness, has great pleasure in the children, and the latter in their mother; she also directed little Albert's music studies with success, for example. A week ago today little Albert was here to visit us for half of an afternoon—we talked about all sorts of things, such as, about the natural wonders in the travel book on Celebes by the Sarasin brothers;[5] about an arithmetical problem, whereby it was revealed that algebra comes quite easily to him already—clear and alert in everything, which makes it a real pleasure.– Unfortunately, our meetings cannot be repeated as often as I would like, since one never knows how it affects the mother's condition.

–I have not thanked you yet for your latest paper on gravitation sent recently.[6] Now I have received 3 copies of it, all told, one of which I gave to Weyl and one to Dällenbach.[7] I must say for myself that I had not come as far as to perceive the gap in the system that is being filled by it; by contrast, Weyl obviously seems to have felt it, since the paper that he has transmitted to you[8] also deals in part with the relation between gravitation equations and the conservation laws.—In this paper by Weyl, it is demonstrated, as he told me, that the "finite circumference of a mass-point" (which I define for myself such that when the space is represented in a Euclidean manner—for masses at rest, this is meaningful also in the third dimension—the representations of the measuring rods are variable, that is, they become so small for the mass-point that the measurement figure approaches a finite limit) *vanishes* through electrical charging of the mass-point (that is, through a very low charge, $\dfrac{\varepsilon}{\mu}$, approx. $\dfrac{1}{20\,000}$ of this ratio for an electron). Is this connected to the fact that the negative gravitational energy of the mass point's field is counterbalanced by the electrical field energy (at this low charge already)? You are going to say: Lazybones, figure it out for yourself! But my thinking machine has become so resinous, ought to think out several things—and therefore refuses this enterprise, quite as a matter of principle. But this does not prevent me from telling others about what I do not know myself. Thus I want to offer an aperçu in the phys. colloquium on earlier attempts to explain perihelion motion,[9] likewise on the papers by Wiechert[10] and Flamm.[11] Regarding those explanation attempts, I have found interesting material by Zenneck on gravitation in the *Enzyclop. der mathem. Wiss.*[12] I have also thought about Gerber's idea:[13] It can be presented in a way that makes it appear entirely reasonable: The potential applicable to the moving point has a value corresponding to its location at a time sufficient for an effect to be able first to reach the Sun and to return from it to the planet in that interval.[14] Why Gerber identified this effect specifically with the potential and not with the force, for inst., or with an arbitrary function of the potential, is naturally not clear. It is not more unreasonable, though, than many other attempts to straighten out novel issues. On the other hand, it *appears* to me (see lazybones remark above!) that although the correct value should come out for Gerber, the opposite sign for the perihelion motion would have to result—at least if, as a rough comparison of the results suggests to me, with a potential of the form $\dfrac{K}{r}\left(1 + a\dfrac{dr}{dt} + b\left[\dfrac{dr}{dt}\right]^2\right)$, the result turns out proportional to the third term's coefficient.

For the Flamm paper also, I have to consider nothing but things I know nothing about. For an unalterable mass configuration, a practical elimination of space and time occurs again. Does light describe the straightest lines in the

Schwarzschild-Flamm space? Or is it rather (since one can clearly speak of an *infinite* velocity here) that geometry specifies for it the solution to the motion equations? *The latter seems* to me to be the case.– In mapping a spherical space on to a plane space, one can (or must?) proceed such that the diminution of the measuring rod images in the mapping is investigated as a function of the distance from an intersection point, for which the "apparent" circumference becomes $=$ $2\pi \sin \dfrac{\text{apparent radius}}{\text{spherical space radius}}$. Is an "isotropic" representation adequate at all, namely, can it be assumed that the dimensions of the measuring-rod picture are independent of its orientation with regard to the intersection point?–

–This mapping possibility ends for an "apparent" radius $= \dfrac{\pi}{2}$ spherical radius. And there my ability to vizualize it also ends. What happens when the density of the spherical mass or the size of the homogeneous sphere is even greater? Then spaces are involved, the content of which cannot influence the outside world, can it? The interior of nuclear spaces, the interior of the celestial cosmos?

Is the solution for the gravitation equations of a spherical shell known?–[15]

Not even with Wiechert am I quite sure about myself. I lectured unbendingly and firmly that it is quite unforgivable to assume an effect on gravitation by energy and then not immediately also assume that amount of energy corresponding to the equivalency of gravitational and inertial mass. And this is undoubtedly right. But what is not just as clear to me is, to what extent the inertial mass is given or suggested, even with the energy *without* the special theory of relativity; so that you would have another proportionality coefficient available *here* just loosely connected to the electron mass.

The gravitational influence of gravitational energy, or the latter itself, is ultimately only a counter for you, isn't it? On its own, it has no tensor properties; it is negative and therefore (?) has no place in your *empty* space. Or read *correctly*, does just this last paper of yours answer these questions otherwise?

Speaking of papers: I don't have your gravitational *waves* paper. Would you still be able to send me a reprint of it? The problem of whether the coordinate system can, in principle, be chosen so that the apparent solutions do not appear seems so very important to me!–[16] An influenza of sorts is plaguing Anna; she sends warm greetings. Still in expectation of it (the influenza), I am still tolerably well at the moment.– Affectionately yours,

Michele.

284.   To Constantin Carathéodory

Berlin, Sunday. [10 December 1916][1]

Dear Colleague,

I find your derivation wonderful. At first, a small slip of the pen, to be found on the second page, caused me difficulties. But now I understand everything. You ought to publish the theory in this form in the *Annalen der Physik*, since physicists usually know nothing of this subject, as was the case with me as well. With my letter,[2] I must appear to you like a Berliner who has just discovered the local Grunewald woods and asks whether people have ever been in it yet.[3]

If you would like to take the trouble to explain the canonical transformations as well, you will find a grateful and careful listener. If you solve the problem of the closed time lines,[4] though, I shall place myself before you with hands folded in reverence. . . . . Behind this is something worthy of the sweat of the best of us.

Best regards, yours,

A. Einstein.

285.   From Constantin Carathéodory

Göttingen, 31 Friedländer La., 16 December 1916

Dear Colleague,

The main point in the theory of canonical substitutions can, in my opinion, be most simply derived as follows.[1] If

$$\int_{t_0}^{t} L(x_k; \dot{x}_k; t)dt \tag{1}$$

is the Hamiltonian integral and we set

$$y_k = L_{\dot{x}_k} \qquad H(x_k, y_k, t) = -L + \sum_k y_k \dot{x}_k \qquad (k = 1, \ldots, n), \tag{2}$$

then the differential equations of mechanics read

$$\dot{x}_k = \frac{\partial H}{\partial y_k} \qquad \dot{y}_k = -\frac{\partial H}{\partial x_k} \qquad (k = 1, \ldots, n). \tag{3}$$

Thus

$$x_k = \bar{x}_k(\alpha_1 \ldots \alpha_{2n}, t) \qquad y_k = \bar{y}_k(\alpha_1 \ldots \alpha_{2n}, t) \tag{4}$$

are the general integral of (3) and $\alpha_1 \ldots \alpha_{2n}$ integration constants. I set

$$\int_{t_0}^{t} L\left(\bar{x}_k, \frac{\partial \bar{x}_k}{\partial t}, t\right) dt = \bar{\Omega}(\alpha_1 \ldots \alpha_{2n}, t), \tag{5}$$

or if I regard the α_j's from (4) as functions of x_k, y_k,

$$\bar{\Omega}(\alpha_1 \ldots \alpha_{2n}, t) = \Omega(x_k, y_k, t).$$

Similarly, if we write

$$\bar{H}(\alpha_1 \ldots \alpha_{2n}, t) = H(\bar{x}_k, \bar{y}_k, t),$$

then (with the aid of (3) and (4))[2]

$$\left.\begin{aligned}
\frac{\partial \bar{H}}{\partial \alpha_j} &= \sum_k \frac{\partial H}{\partial x_k}\frac{\partial \bar{x}_k}{\partial \alpha_j} + \frac{\partial H}{\partial y_k}\frac{\partial \bar{y}_k}{\partial \alpha_j} = \sum_k -\frac{\partial \bar{y}_k}{\partial t}\frac{\partial \bar{x}_k}{\partial \alpha_j} + \frac{\partial \bar{x}_k}{\partial t}\frac{\partial \bar{y}_k}{\partial \alpha_j} \\[2mm]
&= \frac{\partial}{\partial \alpha_j}\left(\sum_k \bar{y}_k \frac{\partial \bar{x}_k}{\partial t}\right) - \frac{\partial}{\partial t}\left(\sum_\mu \bar{y}_k \frac{\partial \bar{x}_k}{\partial \alpha_j}\right)
\end{aligned}\right\} \tag{6}$$

is the result. However,

$$\sum \bar{y}_k \frac{\partial \bar{x}_k}{\partial t} - \bar{H} = L(\bar{x}_k \bar{y}_k, t) = \frac{\partial \bar{\Omega}}{\partial t}.$$

Thus, instead of (6) one can write

$$\frac{\partial}{\partial t} \sum_k \bar{y}_k \frac{\partial \bar{x}_k}{\partial \alpha_j} = \frac{\partial^2 \bar{\Omega}}{\partial \alpha_j \partial t},$$

and from this follows

$$\sum_k \bar{y}_k \frac{\partial \bar{x}_k}{\partial \alpha_j} = \frac{\partial \bar{\Omega}}{\partial \alpha_j} + A_j(\alpha_1 \ldots \alpha_{2n}) \qquad (j = 1, \ldots n).$$

If to these n equations

$$\sum_k \bar{y}_k \frac{\partial \bar{x}_k}{\partial t} = \frac{\partial \bar{\Omega}}{\partial t} + \bar{H}$$

is added, they are then equivalent to the equation

$$\sum \bar{y}_k d\bar{x}_k = d\bar{\Omega} + \sum_j A_j d\alpha_j + \bar{H} dt. \tag{7}$$

On the other hand, however, the canonical differential equations follow from (7); for from (7)

$$\frac{\partial}{\partial t} \sum \bar{y}_k \frac{\partial \bar{x}_k}{\partial \alpha_j} = \frac{\partial^2 \bar{\Omega}}{\partial \alpha_j \partial t}$$

$$\frac{\partial}{\partial \alpha_j} \sum \bar{y}_k \frac{\partial \bar{x}_k}{\partial t} = \frac{\partial^2 \bar{\Omega}}{\partial \alpha_j \partial t} + \frac{\partial \bar{H}}{\partial \alpha_j}$$

follow, and from this[3]

$$\frac{\partial \bar{H}}{\partial \alpha_j} = \sum_k \frac{\partial \bar{y}_k}{\partial \alpha_j} \frac{\partial \bar{x}_k}{\partial t} - \frac{\partial \bar{x}_k}{\partial \alpha_j} \frac{\partial y_k}{\partial t} \qquad (k = 1, \ldots n).$$

In combination with the identity

$$0 = \sum_k \frac{\partial \bar{y}_k}{\partial t} \frac{\partial \bar{x}_k}{\partial t} - \frac{\partial \bar{x}_k}{\partial t} \frac{\partial \bar{y}_k}{\partial t},$$

these last equations can be written:

$$d\bar{H} = \sum_k \frac{\partial \bar{x}_k}{\partial t} d\bar{y}_k - \frac{\partial \bar{y}_k}{\partial t} d\bar{x}_k + \frac{\partial \bar{H}}{\partial t} dt,$$

from which the canonical differential equations can easily be drawn.

Hence, this theorem is valid:

The functions (4) are solutions of the canonical differential equations, in every case where an equation

$$\sum y_k dx_k = d\Omega + \sum_j A_j d\alpha_j + H dt \tag{8}$$

in which the $d\Omega$'s are a total differential and the A_j's are independent of t is valid.

From this theorem, everything else follows immediately:

I. *Canonical transformations.* The new variables

$$\left. \begin{array}{l} \xi_k = \xi_k(x_1 \ldots x_n, \, y_1 \ldots y_n, \, t) \\ \eta_k = \eta_k(x_1 \ldots x_n, \, y_1 \ldots y_n, \, t) \end{array} \right\} \tag{9}$$

are such that a function $\Psi(x_1 \ldots x_n, \, y_1 \ldots y_n, \, t)$ exists for which

$$\sum_k (\eta_k d\xi_k - y_k dx_k) = d\Psi + p(x_1 \ldots x_n, \, y_1 \ldots y_n, \, t) dt \tag{10}$$

is satisfied identically. Then you have according to (8), if you substitute x_k, y_k for the general solution of (3),

$$\sum_k \eta_k d\xi_k = d(\Omega + \Psi) + \sum_j A_j d\alpha_j + (H + p)dt \tag{11}$$

and therefore, if you introduce ξ_k, η_k as an independent variable and set

$$\mathrm{H}(\xi_k,\, \eta_k,\, t) = H(x_k,\, y_k,\, t) + p(x_k,\, y_k,\, t),$$

equations (3) transform themselves into

$$\dot{\xi}_k = \frac{\partial \mathrm{H}}{\partial \eta_k} \qquad \dot{\eta}_k = -\frac{\partial \mathrm{H}}{\partial \xi_k}. \tag{11^{a}}$$

II. If the y_k's can be eliminated from the n's of the first equations (9) and, consequently, the x_k, ξ_k's can be chosen as independent variables and we write

$$\Psi(x_k,\, y_k,\, t) = -\Phi(\xi_k,\, x_k,\, t), \tag{12}$$

it thus follows from (10)

$$\frac{\partial \Phi}{\partial \xi_k} = -\eta_k, \qquad \frac{\partial \Phi}{\partial x_k} = y_k, \qquad \frac{\partial \Phi}{\partial t} = p. \tag{13}$$

If, conversely, $\Phi(x_k,\, \xi_k,\, t)$ is an arbitrary function and if it is possible to eliminate the x_k's from the first equations (13) and the ξ_k's from the middle equations (13), then equations (13) deliver a canonical transformation for which

$$\mathrm{H} = \frac{\partial \Phi}{\partial t} + H \tag{14}$$

applies.

III. *Jacobi's method of integration.*[4] If Φ is a solution of the Jacobian differential equation, then according to (14),

$$\mathrm{H} = 0$$

and according to (11^{a}) for the solutions in the new coordinates,

$$\xi_k = \text{const.} \qquad \eta_k = \text{const.}$$

The entire theory is thus a consequence of the transformation that led to equation (8).

By the way, compare the explanation in Whittaker. *Analytical Dynamics*, p. 282 and f.[5]

With best regards, yours very truly,

C. Carathéodory.

286.   To Hermann Weyl

[Berlin,] 3 January 1917

Highly esteemed Colleague,

Many thanks for your recent kind letter. I have examined your earlier letter thoroughly as well and admired very much your elegant solution for the case of an electrically charged singular mass-point.[1] The issue of whether the electron can be treated as a singular point, or whether actual singularities are admissible at all, in the physical description, is anyway of the greatest interest. In Maxwellian electrodynamics, a finite radius was chosen to explain the electron's *finite* inertia or to obtain a finite energy for the electron.[2] They did *not* want to tolerate the integral

$$\lim_{\varepsilon=0} \left\{ \int_{r=\varepsilon}^{r=0} (\text{energy density}) d\tau \right\} \text{ becoming } = \infty.$$

It would be very interesting to see how this comes out in your solution. I have already made some efforts in this direction but, with my unreliable calculation work, have not yet arrived at any secure results. The question is whether

$$\int_{r=\varepsilon}^{r=\infty} (\mathfrak{T}_4^4 + \mathfrak{t}_4^4) dV \qquad \text{(three-dimensional)}$$

in the limit for $\varepsilon = 0$ becomes finite or infinite. Owing to the field equation[3]

$$\frac{\partial}{\partial x_\alpha} \left(\frac{\partial \mathfrak{G}^*}{\partial (g_\alpha^{\mu\sigma})} g^{\mu\nu} \right) = -(\mathfrak{T}_\sigma^\nu + \mathfrak{t}_\sigma^\nu),$$

this leads us to the question of whether the contribution[4] of the quantity $r^2 L_\alpha$ becomes infinite or not for $r = 0$, where

$$L_\alpha = \frac{\partial \mathfrak{G}^*}{\partial g_\alpha^{44}} g^{44}.$$

The computation will tell us whether the singularity is to be understood as the carrier of either an infinitely large, finite, or a *nonexistent* mass. Only if the latter applies do I believe that the point electron must be taken seriously physically. It would then have no singularity, *energetically*. If you do the calculations, please let me know the outcome.

I have just conveyed your greetings to Freundlich & ask you to give Dällenbach my best regards. I am very glad that he is able to benefit from your suggestions.[5]

Cordial regards and best wishes for the New Year, yours,

A. Einstein.

287. To Hans Albert Einstein

[Berlin,] 8 January 1917

My dear Albert,

Your last letter pleased me greatly; it was very nice and detailed. Now I can imagine the ship well.[1] It was also my favorite hobby at your age. I did not have such a fine body of water at my disposal, though, just a round metal drum, 1 m in diameter. That's why the ships had to be small. What did you do with the Christmas money? The Mozart sonata you have been playing is a pretty one; I know it well. Do keep to Mozart sonatas; your Papa also became properly acquainted with music through them.[2] I am glad that now Mama is feeling better again;[3] you may always show her what I write you.[4] I am coming to Switzerland again in the spring and am eager to see you both. I intend to stay at a small inn close to Zurich and have you both stay with me for a couple of days. We might make another small excursion again then, like last year.[5] I am not going to stay in Zurich itself. Do you have a lot of work to do for school? If you have a particular interest in a subject, I'll send you an appropriate book. If you have such wishes, write me about it.

Greetings and kisses also to Tete from your

Papa.

Kind regards to Mama.

288. From Alexander Moszkowski[1]

[Berlin,] 5 Fasanen St., 18 January 1917

Most highly esteemed Professor,

A quite voluminous book authored by me and Artur Fürst recently appeared, *Das Buch der 1000 Wunder* [*The Book of 1000 Wonders*] (published by Albert Langen), which in some parts also covers the wonders of theoretical physics.[2] The rel[evant] sections were written by me; I have made an effort to convey to a wider audience, in a very popular form, some facts about your major achievements. I know very well that I have only brushed the surface with it, but at least a vague idea is given to a considerable readership who would otherwise never hear about the topic. It goes without saying that I did not miss this opportunity to pay tribute to you.[3]

I presented the rel[evant] sections to Dr. Fritz Reiche before it went into press, to be sure that the topic itself is error-free, despite the light treatment. To my relief, I received no censure from him.[4]

I considered it my duty to notify you of this publication; I do not enclose the book, however, because that could look like an attempt to press you to read it. I am satisfied if you are kind enough to devote this moment to me.[5] As things stand, I may not hope for more—or perhaps I may? Sometime in the future?

In expressing my deepest admiration, I am most sincerely and devotedly yours,

Alexander Moszkowski.

289.   To Georg Nicolai

[Berlin, ca. 22 January 1917][1]

Dear Nicolai,

After you had left, I continued to mull over your plan in my head.[2] I consider the affair entirely hopeless and apt to result in much trouble and disappointment for you. Imagine for yourself how we are going to look to the people whom you want to approach now if the whole business misfires. Consider how valuable relationships will suffer by it! We must not let ourselves be deceived about the prospects of the enterprise. In these agitated times, a tiny few concern themselves with writers of the past; the latter remain undisturbed in the libraries. Added to this, I believe that Mr. Buek[3] ought not to be burdened with a task to which even a more energetic person would not be equal; agreeable though the man is otherwise, with his shy and meek demeanor, he is absolutely unsuited to execute such a mission.

Do not be annoyed by this frank expression of my opinion. I feel so firmly convinced, that silence would be disingenuous.

Cordial greetings, yours,

A. Einstein.

290.   To Willem de Sitter

[Berlin, 23 January 1917]

Dear Colleague,

It is a fine thing that you are throwing this bridge over the abyss of delusion.[1] You will receive the requested paper, along with some others for our colleague, simultaneously with this postcard.[2] When peace has returned, I shall write to him. If you write him yourself, please tell him that his intuition about my paper of 1914 has not deceived him; the following two errata were in it:

1) The consideration in §12 is incorrect because the events can be determined uniquely, without the functions describing them being unique.[3]

2) In §14 at the top of page 1073 there is a flawed consideration.[4]

I noticed my earlier mistakes by calculating directly that my former field equations were *not* satisfied for a rotating system in a Galilean space.[5] Hilbert detected the second error as well.[6]

Cordial greetings, yours truly,

A. Einstein.

291.   To Władysław Natanson

[Berlin,] 28 January 1917

Dear Colleague,

I return most cordially your good wishes for the now not quite so new year. May more reason and goodwill develop among people than in the last few years. I am also sending you a couple of short papers.[1]

With cordial greetings to you and your family, yours,

A. Einstein.

292.   From Alexander Moszkowski

[Berlin,] 5 Fasanen St., 1 February 1917

Most highly esteemed Professor,

Your kind words occasioned by the book on wonders[1] made me overjoyed; I thank you heartily!

From time to time I have made a number of attempts at describing certain things connected with the princ. of relat. for our large daily newspapers, that is, of necessity, in popularized reduction. Since these matters have met with a positive response by the public, the editors approached me about covering the topic more thoroughly; but then I had to deal with the concern about whether the descriptive devices employed in my journalistic capacity could suffice. An informal cue from you, the master, on this issue of conscience would be very valuable to me. Your amicability gives me the courage to bring this up, the more so since months ago you sent me away with the promise of allowing me another visit to your flat.

May I therefore deliver my other book, which has just appeared,[2] personally with the prospect of a brief meeting?

In this case, I request, most respectfully, notification by you of the appointment time by telephone. Otherwise, I would permit myself to send you this book in the mail, like the previous one.

Regardless of what happens, I would like to continue the "cult"; for you it is secondary, for me it is of paramount importance in life.[3] Additionally, I have the encouraging feeling that, with my modest writing abilities, I may also serve the cause once in a while.

In expressing my greatest admiration, I am truly and devotedly yours,

Alexander Moszkowski
5 Fasanen St.
Telephone: Steinplatz office, No. 10185.

293.   To Willem de Sitter

[Berlin,] 2 February 1917

Dear Colleague,

You will receive the requested paper simultaneously with this postcard; I am very happy that you have taken an interest in it. The P[otsdam] appointment affair has come to a standstill, which seems very suspicious to me. An intrigue must be at play.[1] I shall naturally hear nothing about it until it is too late. Presently I am writing a paper on the boundary conditions in gravitation theory.[2] I have completely abandoned my idea on the degeneration of the $g_{\mu\nu}$'s, which you rightly disputed.[3] I am curious to see what you will say about the rather outlandish conception I have now set my sights on.[4]

With cordial greetings, yours,

A. Einstein.

294.   To Paul Ehrenfest

[Berlin,] 4 February 1917

Dear Ehrenfest,

The paper by Mr. Burgers delighted me.[1] The adiabatic method and the Schwarzschild-Epstein method now mutually support each other.[2] It is a fine success; the paper is also very well written. I have perpetrated something again as well in gravitation theory, which exposes me a bit to the danger of being committed to a madhouse.[3] I hope there are none over there in Leyden, so that I can visit you again safely. It is a pity that we do not live on Mars and just observe man's nasty antics by telescope. Our ⟨Lord⟩ Jehova no longer needs to send down showers of ash and brimstone; he has modernized and has set this mechanism to run on automatic. Cordial greetings, yours,

Einstein.

295.   From Max Planck

Grunewald, 4 February 1917

Dear Colleague,

Now that I have gained a bit more clarity on the dynamic significance of the Jacobian action integrals, I would like to report on it to you.[1] Let E be the energy as a function of the coordinates q_1, q_2, ... and of the momenta p_1, p_2, ...; and if we set in this function

$$p_1 = \frac{\partial W}{\partial q_1}, \qquad p_2 = \frac{\partial W}{\partial q_2}, \ldots \tag{1}$$

the Hamilton-Jacobi differential equation is then:

$$\frac{\partial W}{\partial t} + E = 0.$$

The same is satisfied by

$$W = -\alpha_1 t + V,$$

where V is a function of q_1, q_2, ... and of the constants α_1, α_2, ..., among which the first α_1 represents the value for E. Then the $2n$ motion equations are, first, equations (1); second, the equations:

$$\frac{\partial W}{\partial \alpha_1} = -t + \frac{\partial V}{\partial \alpha_1} = \beta_1, \qquad \frac{\partial W}{\partial \alpha_2} = \beta_2, \ldots \tag{2}$$

(This in preparation)

Now,

$$E = \sum p_1 \dot{q}_1 - H \qquad (H = T - U),$$

consequently:

$$H = \sum \frac{\partial V}{\partial q_1} \dot{q}_1 - E$$

and, if you integrate over t from 0 to t:

$$\int_0^t H \, dt = V - V_0 - Et = W - W_0.$$

The integral $\int_0^t H \, dt$ is thus *not* $= W$, that is, equal to that function which occurs in motion equations (1) and (2). In fact, equations (2) would become incorrect if $\int_0^t H \, dt$ were substituted for W. For W_0 also depends on the constants α_1, α_2, ...

Until we meet again on Wednesday at the colloquium![2] Yours,

Planck.

296.   From Moritz Schlick

Rostock, 23 Orléans Street, 4 February 1917

Esteemed Professor,

On the occasion of my last visit, you were so kind as to declare your willingness to look over an essay on relativity that I had promised to submit to the *Naturwissenschaften*. Overwork and other interruptions prevented me from completing the essay until now, but I have finally gotten around to it after all and take the liberty of sending you the manuscript now with the cordial request that, time permitting, you subject it to your scrutiny.[1] I would be exceedingly grateful to you if you would draw my attention to deficiencies in the paper, and request that you kindly point to any errors, inaccuracies and anything else (with comments on the reverse side of the sheets, perhaps). The topic had been proposed in that form by the editors, as the caption indicates,[2] and thus the essay is less an explanation of the general theory of relativity itself than an in-depth commentary of the theorem that space and time have now lost all trace of concreteness in physics.[3] My main goal was to make the explanation as easy to understand as is possible; whether this has succeeded to the degree intended seems questionable to me, of course. It really is very desirable that the ideas of the general principle of rel. be known and understood everywhere very soon, not merely for physical but also especially for philosophical reasons—and I would consider myself privileged if the essay could contribute tangibly toward this. Since it truly does involve promoting the subject, I also do not hesitate to make use of the permission you had granted earlier and to present you with the paper for evaluation prior to publication. Thus I await your verdict and remain, with best wishes for your well-being and in greatest respect, yours very truly,

M. Schlick

P. S. If, contrary to expectation, no significant changes are needed on the manuscript, may I possibly ask that in this case you be so good as to send it on directly to the editors of the *Naturwissenschaften*, 23/24 Link Street?

297.   To Moritz Schlick

Berlin, 6 February 1917

Esteemed Colleague,

Your exposition is of matchless clarity and perspicuity.[1] You did not dodge any problems but took the bull by the horns, said all that is essential, and omitted all that is unessential. Whoever does not understand your exposition is totally incapable of grasping such trains of thought. I liked particularly that you did not present the general theory of relativity *a posteriori* as epistemologically *necessary* but only as highly *satisfactory*. This incorruptibility pleases me especially. I have absolutely nothing to criticize but can only admire the pertinence of your way of thinking and expression. I am sending the paper back to you, nevertheless, because of a small inaccuracy on each of pages 27 and 28, which must still be corrected.[2]

Even your essay on the special theory of relativity is outstanding.[3] Do you still have copies of it? Unfortunately, the one you had given me has gone astray on loan, and I would very much like to have it among my possessions. May I be so forward as to ask you for 2 or, if possible, 3 copies of this new paper of yours? I would like to let my friends in Zurich have one.[4]

Best regards, yours,

A. Einstein.

298.   To Paul Ehrenfest

[Berlin,] 14 February 1917

Dear Ehrenfest,

Very, very unfortunately, I can't come this time, as much as I would like to be present at your celebration[1] and to see all of you again. For I am quite infirm from a liver condition,[2] which imposes upon me a very quiet lifestyle and the strictest diet & regimen. This protracted thing was the cause of my constant sickly appearance.[3] Tell Lorentz and de Sitter directly, so that they do not plague me in vain with letters. De Sitter's condition concerns me; please write me about it in a bit more detail. I hope it does not involve a tuberculous infection.[4] I am sending you my new paper.[5] My solution may appear adventurous to you, but for the moment it seems to me to be the most natural one. From the measured stellar densities, a universe radius of the order of magnitude 10^7 light years results, thus unfortunately being very large against the distances of observable stars.[6] The odd thing is that now a quasi-absolute time and a preferred coordinate system do reappear in the end, while fully complying with all the requirements of relativity. Please show the paper also to Lorentz and de Sitter. Has Lorentz received the letter by Waldeyer?[7]

Warm greetings also to your family, yours,

Einstein.

299.   To Walter Dällenbach

[Berlin, after 15 February 1917][1]

Dear Dällenbach,

Your observations are, in my view, to a high degree legitimate.[2] Strictly speaking, even the concept of the ds^2's evaporates into an empty abstraction, in that ds^2 cannot be construed strictly as a measurement result, not even in the absence of electromagnetic fields. You have indicated the reasons for this entirely correctly. Even so, in a reasonable didactic exposition of the theory, ds^2 will be treated as though it were strictly measurable. The issue here runs analogously to that in electrical science, where the definitions for $\mathfrak{e}$ and $\mathfrak{h}$ are given, even though these definitions do not hold out against strict criticism.

A logically more satisfactory description is obtainable (*a posteriori*) by relating the theory's more complex individual solutions to observed facts. A standard could then be correlated with a certain type of atomic system that could not claim a privileged position in the theory. Thus a four-dimensional continuum *can* still be maintained and, in upholding the postulate of general covariance, it then has the advantage of circumventing the arbitrariness in the choice of coordinates.

However, you have also understood correctly the disadvantage attached to a continuum. If the molecular interpretation of matter is the correct (practicable) one, that is, if a portion of the world must be represented as a finite number of moving points, then the continuum in modern theory contains *much too* multifarious possibilities. I also believe that this multifariousness is to blame for the foundering of our tools of description on quantum theory. The question seems to me to be how one can formulate statements about a discontinuum without resorting to a continuum (space-time); the latter would have to be banished from the theory as an extra construction that is not justified by the essence of the problem and that corresponds to nothing "real." But for this we unfortunately are still lacking the mathematical form. How much I have toiled in this direction already![3]

Indeed, I see principal difficulties here as well. Electrons (as points) would be final conditions (building blocks) in such a system. Do final building blocks exist at all? Why are they all similar in size? Is it adequate to say: God in His wisdom has made them all the same size, each like any of the others, because He so wished? If it had suited Him, He could also have made them dissimilar. Thus we are better placed with the continuum conception, because elementary building blocks to not have to be provided from the outset. The old question of the vacuum notwithstanding! But these reservations must pale against this glaring fact: The continuum is more comprehensive than the objects to be described . . . .

Dear Dällenbach, what help is all this argumentation, when we do not come through with a satisfactory interpretation? But this is devilishly difficult. There will have to be a hard struggle before the step we have in mind is actually taken. So, tax your brain, maybe you can master it.

I am sending all of you herewith a copy of the quantum paper[4] and one on the cosmic gravitational field.[5] This latter subject is a bit daring but, without a doubt, worth consideration.

Best regards, yours,

Einstein.

P. S. In principle, it certainly is correct that the local system defining the ds's is chosen from such motion constraints as cause the gravitational field forces to vanish. For we obviously do not know whether the field forces $\Gamma^\sigma_{\mu\nu}$, or acceleration, do not (in principle) affect the yardsticks and clocks directly. Best regards to Wohlwend.[6] I am very pleased that he wants to write me. ⟨I do not yet know when exactly I'll be traveling to Switzerland. (Healthwise, I am feeling decidedly better). In the spring or⟩ In the summer I shall definitely come to see you in Switzerland; traveling is too repugnant now, especially for a rickety old fellow.[7]

300.   To Erwin Freundlich

[Berlin,] Sunday. [18 February 1917 or later][1]

Dear Freundlich,

I have been turning the elliptic space problem over in my mind and understand it completely. My calculation remains correct, in particular, the relation between λ, ρ, and R; only, the total volume is half as large as for the originally spherical space.[2] The elliptic space is simply a spherical one in which points that are symmetric with respect to the center are identical (not distinguishable). Put another way,

Let G and G' be two geodetic lines that start at P. These same lines intersect each other at P'. According to spherical geometry, P and P' are opposing points; according to the elliptic one, they are identical. An elliptic space can aptly be described as "hemispherical."

The star statistics question has become a burning issue to be addressed now.[3] I suggest that we discuss the problem in a joint paper. If it is necessary or beneficial, I shall seek to have you granted time off from your duties for a specified period. We must act cautiously in doing this, though, so that your position is not jeopardized.[4]

The matter of great interest here is that not only R but also ρ must be individually determinable astronomically, the latter quantity at least to a very rough approximation, and that then my relation between them ought to hold. Maybe the chasm between the 10^4 and 10^7 light years can be bridged after all![5] That would mean the beginning of an epoch in astronomy. The paper by Harzer is very interesting. If only it were known what should be done with the loathsome absorption; it is so abominably up in the air.[6] But managing without this quantity is unfortunately impossible.

Best regards, yours,

A. Einstein.

301. To Kathia Adler

Berlin, 20 February [1917][1]

Dear Mrs. Adler,[2]

Even without the letter from your brother, I had the strong urge to write you but did not know your address. Your misfortune and that of your husband, whom I have always admired, touched my heart deeply like little else that I have witnessed in this hard world.[3] He is one of the most splendid and purest men I have ever known.[4] I cannot judge his deed, since I cannot assess the motive behind it; I do not believe him capable of rash acts; he is much too conscientious for that. If there is something you think I can do for him or for you, think of me and write me. I am sending you in the same post the little paper on Mach.[5] It is not particularly good; my style is heavy and wooden, my knowledge of the literature poor. Prof. Ph[ilipp] Frank at the German University in Prague has published an excellent essay on Mach (in *Naturwissenschaften*).[6] He will surely gladly send you a copy. If he has none to spare, I am willing to lend you mine, if I can find it in the chaos of my papers.

Warm regards, yours,

A. Einstein.

Please convey my cordial greetings to your father-in-law, whom I met in Vienna.[7]

302. From Georg Nicolai

Langfuhr, Danzig, 133 Haupt Street, 26 February 1917

To Prof. Albert Einstein
Wilmersdorf, Berlin
13 Wittelsbacher Street

Esteemed Professor,

I come back to the publisher affair again[1] for two reasons; for one, because this is of practical importance, as you will gather from the following, and secondly, because I would not like to have this matter, in which obviously some misunderstandings have played a part, remain as a permanent impediment between us.

I was prompted to do this by a letter that Mr. Rudolf Moos had directed to one of my financial backers, Captain of the Cavalry Kellner, in response to his inquiry about where he in fact stood on the matter now.[2]

Mr. Moos writes, in accordance with the facts, of course, that he had followed your own decision in agreeing to participate in the publishing house[3] and had also encouraged a friend to participate, and that he had been willing to assist in its creation, both actively and through offering advice, as far as his time permitted. Then, after having had three meetings with us, as he adds verbatim, "some days later, Prof. Einstein let me know that he had reconsidered the publisher matter again thoroughly and had come to the conclusion to withdraw his participation because he did not believe that the envisaged goal was attainable in the manner planned by Prof. Nicolai. Thus my reason for being involved in the affair also fell away."

He then continues: "A few days later I found out that Prof. Nicolai had been informed by Elsa Einstein about this new state of affairs."

To this I would like to note that I have been informed neither by Elsa Einstein nor by you yourself about your having withdrawn participation, but only that you had serious doubts about the plan.[4] To the contrary, upon receiving your cousin's letter, I asked you explicitly whether you were perhaps intending to withdraw; and you did not give me an answer to this direct question but in response merely listed the misgivings you had about the plan in general. Be this as it may, you could now state that you are withdrawing your involvement and then, although the preceding events would not have resulted differently, the matter would nevertheless be clarified for the future.

Before you reach your decision, though, I would like to clarify my position in this affair. I do not dispute for a single moment your right to withdraw from a contract that you have perceived as improper, but I do dispute your right to do this without having tried simultaneously to do everything not to inflict superfluous harm additionally to the other party.

You know very well that during my last brief stay in Berlin,[5] after having straightened out the matter with Moos (that is, specifically with Moos at your or your cousin's instigation), I could justifiably assume the publishing house matter settled and therefore did nothing since about it. You know, furthermore, that from here I am hardly in the position to place the publishing house matter on another foundation; you must thus also concede that, if Moos, who in following your example had consented, is now led by you to withdraw (to which he has full right, since he never gave a binding promise), you would be to blame if the entire arrangement possibly fell flat.[6]

But to this you would have the right, in my opinion, only if upon your initial consent you had either been deceived by me or you could look upon the entire plan not only as an unwise plan, as you write, but as an unethical plan. I am convinced, however, that you do not believe either of these.

Hence I believe I can base myself not only on the right of friendship but also on a much broader and more general right when I make the request that you continue to support the plan, or at least not harm it, even if you consider the plan itself unwise, which, incidentally, all businessmen involved deny, who ought to be more competent judges in this regard than the two of us.

Mr. Moos himself is not entirely adverse to participating despite your advice against it. He writes further: "Nonetheless, or perhaps precisely because of this, I do not have the heart to turn Prof. Nicolai away now upon your inquiry. But I am not yet clear about how he can be helped and what I personally can contribute."

You see from Mr. Moos's words that the matter does indeed depend on you now, and I leave it entirely to you to find a way you consider appropriate. It may well be that, for some reasons unknown to me, the consequences of your kind consent at that time are unpleasant to you today, but you will admit that I am not to blame for that. In any case, I naturally do not want you to continue to be involved personally in the publishing house; I think as matters now stand, that would not be good and even superfluous, because it seems to me that Mr. Rudolf Moos would still participate if you told him that although you yourself did not want to be involved because you do not expect much to come of the business, you do consider the matter itself desirable and worthwhile.

Otherwise, I would like to inform you that the manuscripts on Kant, Herder, Fichte, and Jean Paul are ready for press. I truly believe, if you were to read them, you would be delighted with them and really would support conveying knowledge of these works to the German people.

This objective knowledge about what our classical authors thought about Germany's politics, though under entirely different circumstances, can surely only be useful. It is irrelevant what view one may hold otherwise, so long as truth is seen as a goal worth seeking, and this I believe we both do.

With best regards also from my wife.[7]

303.   To Georg Nicolai

[Berlin,] 28 February 1917

Dear Nicolai,

Nothing is more difficult than turning Nicolai down.[1] The man, who in other things is so sensitive that even grass growing is a considerable din to him, seems almost deaf when the sound is attached to a cancellation. A lame excuse of science in face of this enigma: "attentiveness."

Thus I raise my voice with the force of a bullock just come of age and cere-
moniously, fervently, and energetically call (bellow) it off herewith. (The music
ceases, a pause of two beats, followed by elegiac piano.)

Reason and justification. The inviolable human rights include: a person may
ride a respectable number of hobby-horses, as meets his temperament. But also
included among these human rights is the right to refuse accompanying our dear
fellow-man on his hobby-horse ride. Therefore, you have the plain right to sow
a lot of old nonsense, fertilize it with modern hogwash, and sell it personally on
the market. I have the right to keep my distance and to devote myself entirely
to my own hobby-horses. That is how it should and must remain.

In requesting that you bear this constantly and consistently in mind, best
regards, yours,

A. Einstein.

304. To Georg Nicolai

[Berlin, after 28 February 1917][1]

Dear Nicolai,

On rereading your letter, I regret the unappealing image I have evoked in
a playfully emotional outburst about the endeavor to publish the venerable authors.[2] You should just take this as a coarse joke. To the point, I add that there
is no question of any harm coming to the enterprise through me. I am alienating
no one from the venture who would be involved in it without me. I just hope
that Mr. Moos does not participate *solely for my sake*;[3] such a personal sacrifice
on the part of a relative and friend would be awkward for me.

Kind regards also to your wife from your old

A. Einstein.

305. To Walther Rathenau

[Berlin,] 8 March 1917

Highly esteemed Dr. Rathenau,[1]

I gratefully accept your invitation for Sunday evening but shall only appear
between 8 and 9 o'clock, owing to my delicate constitution.[2]

I was delighted with your book and have already read through it carefully.[3]
What pleased me most was the mentality of representing what a good person

wants as an end in itself. This should take precedence over everything else and needs no threadbare theories in its support. Then I saw with astonishment and joy how extensive a meeting of minds there is between our outlooks on life (even including your estimation of professors). If I must stay somewhere in the opposition, it is in the far-reaching function you assign to the state. I am convinced that the representatives of economic interests should have no military weapons at hand. If with the great new debt burden of the independent states this goal cannot be attained otherwise, I should prefer by far general state bankruptcy. On these grounds, absolutely no legitimacy can be ascribed to the existence of supra-regional states. The state seems justifiable to me only as a supporter of institutions for the common weal, such as hospitals, universities, police, etc. That is why I do not understand why states larger than the province of Brandenburg could be desirable. In my opinion, only in small districts can a nation-state have lasting stability. Switzerland is a model to me in this regard, albeit there the individual states are so small that they are barely able to master the functions indicated above.—I know very well, though, that the world will not take shape according to my wishes!

Best regards, yours truly,

A. Einstein.

306. To Michele Besso

[Berlin,] 9 March 1917

Dear Michele,

It is nice that you took my Albert along to Mr. Bas. I still remember very well his exquisite, intelligent, though somewhat cold playing.[1] It is also splendid that Dällenbach[2] is visiting you. There is surely a lot of communal theorizing going on; I am already looking forward to being able to make my contribution in the summer. I prefer to come in the summer to now, you see, not least because a trip in the fair season is more suited to my somewhat cranky body. I am feeling well, incidentally. The doctor[3] says that I have gallstones. Spa treatment, strict diet. You surely know that our Zangger has arranged for the appropriate fodder for me.[4] I'm feeling very significantly better; no more pain, better appearance.

My little boy's condition depresses me very much.[5] It's impossible that he will become a fully developed person. Who knows whether it wouldn't be better for him if he could depart from us before he really came to know life! I am to blame for his being and reproach myself, for the first time in my life. Everything else I have taken more lightly or did not actually feel responsible for. I just was not familiar with the nature of scrofula; I did not know that this is tuberculosis with hereditary risks for the children.[6] Indeed, to be honest, I knew nothing about scrofula either and did not attach any particular importance to the glandular swelling that my wife had at the time. Now the misfortune is upon us, as it had to be.   Bearing it, not bewailing it, is the solution. We care for the sick and console ourselves with the healthy. I draw strength from the idea of taking Albert out of school and teaching him myself and, where I fall short, supplementing it with private tutoring. The only thing that makes me vacillate in this resolve is the consideration that he would lack the necessary contact with boys of his own age, which would not be an insignificant deficiency with his undeniable predisposition for a certain reserve. What do you think about this? I think that I could offer the boy very much, not just intellectually. Do you think that my wife would consent?

Your suggestion regarding relativity is a good one. However, the booklet is finished[7] and the proofs are almost completed already, so I cannot make use of it anymore. The description has turned out quite wooden. In the future, I shall leave writing to someone else whose speech comes more easily than mine and whose body is more in order.

You will have received the "Cosmic Considerations."[8] It is at the very least proof that general relativity can lead to a system free of contradictions. Before, there had always been the fear that the "infinite" harbored irreconcilable contradictions. There is unfortunately little prospect of testing the advocated view against reality, though. With the aid of the astronomers' inquiries into the density distribution of stars, we arrive at the order of magnitude[9]

$$R = 10^7 \text{ light-years,}$$

whereas visibility approaches only

$$R = 10^4 \text{ light-years.}$$

The question arises, incidentally, whether we should not be able to see stars lying closely enough to our antipodal points. These would have to have a negative parallax.[10] It should not be forgotten, though, that the curvature of space is an uneven one, so light rays traverse a medium full of streaks. The quantum paper I sent out has led me back to the view of the spatially quantumlike nature of radiation energy.[11] But I have the feeling that the actual crux of the problem posed to us by the eternal enigma-giver is not yet understood absolutely. Shall we live to see the redeeming idea? Politically, things look strange. When I speak with people, I sense the pathology of the general state of mind. The times recall the witch trials and other religious misjudgments. It is precisely the responsible and privately most selfless of people who are frequently the most rabid supporters of dogmatism.[12] Social sentiment has gone badly astray. I could not imagine such people existed, if I did not see them before me. Now I can only hope for salvation from an external force.

Affectionate greetings also to Anna and Vero, yours,

Albert.

307. From Friedrich Adler

Vienna VIII, 1 Alser St., 9 March 1917

Dear Professor Einstein,

Now that I have plenty of free time at my disposal, I have taken up my studies on the foundations of physics again, which I had abandoned 7 years ago.[1] I intended to assemble in a book my previously published and unpublished papers on Mach and to expand on them in various directions. The work had already progressed quite far when I started the chapter on relativity. Then something quite unexpected happened to me.

Ever since taking Mach's point of view (1903) I also assumed as a matter of course the relativity of rotation and already advocated it in 1907 in my lecture courses in Zurich.[2] You may still remember as well that around 1909 we had a longer argument about it in our attic rooms on Mousson Street.[3] It remained vividly in my memory because it was very important to me. I was very irritated, you see, that in taking the centrifugal effects into account, you rejected relativity for rotations. I asked myself at the time whether an error existed in Mach's or my argumentation after all. But I could find none. So much the greater was my joy when your general theory of relativity came out. I did not have the time to follow the subject closely, however. It is only now that I sent for the more recent literature in order to be able to present Mach's position on relativity. And there I saw, first in Freundlich's brochure,[4] and then in your own papers,[5] that you have accepted Mach's position entirely, including the centrifugal phenomena.[6] I was in ecstasy when suddenly, 4 weeks ago, a turning point came in my considerations which reveals the whole problem differently from how I had seen it previously. I found, first in a more recent discussion of Foucault's pendulum experiment, and then generally, a criterion that you and Mach do not take into consideration, or at least, not sufficiently, which sheds new light on it all. I believe I have found where the error in the assumptions not only by Mach but also by you lies. I cannot discuss this in detail within the bounds of a letter, but just want to say that, naturally, it does not involve a return to "absolutes" but a criterion of a relativistic nature for preferred reference systems.[7]

In the last 7 years, I was able to follow the course of the literature only very cursorily, of course, and now can catch up only with the most important papers.

I should therefore be *very grateful* to you if you would read my work, as soon as it is finished, and give me your verdict. I shall provide you with a duplicate as soon as possible.

I should also be grateful if you could send me an offprint of your paper from the *Annalen*, 1916, Volume 49, p. 769,[8] as I must return the issue again and still need it. I possess a quite complete set of your earlier papers. On the other hand, if anything has been published since the paper mentioned, I would appreciate your sending it to me.

Healthwise I am fine and, since I have the opportunity rarely offered in life of being able to work in peace, I am completely happy, even though the time still available to me can be counted only by the week.[9]

I often think back with pleasure on the hours I was able to spend with you and send you my cordial greetings as an old friend and in sincere admiration, yours,

Fr. Adler.

308.   To Michele Besso

[Berlin, after 9 March 1917][1]

Dear Michele,

It is very nice of you and Vero to look after my Albert. It has always been a dear wish of mine that he come to see you. I advised him to do so in almost every letter. I am also pleased that my poor little boy is doing all right; but I have no grand illusions.[2] One must be able to look the truth in the eye, hard though it may be. I have difficulty explaining why Miza disapproves of your visiting; but there are more things in heaven and earth . . . . . .

A letter to Dällenbach and you is on the way.[3] Now to the $\lambda = \dfrac{1}{R}$ matter.[4] Quite apart from the correctness or incorrectness issue, we are not dealing with a matter of great scientific significance; I do not delude myself about that. It merely involves this question: If at one place I choose Galilean $g_{\mu\nu}$'s and continue the system in the most suitable manner possible, how do the $g_{\mu\nu}$'s behave when I continue on for a vast distance, spatially and temporally? Is it possible to arrange it that the $g_{\mu\nu}$'s really are determined solely by the matter, as the relativity concept would have it?[5] Now, almost all the objections you are making are legitimate. The arguments provided by me are not actually compelling, as is normally the case in all things concerning real facts. But I do believe that I

have hit upon essentially the right thing and also that I can convince you of it in
person, if necessary, when I come to visit you again.

First the main issue: Take Newton's
theory as a basis. You suggest it should
be assumed that a mass filling the space
uniformly into infinity does not produce a field (for reasons of symmetry).
This is not correct, though. Let there
be no field at point P. Even so, there

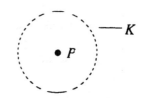

has to be a flux of gravitation running from the spherical surface K according to
Gauss's theorem, arising from the masses enclosed by K. According to Gauss's
theorem, every mass is a point of convergence for lines of force! Even if the world
outside of K is filled with mass, even into infinity, the matter must, nevertheless,
fall toward P; to be more precise, the acceleration is greater, the greater the
distance is from P. Jehovah did not found the world on such a crazy basis.

If the world is to be permanently stable, motion has to obstruct the fall
(centrifugal forces). That is how it is in the solar system, of course. But this only
works when the average density of the matter is brought to zero at infinity in the
appropriate manner; otherwise infinitely large differences in potential occur.

Such an interpretation is unsatisfactory even according to Newton[6]—problem
of reduction of matter and energy. Dispersal at infinity—and is even more unsatisfactory according to relativity theory, because the relativity of inertia is not
satisfied. The latter would be determined for the most part by the $g_{\mu\nu}$'s in infinite space and, to a very small part, by interaction with the other masses. This
interpretation is intolerable to me. The only alternative I find is the hypothesis
of spatial closure, the feasibility of which I have proven.

I do not seriously consider believing that the universe is statistically and mechanically at equilibrium, even though I argue as I do.[7] The stars would all
have to conglomerate, of course (if the available volume was finite). But closer
reflection reveals that statistics can be legitimately applied to the problems of
importance to me. It could also be done without statistical considerations, by
the way. It is certain that infinitely large differences in potential would have to
yield stellar velocities of very significant magnitude that probably would have
ceased long ago. Small differences in potential, in conjunction with infinite extension of the world, require emptiness in the universe at infinity (constancy of
the $g_{\mu\nu}$'s at infinity for a suitable choice of coordinates), in contradiction to a
sensible interpretation of relativity. Only the closure of the universe frees us from
this dilemma; this also suggests itself *in that the curvature has the same sign*

throughout because, according to experience, the energy density does not become negative.

The newly introduced λ has nothing to do with the earlier one. I failed to notice then that on the left-hand side of the field equation an addition of $+\lambda g_{\mu\nu}$ does not disturb the tensor character. I initially ought to have set $\lambda = 0$ in Newton's sense. But the new considerations plead for a λ differing from zero, which tends toward producing a non-zero mean density ρ_0 of matter.[8] Fixed-star astronomy (star counts) points to the order of magnitude $\rho_0 = 10^{-22}$gram/cm^3, corresponding to a world radius of $R = 10^7$ light-years,[9] while the farthest visible stars are estimated at 10^4 light-years. Read the paper with Dällenbach. It will amuse you both.

Dear Michele, back to Earth, which is so ugly only because we are seeing and experiencing it too keenly. Unless my appearance in Zurich does not suddenly become necessary, I'm not coming until the summer. We'll all get more out of it, especially my Albert. I am looking forward to it immensely already. I ask Zangger to set the package aside until that time;[10] then it is sure to arrive safely at its destination. There is no question of my taking Albert away *against Miza's will.*[11] I am no ruthless tyrant, you know. I know that some other things argue against it as well. If the boy is not endangered physically by his environment and is content, then he should very well stay there. Here he has *just me*, while I would have to shield him from other influences. There he has a healthy environment with his same-aged friends at the Swiss school and with you all. I would have to expend a major part of my energy, which is anyway weakened, on him and his education and manage a proper household, which would be very complicated for me. I'll do it only if, according to our unanimous opinion, it is the right thing, not out of blind preference. If I know him to be well cared for and content, I also am satisfied. It is lucky that I got the position here, otherwise I would be completely ruined financially. But this way I have put away handsome savings. Keep an eye out *that Zangger does not have to pay for anything* in his solicitude![12] If the money does not suffice, I shall send more. Ask him, appealing to his conscience. I do not have the courage or the skill to ask him properly. *I have enough*, since I have been living extremely frugally.

Affectionate greetings also to Anna and Vero, yours,

Albert.

309. To Heinrich Zangger

[Berlin,] 10 March 1917
Upon receipt of the typewritten letter.

Dear friend Zangger,

Did the little Zürcher boy not get infected by Tete?[1] It is a truly terrifying thought for me that this should be the reward to these excellent people for the goodwill they have shown toward my family.[2] Your consoling words about my little boy are a real comfort to me. Of course, I approve of his being brought to high altitude for a year. It is only human! I am inwardly convinced that it would be in the public interest to imitate the Spartans' method.[3] And yet, I should blame myself most severely if I really did guide my actions accordingly. Do rational arguments here really not have a place in life, or must they be carried further until they coincide with the impulsive feeling of what ought to be done?

A lesser evil is that one of the 3 packages, into which you have put so much effort, has not arrived.[4] The post office here has refused to recompense us, as I have also read recently in the newspaper. The best upbringing is powerless in the face of hunger.[5] I ask you now please to receive the March packet for me, which you had announced, but to keep it in Switzerland until the time of my stay there.[6] In this way, we are sure it will arrive at its destination. I urge you please to reimburse yourself from my last remittance, at least for the direct disbursements you make for the packages, even though this is nothing in comparison to the time and effort you have spent. Boas has also advised me to take the oil;[7] I did not want to keep any secrets about medically related conduct.

You write very gloomily about the economic future of Switzerland. I tremble at the thought that it could really be so bad. If so, then Europe must truly be pitied. You know my views on political matters, of course. They have only become stronger since. I have the ardent wish to discuss these things with you again. But I shall rein back my craving until the summer, if you do not consider it desirable that I travel to Switzerland earlier, for my family's sake. If, for ex., the household has to be dissolved, I shall come out. I am not as useless in these matters as you think, thanks to the unsettled life I now have behind me. And it should not happen that you and others sacrifice *too much valuable time* on my private affairs. My health, which, thanks to good care, has improved very much, permits me to undertake such a thing. I have had no more pain,[8] and my appearance and subj[ective] well-being have also improved substantially.

Regarding Albert's well-being, I have not taken a decision yet and shall not do so without private consultation with you and Besso. One always sees least clearly in one's own affairs. Also, everything should occur, if possible, in agreement with my wife, who is troubled so much already. Obstinacy need not be added unnecessarily to the unavoidable hardships. I have already written you, of course, how I imagine it. I believe that Albert is mature enough for me to be able to be a good teacher and companion to him for a few years.[9] The little one should be cared for in a suitable climate as soon as you consider it appropriate. But who should take him there? Shouldn't I come, and go with him to the place and stay there a while so that the boy does not feel so abandoned? I shall do everything you consider proper or desirable and can always arrange it. These responsibilities now have priority over all others.

I shall take care that Albert is not spoiled here, just as I also always have kept myself independent of the mentality of others. I have come to know the mutability of all human relationships and have learned how to insulate myself against heat and cold, so the temperature is quite steadily balanced. Especially in these times of general excitability, one never knows what the next day will bring, especially when one's own judgment and values contrast so terribly with those of the surroundings. Cordial greetings, yours,

Einstein.

310.   To Heinrich Zangger

[Berlin, after 10 March 1917][1]

Dear Zangger,

The day before yesterday I got a scientific letter from my old friend Fritz Adler.[2] You surely know what the man has perpetrated. I know you never thought particularly highly of him because of his scientific orientation. But as a person, he is an exceptional fellow and extraordinarily selfless, which has gotten him into this stew, as a matter of fact. My compassion for him has grown so strong that I really would like to do something for him. He does not know a soul here, but surely he must have much sympathy in Zurich, the place of his former activities.[3] I beg you now please to arrange a *prompt* action in my name at the Zurich Physical Society and possibly also at the Zurich Scientif. Society, for having a ⟨personal⟩ pardon *plea* submitted to the solely responsible authority. It is unclear whether the associations themselves ought to take the step or a large number of their members; but it seems to me that the associations themselves ought to do it, at least the Physical Society,[4] since it makes more of an impression, especially considering the small number of members in the Phys. Society. Besso and Dällenbach will take over all the time-consuming aspects. However, it is *you* who must assume the "direction" so that something comes of it.

Dear Zangger, I plague you with a second matter as well. I had magnanimously refused the March package[5] you had promised. But now I ask you please to send it after all, if it is still possible. Rice, macaroni, ⟨noodles,⟩ cornflour, ⟨rolled oats,⟩ or semolina are the most important. No more rusk, since this can be substituted by toasted bread.[6]—The third package with the bottle of fine oil has now arrived as well; hearty thanks. I have not started to take the oil yet. Although Boas[7] fully agrees with you on its curative effect, he said: Hoard this rich store for a time of even greater need, if you still have a tolerable amount to eat now! God forbid that, where you are, it ever gets to be like it is here.[8] From all that I am hearing from informed quarters, I still consider it *out of the question*. I can assure you, it is highly probable that you are overestimating the danger. Boas also insists adamantly that I go to Tarasp.[9] But I do not want to go without my boy, and with him it would be much too costly. I want to save as much as possible; I think, in my circumstances this is very much in order. I naturally fully approve of having accommodations made for Tete at high altitude for a whole year or longer. Your consoling words about him were a comfort. I do not want to know so precisely how you assess the case, but I merely want to do what I can and humbly take what comes.[10] We shall see with Albert. Nothing that would make my wife even more unhappy than she already is should happen *without necessity*.[11] In any case, I am relieved that you see no danger to his health as things currently stand.

Scientific life has dozed off, more or less; nothing is going on in my head either. Relativity is complete, in principle, and as for the rest, the slightly modified saying applies: And what he can do he does not want; and what he wants he cannot do. I am not coming to visit you until the summer, but words cannot describe how much I'm looking forward to it. It is quite nice for me here and I am floating right "at the top," but on my own, like a drop of oil on water, isolated by mentality and outlook on life. Your suggestion about the books is a good one *per se* but is surely only practicable if we have a market for these books.[12] You should not take on even more bother on my account than is unfortunately already burdening you.

Dear Zangger, start up the Adler campaign soon, otherwise it could be too late! Affectionate greetings, yours,

Einstein.

Besso's note: "I returned Beck's telegram to him today. 5 May 1917."

311.   To Willem de Sitter

[Berlin, before 12 March 1917][1]

Dear Colleague,

I am terribly sorry that you have health complaints and are confined to bed.[2] I hope you will soon recover. There is something amiss with me too,[3] but at least I am allowed to go about my normal business. Furthermore, it is bad that they have chosen M. instead of K. for Potsdam, in spite of the Academy's recommendation![4] All who mean well in the matter are unhappy about it. It is unclear what forces are to blame in this. There is talk of von Seeliger.[5]

Now to our problem! From the standpoint of astronomy, of course, I have erected but a lofty castle in the air.[6] For me, though, it was a burning question whether the relativity concept can be followed through to the finish or whether it leads to contradictions. I am satisfied now that I was able to think the idea through to completion without encountering contradictions. Now I am no longer plagued with the problem, while previously it gave me no peace. Whether the model I formed for myself corresponds to reality is another question, about which we shall probably never gain information. On the value of R, I contemplated the following.[7]

Astronomers have found the spatial density of matter from star counts up to the nth size class, fairly independent of the class to which the count extends, at about

$$10^{-22} \text{g/cm}^3.$$

From this, approximately

$$R = 10^7 \text{ light-years}$$

results, whereas we only see as far as 10^4 light-years. One thing seems strange to me, though. Stars close to our antipodal point should be emitting a lot of light to us.[8] It is doubtful, however, that they could appear point-shaped, since the light velocity varies irregularly. If such a thing were visible in the heavens, it would be noticeable through its negative parallax.[9] We should at least keep an eye out whether any objects with a negative parallax exist in the sky. But now, enough of this, or else you will laugh at me.

Of course, the question of the three-dimensional composition of the world only makes sense when a *static* (approximated) conception is possible (independence of all functions of a suitably chosen x_4). There is no violation of the relativity postulate.

Our problem can be illustrated with a nice analogy. I compare the space to a cloth floating (at rest) in the air, a certain part of which we can observe.

This part is slightly curved similarly to a small section of a sphere's surface. We philosophize on how we must construe the continuation of the cloth so that an equilibrium is reached in its tangential tension, whether it is fastened in position at the edges, extends infinitely, or has a finite size and is a closed unit. Heine has provided the answer in a poem:

"And a fool waits for an answer."[10]

So let us be satisfied and not expect an answer, and rather see each other again as soon as possible in acceptable health in Leyden!

Sending you cordial greetings and wishing you a speedy recovery, yours,

A. Einstein.

312.   From Willem de Sitter

Leyden, 15 March 1917

Dear Colleague,

Many thanks for your kind letter.[1] Well, if you do not want to impose your conception on reality, then we are in agreement. I have nothing against it as a contradiction-free chain of reasoning, and I even admire it. I cannot concur completely before having calculated with it, which is not possible for me to do right now.

My paper in English is now completely in print, the second piece as well.[2] Unfortunately, I cannot send an offprint, because I have not received any; probably torpedoed.[3] It does not contain anything that we did not already discuss in Leyden anyway. Just this: The simple fact *that we can identify spectral lines* proves that in all stars and nebulae, no matter how far away they may be, the potential is of the same order of magnitude as here.[4] This proof is stronger than that of the low stellar velocities.[5] Also, the fact that there is no systematic shift toward the *violet* in stellar spectra leads to the conclusion that an upper limit exists for the total mass of all stars within a given distance.[6]

Yours truly,

W. de Sitter.

I regret the news about Potsdam very much, for K. and for Potsdam.[7] deS.

313.   From Willem de Sitter

Leyden, 20 March 1917

Dear Mr. Einstein,

I have found that the equations

$$G_{\mu\nu} - \lambda g_{\mu\nu} = 0,$$

thus your equations (13a) *without matter*,[1] can be satisfied by the $g_{\mu\nu}$'s, which are given by[2]

$$ds^2 = \frac{-dx^2 - dy^2 - dz^2 + c^2 dt^2}{(1 - \mu h^2)^2} \tag{1}$$

$$\mu = \frac{\lambda}{12}, \qquad h^2 = c^2 t^2 - x^2 - y^2 - z^2. \qquad x, y, z, t \text{ can become } \infty.$$

At infinity (either spatial or temporal or both) the $g_{\mu\nu}$'s become

$$\left\{ \begin{array}{cccc} 0 & 0 & 0 & 0 \\ 0 & 0 & 0 & 0 \\ 0 & 0 & 0 & 0 \\ 0 & 0 & 0 & 0 \end{array} \right\}$$

Here we thus have a system of integration constants, or boundary values at infinity, which is invariant under *all* transformations. For relatively small values of h, i.e., in our spatial *and* temporal proximity, we have the $g_{\mu\nu}$'s of the old theory of relativity if only $\mu(\lambda)$ is small enough. This is achieved *without* supernatural masses only through the introduction of the undetermined and *undeterminable* constant λ in the field equations.[3]

I do not know if it can be said that "inertia is explained" in this way. I do not concern myself with explanations.[4] If a single test particle existed in the world, that is, there were *no* sun and stars, etc., it would have inertia. *Also in your* theory, as I see it, if *physical* masses (sun, etc.) were not there. Conjecturing that supernatural masses did not exist is just as impossible in your theory as saying "assume the world did not exist."[5]

(1) can also be interpreted in such a way that the *four*-dimensional world is finite, with a radius given by $\lambda = \dfrac{3}{R^2}$. The analogy with your solutions emerges from the following comparison:[6]

Three-dimensional With supernatural masses.	*Four-dimensional* Without any masses.

$$\lambda = \frac{1}{R^2}$$

$$\lambda = \frac{3}{R^2}$$

Coordinate System I

$x_1, x_2, x_3, ct : x_1^2 + x_2^2 + x_3^2 \le R^2$

$x_1, x_2, x_3, x_4 = ict' : x_1^2 + x_2^2 + x_3^2 + x_4^2 \le R^2$

$$g_{44} = 1 \qquad g_{i4} = 0$$

$$g_{\mu\nu} = -\delta_{\mu\nu} - \frac{x_\mu x_\nu}{R^2 - (x_1^2 + x_2^2 + x_3^2)}$$

$$g_{\mu\nu} = -\delta_{\mu\nu} - \frac{x_\mu x_\nu}{R^2 - (x_1^2 + x_2^2 + x_3^2 + x_4^2)}$$

$$\mu \text{ and } \nu = 1, 2, 3, 4.$$

Coordinate System II (Hyperspherical coordinates)

$$ds^2 = c^2 dt^2 - R^2[d\chi^2 + \sin^2 \chi (d\psi^2 + \sin^2 \psi d\vartheta^2)]$$

$$ds^2 = -R^2[d\omega^2 + \sin^2 \omega \{ d\chi^2 + \sin^2 \chi (d\psi^2 + \sin^2 \psi d\vartheta^2) \}]$$

$-\infty < t < +\infty \; 0 \le \vartheta \le 2\pi \; 0 \le \chi, \Psi \le \pi$

$0 \le \vartheta \le 2\pi \; 0 \le \omega, \chi, \Psi \le \pi$

Coordinate System III (Cartesian) $\begin{cases} \text{is obtained from } II \text{ through a} \\ \text{"stereographic projection"}^{[7]} \end{cases}$

$$ds^2 = c^2 dt^2 - \frac{dx^2 + dy^2 + dz^2}{\left[1 + \frac{1}{4R^2}(x^2 + y^2 + z^2)\right]^2}$$

$$ds^2 = \frac{-dx^2 - dy^2 - dz^2 + c^2 dt^2}{\left[1 + \frac{1}{4R^2}(x^2 + y^2 + z^2 - c^2 t^2)\right]^2}$$

At infinity: $g_{\mu\nu} = \left\{ \begin{matrix} 0 & 0 & 0 & 0 \\ 0 & 0 & 0 & 0 \\ 0 & 0 & 0 & 0 \\ 0 & 0 & 0 & 1 \end{matrix} \right\}$

At infinity: $g_{\mu\nu} = \left\{ \begin{matrix} 0 & 0 & 0 & 0 \\ 0 & 0 & 0 & 0 \\ 0 & 0 & 0 & 0 \\ 0 & 0 & 0 & 0 \end{matrix} \right\}$

Invariant under all transforms.
with $t' = t$.

Invariant under *all* transforms.

$$G_{44} = 0 \;\; G_{ii} = \frac{2}{R^2} g_{ii} \;\; i = 1, 2, 3$$

$$G_{ii} = \frac{3}{R^2} g_{ii} \;\; i = 1, 2, 3, 4$$

To find the relations between λ and R^2, I have the field equations

$$G_{ij} - \lambda g_{ij} = -\kappa(T_{ij} - \frac{1}{2} g_{ij} T).$$

I have only one test particle and no other *physical*[8] masses (sun, etc.) [From *physics* I know *only* that I must find $g_{\mu\nu}$'s very close to those of the old theor. of rel. for finite distances], but perhaps I am going to need supernatural masses, which I assume to be static, thus $T_{44} = \rho$, all other $T_{ij} = 0$. Then the equations become: (in coordinate system III)

$$G_{ii} - (\lambda + \frac{1}{2}\kappa\rho)g_{ii} = 0 \qquad i = 1, 2, 3$$

$$G_{44} - (\lambda + \frac{1}{2}\kappa\rho)g_{44} = -\kappa\rho$$

From this follows: for the ⟨four-dimensional system⟩ both cases:

$$\lambda + \frac{1}{2}\kappa\rho = \frac{2}{R^2}, \qquad \lambda = \frac{1}{2}\kappa\rho$$

$$\lambda = \frac{3}{R^2}, \qquad \rho = 0$$

Therefore, the supernat. masses are needed.

Supernat. masses do not exist.

I am curious about whether you can agree with this approach and whether you prefer the *three*-dimensional or *four*-dimensional system.

I personally much prefer the *four*-dimensional system, but even more so the original theory, without the undeterminable λ, which is just philosophically and not physically desirable, and with noninvariant $g_{\mu\nu}$'s at infinity. But if λ is only small, it makes no difference, and the choice is purely a matter of taste.

I hope that your health has improved[9] and you will soon be back to normal again. With cordial greetings, yours truly,

W. de Sitter.

314. To Moritz Schlick

[Berlin,] 21 March 1917

Esteemed Colleague,

Upon rereading your fine essay in *Naturwissenschaften* I do find another small inaccuracy. I am informing you of it in case your article is reprinted elsewhere.[1]

The derivation of the law of the motion of a point provided on page 184 assumes that, seen from the local coordinate system, the point moves in a straight line. Nothing can be derived from this, however. The local coordinate system is generally of importance only at the infinitesimal level, and at the infinitesimal level every uninterrupted line is straight. The correct derivation proceeds as follows: In principle, *finite* (matter-free) portions of the world can exist for which, on suitable choice of frame of reference,

$$ds^2 = dX_1^2 + \cdot + \cdot - dX_4^2$$

applies. (If this were not the case, the Galilean law of inertia and the special theory of rel. could not be upheld.) In such a segment of the world, with this reference system choice, the Galilean law of inertia is valid, and the world line is a straight one hence, for an arbitrary choice of coordinates, a geodetic line.

That the point's world line is a geodetic line otherwise as well (if others do not exert gravitational influence) is a hypothesis, albeit one that suggests itself.–

You are right with your criticism on page 178 (footnote). On close examination, the demand for causality is merely not sharply defined.[2] The causality requirement can be fulfilled to various degrees. It can only be said that the general th. of r[elativity] has been more successful than classical mechanics in satisfying it. It might be a rewarding task for an epistemologist to think this through carefully.

Cordial greetings, yours,

A. Einstein.

P. S. I am sending you a new paper which treats a principal point of the gen. th. of rel.[3]

315.   From Hendrik A. Lorentz

Haarlem, 22 March 1917

Dear Colleague,

We were very sorry to hear from Ehrenfest that recently your health leaves something to be desired[1] and that for this reason you will not be able to attend the Congress of Dutch Scientists and Physicians to be held during the Easter holidays.[2] I understand very well your wish to avoid the fatigue attached to such gatherings; and as much as we would enjoy your participation, it is much more important that you convalesce properly as soon as possible. Should you occasionally find it hard to be easy on yourself as precaution would require, I beg you to think of your many friends and colleagues, who dearly wish to see you healthy and vigorous.

After the exhausting work of the last few years, you surely have earned some rest. Could you not leave Berlin now for a time to seek relaxation in Switzerland or here somewhere in the countryside? There are several places in our eastern provinces as well as on the seashore where one could stay comfortably for a few weeks; and you could make your selection such that you are not entirely shut off from scientific interaction but do not have to join the daily discussions.   If we can be of any assistance in this regard, I ask you please to tell me. But perhaps the Swiss air is better for you after all, and naturally you will be attracted by the prospect of having your children nearby.[3]

In the meantime, . . . I have set up a "Lagrangian function," dependent only on the present-day distances r and velocities ν, for an arbitrary number of celestial bodies. Precisely in the second-order terms (in which form they must be considered in the explanation of Mercury's perihelion motion) this function assumes the following form:[4]

$$
L = \frac{\pi}{\kappa c} \sum (i) \left\{ 2\alpha_i \nu_i^2 + \frac{1}{2c^2} \alpha_i \nu_i^4 - \frac{\alpha_i^2 \nu_i^2}{R_i} \right\} +
$$
$$
+ \frac{\pi}{\kappa c} \sum (\overline{ij}) \{3(\nu_i^2 + \nu_j^2) - 7(\nu_i \cdot \nu_j) + \nu_{ij}\nu_{ji}\} \frac{\alpha_i \alpha_j}{r_{ij}} +
$$
$$
+ \frac{\pi c}{\kappa} \sum (\overline{ij}) \left\{ 2\frac{\alpha_i \alpha_j}{r_{ij}} - \frac{3}{5}\left(\frac{\alpha_i \alpha_j^2}{R_j} + \frac{\alpha_j \alpha_i^2}{R_i} \right) \frac{1}{r_{ij}} - \frac{\alpha_i \alpha_j^2 + \alpha_j \alpha_i^2}{2r_{ij}^2} \right\}
$$
$$
- \frac{\pi c}{\kappa} \sum (\overline{ijl}) \alpha_i \alpha_j \alpha_l \left(\frac{1}{r_{ij}r_{jl}} + \frac{1}{r_{jl}r_{li}} + \frac{1}{r_{li}r_{ij}} \right)
$$

After the conversations you had with our Berlin colleagues last fall, I received very friendly letters from Planck and Waldeyer.[5] I have not yet found occasion to take up the matter again. I do not want to be presumptuous, considering that the issue involved is primarily of importance to Germans: Nevertheless, the attitude of the gentlemen mentioned gives me full confidence that, after the war, something will happen in the direction I wish.

With cordial greetings from both of us as well as from de Haas and my daughter, and with best wishes, devotedly yours,

H. A. Lorentz.

316.   From Friedrich Adler

Vienna, 1 Alser St., 23 March 1917

Dear friend Einstein,

Your letter was a welcome joy, above all, because it proved to me what I had hoped and suspected, that our earlier personal relations have remained the same. I am very glad that you are prepared to read my paper.[1] Unfortunately, I am only going to be able to send it to you around Easter, or perhaps even only after Easter. Since, although the first part is already finished, which would already show you what I mean, external difficulties also exist currently regarding the feasibility of transcription and dispatch. However, I hope that it will work by Easter. I cannot write you about the content within the limited bounds of a letter—which are particularly constrained this time, as this is the remains of my weekly paper supply.[2] Today I just want to turn your attention to a paper not directly related to it. Are you familiar with: Paul *Gerber* "Die räumliche und zeitliche Ausbreitung der Gravitation" ["The Spatial and Temporal Propagation of Gravitation"] in *Zeitschrift für Mathematik und Physik* (Schlömilch) 43rd Volume, Page 93 (1898).[3] He calculates the velocity of the gravitational potential from Mercury's perihelion motion and finds it equal to the velocity of light. I do not have time to work on this now, but it seems to me that it ought to be examined whether the decisive thing in your explanation of the perihelion motion is the *introduction of the velocity of light* and not the *overall* structure of the general theory of relativity. For, Gerber obviously comes to his result using Euclidean geometry. Therefore, I think that it ought to be possible to explain the perihelion motion of Mercury also using the *old* tools, thus that the verification of the gen. theory of relativity through this result is *not as far-reaching* as you assume it to be. I shall be very pleased if you write me again. Cordial regards, yours,

F. Adler

P. S. Many thanks for the offprints, which I have received.[4]

317. To Willem de Sitter

[Berlin,] 24 March 1917

Dear Mr. de Sitter,

I find your new results[1] very interesting. Also, I am very pleased that you are up and about again and (I hope) completely restored. I also am feeling considerably better.[2] Regarding your solution, I think the following:

1) Your world is spatially finite. Because, the naturally measured length of the positive x-axis is[3]

$$(-i) \int ds = \int_0^\infty \frac{dx}{1 + \mu x^2} = \text{finite.}[4]$$

It certainly would be most natural to describe this world as spatially closed and finite.

2) The hyperboloid surface

$$1 - \mu h^2 = 0$$

is a singularity. On crossing it, the $g_{\mu\mu}$'s jump from $-\infty$ to $+\infty$, and the g_{44}'s from $+\infty$ to $-\infty$, resp.[5]

3) Such singularities are to be ruled out in the physically finite. The intersection of the positive t-axis with the singul[arity's] surface is at $t = \dfrac{1}{\sqrt{\mu c}}$. The naturally measured (time-like) distance of the intersection point from the origin $x_1 = x_2 = x_3 = t = 0$ (measured on the time axis) is

$$\int ds = \int_0^{\frac{1}{\sqrt{\mu c}}} \frac{c\,dt}{1 - \mu c^2 t^2} = \frac{1}{\sqrt{\mu}} \frac{\pi}{2} \, (\text{finite}).^{[6]}$$

The surface with singular qualities (discontinuities) thus lies in the physically finite.

———————

It seems to me for this reason that your solution does not correspond to a physical possibility. The $g_{\mu\nu}$'s and $g^{\mu\nu}$'s (together with their first derivatives) must be continuous everywhere.

In my opinion, it would be unsatisfactory if a world without matter were possible. Rather, the $g^{\mu\nu}$-field should *be fully determined by matter and not be able to exist without the matter.* This is the core of what I mean by the requirement of the relativity of inertia.[7] One could just as well speak of the "matter conditioning geometry." To me, as long as this requirement had not been fulfilled, the goal of general relativity was not yet completely achieved. This only came about with the λ term.

Cordial greetings, yours,

Einstein.

P. S. The fact of low stellar velocities is an argument whose cogency exceeds that of the nonexistence of the violet shift only insofar as one is not limited here exclusively to the effect of the visible portion of the stars.[8] *Under the assumption of a mechanically quasi-stationary behavior of matter,*[9] the low stellar velocities prove that great differences in gravitational potential do not occur at all in the universe.

Assuming that the correctness of the general theory of relativity were established, the λ-problem could be decided through observation of the spectral lines of remote stars. For if $\lambda = 0$, the average stellar density would suffice to produce a difference in potential between us and distant stars, which would entail a substantial violet effect.

318.   From Max von Laue

[Frankfurt-on-Main,] 24 March 1917

Dear Einstein,

Well, it has happened! My revolutionary views about wave optics are in print and, no doubt, at this very moment are arousing utmost disgust in every peace-loving physicist.[1] Nonetheless, I stand by my reprehensible position and, in corroboration of it, have written a long letter to Planck today with the request that he show it to you. Hence I should not have had to write to you at all, had I not forgotten a point of some importance. Instead of writing to Planck again tomorrow, I am sending you this with the request that you make it available to Planck. But first read my letter to Planck.

Exner, whom I cite in my paper in the *Berliner Berichte*,[2] had accurately observed the filaments in the diffraction phenomena in question of white light, but only a pointlike granulation for narrow spectral lines. Presently I can only confirm his observations. It should actually have struck him, though, that these filaments are completely white, and not colored in the least. For if they are explained as monochromatic light from granulation, according to the old optics (the theorem that the diffraction pattern is proportional in all dimensions to the wavelength), they can be expected to be blue in color inside and red outside. Now, I do not know: Was Exner absentminded enough not to notice this or, on the contrary, is there hidden behind his complete silence on this point the profound knowledge of the impossibility of making any rational conjectures at all in 1879? For me, in any case, this is a new and very compelling proof against the classical theory.[3]

With cordial regards, yours,

M. von Laue.

N. B. Do you know, perhaps, which Exner this is? Does he possibly still live in Vienna?[4] You are familiar with the Austrian situation since Prague, aren't you.

319.   To Felix Klein

[Berlin,] 26 March 1917

Esteemed Colleague,

Many thanks for your papers and your friendly card.[1] It is really astonishing that such different points of departure as mathematics and physics give occasion to establishing the same thought structures in the end. I have already read with pleasure the marked places in your papers; the other will follow. As I have never done non-Euclidean geometry, the more obvious elliptic geometry had escaped me when I was writing my last paper.[2] Mr. Freundlich has already made me aware of this point.[3] My observations are just altered thus, that the space is half as large; the relation between R (radius of curvature) and ρ (mean density of matter) is retained.

The new version of the theory means, formally, a complication of the foundations and will probably be looked upon by almost all our colleagues as an interesting, though mischievous and superfluous stunt, particularly since it is unlikely that empirical support will be obtainable in the foreseeable future. But I see the matter as a necessary addition, without which neither inertia nor the geometry are truly relative. But someone who does not find it disturbing when the existence of a $g_{\mu\nu}$-field can result from the theory without field-producing matter and when a single mass (imagined as being alone in the universe) can have inertia, cannot be convinced of the new step's necessity.[4]

I am very much looking forward to seeing Debye in the next few days and ask you please to inform me when you are coming to Berlin again so that I can visit you. It would certainly be desirable if between Göttingen and Leyden there were an exchange of papers addressing the area of general relativity. In that way, much intellectual effort would be spared. For ex., a precise treatment of the point problem, as Hilbert provides in his last article,[5] already exists in the dissertation by the Dutchman Droste.[6] Furthermore, Lorentz has worked on many things.[7]

With kind regards, yours truly,

A. Einstein.

Document description: "*Einstein to Klein.* To Messrs. Baade and Fréedericksz for your o[bliging] perusal and return on Saturday. K[lein]."

320.   To Moritz Schlick

[Berlin,] Sunday. [1 April 1917]

Dear Colleague,

Many thanks for kindly sending the separates.[1] Your splendid work has already helped a good many people come to understand the theory, as I have been convinced. I approve of the small change you are planning.[2] I shall be very pleased if you call on me again sometime. Then we can also discuss the problem of the constitution of space. I recommend to you my old acquaintance Hopf, a capable physicist who is also employed in physics in Adlershof.[3]

Best regards, yours,

A. Einstein.

321.   From Willem de Sitter

Doorn, Dennenoord Sanatorium, 1 April 1917

Dear friend Einstein,

Your letter pleased me very much because from it I can gather that you are feeling much better.[1] As for me, it is still the same; I am sitting—or rather lying—here now in this sanatorium and am slowly recovering. It is not a serious matter at all, just boring.[2]

I understand your position now fully. Mine was different.[3] [When I say "I," you must not misunderstand me; I personally take the position that all extrapolation is uncertain—it can be useful at times to get a better overview of the picture we make of the universe—but most of the time it is dangerous, because it leads us to assume that we have solved a puzzle, when we have just clothed it in other words.][4] But in working out this four-dimensional world, which you call mine, I had only taken the position of those (among whom I believe Ehrenfest belongs) who require that the boundary values for $g_{\mu\nu}$ be invariant, that is, that the world as a whole can perform random motions without us (within the world) being able to observe it. [These "motions" are naturally *not* physical, conceivable phenomena, only a pseudo-physical name for a purely mathematical concept.] From the point of view of this *mathematical relativity condition*, "my" four-dimensional world is perhaps better than your three-dimensional one.[5] Your condition, the *material relativity requirement*, will obviously just be satisfied by the three-dimensional, finite world.[6] May I cite this material relativity requirement from your letters in a postscript to the communication I am submitting to the Amsterdam Academy?[7]

It is obvious that "my" world is finite like yours;[8] I have forced it into a Cartesian Euclidean coordinate system merely to show that the finiteness in natural measure is equivalent to the requirement of the $g_{\mu\nu}$'s becoming zero at Euclidean infinity.

As concerns the singularity, I have nothing here, no books, etc., thus I cannot do the calculations. But I think that it is just apparent and that it stems from my description of my actually hyperbolic world as spherical.[9] Thus the hyperboloids' infinity is dragged into the finite realm, where it naturally must appear as a singularity. If the $g_{\mu\nu}$'s on the hyperboloid are finite at infinity, the $g'_{\mu\nu}$'s must necessarily become infinite there, when this infinite part of the hyperboloid is mapped onto a finite region of a flat space.[10]

I cannot agree with the second part of your letter. I must emphatically contest your assumption that the world is mechanically quasi-stationary.[11] We only have a snapshot of the worl, and we cannot and must *not* conclude from the fact that we do not see any large changes on this photograph that everything will always remain as at that instant when the picture was taken.

I believe that it is probably certain that even the Milky Way is *not* a stable system. Is the entire universe then likely to be stable? The distribution of matter in the universe is extremely *inhomogeneous* [I mean of the stars, not your "world matter"], and it cannot be substituted, even in rough approximation, by a distribution of constant density. The assumption you tacitly make that the mean stellar density is the same throughout the universe[12] [naturally, for vast spaces of, e.g., $(100,000 \text{ light-years})^3$] has no justification whatsoever, and all our observations speak against it. I do not have the data here to do anything other than express my conviction; I do not intend to try to prove it today.

With cordial greetings, yours truly,

W. de Sitter.

322.   To Hendrik A. Lorentz

Berlin, 3 April 1917

Dear Colleague,

Your warm words[1] were most comforting to me. In you, nature was seized by the rare impulse of combining a keen mind with warm sentiment. If only this were more often the case; the public at large would be considerably better off! My condition is probably not connected appreciably to work; it is probably attributable to a constitutional flaw.[2] I have not been working much at all, and that under ideal external circumstances.

My burning desire to see you and my other colleagues and mindmates more often will assure my coming back to Holland soon.[3] This summer I must first go to Switzerland, though, to be together with my children, perhaps also to spend time at a health resort in Tarasp.[4] However, my next trip will certainly take me to all of you.

I am pleased that Planck and Waldeyer have written you.[5] The latter still hails from the good old pre-Treitschke days.[6] Planck did not tell me that he had written you at all; that is why I feared he had neglected to do so. The younger ones are considerably worse than they, however. I am convinced that a kind of mental epidemic is involved. For otherwise I cannot grasp why people who are thoroughly decent in their personal attitudes take on such an entirely different view with respect to public affairs. I can only draw comparison with the times of martyrdom, crusades, and witchcraft trials. Only very uncommon independent characters can escape the pressure of prevailing opinion. There seems to be no such person at the Academy.[7]

Your joint paper with Droste is very interesting.[8] It is splendid that in L only first derivatives occur according to time. This is analogous to Hamilton's function of the gravitational field. It is a pity that the deviations from Newton are so small; but ultimately, we ought to be glad that Mercury's case has presented itself, at least.[9]

My last paper will not appeal to you.[10] It is convincing only if the relativity of inertia is considered a requirement, that is, if one is convinced that inertia can be attributed entirely to an interaction with the observed mass.[11]

Healthwise I am feeling considerably better again, thanks especially to the meticulous diet which my relatives here have been able to provide with the help of their southern German connections.[12] Without this help I would scarcely be able to stay here; I do not know either whether it can continue on like this.[13]

Affectionate greetings also to your wife and the two de Haases, yours,

A. Einstein.

323.   To Felix Klein

[Berlin,] 4 April 1917

Highly esteemed Colleague,

Thank you very much for your kind and interesting letter. I shall not fail to read your mentioned papers in the original.[1] That which you call agnostic in your position is present also in mine, specifically, in the following form: No matter how we draw a complex from nature for simplicity's sake, its theoretical treatment will ultimately never prove to be (adequately) right. Newton's theory, for ex., seems to describe the gravitational field completely with the potential φ. This description proves to be insufficient; the $g_{\mu\nu}$ functions take its place. But I do not doubt that the day will come when this approach will also have to give way to a principally different one for reasons that we do not anticipate today. I believe that this process of securing the theory has no limit.

I am sorry that you have health complaints; I know how dependent on bodily well-being the mental capacity for work is. It's a similar story with me as well.[2] I am sending you my last paper together with these lines.[3] The gist of its contents is, in particular, that the size of the universe seems to be linked to the mean density of matter. It is not at all out of the question that in the foreseeable future the statistics of fixed stars will confirm or refute the theory.[4]

With best regards, I am yours very truly,

A. Einstein.

324.   To Friedrich Adler

[Berlin,] 13 April [1917]

Dear Adler,

I hope you have received my package.[1] I now have an unusual request: When your affair is deliberated in court,[2] I would like to be summoned there as a witness; you should apply for this. Do not think this is senseless; for not only the circumstances directly connected to the event are clarified through witness accounts, but statements are also brought forward that shed light on the character of the perpetrator.[3]– How much I would like to discuss the relativity problem with you! I hope that we can make up for this later. I am curious about your arguments. Easter has passed, of course, so I may expect them shortly.[4]

With best regards, I am yours,

A. Einstein.

325.   To Willem de Sitter

[Berlin,] 14[13] April 1917[1]

Dear de Sitter,

I hope the quiet there agrees with you very well, so that you are back to normal soon.[2] I am considerably better, thanks to good care.[3] We should not devote all that much time to our difference of opinion,[4] which is only a difference in creed, so to speak, since it would not be fruitfully spent. We should see the possibilities without *wishing*. Of course you can quote my formulation, provided that my rendition was not too vague and imprecise.[5] This you can judge for yourself, of course.

I still do not understand the remark about the manifold you considered,[6]

$$ds^2 = \frac{\sum dx_\nu^2}{\sum x_\nu^2},$$

according to which the singular behavior in finitude is only apparent, caused by the choice of coordinates.[7] But your new article[8] will clarify this for me.

In any case, one thing stands. The general theory of relativity *allows* the addition of the term $\lambda g_{\mu\nu}$ in the field equations.[9] One day, our actual knowledge of the composition of the fixed-star sky, the apparent motions of fixed stars, and the position of spectral lines as a function of distance, will probably have come far enough for us to be able to decide empirically the question of whether or not λ vanishes.[10] Conviction is a good mainspring, but a bad judge!

Cordial regards, yours,

A. Einstein.

326.   From Otto Neurath[1]

Vienna XIX, 6 Billroth Street, 15 April 1917

Highly esteemed Professor,

I thank you sincerely for the kind note you sent my way last fall. Forgive me if I bother you with an inquiry connected to my research.

For a long time I have been contemplating the idea of tracing the parallel course of development of acoustics and optics. This seems to me to be a fascinating task in more than one respect. As you know, for a time, optics and acoustics followed approximately the same course of development; in particular, the idea of interference appears approximately simultaneously in both fields. They have undoubtedly influenced each other as well.

I should be sincerely obliged to you if you would do me the great favor of informing me whether you know of anything related to this question or even just of a more or less instructive contribution to the history of interference in acoustics. Whereas I have found with relative ease valuable essays in the optical literature,[2] all that has come into my hands up to now on acoustical interference is fragmentary and scattered. Before I set about searching out the essentials from the abundance of primary material, I would first like to get one or two important pointers. Mach, with whom I have exchanged a couple of brief letters on this, could not refer to anything either, unfortunately.[3]

I do not want to trouble you in any way with this inquiry. But it really is possible that by chance you might be able to give me some information on this point. I might be holding a lecture in Berlin in a while and, if this would not disturb you, would like to come to see you.

In expressing my special admiration, I am yours very truly,

Dr. Otto Neurath.

327.   From Willem de Sitter

Doorn, 18 April 1917

Dear Einstein,

The main point in our "difference in creed"[1] is that you have a specific belief and I am a skeptic.

Observations will never be able to prove that λ vanishes, only that λ is smaller than a given value. Today I would say that λ is *certainly* smaller than 10^{-45}cm^{-2} and is probably smaller than 10^{-50}. Maybe one day observations will also provide a specific value for λ, but up to now I have no knowledge of anything pointing to this.[2]

The problem of the discontinuity in $ds^2 = \dfrac{-dx^2 - dy^2 - dz^2 + dt^2}{\left(1 + \frac{x^2+y^2+z^2-r^2}{4R^2}\right)^2}$ is actually not interesting,[3] because the hyperboloid $4R^2 + x^2 + y^2 + z^2 - t^2 = 0$ intersects the t-axis at $t = \pm 2R$, and the natural distance of these points to the origin is $\displaystyle\int_0^{2R} \frac{dt}{1 - \frac{t^2}{4R^2}} = \infty$ and *not* finite as you initially thought.[4] The length of the x-axis, on the other hand, is $\displaystyle\int_0^{\infty} \frac{dx}{1 + \frac{x^2}{4R^2}} = \pi R$, hence finite.[5] "My"[6] four-dimensional world also has the λ term, but *no "world matter."*[7]

With cordial regards,

W. de Sitter.

328.   To Felix Klein

Berlin, 21 April 1917

Highly esteemed Colleague,

If presently I weren't the slave of a book taken out of the library, I would have already looked at the papers you mentioned, not only your own but the other two as well.[1] Thus, however, I must allow another couple of weeks to pass. Grossmann (I believe) had the little book by Wright when we were working together on relativity 4 years ago.[2] I have already had Batman's paper in hand but must confess that I cannot quite imagine it possible that any physical meaning be afforded to substitutions of reciprocal radii.[3] Not only does the Lorentz transformation leave the velocity of light invariant, but the coordinates of coordinate systems related in such a way have a simple physical meaning as measurement results from measuring rods and clocks. This latter quality is certainly lost in Bateman's transformations.[4]

I am very much looking forward to your lectures, which you have promised me.[5] I shall pass them on to Sommerfeld as soon as I have received and studied them. Probably—knowing me—nothing will come of my article for the encyclopedia. I was cautious enough, if I recall correctly, not yet to give a positive answer to our colleague Sommerfeld.[6]

Hilbert's second relativistic paper interested me very much.[7] Just the geodetic reference system introduced there, which is characterized by $g_{14} = g_{24} = g_{34} = 0$ and $g_{44} = 1$, does not seem acceptable to me. For it can be demonstrated through examples that this system does not preserve the uniqueness of the assignment of coordinates to space-time points. This stems from the temporal geodetic lines intersecting one another. However, it would not be inconceivable that otherwise definable coordinates existed, which would simplify the relations. Although the coordinate choice I made in some papers according to the condition $g = -1$ is natural in some cases, it limits the choice just a little (only *one* limiting condition rather than *four*).[8]

With best regards, I am yours truly,

A. Einstein.

329.   From Friedrich Adler

Dear friend Einstein,

I received your printed-matter package as well as your postcard of the 13th
inst.[1] A duplicate of my paper was sent off to you the day before yesterday.[2]
You can keep that copy. I have been able to complete only about 2/3 of the tract
now, as other business will be occupying me once again in the next 4 weeks.[3]
I believe, though, that two points essential to me become clear in this section
already:

1) Physical problems exist only with respect to "Copernican coordinate systems."[4]
Therefore, there is no reason to formulate the theory more generally than as relates to *this* manifold. The "Copernican coordinate systems" are preferred but
not equivalent. The same natural laws apply in all of them, but in each they are
"primarily applicable" to different areas.

2) In the equations of *classical* mechanics there can be no dependence of inertia on
other *simultaneous* parameters (gravitation). In classical mechanics, inertia char-
acterizes the *prehistory* of the object, it is evidence of simultaneous dependencies
of the past; it appears as an *initial condition.*

I fell into a very extensive study of the literature, which has held me up
very much, but now I am quite finished and will be able to settle the remainder
quickly at the end of May. I wanted to get the assurance, though, that what I
am saying has not already been said. Yet as close as the various sides had come
to my standpoint, none have all the aspects that I believe must be combined
in order to arrive at a tenable solution. As you will see, to me it is primarily
a problem of classical mechanics. I also touch briefly on your general theory in
the introduction, and you will see that I could not quite clear up what I call in
my terminology "independent motion" with respect to your theory. I would be
interested to know with which interpretation you agree. This point, which seems
to me very essential, is ambiguous also in Schlick.[5]

Concerning the entire paper, I naturally count on your giving me your true
opinion without sentimentality and also about whether you find publication worthwhile. In any case, I am already very pleased with what I have learned for myself
in the two months I spent working on it.

I shall speak with my lawyer about your kind offer in your last postcard.[6]

As I could not review some pages of the duplicate—so as not to delay the
sending any further—a few errors distracting to the content have remained:

Page 83, line 10 from the bottom, instead of "continuous" [*unausgesetzte*] it should read "assumed" [*vorausgesetzte*].

Page 15, point 4, the words "determined by gravitation and inertia" are missing; and in the note at the top, instead of "geodetic line" it should read "gravitation + inertia."

All in all, the paper has become a bit inflated, but I did not have the time to be as brief as I would have liked. Cordial regards, yours,

Fr. Adler

The paper by Schlick will be returned to you. Many thanks for it!

330.   To Eduard Hartmann[1]

[Berlin, 27 April 1917]

Einstein repudiates an objection to his general theory of relativity that had been raised by Ernst Gehrcke

. . . the difficulty of answering it lies in that, in order to give the case full treatment, the cosmos would have to be drawn in, the overall structure of which theory has not given any definite information until now. Not even in the question of whether the universe ought to be conceived spatially or as finite or infinite is there any transparency. I wrote a paper on this point recently[2] . . . it is easy to see that Gehrcke's objection has no weight. It predicts that gravitation lines (on the positive side) could only end in masses, from which the absurd inference then immediately results.[3] This condition is borrowed from Newton's gravitation theory, which can obviously only claim validity in the static case. The gravitational field perceived by the accelerated observer is not produced *statically*, though, but *dynamically* by the distant masses, through their acceleration.[4] In analogy, the generation of electrical (electromotive) force in Faraday's induction could be named, which originates from a temporally variable current.

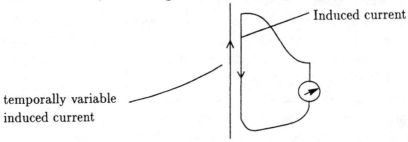

Induced current

temporally variable
induced current

In fact, the following results are obtained when calculating to first-order approximation in the new theory.

A hollow sphere *at rest* generates no motive gravitational field within its interior. But if it is accelerated (in the direction of the arrow), it then generates within its interior a gravitational field in the same orientation, which would be capable of setting masses into motion.

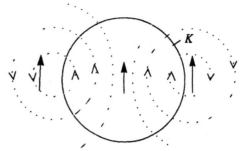

However, this escapes observation owing to its minuteness. On the other hand, the resting hollow sphere K produces an increase in inertia of a test object encased within it, which is again also undetectably small.[5]

. . . Now, in order to derive the gravitational field perceived by an accelerated observer, one would have to set, instead of the hollow sphere, the entirety of all the masses contained in the universe, which effort founders against our ignorance of the large-scale structure of the world . . .

Einstein cannot take a position on the recipient's paper, as he is unclear

about what is to be understood by a realistic conception of the world[6] . . . From reading philosophical books I had to learn that I stand there like a blind man before a painting. I grasp only the inductive method at work . . . but the works of speculative philosophy are inaccessible to me . . .

331. To Michele Besso

[Berlin,] Sunday. [29 April 1917][1]

Dear Michele,

I let you both wait so long for a message because I myself was undecided about whether a petition ought to be made regarding A. or not because I doubted its usefulness and because, as I have repeatedly heard, sympathy for A. is supposedly very great in his hometown.[2] However, it cannot *harm* the cause, and that is why it should happen. If you prefer to have the letter directed elsewhere to where I had initially thought, I remain in agreement and shall be happy to confess to being the instigator before the Society and, if need be, also in the petition. If you prefer not to mention this explicitly, I am also satisfied. Come to that, I give you both the authority to do as you see fit in my name.

Emphasis should be given to the fact that, during his years of teaching,[3] A. showed himself to be a selfless, composed, diligent, kindhearted, and scrupulous person who enjoyed the respect of all,[4] and that it is therefore a wish close to our hearts to put in a good word for him.—For your personal information: A. is a quite sterile rabbinical mind, inflexible, with no sense of reality; ultra-selfless, with a heavy self-tormenting, even suicidal streak. A true martyr type. When it was a question of which of us (he or I) ought to be appointed to the Zurich chair, Privy Councillor Ernst would have liked to call him, since he was a fellow party member. But his statements to Ernst about himself and about me made it impossible for Ernst to advocate *his* nomination.[5] This I know from Ernst himself, who told me so during my visit. You cannot say too much ⟨positive⟩ about his scientific qualities; nonetheless, you can say that he was a careful thinker who sought (with success) to work his way through to lucidity. I just received a manuscript from him, completed a few days ago, on relativity[6] in which he presents extremely expansively and with prophetic conviction quite worthless hypercritical points, so that I am in the awkward predicament of what I should say about it. I am constantly racking my brains over it. He rides Mach's hack to exhaustion. You both will certainly know better than I how to strike a responsive chord, wise as you are in the ways of man. I convey my cordial thanks to Mr. Beck for his willingness and likewise to Mr. Erismann, who has written me an express postcard.–[7]

I received a very cheerful letter from my Albert, which pleased me tremendously. The boy is bursting with a zest for life, thank heavens, despite his unfortunate lot, which looks tragic from the outside. This is happy childhood and the healthy environment in Zurich! Yesterday I presented a little thing on the Sommerfeld-Epstein formulation of quantum theory before the thinned ranks of our Phys. Society. I want to write it up in the next few days.[8] L. Civita wrote a critical paper on gen. rel.[9] I think, though, that he is mistaken. I'll tell you about it in the summer. When should I come to see you all? It is still uncertain whether Albert wants to come with me; he expresses himself very comically and guardedly about this in his letter: "How about if *just* us old chums went on a little excursion together?" What's the matter with Zangger? He seems so depressed.

Warm regards, yours,

Albert.

332.   To Emil Beck

[Berlin, 30 April 1917]

Upholding application. Letter Besso.[1] Greetings—Einstein.

Document description: "1917 10 o'clock."

333.   From Michele Besso

Zurich, 4 May 1917

Dear Albert,

The motion was discussed this evening by the Physical Society and was passed in the sense that his former colleagues and friends bear witness, from their own experience, to the personal qualities and scientific efforts of the former member (entrusted to the hands of the defense attorney, who should make use of the attestation at his discretion).–[1]

Our dear friend Zangger reported about your family for you: Miza and the little one are—very well provided for—at the private hospital;[2] your little boy is so well that with his vitality he does not quite fit into the surroundings: But now it will be possible, after all, despite the late spring, to bring him to Arosa, where it is hoped that an extended stay will secure him lasting health.[3]

Little Albert is also there now, since Sunday, because he had a sudden bout of fever: very many such unexplained minor sicknesses of short duration have appeared among the children here, and it is obviously nothing grave in his case either.– Then friend Zangger will take him in temporarily: He had been there for one or two days already before he was overcome by the fever attack (which just set in exactly when he was visiting his mother) and felt very comfortable there.[4] I hope to be able to meet with him more often now as well.– Then it will be revealed whether a few months of absolute rest can help Miza.—Life is a strange thing. I believe it is—by definition—surely always worth living through: but the worst it can hold and the best are not so terribly far apart from one another and ⟨not substantially⟩ different to the most absolute nonexistence we can imagine for ourselves.—Written down, it is empty rhetoric; but you know how I feel about it! Your old

Besso.

334. From Michele Besso

Zurich, 5 May 1917

Dear Albert,

You do not need to extol our wisdom in the ways of man in the least;[1] it embarrasses me very much, because I appear to myself like a wanderer with an inpenetrable wall of fog before him who, at most, can see thinned spots upon looking back, which reveal to him—how much *this also* is an illusion, who can say?—what he ought to have done and what abysses he had shuffled past in his sleep. Thus *now* it is clear to me how I ought to have dealt with the Phys. Soc. affair. But I just did it as best I could. A stormy evening was its price, and we shall have to wait and see what else will come of it.[2]

As regards Mach's little steed, let us not berate it so much; did it not serve trustily in the hellish ride through the relativities?[3] And who knows whether it does not carry its rider, Don Quixote de la Einsta, through the nasty quanta as well!

Well: We are all very much looking forward to your arrival here, Zangger, Dällenbach, Besso the elder, and Besso the younger:[4] Vero took the school-leaving examination this spring but is remaining in Glarisegg for another half-year and is going to try to give his former classmate extra help in studying optics. I am curious how he will make out! If the Lord means fairly well by us, an agreeable hiking tour, at least by the two of us, but if possible also with our two boys and with Dällenbach, should not be denied us.

Zangger is much troubled in the first place by his book,[5] in the second by a referee report, malicious among many others (I mean, one of the most malicious imaginable);[6] added to this, he also had physical complaints (extensive periostitis of the lower jaw, probably from his experimental examinations of commercial poisons) and is also being tormented by his dear colleagues.– I, on the other hand, am tormented neither by colleagues nor by pupils: I had to persuade our former colleague at the Office Furrer[7] to attend the "lecture" in order to be able to hold it at all.[8] On the other hand, I also am riding up against my modest little windmills with some forensic expert opinions, among these, two which Zangger just referred me to, precisely that nasty one, which also has a technical aspect, and a simple question regarding elevator design but which regularly claims accident victims in Zurich without anyone being able to bring himself to take any official measures until now.

Come soon, we are all waiting eagerly! Yours,

Michele

P. S. Oseen: The electron theory of metals calculates out once again that establishing Jean's radiation equivalency would require enormous amounts of time.[9] I know that we had brought up this point long ago; if I am not mistaken, it is Wien's displacement law in combination with the universality of black body radiation, which reveal that there is nothing to it of substance.   The fact that Planck's radiation equivalency can be established instantaneously is connected to the quantity of short-wave quanta, isn't it?

335.   To Michele Besso

[Berlin,] ⟨8⟩7 May 1917

Dear Michele,

I am sending you some papers with the request that you forward them to Dr. L. Silberstein,[1] 4 Anson Road, Cricklewood, London N.W. 2, who has asked me for them. He is the brother of the woman at whose home we were together.[2] Did the A. affair come through well?[3] I would like to lodge my Albert at my sister's, since I do not consider it good for him to grow up here, after all.[4] Other options might be: Tanner, cantonal schoolteacher in Frauenfeld, a former student of mine, and Glarisegg.[5] How much does it cost there? Do talk to Maja & Paul once about whether they would like to take in Albert. I would pay them for room and board a total of Fr 2,000 annually. But don't say that the inquiry comes from me so that they can refuse without embarrassment if the burden seems to them too great. My wife should not be taken into account in this matter, but rather only what is advisable for the boy. I am very much looking forward to our reunion in July. My doctor here and Zangger want me to go to Tarasp to take a cure.[6] But I have difficulty mustering up the necessary superstition for it. I should be very glad if Albert were finally taken care of very soon. Help me to relieve Zangger very soon;[7] he is saddled with far too much already.

Heartfelt greetings also to Anna and Vero, yours,

Albert.

336.   From Friedrich Adler

Vienna VIII, 1 Alser St., 7 May 1917

Dear friend Einstein,

Hearty thanks for your long letter of the 1st inst., which I received today.[1] As you know, too many demands are being made on me otherwise at the moment[2] to be able to answer you thoroughly. Hence only a couple of allusions regarding the relativity postulate.

1) We may be of one opinion on the conditions of the "reality" of Copernican coordinate systems, as you will see when I send you the next chapter (on "inertial systems"), which is essentially finished but not yet developed enough for transcription. Incidentally, it seems to me, the sense in which "real" is to be understood has been clarified long ago by Lange and especially by Seeliger (Munich *Berichte* 1906).[3] Although the center of gravity is not always known, it is *always traceable*, no matter what displacements occur within the system. The center of gravity is unique in this respect as well.

2) I share your opinion entirely and at various places in the paper have also stated that the legitimacy of all Copernican systems also entails a "relativity postulate," but a limited one. *All* coordinate systems are "Copernican"; they differ only in the range in which they are "Copernican." Thus it is possible to proceed initially from *any* coordinate system, but determinations of mass are possible only to the extent that the objects belong within the "inner region."

3) It is completely correct that masses, accelerations, and the coordinate system are needed to ascertain the center of gravity. But the crucial point is that one never has the masses without the related Copernican coordinate system. If, therefore, one wants to apply dynamics to a system of *material bodies*, thought *experiments* must be performed in which an approximation of the Copernican coordinate system is sought. The point of origin of the Copernic. coordinate system and the *possibility* of examining the masses of this system of bodies are found *at the same time*. This becomes clearly apparent in the astronomers' efforts to find the "Copernican system" of the fixed-star system. Only once this has been achieved can there be a true *dynamics* for fixed stars. A decisive counterargument to my conception would be if a method were successfully found according to which the masses of the fixed stars could be indicated without using the connected Copernican system. I know of no such possibility, though.

Cordial greetings, yours,

Fr. Adler.

337.   From Walther Rathenau

[Berlin,] 10–11 May 1917

Dear and esteemed Mr. Einstein,

I have been immersed in your ideas for weeks; I had barely finished the evangelist Schlick when the *verba magistri* arrived, which are now before me and for which I thank you wholeheartedly.[1]

First a preliminary remark, which is not meant to be a platitude: the prophet is clearer than the evangelist. I would not have thought it possible to force such a radical rearrangement of ideas through, the way you do, with such simple means and using such classical architectonics—I underscore the word classical, in contrast to your "bumpy."[2]

I have read to p. 39[3] and do not say that it comes easily to me, but certainly relatively easily—as everything that you touch becomes relative. Perhaps I am complicating the matter for myself because all sorts of rudimentary ideas from various sources have led me within the proximity of your force field, and because now I must take in the radiating effects and assimilate them within the existing chains of reasoning.

Shall I tell you a bit about such rudimentary thoughts? Within the light of your halo they will appear as pitiful unmasked ghosts—but maybe I can extend my bizarre thanks to you for my joy and admiration by making you smile for a minute or two. Will the things still occur to me now, around midnight? Let us enumerate, then it will work.

1. Gyroscopes always seemed senseless to me. When built with precision, how does it know that it is rotating? How does it distinguish the direction in space in which it does not want to let itself be tilted? Even if I put it in a box and make it blind, it knows where the polestar is. I have always had the secret feeling that it rotates only when it has a spectator. But if so—it would then have to protect itself with counterforces against the approach of such spectators from infinity. Are there such forces?

2. Ever since the arrival of vestibule trains, I have found walking in the corridors not only an ordeal but also a problem, hence a pleasure. Often I imagined a vestibule train that extended from Berlin not quite all the way to Paris but only to St. Quentin, and within it a smaller one to Verviers, etc. Then one would arrive quite quickly in Paris.[4] Well, there is an end to this anyway—so I do not have much to lose.

3. The smaller insects are, the faster they move. It is customary to feel sorry for dayflies. I told myself sometimes: Maybe it is not so bad; in the end, time diminishes with mass.—Or is it only the sense of time?—If one had to play the

allegro to the 5th [symphony] for such a gnat, it would be over in one minute, otherwise it would mistake it for a funeral march in C major.[5]

Well, time really does depend upon motion! But we do not notice it! And motorcar driving was a fleeting pleasure nonetheless.

(Now it is getting more and more nonsensical and muddled; I think you should not read on. But ultimately, all thoughts arise from elements of lunacy; it is only the critical cement which is lacking here to glue them together.)

4. One thing which is entirely unrelated and yet which seems to me to have a remote connection: I have always had an emotional resistance to the entropy business. It seems to me as if it were correct only inside a bathtub. And moreover: What happens to rays of light that go out into the distance without ever meeting an object? As long as there is a medium: fine. But why should it not come to an end—or be interrupted?

5. A metaphysical frivolity: My senses tell me that everything is at rest in the absolute. Illustration—somewhat in our Lord's spirit: a trip to Italy. I travel the length and breadth of the land, experience it minute by minute, register it, and cut it up into sections. I come home and have a concept of: Italy. An impression like the taste of a fruit or the character of a woman. I can scroll through it again (using the memory sections) and do so when I want to answer individual questions. But otherwise, Italy *rests* within me, is present, alive, and yet motionless. I possess the whole (unfortunately only figuratively speaking, since travel and life are finite)—or I possess at least *a* whole.

Your illustration of the two flashes of lightning and the train really gripped me here[6] (incidentally, I turn it into two dynamite explosions and a czar train). What startles the czar twice is only *a single* matter for the assassin. His rest (in both senses) is greater. Now I expand further. The assassin stands outside the train. Now I place him beyond the Earth's rotation; then beyond the Earth's orbit; then beyond the translation; then beyond—etc. Will the man not be surrounded by an ever deeper silence?

Another sidetrack: Time (epistemologically) dissipates. It only comes into existence, so to speak, through motion. Motion, however, has time as its precondition again, for it is $\frac{s}{t}$. Would we not have to find a way out (epistemologically) by conceiving of motion as a manifestation of force (thus obversely to normally), so we merely note that elements containing varying amounts of force (metaphysical charge) exist—thus ultimately a kind of monadology would emerge?

Now, enough of this. Instead of cheerful thanks I have given you a *pathologia mentalis*, which must have appalled you. Nevertheless, I am sending the letter off, because it has to prove one thing to you *in corpore viti*:[7] the forceful effect of your ideas on a poorly shielded brain. A steel helmet [*Stahlhelm*] is necessary, or at least a sun hat, in order to hold the balance.

My regards in amicable admiration, yours,

Rathenau.

338.   To Paul Mamroth

[Berlin,] 11 May 1917

Esteemed Dir. Mamroth,[1]

Thank you very much for the two essays you sent me, both of which I have read with the greatest interest. The funeral oration is a masterpiece that stands alone.[2] The Judaism–Christianity issue must be answered variously, depending on the political view held by the (free-thinking) Jew examining the question. Rathenau himself wrote unofficially on the issue again just a few weeks ago in eminently witty and fine style. Perhaps he will give you a duplicate copy of his reflections, which are in the form of a letter.[3] If I were disposed toward the opposition and I saw in the state church an objectionable means of encouraging people to maintain a mentality convenient for the ruling caste, then naturally I would not support this established church. But if I loved the established church as a preserving element of the state which was according to my taste (not *mine*), then I, as a free thinker, might safely join it . . . .

Thanking you exceedingly once again for the kind communication, I am yours truly,

sig. A. Einstein.

339.   To Michele Besso

[Berlin,] 13 May 1917

Dear Michele,

Thanks from the bottom of my heart for handling the A. affair, also to the Phys. Soc.[1] The general view is that there is no serious threat to the man. Now I am firmly resolved to place Albert with Maja.[2] Talk to Zangger about it sometime; I am writing you first so that Albert is not worried prematurely. It is very charming that Vero is tutoring in Glarisegg.[3] What an endearing little fellow he is; I am thrilled with him. I'm not complaining about Mach's little mount; you know what I think of it. However, it cannot bring forth any living thing but only exterminate harmful vermin. If you had savored A's longwinded and elaborate exposition, you would understand quite easily my metaphor of the nag ridden to its death.[4] My doctor[5] is absolutely demanding that I go to Tarasp for a cure; I still cannot believe that I should allow my holidays to be spoiled this way; maybe Zangger's powerful verdict will save me from it. By contrast, I am committing myself to do everything else—which is unbelievable— to abstain from drinking, etc., in short, to perform the rites of medicine loyally and piously. I regret Zangger's troubles very much.[6] If he had my thick skin against human concerns, it would be less serious. The periostitis I like least of all; even colleagues, etc., are more pleasant than that. I am sorry that you have no students, also for the students who don't have you.[7] People are simply overburdened with obligations. I am coming at the beginning of July; earlier is impossible for me because of lectures, Phys. Society, and also because of a toe, which I had the misfortune of breaking. Nernst lost his two sons in the war . . . is old Jehova still alive?[8] These people's psychology is peculiar. I have forgotten how to hate. Warm regards, yours,

Albert.

What is happening to the apartment? I owe it to my children to be thrifty. Talk to Maja.[9]

Best regards to Anna, Vero, and Dällenbach.

340. To Michele Besso

[Berlin,] Tuesday. [15 May 1917]

Dear Michele,

Yesterday I learned that certain supernumerary earnings I had in the recent past are about to cease, so I must think in earnest of arranging my family's and my own livelihood as frugally as possible. After subtracting the tax, my annual income totals about 13,000 M.[1] Of it, what I give *regularly* for my family's support is $5,600 \cdot \dfrac{125}{100} = 7,000$ M,[2] thus over half already. If substantial irregular expenses are added to this, the case can easily arise that I am forced to use up my savings, which would be irresponsible toward the children. I beg you to consider this with Zangger and to take it into account in your decisions. This aspect has to be borne in mind apart from the purely medical one. For inst., I intend to omit the planned cure at Tarasp as well and to seek relaxation in Lucerne instead.[3] My mother also is dependent on my support (ca. 600).

I mean to come to Switzerland at the beginning of July and am looking forward to it very much. Albert will be accommodated at my sister's by then, I hope.[4] Thanks again for your efforts at the Phys. Society;[5] it is to be hoped they will be of some use.

Affectionate greetings, yours,

Albert.

341.   To David Hilbert

[Berlin,] 19 May 1917

Dear Colleague,

I would very much like to participate in your interesting conferences.[1] However, I have a serious liver condition[2] and must avoid all temptations away from maintaining a hypochondriacal lifestyle, entirely on doctor's orders, although it has improved again somewhat compared to 4 months ago.

Wishing you a productive meeting, though, and favorable work, with best regards, yours,

A. Einstein.

It seems that we simultaneously found the same generalization of the quantum law.[3]

342.   From Heinrich Zangger

Zurich, 20 May 1917

Dear friend Einstein,

I received letter & postcard. At such distance, you have many a worry you would not have here.

It is good news that job & income are continuing, but here there would also be position & money & everything without having to be Kappeler, to whom you allude, comfortably arranged without all the obligations.[1]

Albert is very cheerful & stable, a good, careful, mentally alert, dear fellow. He has been here for almost ⟨a fortnight⟩ 4 weeks now. The day following his move he fell sick at his mother's with diarrhea, that's all.[2] Now he can eat everything again. For a time we both followed the same diet together. He stayed in bed for 4 days.

For the little one, I have arranged a place in Dr. Pedolin's children's sanatorium in Arosa, in agreement with Prof. Bernheim,[3] albeit for Fr 10 per day; but the sanatoria are very full now because of the foreign children, so that one has to be happy to have anything first-class at all. Unless you report otherwise, I shall take the little boy there as soon as the weather is good & then report address, etc., to you from there. Next week I am going to Berne & shall try to procure a Swiss remedy for your stomach.[4]

This will interest you: The Swiss government has indicated it will stand by me, so that I did not even have to run out all my cannons & did not have to fire any of them; so these are welcome successes.–[5]

Your sister wanted to visit me; I was away.

Albert likes staying here very much—Papa has little to say here; come here to give orders. Regards,

Zangger

Incidentally: My wife simply went there & picked up A. with no protests.

Postscript to the letter.

Meanwhile I was in Berne. You should "lodge a package complaint" in case they do not arrive on time. At all events, they are going to arrive between 5–10 July. You should not subject Albert to the disappointment of expecting you & then not coming for the third time; promises to children . . .

Your little boy will certainly be brought to Arosa as soon as the weather is stable; he is looking forward to going "on holiday," like the other children. You obviously wrote your book with Albert very much in mind;[6] if you were closer, you would have adopted a different tone in a number of instances. It is a pleasure for me to read, & particularly for another special reason you will not guess . . . I have often considered how this could be presented to the solid, naive thinker. I have also told Besso that I am considering describing relativity as I see it and have & am experiencing it.[7]

Should I not give the little book to Albert? He is a bit thin, but his mother says that he always is a bit lanky in the summer. He is eagerly awaiting the holiday excursion with Papa. Now he is cutting flowers in our garden, to bring to his mother in the evening. Today I was at the hospital for a longer period of time. The little one is doing decidedly much better, but we do not know how times will be; let us all invest what capital is available on health before the forties.

Albert is going to write in the next few days.

343.   To Moritz Schlick

[Berlin,] 21 May 1917

Esteemed Colleague,

Time and again I look over your little book and delight in the splendidly clear discourse.[1] I also find the last section, "Links to Philosophy," superb. If something did catch my eye while occupied with this ruminating business, I let you know, so that you can make some corrections for an even[tual] new edition.

The discussion on the invalidity of Euclidean geometry at the top of page 33 is misleading. It cannot be said that Euclidean geometry is not valid for two systems rotating relative to each other. Rather, the following can be deduced: Assuming system K is Galilean, or that a system K exists, in which the possible positionings of practically rigid bodies (at least in a certain region) at rest rel. to K are governed by Eucl. geometry, then the same surely is not the case for a system K' rotating relative to K.—(In this proof, systems K and K' thus play entirely different roles.) From this it is deduced first that the existence of a gravitational field precludes the validity of Euclid. geometry (a field exists, of course, rel. to K'). Finally, from the circumstance that, upon closer examination, gravitational fields are absolutely never absent, it is deduced further that in reality Galilean coordinate systems do not exist at all within finite areas, that therefore Eucl. geometry is never valid within finite spaces.—[2]

The second point I would like to make regards the concept of reality.[3] Your interpretation compares to Mach's according to the following scheme:

Mach: Only perceptions are real.

Schlick: Perceptions and events (of a phys. nature) are real.

Now, it seems to me that different senses of the word "real" are being used, depending either on whether perceptions are being discussed or events and facts in the physical sense.

When two different nations engage in physics independently from each other, they will create systems that certainly concur concerning the perceptions ("elements" in Mach's sense). The theoretical constructions that the two devise to connect these "elements" can be considerably different. Both constructions do not need to agree with each other either regarding the "events"; because they clearly belong to the conceptual constructions. Certainly only the "elements" are real, in the sense of "irrefutably given in experience," but the "events" are not.

If, however, we denote as "real" what we have classified in the space-and-time scheme, as you have done in your *[General] Epistemology*,[4] then no doubt, the "events" are real in the first place.

What we describe as "real" in physics is unquestionably that which is "spatio-temporally classified," not what is "immediately given." What is immediately given can be an illusion, what is spatio-temporally classified can be a sterile concept contributing nothing toward illuminating the connections with what is immediately given. *I would like to suggest a clear conceptual distinction here.*[5]

Best regards, yours,

A. Einstein.

344.   To Paul Ehrenfest

[Berlin,] 25 May 1917

Dear Ehrenfest,

Don't swear at me for answering your kind invitation card only now! Unfortunately, I cannot think of a trip to Holland foɪ now.[1] At the beginning of July I'm going to Switzerland to see my children. The little one is very sickly and must go to Arosa for 1 year.[2] I'm going to go there with Albert. My wife is also ailing.[3] Worries and more worries. Nonetheless, I have found a nice generalization of the Sommerfeld-Epstein quantum law.[4] *Droste's dissertation is extraordinarily fine.*[5] This man must get a proper position soon.[6] I should discuss the relativity of inertia, etc., urgently with de Sitter and you;[7] but it is no good in writing. My health is in decent shape, but I have to follow a strict diet.[8] Pardon my scribblings to you here and cordial regards also to your family, yours,

Einstein.

345.   From Max Planck

[Berlin,] 26 May 1917

Dear Colleague,

If anything had been able to entice me yesterday to the Physical Society, it would have been the suggestion you relayed to my daughter[1] by telephone. However, I must tell you that I still feel physically incapable of work these days. I can take only absolute necessities in hand. My grief over my second daughter,[2] who passed away in my arms last week from pulmonary embolism, still gnaws too persistently at my thoughts for them to be able to exert their normal freedom of action. After Pentecost it will improve again. Until then, please make allowances for your devoted colleague

M. Planck

I console myself with the thought that, if subelectrons really do exist, they will also make themselves noticed otherwise than in the lecture I had to miss.[3]

346.   From Gustav Mie

curr[ently] Schierke, in the Harz, Haus Tannenheim, 30 May 1917

To Privy Councillor A. Einstein, Berlin, Kaiser Wilhelm Academy

Highly esteemed Colleague,

Privy Councillor Hilbert probably informed you a while ago that I am going to give a few lectures on the 5th, 6th, and 7th of June in Göttingen and that in these lectures I intend to present some thoughts on your theory of gravitation.[1] I consider it necessary to write you about what approximately I plan to say. I am convinced that, of all the theories of gravitation, your new theory is the only one that will induce science to investigate it thoroughly and to test all of its consequences, because it alone is determined entirely by intrinsically convincing principles.[2] The peculiar thing is, though, that these principles can be looked upon in quite different ways, depending on the standpoint one chooses; and at Göttingen I shall try to consider your theory for once from a standpoint other than the one you prefer and which had helped you find your way. For I want to apply to your theory the approach upon which I based my paper on the relativity of gravitational potential.[3] I am only employing considerations of a very general nature, and very much of what I am going to say is long since known to you, in any case. Despite this, however, through my very simple considerations on a very important point, I come to an interpretation different from yours, namely, on the meaning of general relativity, the general transformability of the equations. I believe I can show that this general transformability does not agree with the requirement of a truly finished, self-contained theory of the physical events, and that some means have to be devised to paralyze the uncertainty that comes with it.[4] I can offer some suggestions about these means, and I believe that your introduction, in particular, can lead to further interesting developments in the theory. ⟨For all that, owing to the issue of my appointment[5] and other matters, my time was unfortunately far too⟩ For all that, in the last year I have unfortunately had too little time at my disposal to apply differential geometry as much as is needed for such problems, and for this reason I must content myself with a general indication of the goal. You can imagine that I should be especially pleased if you were in Göttingen and were to participate in the discussion. In any case, I shall take the liberty of making available to you as soon as possible the exact content of my lectures.

In assuring you of my greatest respect, I am yours very truly,

Gustav Mie.

347.   From Wilhelm Wien

Würzburg, 8 Pleicher Ring-road, 1 June 1917

To Prof. Einstein, Berlin, 13 Wittelsbacher St.

Esteemed Colleague,

While reading your interesting paper on quantum theory ($P. Z.$ 1917, p. 121)[1] a thought occurred to me of which I would like to inform you.

You consider the equilibrium between molecules and radiation and come to the result that the balance can only subsist if the radiation is emitted from a particular direction and conveys a kinetic moment to the transmitting molecule.[2] This system can never be complete, however, because the impinging radiation releases photoelectric electrons and these, in their turn, excite radiation. Upon emission of an electron, a momentum of the quantity $\dfrac{2h\nu}{v}$ is conveyed when the relation $\dfrac{m}{2}v^2 = h\nu$ holds. On the other hand, upon collision, the emitted electron will transmit momenta whose quantities will remain unknown, since the electron does not necessarily have to be brought to rest by only a single inelastic impact but can experience changes in direction in the vicinity of the molecules, which likewise convey momentum. It ought to be entirely possible to satisfy the thermal equilibrium condition through suitable assumptions about the behavior of these electrons, without having to make the assumption of singly directed emission, which conflicts with the wave theory of light.

Our last meeting was very brief. I hope next year we can discuss science under more peaceful stars.

With best regards, yours,

W. Wien.

348.   To Gustav Mie

[Berlin,] 2 June 1917

Highly esteemed Colleague,

If my state of health were better, I could not have been dissuaded from traveling to Göttingen for your lectures, particularly since our colleague Hilbert had invited me to it very heartily more than once.[1] It would have been a great pleasure for me to hear your talks, from which I would have learned more, no doubt, than from a mute article. On this occasion, I have the urge to say something to you for which I had always lacked the opportunity. I deeply regretted that Lecher behaved so tactlessly toward you in Vienna at our discussion.[2] I beg you not to allow your feelings toward me to be clouded by this minor faux pas; I probably do not need to impress upon you that the proceedings of that time are in stark contrast with the enthusiastic respect I have for your scientific research.[3]

I am delighted that now you also value the generalized relativity idea as a hypothesis that stringently restricts the possibilities. That which you have called the relativity of gravitation potential,[4] must—if I have understood you correctly— apply to any system of field equations that is covariant under *linear* transformations; invariance under similarity transformations already suffices, of course. Your result that general transformability (covariance) is not compatible with the requirements of a complete theory was my opinion also three years ago. However, my reasons at the time were not sound.[5] I am very curious about what you are going to say regarding this crucially important problem. Like you, I believe that a specialization of the coordinate system *a posteriori* could perhaps enhance the theory very much. But no attempts made with this intention until now are genuinely satisfactory.[6]

At all events, please do give me a few hours of your time when your path leads you through Berlin. A conversation is usually a great deal more useful than mere correspondence. In wishing you happy and inspiring days in Göttingen, I am with best regards, yours very truly,

A. Einstein.

349.   To Wilhelm Wien

[Berlin, 2 June 1917]

. . . My remarks are limited to such absorption processes in which electrons are not released.[1] The fact that such do exist, we learn from Bohr's theory, for ex., and naturally also from direct experience. If, for inst., there is a Bohr monatomic hydrogen molecule, at sufficiently low temperature the third innermost electron orbit becomes improbable (rare) enough against the innermost and second innermost ones, that the only absorption and emission reactions to be taken into consideration are thus transfers between the innermost and second innermost orbits. In this case, the theory would be directly applicable. In cases where a multitude of transfer reactions can take place, it is assumed that any reaction corresponding to a pair of states conserves the thermodynamic equilibrium on its own. . . .

Einstein discusses Felix Ehrenhaft's theory of "subelectrons," providing equations for its testing. The last section is devoted to the question of the compatibility with the wave theory of directed momentum transfer:

. . . Although Pointing's theory of momentum is compatible with Maxwell's equations, it is not a consequence of them. Our inability until now to discover a detailed energy-momentum localization analogous to quanta, should not immediately lead to the interpretation that it is an impossibility . . .

350.   To Paul Ehrenfest

[Berlin,] 3 June 1917

Dear Ehrenfest,

I am corresponding quite busily with Adler because he has written something about relativity theory that is unfortunately only very flimsily supported.[1] But otherwise, he is a terrific fellow. It would be best to send offprints to his wife (Mrs. Adler, junior, for Dr. Fritz Adler, 1 Blümel Alley, Vienna VI). I optimistically hope that nothing will happen to him. He has the sympathy of all discriminating persons, also here more than would be expected. Slowly but surely, a change is generally taking place toward a moderate and natural mentality. But a tremendous amount of fertilizer is needed before the little plant can grow!

The generalization by Sommerfeld is as follows.[2] Let there be a problem in which as many integrals

$$L(q_\nu, p_\nu) = \text{const}$$

exist as degrees of freedom. Then the momenta p_ν can be expressed as (multiple) functions of q_ν. On the other hand, the path's curve fills a certain portion of the q_ν-space entirely, so that it comes arbitrarily close to every point in it. Then the system's path in the q_ν-space generates a vector field for the p_ν's. In a "Riemann foliated" q_ν-space, the p_ν's can be interpreted as *unique* and always constant functions of the q_ν's.

Now we regard the sum

$$d\sigma = \sum_\nu p_\nu dq_\nu,$$

formed for a random line element of the q_ν-space. This sum is invariant under coordinate transformations and is in addition *a complete differential*. The latter can be inferred from Jacobi's law.

The integral

$$\int d\sigma$$

extended over a closed curve in the "Riemann foliated" q_ν-space is now considered. This integral vanishes when the curve can be contracted continuously into a point which, owing to the "Riemannization," is by no means the case for all curves. Now, the quantum rule requires quite simply that for *any* closed curve,

$$\int \sum p_\nu dq_\nu = nh$$

has to apply. Epstein's special case is simply that each p_ν depends only on the related q_ν.[3] Handsome though this is, it is just limited to the special case where the p_ν's can be described as (multiple) functions of the q_ν's. It is interesting that precisely this limitation removes the validity of statistical mechanics. For it presupposes that, at the recurrence of the q_ν's, the p_ν's of a system left to itself gradually adopt all value systems compatible with the energy principle.

It seems to me that real mechanics is such that the existence of the integrals (which preclude the validity of statistical mechanics) is already secured by virtue of its general foundations. But how to implement this??

I shall be glad to write to Onnes.[4] Have you seen Nernst? He was in Holland to find out whether his pilot son, who has crashed, is dead. The other died in action right at the beginning of the war. Old Jehovah is still around.[5] Unfortunately He also smites the innocent, and *He smites the guilty terribly with blindness* so that they cannot feel guilt. So where does He get the right to punish and crush? Also from force, perchance?[6] I have become much more tolerant, without changing my views in principle in the least. I see that often the most power-hungry and politically extreme people could not kill a fly as private individuals. A delusion of epidemic proportions is afield which, after producing immeasurable suffering, will vanish again, so the generation after the next can gasp at it as something completely monstrous and incomprehensible. I am very eager to see Burgers's new paper.[7] You yourself complain again about yourself and are not satisfied with yourself. Just think how insignificant it will be 20 years from now how one knocked about on this Earth, provided no wrong was done. It is all the same whether you or someone else has written this or that paper. You are certainly never mindless, except for when you happen to be contemplating whether you are mindless or not. So, away with the hypochondriac! Enjoy yourself with your family in the fair land of the living!

Affectionate greetings, yours,

Einstein.

351.   To Willem de Sitter

[Berlin,] 14 June 1917

Dear de Sitter,

From your postcard and from your correction proof I see that you have misunderstood my paper on one point.[1] According to my conception, *apart from the stars* there is no "world matter," or at least, there does not have to be any. The density ρ in my paper is the density of matter which would result if the matter condensed in the stars were spread evenly throughout the interstellar spaces. "World matter" in my sense is thus *definitely present* in reality.[2] This density yields a cosmic radius of about 10^7 light-years. At most, it could be a question of whether its density does not vanish in parts of the universe. I have merely examined the *simplest* case conceivable, namely, the one where ρ is constant throughout. (This can naturally be valid only as a mean value for ρ in spaces that are large compared to the average distance of neighboring fixed stars.)

I cannot make sense of your four-dimensional structure[3]

$$ds^2 = f(h)[-dx^2 - dy^2 - dz^2 + dt^2]$$

mainly for the following reasons:

1) It could only exist without "world matter," that is, if there were no stars.

2) Your four-dimensional continuum does not have the property that all its points are equivalent; rather, it has a spatio-temporal center $x = y = z = t = 0$. This is the center of the conic section

$$1 + h^2 = 0$$

(with the above coordinate choice[4]). The spatial extension of your universe (in natural measure) depends in a peculiar way on t. For sufficiently negative $[t]$,[5] a rigid circular hoop can be placed in your world, for which there is no room in your world at $t = 0$.[6]

3) It seems to me that a reasonable interpretation of the present universe necessarily requires the approximate spatial constancy of g_{44}, owing to the fact of the small relative motion of the stars.–[7]

I would also like to emphasize particularly that the identifiability of the lines of the visible stars does not prove the approximate constancy of the $g_{11} \ldots g_{33}$'s but only the approximate constancy of g_{44}.[8]

With cordial regards and best wishes for your health, yours,

A. Einstein.

P. S. I shall be away on a trip to Switzerland in July and August ⟨and September⟩. My health is still shaky too.[9]

352.   From Paul Ehrenfest

[Leyden,] 14 June 1917

Dear Einstein,

Thank you very much indeed for your detail[ed] letter. In connection with your comm[ent] on quantum formulations, the enclosed summary of a calculation made by Burgers a few weeks ago may be of interest to you. I shall send you his calculations on band spectra very soon.[1]

—o—

Yesterday during an intermission between two talks at our physics colloquium, I made public the engagement of Nordström with the younger of the two van Leeuwen physics students. (Her name is Nelly van Leeuwen;[2] her older sister, H. van Leeuwen, is assistant to Lorentz and me and will soon take her doctorate under Lorentz.)[3]

The day before yesterday it was also a great surprise to my wife and me when the two came to see us. We know them both about equally well, had noticed— even though they were in front of our eyes almost daily—nothing in the least, were extremely astonished, but also absolutely pleased without restraint. This woman is a very capable, warmhearted, vivacious person and is an *excellent* match for Nordström. The entire big family of Leyden physicists and mathematicians (you know—we all hang together here very congenially) is celebrating the couple's joy merrily and boisterously, except perhaps for one or the other of the younger ones who was in love with Nelly ($n > 0$). It was incidentally very evident on this occasion in particular that we are largely united here, not only professionally but also personally. (N.B., in the last few months, an engagement epidemic has been surging through our building, and the first stop is always to my wife and me—but never were we as unreservedly pleased as this time!)

—o—

Nordström is preparing to write you—but don't wait for it!
Fond greetings, yours,

Ehrenfest

Onnes and de Sitter are now much better.[4]

353.   From Erwin Freundlich

Neubabelsberg Roy. Observatory, 17 June 1917

Dear Einstein,

In the following I would like to sketch my research plans briefly once again.[1]

The two most important experimental projects I would like to work on in connection with your theory of gravitation are:

1) Verification of the deflection of light rays in a gravitational field.

2) Verification of the relative shift of spectral lines in gravitational fields of differing intensity.

As concerns the execution of both projects, essentially no special instruments are required, rather only large astronomical telescopes with the latest accessories, such as those available at the Babelsberg, Berlin, Roy. Observatory. I have merely sought to refine details of the measuring instruments and the observation methods employed in processing the observation data gained. Thus at the beginning, the research work will cover the theory underlying the new measurement methods and the efficiency of the observation methods.

1) The first project, that is, the investigation on the deflection of light rays in gravitational fields, can be carried out by photographing the background sky during a total solar eclipse and comparing the stellar distances on such plates with corresponding shots of the same region of the sky obtained some months before & after the solar eclipse.

For the expedition undertaken by Dr. Zurhellen and me in 1914, which yielded no results because of the outbreak of the war,[2] we had devised a series of improvements toward obtaining sufficiently precise data, which are of some consequence to astrophotography and the examination of which would be *important preparatory work* for the coming unusually favorable solar eclipse in 1919.

In order to become independent of the rare moments of total solar eclipse in studying these problems, however, I have also focused my attention on the following two methods:

a) The exposure of stars in the proximity of the Sun in daytime. My initial efforts in this direction could never be taken up practically, since neither funds nor instruments were made available to me.[3] The fact that such exposures have already been tried out with success in England by Mr. Lindemann a short while ago proves that this method is not without prospects.[4]

b) Photographic exposures of stellar occultations by Jupiter using Kapteyn's parallax method.[5] I have described this method in more detail in an unpublished paper "Über die Möglichkeit und die Methoden, welche die Astronomie zur Begründung einer Theorie der Gravitation bieten kann" ["On the Possibilities and Methods Astronomy Can Offer toward Establishing the Theory of Gravitation"]. It requires a photographic telescope with a large focal length. As the precision called for is very high, I presented my plan to the American astronomer Slocum to get an expert opinion, since Slocum in particular is most experienced with Kapteyn's parallax method.[6] He not only pronounced my plan feasible, but at that time invited me in the name of the director of the Yerkes Observatory[7] to take up the research over there. In order to obtain greater precision in the plate measurements, in 1914 I submitted to the Zeiss Company an idea for a new stellar-plate measuring instrument, the construction of which was assured me; at the same time, I took up the application of photoelectric methods for determining stellar positions on photographic exposures and am currently having a measuring instrument built by the Toepfer Company in Potsdam along these lines.[8] The development of this method thus requires a series of preparatory researches which are anyway of interest in astrophotography.

2) Verification of the relative shift of spectral lines on the basis of the general theory of relativity can also be tackled in various ways.

In the case of solar lines it is not definitely establishable whether an actually identified redshift of its lines speaks for the new theory (see E. Freundlich, *Phys. Zeitschrift* 1914, pg 369–371),[9] since various effects overlap one another. I have therefore attempted to apply the more powerful gravitational fields of larger fixed stars to the study of this question. In a statistical analysis published in the *Astr. Nachrichten*, No. 4826, I demonstrated that for the B (Orion) stars, an effect of the expected type is indeed indicated.[10]

A careful implementation of this method is of extreme interest on a purely astronomical basis as well.

Another method of investigating this problem is provided by spectroscopic binary stars in whose spectra both components are separately distinguishable.[11] The necessary precision for obtaining reliable results is anticipated with the introduction into astrophysics by P. P. Koch of a measuring instrument employing photoelectric cells.[12] A measuring appliance of this type, developed and completed for the purposes of such astronomical examinations, is being built for me at this time. It is projected to allow determining the spectral lines in fixed-star spectra and ascertaining the energy distribution within the spectra with an accuracy not even remotely achieved previously and promises to deliver important results for countless other types of problems.

In addition to these possibilities toward testing the general theory of relativity, there are other important theoretical points of attack as well, such as the spherically shaped star clusters, the Milky Way, etc., where astronomy also is called upon to provide important assistance in testing new physical theories.

Yours truly,

E. Finlay Freundlich.

354.   From Max von Laue

Würzburg, 18 June 1917

Dear Einstein,

With great regret I heard yesterday in Frankfurt that your doctor only wants to grant you such a short stay in Frankfurt.[1] Wachsmuth and I have decided to monopolize you at least for the evening following the talk, and Wachsmuth will probably give you the details soon.[2] But I would like to have you for a bit longer and for once not in a big throng of people. That is why I would like to know the following: On which train are you arriving, and when do you have to leave?

Additionally, I have a request: Do commend our colleague Born as much as possible in Frankfurt. This can be done with a good conscience, of course. For if all else fails, I should like to try to swap positions with him;[3] and the only problem that could possibly arise in this, as far as I can see, lies in Frankfurt.

With cordial regards and hoping that you will soon be feeling better, yours,

M. Laue.

355.   From Willem de Sitter

Doorn, 20 June 1917

Dear Einstein,

From the smallness of the relative stellar velocities, one can also actually just draw conclusions about g_{44} and not about the spatial g_{ij}'s.[1] Actually, nothing can be inferred from the *velocities*, rather, the line of reasoning is: If the accelerations were not small, the velocities could not remain small, and if the accelerations are small, $g_{44} - 1$ must be small. Let us imagine that the velocities are of the order α; then if $g_{44} - 1$ is of the order γ and $g_{ij} + \delta_{ij}$ of the order β, the accelerations thus contain terms of the orders γ, γ^2, $\beta\gamma$, $\alpha^2\beta$, $\gamma\alpha^2$, etc.—but *not β alone*. Thus only the smallness of γ, not β, can be inferred from the smallness of α and the accelerations.[2]

Lately, I have been working on the problem of computing the field of a small sphere in the three systems.[3] This has long since been done for system C, of course (Schwarzschild, etc.).[4] In system A, I thus imagine the world matter condensed into a sun at *one* point only ($x = y = z = 0$, or $r = 0$), not anywhere else. Then I am surely permitted to interpret the problem such that I set $\rho = \rho_0 + \rho_1$, where ρ_0 is the density of the cosmic matter and a constant: $\kappa\rho_0 = 2\lambda$; and ρ_1 the density of the "normal matter," hence $\rho_1 = 0$ for $r > \mathbf{r}$, if $\mathbf{r}$ is a *small* positive number. Then $\rho_0 = 0$ throughout system B. In A, $\lambda = \dfrac{1}{R^2}$, in B, $\lambda = \dfrac{3}{R^2}$, and in C, $\lambda = 0$ (and $\rho_0 = 0$). I now imagine a stationary state, thus the normal matter and the cosmic matter are at rest. If $g_{\mu\nu}^0$ stands for the $g_{\mu\nu}$'s without normal matter, thus[5]

A: $ds^2 = -dr^2 - R^2 \sin^2 \dfrac{r}{R}(d\psi^2 + \sin^2 \psi d\vartheta^2) + dt^2$

B: $ds^2 = -dr^2 - R^2 \sin^2 \dfrac{r}{R}(d\psi^2 + \sin^2 \psi d\vartheta^2) + \cos^2 \dfrac{r}{R}dt^2$

C: $ds^2 = -dr^2 - r^2(d\psi^2 + \sin^2 \psi d\vartheta^2) + dt^2$,

then the $g_{\mu\nu} - g_{\mu\nu}^0$'s $(= \gamma_{\mu\nu})$ must be small. The $T_{\mu\nu}$'s are: $T_{\mu\nu}^0 = 0$, apart from $T_{44}^0 = g_{44}\rho$; $T_{\mu\nu} = T_{\mu\nu}^0 + T_{\mu\nu}'$ and $T_{\mu\nu}'$ is of the order $\gamma_{\mu\nu} \times \rho$.

I did *not* take the $T_{\mu\nu}'$'s into account. They may be disregarded, though, [in an approximate solution up to the *first* order in $\gamma_{\mu\nu}$ (or κ)],[6] only when $\kappa\rho_0$ is of the same order as $\kappa\rho_1$, that is, when λ is of the same order as $\kappa\rho_1$, within systems B and C, $\rho = 0$ outside of the "sun," and a complete solution can be obtained with $T_{\mu\nu} = T_{\mu\nu}^0$. In system A, however, one cannot get beyond the first approximation without an assumption about the $T_{\mu\nu}'$'s, that is, about the pull and pressure effects in the "world matter" as a consequence of the presence of the "sun." But even the first approximation leads to interesting results.[7] For I find in system A:[8]

$$g_{44} = 1 - \mathrm{a}\cot\frac{r}{R} \qquad \mathrm{a} = \int_0^r \kappa R^{(-)}\rho_1 \sin^2 \frac{r}{R}dr.$$

For a spherical world, at the antipodal point of the sun, we would have ($r = \pi R)g_{44} = \infty$. For $r = \dfrac{1}{2}\pi R$, g_{44} becomes $= 1$. Hence, it would *not* be permissible to assume a spherical world, but the world must be imagined as "elliptical,"[9] that is, the greatest possible distance between two points is $\dfrac{1}{2}\pi R$, two (straight) lines intersect each other only at *one* point, not two, etc. Schwarzschild already concluded the same in 1900 (*Vierteljahrsch. der Astron. Gesellsch.*, p. 337)[10] from the improbability that all of the light rays emitted from the sun intersect one another again at the antipodal point.—In this elliptic space, however, there are no negative parallaxes.[11] For the parallax is $\pi = a \cot \dfrac{r}{R}$ (better, $\tan \pi = \sin \dfrac{a}{R}$ $\cdot \cot \dfrac{r}{R}$). Also in system B it would have to be assumed that $r \leq \dfrac{1}{2} \pi R$. In this system, the parallax has a minimum $\pi_0 = \dfrac{a}{R}$. The establishment of a [lower] limit for R from the nonexistence of negative parallaxes is thus also eliminated.

With cordial regards, yours,

W. de Sitter.

356.   To Willem de Sitter

[Berlin,] 22 June 1917

Dear de Sitter,

Your detailed letter pleased me very much because I see how deeply you have been thinking about the problem of mutual interest to us. I have just a few comments to add.[1]

1) My opinion is not that the sphere must approximate the world *well*. The system could actually be quite irregularly curved, also on a large scale, that is, it could relate to the spherical world like a potato's surface to a sphere's surface. Large parts of it could then really be void (without matter). The sphere only serves to show, through an idealization, that a spatially closed (finite) system is possible.[2] If, therefore, coordinated Milky Way systems really do exist (a view which, as far as I know, not all astronomers share), there does not have to be matter in the space between them, by any means. It is not necessary to assume that matter in any form other than that of the stars exists. However, it is necessary to assume that the world is considerably bigger than the Milky Way's 10^4 light-y[ears].

2) I do not quite know what you mean by *finitude* and *boundedness*; I think of it as follows. Between two points, A and B, there are space-like geodetic paths; the length of the shortest of these paths is l_{AB}. Now, if a number G exists such that for any choice of A and B

$$l_{AB} < G,$$

I call the world (spatially) finite. It is probably always possible to regard such a continuum as *closed* in my sense. The naturally measured volume is then finite (strictly speaking, one can only talk about the world being closed if it is "static").

3) When I say that your world has a preferred center,[3] I mean that the points are not of equal value, apart from in their coordinate system arrangement.

For each point there is an infinitesimal light cone as well as a "light surface" formed by the totality of the geodetic continuations of the cone's generators.

Let a light surface be formed for every point in your world (in your most transparent coordinate choice,[4] they are cones). Furthermore, the hyperboloid[5]

$$h^2 = 0$$

has meaning independently of the coordinate system only in your world. It signifies the temporally infinitely distant (infinitely distant, measured in $\int ds$[6]). The light surface of an arbitrary point of your world intersects this surface of the temporally and geodetically infinitely distant. The light surface of just *one* point (your origin) does not intersect this surface but approaches it asymptotically.[7] Thus this point is *de facto* privileged;[8] your world is thus *not* homogeneous.

This, naturally, does not constitute a refutation; but this state of affairs irritates me.–

What pleases me especially is that you did take an interest in the problem and *also* feel the need to discover what kinds of conceptions of the large-scale structure of the $g_{\mu\nu}$-field are theoretically possible.–[9]

From Ehrenfest I hear to my great joy that your health has improved.[10] In wishing you a quick, complete recovery, I am with cordial regards, yours,

A. Einstein.

At the end of the month I am departing for Switzerland for 6–8 weeks. My address there is: c/o Mr. P. Winteler, 16A Bramberg St., Lucerne.

357. To Michele Besso

[Berlin,] 24 June [1917][1]

Dear Michele,

I hope to be able to arrive in Zurich between the 5th & the 7th of July. Words cannot express how eager I am to go, even though the journey itself is expected to be a great strain now.[2] I am staying in Zurich a few days only, because it is too stressful for my creaky frame.[3] Then I am going to Arosa with Albert, after that to Lucerne to my sister's, where I'll be staying for a number of weeks.[4]

I am very much looking forward to our conversations, also to Dällenbach if he is there. I hope to see Weiss again as well, who is supposedly back in Zurich.[5]

Do not be annoyed that I write you so little; there is no other way, it seems! Good intentions don't work in the long run!

Best regards also to Anna and Vero, yours,

Albert.

I am leaving on Friday & and am at my mother's until my departure for Switzerland.[6]

358. From Max von Laue

Würzburg, 25 June 1917

Dear Einstein,

I am arriving in Frankfurt on the 30th, probably not before the afternoon;[1] to be more precise, the train is arriving at 3 o'clock but is usually between a $\frac{1}{4}$ to $\frac{1}{2}$ an hour behind schedule. From the train station, I am going directly to the Basler Hof Hotel. Then we still have a couple of hours ahead of us, of course, even if I leave you undisturbed for the last hour before your lecture.[2]

I have some questions, namely, I have heard of a generalization of Epstein's quantum approach for not necessarily periodic systems,[3] and am very curious about it.

With cordial regards, yours,

M. von Laue.

359. To Willem de Sitter

Berlin, 28 June [1917][1]

Dear de Sitter,

I agree with your first remark about the orders of the $g_{\mu\nu}$'s.[2] This was and is exactly my opinion as well. Experience seems to indicate only the approximate constancy of g_{44}, where the argument of low velocities and of the nonexistence of systematic line shifts increasing with distance are, in principle, equivalent.

The second question is whether to choose elliptical or spherical in the sense proposed by me. When I was writing the paper,[3] I did not yet know about the elliptical possibility and did not think of it myself either. It is obtained from what I am advocating through identifying antipodal points. This (elliptical) possibility was pointed out to me very soon after publication of my article.[4] This possibility seems more likely to me as well.[5]

I cannot accept your argument in support of it, though.[6] It is incorrect, in principle, to force more mass into a world that is at equilibrium with its world matter, without altering the world's volume somewhat.   In fact, the mass surplus in the preferred mass-point corresponds to a mass deficit in the continuously distributed world matter, causing the lines of force originating from the preferred mass-point to terminate eventually within space before having traversed a quadrant, in the elliptic case, and a semicircle, in the properly spherical one. A termination of lines of force at the antipodal point in the absence of masses is an absurdity, of course. In the elliptic world, it corresponds to the impossibility of the appearance of lines of force of the length $\frac{\pi}{2}R$.

I am traveling to Switzerland now[7] (address: 16A Bramberg St., Lucerne, for 6 weeks). Cordial regards, and best wishes for your health![8] Yours,

A. Einstein.

360.   From Friedrich Adler

Vienna VIII, 1 Alser St., 4 July 1917

Dear Friend,

I thank you heartily for your efforts with Barth. It will work without him as well. My wife is back in Zurich now (address: Kathrin Adler, Zurich 7, 46 Carmen St., No. II), but I hope she will return in a few weeks. As the children, or the two older girls,[1] also came here to visit for a week and she had to transport them back again, which is not so simple, you know, under the current less than civilized conditions. As I have heard from Prof. Jerusalem,[2] Mach's son, Dr. Ludwig Mach, is now also in Berlin (Nieder Schönenweide, 26 Brücken St.). He intends to send me the correction proofs of Mach's *Optics*,[3] but I have received no messages from him directly yet. Now I am back to work on the coordinate systems and hope to come to a close soon and return again to my main research on Mach and Kant.[4] Recently I received a postcard from Prof. Ehrenfest in Leyden, which delighted me.[5] Judging from the current post, I am not sure if he also received my answer, though. If you find time to write me a few words again, I would be very pleased.

Most cordial regards, yours,

Fr. Adler.

361.   From Hans Thirring[1]

[Vienna, 11–17 July 1917]

As Prof. Frank[2] may have told you, the young Viennese school is examining the theory of gravitation closely. My friend Flamm had a little notice appear in the fall in the *Phys. Zeitschr.*, is presently working on a new publication,[3] and I likewise have been doing a few calculations in the short leisure hours left to me beside my technical duties in the military.[4] Before publishing my results, I would like to send you a little report in the hope of perhaps receiving another stimulus directly from you—we have been orphaned here in Vienna, as you know, ever since Hasenöhrl's death and must help ourselves as best we can.[5]

I am primarily occupied with the relativity of rotational motion—a problem that naturally must be regarded as solved by your theory. For ex., the fact that for an observer positioned on the Earth's surface the same laws of motion are valid both in system (I): rotating Earth, fixed-star sky at rest, and in system (II): Earth at rest, rotating fixed-star sky, is guaranteed, of course, by the general invariance of the field and motion equations.[6] The matter thus actually should be settled. But there are many physicists (particularly experimentalists) for whom this result is too mathematically abstract and who would like to have proven by a concrete example that according to your theory the rotational motion of distant masses is capable of causing a kind of force like centrifugal force on a mass-point at rest. In order to show this, I proceeded as follows: I imagine an ∞ thin hollow sphere of radius a (the fixed-star sphere) at $\angle$ velocity ω rotating around the Z-axis,[7] and calculate the field at the origin $x = b$; $y = z = 0$; $b \ll a$, according to the method of approximative integration of the field equations (*Berl. Ber.* 1916, p. 688).[8] Excepting the higher-order terms in b/a, I obtain:[9]

$$g_{44} = -1 + \frac{\kappa}{4\pi} \frac{M}{a} \left\{ 1 + \omega^2 a^2 \left[1 + \frac{1}{10} \frac{b^2}{a^2} \right] \right\}$$

M is the mass of the hollow sphere. The significant term (all the rest are obviously constant) is: $\dfrac{\kappa}{4\pi} \dfrac{M}{a} \dfrac{1}{10} \cdot \omega^2 b^2$. Through differentiation $\dfrac{d^2 x}{dt^2} = \text{const. } \omega^2 b$ results for the acceleration.[10] This completely corresponds to the centrifugal force, by which the equivalency of systems I and II from before are thus proven with a simple example. Now, the force on a point outside of the equatorial plane was of further interest to me. For the test point with the polar coordinates r and ϑ, $(\varphi = 0)$:[11]

$$g_{44} = -1 + \frac{\kappa}{4\pi} \frac{M}{a} \left\{ 1 + \omega^2 a^2 \left[1 - \frac{1}{5} \frac{b^2}{a^2} \left(1 - \frac{3}{2} \sin^2 \vartheta \right) \right] \right\} =$$

$$-1 + \frac{\kappa}{4\pi}\frac{M}{a}\left\{1 + \omega^2 a^2\left[1 + \frac{1}{10}\frac{b^2}{a^2}(1 - 3\cos^2\vartheta)\right]\right\}$$

results. The significant term becomes:

$$-\frac{\kappa}{4\pi}\frac{M}{a}\frac{\omega^2 b^2}{5}\left(1 - \frac{3}{2}\sin^2\vartheta\right) = \frac{\kappa}{4\pi}\frac{M}{a}\frac{\omega^2}{5}\left(\frac{x^2}{2} - z^2\right) = \frac{\kappa}{4\pi}\frac{M}{a}\frac{\omega^2 b^2}{10}(1 - 3\cos^2\vartheta).$$

Well, to my astonishment, for a point situated outside of the equatorial plane, another axial force occurs which draws it into the equatorial plane.[12] If the appearance of this force term is traced to the integrals of the field equations, then the following emerges: Those surface elements of the hollow sphere lying in the vicinity of the equator also have, by virtue of their higher velocity, a larger gravitational mass. The field of a rotating hollow sphere of constant surface density is hence equivalent to that of one at rest with a density distribution corresponding to the one (exaggeratedly drawn) in the fig.:[13]

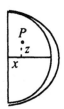

It is not surprising, of course, that here a radial and an axial force will be exerted on a point P. Evidently, for relativistic reasons, the following corollary law must now be required: For a mass-point within a static hollow sphere of constant density to describe in constant motion, in the absence of other masses in the universe, a circle whose plane does not go through the center of the sphere, the following are required: (a) a force against the center of its orbit, (b) a force perpendicular to the plane of motion.

Now, it suggests itself to ask: How is it that we have always only observed the radial force and never the axial force in rotational motion? In all cases the error lies in the incorrect approximation of the cosmos with a hollow sphere. A more satisfactory result would be obtained from calculating the field within a rotating solid sphere. But, as I have ascertained upon cursory examination, this leads to integrals, the integrands of which are elliptic integrals. I have abandoned the calculation in the meantime, for I think it is not worth the effort, since it clearly merely involves demonstrating by a simple textbook example that this general relativity, which is, of course, the basis of the whole matter, actually does hold.[14]

It would be of more practical interest to examine what influence solar rotation has on the planetary orbits (or the rotation of the planets on satellite orbits).[15] Toward this end I have calculated the field of a rotating solid sphere for externally located remote initial points. The result is:

$$g_{44} = -1 + \frac{\kappa}{4\pi}\frac{M}{b}\left\{1 + \frac{3a^2\omega^2}{5}\left[1 + \frac{a^2}{14b^2}(1 - 3\cos^2\vartheta)\right]\right\}$$

M and a are the mass and radius of the sphere,
the initial point has the coordinates: $r = b$, $\varphi = 0$, $\vartheta = \vartheta$. $b \gg a$.

Potentials higher than a^2/b^2 and ω^2 are negligible.

For very remotely lying points the influence of the rotation consists only in an increase in the apparent mass; in the case of initial points where a^2/b^2 cannot be disregarded anymore, a term is added, which breaks the field's central symmetry. Do you consider it possible that an influence on Jupiter's innermost moon were observable? Of all the planets, Jupiter has the largest ω and the largest a—I fear, though, that even so, the effect is too small against the disturbances of the moons amongst one another and against the perihelion motion.

To close, a little comment on your cosmological paper.[16] The "average" line element of the spherical world (eq. (12), p. 150) can be expressed very nicely in polar coordinates which are orthogonal coordinates for this space. It then reads:[17]

$$ds^2 = -\left(1 - \frac{r^2}{R^2 - r^2}\right) dr^2 - r^2 d\vartheta^2 - r^2 \sin^2\vartheta d\varphi^2 + dt^2.$$

362. To Paul Ehrenfest

Arosa. [22 July 1917]

Dear Ehrenfest,

Mr. Grommer, a marvelous mathematician, whom you know from Göttingen[1] (to aid your memory, I remind you of the enormous proportions of his head and hands), would very much like to obtain a mathematician's position in Russia, if only a modest one. The man is a Jew and a true Russian. The dissertation he wrote on whole transcendental functions[2] is acknowledged as superb, and he has a command of the gen. theory of rel. I have authorized him to offer a supplementary session to my lecture course on rel. th. in my absence.[3] It would be nice if you would recommend the man to your Russian friends.—[4]

I am spending happy days with my boys here in Switzerland and am on the way to a proper recovery again.[5] Wishing you pleasant holidays, yours,

A. Einstein

Affectionate greetings also to your family & to Lorentz.

363.   To Willem de Sitter

Arosa, Saturday. [22 July 1917]

Dear Colleague,

Your interesting letter reached me with some delay, because I was traveling around in Switzerland before your letter could be forwarded to me from Lucerne. Your new expression for the line element[1]

$$ds^2 = -dr^2 - R^2 \sin^2 \frac{r}{R} \left[d\psi^2 + \sin^2 \psi d\vartheta^2 \right] + \cos^2 \frac{r}{R} dt^2$$

is very instructive.[2] *Spatially*, all the points in your world are equivalent,[3] but the rate of a standard clock at rest equals

$$\frac{ds}{dt} = \cos \frac{r}{R},$$

hence, depends on position and reaches the value zero at an "equator."[4] This is a position of minimal gravitational potential & of vanishing light vel. Such a singularity ($g_{44} = 0$) in a finite world ⟨obviously⟩ is, in my opinion, to be discounted as physically out of the question. Masses would have the tendency to aggregate at the "equator."[5] The total energy of $g_{44} \dfrac{dx_4}{ds}$ of a mass-point would vanish at the "equator"!

Don't you also have the feeling that, in reality, such cases do not come into consideration? Nevertheless, I cannot give a precise, general formulation of what, in my opinion, must be viewed as ruled out.–[6]

My sojourn in Switzerland is physically and mentally very reinvigorating. I am staying until mid-August. If you inform me, at 16A Bramberg St., Lucerne, of how you now consider the matter, I would be very pleased.

With cordial regards, I am yours,

A. Einstein.

364.   From Franz Selety[1]

Vienna I, 11 Zedlitz Alley, 23 July 1917

Esteemed Professor,

It may not seem of importance to you if I inform you herewith that I had already corresponded with Professor Philipp Frank three years ago in the spring of 1914 about Friedrich Adler's idea, to which he himself attached such great significance. The former did not find it noteworthy, and so I did not concern myself with it further. I am not a physicist, rather a philosopher, but frequently study physical problems. Upon discussing this with Professor Frank, he now confirmed to me that my proposal of that time coincides with Adler's idea, which I am familiar with only through newpaper reports; yet, he continues as before not to consider the matter worth notice.[2] I would like to present in the following my current conception of that idea, which I have now taken up again and which in my view contradicts neither the special nor the general principle of relativity, since different issues are involved here. The principle of relativity is concerned with the formulation of *differential equations* through which the *natural laws* are expressed; the latter observations of mine, on the other hand, concern the description of real processes in nature. In describing these processes with the given constants, obviously not all possible systems are equally suited; certain ones portray the phenomena *more simply*. The general principle of relativity, in particular, compels us to give due consideration to this drawback in general relativity as well. Although the general principle of relativity allows the depiction of the phenomena in the Ptolemaic system as well, there is no doubt that despite the general covariance of the differential equations, the real phenomena, which are described by doubly integrated equations, are presented more simply in the Copernican system than in a system in which the geometry deviates so enormously from the Euclidean and in which the light ray of a star comes to us along a spiral path that revolves around the Earth in 24 hours.[3] I often try to imagine as intuitively as possible the Ptolemaic system according to the general theory of relativity, with all the peculiarities that result from it. How ought one to imagine a world system in which a billiard ball, brought into rotation through an eccentrical stroke, is viewed as at rest? In general, it can be said (if all systems, moving uniformly and in a straight line with reference to one another, are initially regarded as *one* system) that the phenomena are most simply described in a system whose geometry is most nearly Euclidean. After having chosen the three spatial axes in accordance with these aspects, we must now determine the time axis, that is, the spatial coordinate system's state of motion. First of all, concerning the relative rotation, could the most simple system not be defined by having its total angular momentum equal to zero? If we place the coordinate system's origin at the Earth's center, this would most closely approximate the Copernican system (with an infinitesimal rotation of this system). This system would probably also be one in which Foucault's pendulum is at rest. I state these considerations only with reservations, since I cannot judge whether, after more precise calculation, they can still be seen as correct. Concerning its rotation, the system is sufficiently defined that way, but the origin of the spatial coordinate system can still describe an arbitrary world line.

Then surely the most closely Euclidean system would be the one in which no arbitrary gravitational field exists apart from those generated by known masses, that is, *no lines of force without a place of origin.* If, for ex., we move over from the commonly standard systems to one which is uniformly accelerated in a straight line relative to it, we then obtain a homogeneous gravitational field permeating the entire universe, for which no source could be found.[4] Thus the system is fully determined except for the uniform velocity. That variations in the simplicity of describing real processes appear here as well is even more foreign a thought to the pure physicist who, in contrast to the astronomer, concerns himself merely with general differential equations and not with concrete integrated equations. In order to understand that not every coordinate system reveals the phenomena in an equally simple manner, also with regard to uniform motion, one must abandon precisely this standpoint of the pure physicist and assume that of the astronomer. On assuming this point of view, it becomes apparent that the one system with the fixed-star system's center of mass at rest is preferable. This is just the idea I had pointed out three years ago and which Friedrich Adler has also had since. In my letter to Professor Frank at the time, I indicated that such a system is referred to implicitly when it is stated that the Sun moves at 25 km per sec against the Hercules constellation. For even if, somewhat like a molecule in a more comprehensive world system, our fixed-star system moves with reference to its center of mass such that by chance our Sun happens to be at rest in that system, we should still talk of such an independent motion by the Sun, using as a reference the center of mass of the larger system to which the Sun directly belongs. My reflections related to the *general* principle of relativity were formed only recently; naturally, they could not have developed earlier than the general relativity principle itself. I cannot judge, of course, whether my considerations are admissible, yet it seems to me that a choice related to rotation, similar to one related to motion, suggests itself.[5] Here we select the system such that the sum of all the momenta is zero; there, such that the sum of all the angular momenta is zero. In the case of rotation, however, we must take the entire cosmos into account and cannot disregard even the remotest masses, which become infinitely large in the "Ptolemaic" system, because of their velocity. To describe the phenomena on the Earth *in the simplest way,* we are not allowed to choose the system in which the Earth is at rest, because a whole series of processes on Earth: the deflection of the winds and of falling bodies, and the rotation of [Foucault's] pendulum plane, demonstrate a link with the fixed-star sky and with the masses of the whole universe. Regarding the rotation, only if a system in which the fixed-star sky does not rotate is taken as a basis, can a system in which the Earth's center is at rest possibly be chosen for the simplest description of the phenomena.

All these considerations on the selection of a system in which the phenomena can be represented in the simplest way, might possibly appear to the physicist as entirely inappropriate to his field, especially the preference of a system with respect to uniform velocity. Physicists can merely deal with an abstraction of the real processes, which alone can be generalized (in the form of a differential equation) and which requires "arbitrary" initial conditions for its integration. Could this ever be otherwise, or is this founded on the nature of the world? Does chance, in this sense, incapable of being subjected to laws, really exist in the universe? Or will it perhaps be possible one day to discover laws about the distribution of actually existing forms in the four-dimensional world, such as, in the case of an infinite universe, of a periodic nature in space and time? In my cosmological paper, which I am sending you,[6] the sole evidence is the probability concept; however, it is not [un]thinkable, in principle, although altogether improbable, that it would be possible to find an integrated formula of the world, without arbitrary constants, that describes the universe as a whole and allows us to identify the patterns in its four-dimensional structure. As this is currently not the case, my reflections on the coordinate system may be relegated to epistemology or to cosmography and astronomy. Nonetheless, my and Adler's idea does not seem totally irrelevant, because it specifically points out that the special or general covariance of the natural *laws* does not exclude that, even with regard to uniform motion, a specific system is preferred for the description of the world of the fixed-star system known to us. If one wants to call this simplest system the "true" one, no objections can be raised, even from the standpoint of the physical theory of relativity, but only from an epistemological and cosmological point of view. However, the possibility of denoting a system as ultimately "true" only remains if the material cosmos is finite, since otherwise, in considering increasingly comprehensive worlds, we are going to be forced to change over to other preferred coordinates, after all. As I myself do not believe in the boundedness of the world, my ideas do not in any way guide me toward a contradiction to the epistemological principle of relativity. You probably gather from all of the above that I furthermore take a purely phenomenalistic stand and therefore do not believe that any coordinate system is the *true one*, in a metaphysical sense. But the *epistemological* principle of relativity does not necessarily follow from the validity of all your physical theories. Essentially, the *physical* theory of relativity actually merely involves (1) theoretically deriving new and unforeseen facts, determined by natural laws, from certain general assumptions on the representation of the phenomena from various points of view (systems); (2) providing us with the means to calculate directly the phenomena for an observer moving with reference to *one* observer for whom the phenomena are known to us.– As long as no difference in simplicity can be found in the points of view of various observers, the theory of relativity has a right to question the quite arbitrary definition of a particular one as true. If, on the other hand, one observer's system is demonstrated to be simpler, if only with regard to the integrated equations, then the theory does not necessarily have this right anymore. To this extent, emphasis on the fact that the phenomena are represented more simply in a specific coordinate system does seem to me to be of certain epistemological importance. It has frequently been pointed out, of course, that Lorentz's system of the "ether at rest" can never be refuted; but this would have to be rejected, after all, without further ado if no preference for a particular system could be found; but if this does happen, the epistemological situation is somewhat different. I would like to interpret my and Adler's ideas in this sense.

Yet all of this is less important to me than that I now have the opportunity of establishing contact with you, to send you two philosophical articles which, as I hope, will interest you. The first of these, on real facts, etc., treats the problem that seems to me to be fundamental to knowledge as a whole, of how the immediate elements of consciousness are actually composed.[7] And since you take a phenomenalistic point of view, based on Mach's epistemology, I believe that this problem will appear also to you as a, or, *the* fundamental problem in philosophy. I believe to have found a simple answer to this problem. Although my paper has been published for a long while now, as far as I know, it has not been given much notice by philosophers. In the *Zeitschrift für positivistische Philosophie*, which you also know,[8] there is a detailed but critical discussion of it.[9] For the basic idea is a bit too mathematically abstract for philosophy, and deviates from normal conceptions. Although philosophers believe they can understand very abstract things, in reality, they only think that they can understand thoughts that had been conceived in a confused way by their authors; they are mostly quite helpless when faced with really clear, abstract ideas, as has been seen especially clearly with your principle of relativity. My interpretation of what is given differs from Mach's in many points. I fully agree with his epistemological conception of science as an economical description of facts established through experience. Mach says in his *Analysis of Sensations*: One does not emphasize the unity of consciousness.[10] But how can one neglect precisely this quality of it, which is the most essential *formal* quality of the essence of all consciousness in the world: the unity of consciousness, on the one hand, and its "disjection," on the other! This unity is surely of greater fundamental significance than any qualitative property of consciousness, thus also than all natural laws, which merely discuss regularities, that is, similar qualities in various elements of consciousness. Although I did not expect it, I heard that Mach also took an interest in my paper, which I sent him in 1914.

To elaborate clearly on consciousness, my paper must address the concept of time in particular, with which you have been so much concerned, albeit in another sense. When I mention your name at the end, I am admittedly only pointing to a very remote analogy of thought.[11] Bergson has shown that the immediately given duration of consciousness has no metric qualities.[12] He is unfamiliar with this mathematical form of expression, though. In my paper, I show much more extensively, through the fictitious reversal of the accepted course of consciousness, (without even changing the contents, which we call after-effects), which is particularly difficult to imagine, that the apparent asymmetry of the order of time is merely based on the qualitative contents, and that the concepts of earlier and later have no absolute meaning independently of these contents. You will understand this and my exposition in the paper, although I fear that most philosophers will not be able to do so. Finally I show, however, that the order of states of consciousness in time can be dissolved altogether, since only the current state of consciousness forms a unit. If nothing were to exist outside of this state, it would not notice anything. Because the state of consciousness would not notice at all if nothing apart from itself existed, no order of states external to it and independent of it can be immediately given for it either, no more than for another state. All consciousness of the world, regarded purely as such, is therefore none other than an *aggregate* of such states of consciousness *without any ordering relation for them.* A temporal order for states of consciousness is just as inappropriate to pure consciousness as a spatial order, which has long since been abandoned, of course. It is not assumed anymore, that anything psychic exists somehow in space.– My theory can be understood in a double sense, either merely as a description of the directly given reality of consciousness, or as a worldview, if it is presumed that the reality of consciousness, which alone is really known to us, is the only kind of reality that exists, and that physical objects, if anything at all, correspond to something analogous in our consciousness to real being. If the theory is taken merely in the first sense, this nevertheless does not prove that metaphysical, temporal, or spatial ordering of elements of consciousness do not exist; rather, it just reveals, despite the apparent "flow" of our consciousness, that the same relations exist for the temporally external order of the states of consciousness as for the spatial one, that both, if they exist, should not be considered, by any means, to belong to pure consciousness as such, and that if one wants to describe the consciousness of the world purely as such, one must speak of a quantity of unitary states of consciousness through *abstraction of any ordering relation.* The temporal order is merely fabricated by us on the basis of the qualitative contents; on this basis, however, the unity of consciousness can be *imagined* as also ordered in the spatial continuum, at the place where we set the corresponding brain in the construction of the physical phenomenal world.

You will probably find it comprehensible that I hope to encounter more understanding for this theory among mathematicians and physicists with an interest in philosophy than from philosophers proper. Certain obstacles in my theory are frequently particularly difficult to surmount for those not versed in the mathematics. The set theorist is accustomed to abstract from the sets of ordered relations in an analysis of cardinal numbers, and it is thus not difficult for him to conceive of a set in which absolutely no ordering relations exist between the elements. My dissolution of the temporal continuum presents a particular problem for nonmathematicians, however. Philosophers usually are completely unaware that, for a mathematician, the linear continuum is no more than a set of separate individuals, numbers, or points, e.g., of a specific cardinality and with a specific order type, although the former also must be familiar with the continuum of real numbers, from middleschool. The continuum appears to them as something mysteriously uniform; and probably extremely few philosophers would take the trouble to study Bertrand Russell's *Principles of Mathematics*, where they could find clarification of this point.[13] Once the temporal continuum is conceived as a linearly ordered set of elements of consciousness, of which each is actually a world of its own, since they would notice nothing if all the rest of the world did not exist, it easily follows from this that this order is not given in the pure consciousness of the elements, rather that *we* merely consider it appropriate to imagine these elements in that order according to certain inner properties and that, therefore, this order must be dissolved in a consideration of pure consciousness.

Philosophers often deny that a moment in time or a state of consciousness even exists, because they do not know how a continuum can be constructed out of elements and therefore do not consider it possible; it will seem even self-evident to the modern mathematician, and primarily to the set theorist, that in a study of a continuum, one must proceed from the elements. If philosophers say, however, that the individual state of consciousness does not "exist" but that only the "flow" of consciousness does, one can reply that a momentary element of consciousness can perhaps doubt whether anything beside itself exists, but the existence of at least *one* of such closed elements is the surest thing that can be known. It is not certain, however, that what we call a finite stretch of time of a person's consciousness corresponds to a set of consciousness elements of the continuum's cardinality or even to a countable infinite set.– Knowledge of set theory is not absolutely necessary to understand my theory, though; I myself had found this theory 6 years ago, while still in secondary school [*Gymnasium*], inspired by Mach's *Analysis of Sensations*, and moved to criticize it. Even while writing the paper, I did not yet know about set theory, so it contains the inaccuracy that no distinction is made between an uniformly dense, denumerable order and a continuum, in that I call the time order continuous where what we call time corresponds to an infinite set of elements of consciousness. I also did not know that the concept of an infinitesimal distance is altogether inadmissible in an exact description of the continuum, so that my paper contains an error also in this regard which, however, is insignificant to the main issue.[14]

I hope to find better appreciation from you, a physicist to whom epistemology is a part of life and to whom abstract conceptions do not present any difficulties. Admittedly, my theory does go very much against a person's senses; many balk at this absolute disjectioning of consciousness into separate "monads,"[15] and Mach's world, with disjointed sensations, will be more attractive particularly to the scientist. However, this perhaps unpleasant quality of the world of consciousness *cannot be ignored.* One may be an instantaneous solipsist, if one wishes, for whom nothing exists other than the momentary state of consciousness; but if anything outside of this element is assumed, there is no way of getting around the fact that the other consciousness is a foreign one, disjected from the given state, that is, not only that of other persons, but also that which we call our own past or future consciousness and, as peculiar as this interpretation is, in an entirely equivalent way. Thus my system agrees entirely with Mach's denial of the identical self, but must be in conflict with the dissolution of the immediately given element of "sensations," which are united simultaneously within a state.–

If you point out that in physics we only observe "spatio-temporal coincidences,"[16] you yourself are resorting to the unitary state of consciousness, which comprises simultaneously with the observation a small spatial area. However, a dissolution of the physical space-time continuum into the individual space-time points, analogous to the philosophical theory, and a relativization of its order, thus irregular transformations, probably would not be possible.

In the second paper I am sending you, I address a not quite so abstract problem of a cosmological nature.[17] In it, I argue from certain simple and frequently made, though unprovable assumptions: the infinity of the universe in space and time;[18] the similarity, in principle, of phenomena in this universe based on natural laws; and the exclusion of singular points; of these assumptions, I attempt to prove that for every existing object and process, there are infinitely many indistinguishably similar ones, to an arbitrary degree of precision, and what is more shocking to most, that everything in anyway possible, even the most improbable things, must be realized somewhere. Many people feel strongly opposed to the consequences of my second paper as well, in that they either do not want to believe the conclusion, or merely because of this result, think they must reject the premises, which would otherwise not be disagreeable to them. I hope to meet with better appreciation on your part for these thoughts as well than I have with others, who shy away from bold conclusions.

Thus, please forgive me for being so forward as to send you these papers, both of which were published a while ago, and for asking you to read them. I am fully aware that I am directing my letter and my request to one of the greatest minds of all time.   I do so, since I dare to consider my philosophical ideas worthy of your attention. To me, your personal attention naturally is equivalent to a large audience.

Yours truly,

Dr Franz Selety.

365.   To Heinrich Zangger

[Lucerne,] Sunday. [29 July 1917][1]

Dear friend Zangger,

I hope you are more or less recovered from your nasty operation[2] so that you can go on holiday soon with your family.   Take the Henle along,[3] with which you brightened my stay in Arosa; I placed it on the table in the veranda room.   I am feeling surprisingly well.[4] The constant outdoor air and the good care, in conjunction with the comfortable, peaceful existence, are having their effect; it is such a wholesome feeling, that you almost slip away into oblivion. I read your correction proofs and enjoyed very much the parts that deal with concrete cases.[5] But I did not like some of the abstract parts; they often seem to me to be unnecessarily opaque (general) and in the process are not worded clearly and pointedly enough (not every word is placed clearly and consciously). Nevertheless, I understood everything; it may well be possible that my perpetual ride on my own hobbyhorse and the conventions at the Patent Office have driven my standards to exaggerated heights in this regard. In any case, I hope that your urgent warnings may have a rehabilitative effect; the examples from your practice show how imperative it is.– Albert is enjoying himself very much. Today he is climbing the Schwalmis with my brother-in-law;[6] the two of them have become good friends.

With heartfelt wishes for a prompt recovery and enjoyable holidays, yours,

Einstein.

I subsequently marked at random on the correction proof some places that seemed unclear to me so that you see what sorts of things I thought objectionable (pp. 268, 69, 72).[7]

366.   To Willem de Sitter

Lucerne. [31 July 1917]

Dear de Sitter,

No matter how I piece it together, I cannot grant your solution any physical possibility.[1] The trouble is simply linked to the fact that in the finite realm (in natural measure) the $g_{\mu\nu}$'s assume singular values. Let us base ourselves, for ex., on the clearest form

$$ds^2 = -dr^2 - R^2 \sin^2 \frac{r}{R}(d\psi^2 + \sin^2 \psi d\vartheta^2) + \cos^2 \frac{r}{R}dt^2.$$

The energy of a mass-point is generally

$$m \sum g_{4v} \frac{dx_v}{ds},$$

where m stands for the mass constant (mass in natural measure). The energy of a point at rest is thus

$$m\sqrt{g_{44}}.$$

This vanishes for $\frac{r}{R} = \frac{\pi}{2}$. For $r = \frac{\pi}{2}R$, the mass-point has no energy; it does not exist there at all anymore but evaporates entirely along the way there. To me the admission of such cases appears to be nonsensical. This will always be true, of course, regardless of how we may choose the variables.

You see from my more physical approach, though, that I am incapable of formulating exactly (in an invariant way) the condition that the four-dimensional world has to satisfy in order to avoid such singularities.[2] Would you consider trying something in this direction?

Cordial regards, yours,

A. Einstein.

367.   To Michele and Anna Besso-Winteler

[Lucerne, 1 August 1917]

Dear Michele and Anna,

You looked after me with such loving care that I don't know how to recount it all. The operating instructions for my stomach are being followed and honored conscientiously, and my entire lifestyle is medically correctly arranged.[1] Maja and Paul create a sanatorium as comfortable as it is rigorous, so that nothing possibly can be lacking. Albert returned to Zurich yesterday in accordance with Miza's wish.   He has developed well but often behaves a bit roughly toward me out of ingrained habit. Paul got along very well with him and climbed the Schwalmis with him on Sunday. I happened to meet Stocker (Patent Office) here; we reminisced[2] and were pleased to see each other. When is Vero coming? We would all be very glad to see him. I shall try to meet Mrs. Adler before I travel to Berlin; thank you for your help in the matter.[3] I haven't been getting much studying done. I have some worthwhile things here by Levi-Civita on gravitation.[4] He is also interested in the λ term.[5] I find the papers on the foundations of crystallography, which Schoenflies sent me at my request, not particularly clear.[6] Where in the countryside are you going and for how long? I would like to know because of my Zurich stop at the end of my stay.

Warm regards to all three of you, yours,

Albert.

368.   To Tullio Levi-Civita

Lucerne, 2 August 1917

Dear Colleague,

I shall make use of this holiday trip to my homeland to send you *direct* greetings, for once.[1] I recently discovered your fine new papers at Hurwitz's[2] and immediately took them along in order to study them here at leisure.[3] I admire the elegance of your method of calculation. It must be nice to ride these fields on the cob of mathematics proper, while the likes of us must trudge along on foot.

It pleased me especially that you have graciously accepted the new λ term,[4] without which it does not seem possible to form an actually relativist notion of the world as a whole. It is true that nature may also have chosen an entirely different path; this one is just mathematically the simplest conceivable. Stellar statistics has not legitimized it yet, though.[5]

Your objections to my interpretation of the energy components for the gravitational field still do not make sense to me. I would like to repeat what lets me cling to my reading.[6]

1) I proceed from a Galilean space, that is, one with $g_{\mu\nu}$ constant. Through a simple change of frame of reference (introduction of an accelerated reference system K'), I then arrive at a gravitational field. If a pendulum clock, driven by a weight, is installed at rest in K', this gravitational energy is converted into heat, whereas a gravitational field, or in other words, gravitational field

energy, is certainly not present relative to the original (Galilean) system K. As absolutely all components of the energy "tensor" in question vanish altogether with reference to K, this vanishing of all components must also take place relative to K', provided the gravitational energy really can be expressed as a tensor. The pendulum clocks example speaks against it. Altogether, it seems that the energy components for the gravitational field should just be dependent on the 1st derivatives of the $g_{\mu\nu}$'s, because this is also valid for the *forces* exerted by the fields. There are no 1st order *tensors* (dependent only on $\dfrac{\partial g_{\mu\nu}}{\partial x_\sigma}$ & $g_{\mu\nu}$), however.[7]

2) You think that the field equations

$$G_\mu^\sigma + \mathfrak{T}_\mu^\sigma = 0,$$

where G_μ^σ depends only on $g_{\mu\nu}$ & the 1st and 2nd derivatives and $\mathfrak{T}_\mu^\sigma$ on the mass, ought to be understood as energy equations, hence that G_μ^σ would be the energy components. But with this approach it is completely incomprehensible why such a thing as an energy law even exists in spaces in which gravitation can be disregarded. Why, for ex., should a body not be able to cool down without emitting heat outwards? The equation

$$G_4^4 + \mathfrak{T}_4^4 = 0$$

does allow $\mathfrak{T}_4^4$ to diminish throughout, whereby this change is compensated by a reduction in the absolute value of the quantity G_4^4, which does not fall within physical observation. That is why I contend that what you call an energy law has nothing to do with what is otherwise known as such in physics.

3) I set up a law in the form

$$\sum_\sigma \frac{\partial \mathfrak{T}_\nu^\sigma + \mathfrak{t}_\nu^\sigma}{\partial x_\sigma} = 0$$

and draw consequences from it.[8] These consequences are correct independently of whether one concedes that the $\mathfrak{t}_\nu^\sigma$'s are "real" energy components of the gravitation. Since, with the vanishing of the $\mathfrak{T}_\nu^\sigma$'s and $\mathfrak{t}_\nu^\sigma$'s at (spatial) infinity, the theorem

$$\frac{d}{dx_4}\left\{\int (\mathfrak{T}_4^4 + \mathfrak{t}_4^4)dV\right\} = 0$$

is always valid, whereby the integral is extended over the entire three-dimensional space. For my deduction, it is necessary only that $\mathfrak{T}_4^4$ be the energy density of the mass, which neither of us doubt.

4) Finally, in an appeal to your physical sense, I would like to call your attention to the following case. Let the space be Galilean (field-free) except for the field

that generates few gravitational masses (Newton's case). According to Newton, the number of lines of force sent forth by this space into infinity depends only on the total *mass* of the bodies, according to the results of the special theory of relativity, therefore from the total energy.   Now, it seems to me beyond doubt that (in the static case) the field at infinity must be fully determined *by the energy of the mass and of the gravitational field together*. This fits with my interpretation of the t_{ν}^{σ}'s, as you can easily prove from my field equations in mixed form.[9]

If you like, I shall be glad to present any of these points in more detail than has been done here. I have no supporting literature with me. With cordial regards, I am yours,

A. Einstein
cur[rently] 16A Bramberg St., Lucerne.

369.   To Hans Thirring

Lucerne, 16A Bramberg St., 2 August 1917

Dear Colleague,

Thank you for your kind and interesting letter,[1] which was forwarded to me here. To your example of the hollow sphere it must just be added that, aside from the centrifugal field, whose axial components you interpreted so nicely, a Coriolis field also results which corresponds to the components g_{41}, g_{42}, g_{43} of the potential and is proportional to the first power of ω.[2] This field has a vertically repulsive influence on moving masses, and in the Foucault experiment, for ex., causes a rotation of the pendulum plane. I have calculated this trailing rotation for the Earth; it remains far below any observable quantity. This Coriolis field is also produced by the rotation of the Sun and Jupiter and causes secular changes in the orbital paths of the planets (or moons) which, however, remain far below the margin of error.[3] All in all, the perihelion motion of Mercury seems to be the only case within the field of celestial *mechanics* where deviations from classical theory are perceptible nowadays.

Nevertheless, the Coriolis fields still seem more accessible to observation than your supplementary terms to g_{44}, because the latter influences have the same symmetry properties as field distortion from oblateness.[4]

With best wishes for the holidays and for success in your work, I am with best regards, yours truly,

A. Einstein.

370. To Willem de Sitter

[Lucerne,] Wednesday. [8 August 1917]

Dear Colleague,

I am delighted with your detailed and clear arguments, which I fully understand. A bridge can be constructed between cases A and B as follows:[1]

B is the $g_{\mu\nu}$-field of a world in which all matter is concentrated at the "equator,"[2] whereas it is uniformly distributed in case A.[3]

With this view, matter must be considered the cause of the falling off of g_{44} (to zero) upon approaching $r = \dfrac{\pi}{2} R$, just as it is for a drop in the g_{44}'s upon approaching the Sun, for ex.[4] The inequivalence of the different space points, in this interpretation, is not an "independent" property of the space. I am of the opinion that it is reasonable to demand that in the finite realm (in natural measure), singular values for $g_{\mu\nu}$ do not occur as long as the energy density is finite everywhere.[5]

With cordial regards, I am yours,

A. Einstein.

371. To Michele Besso

[Lucerne,] Wednesday. [15 August 1917]

Dear Michele,

Many thanks for the letter and the explanations. It's good that you have obtained the information. Ultimately, they will just have to make the best of it in Zurich.[1] Are they living at home again?[2] The pain went away again soon after we parted, so that by evening already there were no signs of it left.[3] I am coming around the 25th (at the earliest, in the evening of the 24th). Let's discuss your suggestions then. Reason versus passion! The latter always wins whenever they are pitted against each other (proof is the experience everyone has already made for himself). So I believe that the whole thing can merely lead to a pretty little debate between us.

Best regards to all three of you, yours,

Albert.

372. To Paul Seippel

[Lucerne,] Sunday. [19 August 1917][1]

Highly esteemed Colleague,[2]

Through a trick of fate it had to be that, upon addressing the letter directed to me, you wrote "Zurich" instead of "Lucerne" on the envelope; thus I received it only today, so when the letter reached me you had already left Zurich. I regret this exceedingly, thinking that I really could possibly be of some use to you or to the cause.[3] Unfortunately, my state of health does not allow me to travel from here to meet you somewhere; however, I can be met anytime this week at my sister's house, Mrs. Winteler, 16A Bramberg St., Lucerne (difficult to find; ask!). I should naturally also be glad to go to the train station, if you inform me of your arrival (identifying portrait: gray suit, pale face, large dark-gray felt hat).

In the hope of soon making your personal acquaintance, I am yours very truly,

A. Einstein.

373. From Romain Rolland

Byron Hotel, Villeneuve (Vaud), Tuesday, 21 August 1917

Dear Sir,

Prof. Zangger tells me that you intend to set off again imminently from Lucerne.[1] I had formed the plan of going to shake your hand before your departure; but there is little chance that I shall find the time for that this week. I would like at least to convey to you my kind regards. I know that your health has suffered rather much and that you do not want to take care of it as would be necessary, which is a crime toward science and distressing to your friends. But I know only too well that there are moments in life when each is his own "torturer":–

I hope that, despite everything, your sojourn in Switzerland has restored enough of your strength to weather, without too much strain, another winter of physical and moral privation in Berlin. I scarcely believe you have lost the optimism that had struck me so at the time of the visit you were so kind as to pay me in Vevey two years ago.[2] I have retained a tonic and illuminating memory of it.– For my part, if I suffer in my heart at seeing the West and, above all, my France, bleed itself white, I remain an optimist of things overall and of the future progress of humanity. It may be that some nations deplete and void themselves, as Spain had done in times past. Yet, humanity continues its march with new elements; and I believe (for the near future) in a richer and more far-reaching civilization than that of today—a civilization in which the intellectual elements of Asia will bring to impoverished Europe the potential for new development.

I have written two works inspired by the present: a distressing novel that is a crisis of the soul amidst the madness of the passions,—and a fanciful and satirical piece.–[3] I read with enraptured interest the admirable book by Prof. Nicolai.[4] He is a friend of yours, I believe. Would you please let him know how much I like him. I have been taking nourishment from him these last months.– For all that, it is fine to discover in this terrible epoque some grand liberated and serene souls like his. It is sufficient recompense to me for the immense stupidity which is the new universal Flood. The Ark is afloat. In the end it will safely touch land.

In fond and devoted sympathy, I am, dear Sir, yours sincerely,

Romain Rolland.

374.   To Romain Rolland

[Arosa,] Wednesday, 22 August 1917

Highly esteemed Romain Rolland,

I am touched by the cordial interest you have in a person whom you have seen but a single time.[1] I would definitely not neglect to visit you if my health were a bit more stable; but the smallest undertaking often takes its toll afterwards. The bad observations we have had to make of people's actions in the interim have not made me *more* pessimistic, though, than I was in reality two years ago.[2] I even find that the surge of imperialistic mentality, which dominates influential sectors in Germany, has subsided somewhat.[3] Yet I still find that it would be extremely dangerous to form a pact with Germany as it is today.

Through the military success of 1870[4] and through the successes in the fields of trade and industry, this country has taken up a kind of religion of might, which has found adequate and by no means exaggerated expression in Treitschke.[5] This religion holds almost all intellectuals in its sway; it has driven out almost completely the ideals of Goethe's and Schiller's time. I know people in Germany whose private lives are guided by a virtually unbounded altruism, but who were waiting with the greatest impatience for the declaration of an unlimited submarine war.[6] I am firmly convinced that this straying of minds can only be steered by hard facts. These people must be shown that it is necessary to have consideration for non-Germans as worthy equals, that it is essential to earn the *trust* of foreign countries, in order to be able to exist, that the set goals cannot be achieved through force and betrayal. Fighting against the goal itself with intellectual weapons seems hopeless to me; people like Nicolai are labeled with sincere conviction as "utopians."[7] Only facts can dissuade the majority of the misled from their delusion that we were living for the state and that its intrinsic purpose was the greatest power possible at any price.

The finest alternative to lead us out of these deplorable circumstances seems to me to be the following: America, England, France, and Russia conclude a military arbitration agreement for all time with settlements on mutual assistance and on a minimum and maximum of military readiness. This agreement would have to contain the establishment of most-favored nation tariffs. Any state that has a democratically elected parliament and whose ministers depend on a majority within this parliament ought to be allowed to enter into this pact. This brief sketch will have to suffice.

If Germany, which is dependent on markets for its industrial products, finds itself confronted with such a permanent state of affairs, the view that it is imperative to abandon the present course would soon prevail. As long as German statesmen can hope, though, that sooner or later a shift in the distribution of power will occur, no sincere change in course will be considered. For proof that everything is still as it was, look at the way in which the change in chancellorship was staged lately.[8]

In wishing that in these sinister times you find comfort in your inspired artistic creativity, I send you my cordial regards, yours very truly,

A. Einstein.

375.   From Tullio Levi-Civita

[Rome, 23 August 1917][1]

Dearest Colleague,

It was only yesterday, but with the greatest pleasure, that I received your friendly letter of the second of this month.[2] I am very grateful for your cordiality, which I return with all my heart, and for your very flattering and favorable opinion of the mathematical considerations of my latest papers: in any case, the main credit goes to you for opening up ⟨such wide horizons in natural philosophy⟩ these new fields of research.

Now to our amicable controversy. As I think I also indicated in a note I wrote a few months ago to my colleague Grossmann,[3] I understand very well your reluctance to occupy yourself with the not very fruitful solution represented by the equations[4]

$$T_{ik} + A_{ik} = 0 \qquad (i,\, k = 0,\, 1,\, 2,\, 3) \tag{1}$$

(T_{ik} energy tensor, A_{ik} gravitational tensor). I acknowledge the importance of your objection that in this way the energy principle completely loses its heuristic value, in that it does not *a priori* exclude any (or almost any) physical processes because it would suffice to modify the ds^2 in the appropriate way.

You point out that in abandoning (1), or rather their interpretation, the energy contributed by the field can be understood as something dependent on the form of ds^2, analogous to what is done concerning the concept of field strength. If one writes the equations of motion in the form

$$\frac{d^2 x_i}{ds^2} = -\sum_{jk} \left\{ \begin{array}{c} jk \\ i \end{array} \right\} \frac{dx_j}{ds} \frac{dx_k}{ds} \tag{2}$$

and carries out the necessary verification, the analogy between the right-hand side (which defines *neither* a covariant system nor a contravariant one) and the ordinary concept of force is made explicit; in your view, your t_ν^σ's (which do not constitute a tensor) should be dealt with in the same way. I have no objections to your view; on the contrary, I am inclined to assume that it is sound, as is always the case with the intuition of a genius. But I would need to see, in appropriately explicated step-by-step reasoning, how, starting from (2), one actually arrives at the ordinary concept of force (or at least how one should go about it).

I shall give the matter more thought when circumstance (or inspiration) is favorable, but it is from you in the first place that I expect a solution.

As I am for now in this state of cautious reserve, I would like to defend my tensor A_{ik}, at least with respect to its logical soundness. Thus, I point out that no contradiction such as you believe to find, exists in the example of a pendulum clock,[5] considered in two different systems K and K', of which the first is stationary (in the Newtonian sense of the word) and the second moves with constant acceleration. You say:

a) with respect to K, the energy tensor is zero, *because the $g_{\mu\nu}$'s are constant*;

b) but this is not the case with respect to K'; on the contrary, the physical process shows a transformation of energy into heat.

c) Given the invariant character of a vanishing tensor, the simultaneous occurrence of (a) and (b) implies that the premises are flawed.

I object to (a), because we can very well argue that the $g_{\mu\nu}$'s are constant outside of bodies but not inside your pendulum clock.

Concerning the final point of your letter (response 4), it is not linked to the particular form of your t_ν^σ, if I understand correctly, but holds just as well for my A_{ik}.

In fact, it seems to me that one can also obtain from (1) the behavior at ∞, by making use of the fact that the divergence of A_{ik} vanishes identically;[6] this is thus also the case for the divergence of T_{ik} which, with $g_{\mu\nu}$ approaching $\varepsilon_{\mu\nu}$, reduces asymptotically to

$$\sum_0^3 k\frac{\partial T_{ik}}{\partial x_k} = 0.$$

Please accept my best wishes for a pleasant vacation and my warm regards.

P.S. An indication in favor of my (1) seemed to me the negative value that results from it for the energy density A_{00} of the gravitational field (assuming $T_{00} > 0$). This is in accordance with the old attempts to localize the potential energy of a Newtonian body, and it explains the minus sign as connected to the unique position of gravitation vis-à-vis all other physical phenomena.

376.   From Romain Rolland

[Villeneuve, canton of Vaud,] Thursday, 23 August 1917

Dear Sir,

Thank you for your letter.[1] I see how much you are suffering, and I sympathize with you. You know Germany, and I have no difficulty believing that what you say about it is sadly true. However, you do not know the sufferings "on the other side." Evil spreads like a splotch of oil. All the nations show solidarity even as they fight one another; and a way has not yet been found to halt the moral epidemics—just as the others—at the borders. The current war looks to me like a fight against the Hydra of Lerna. For every head chopped off, two other ones spring back to life.–[2] That is why I do not believe in the effectiveness of these clashes of armies. I am awaiting salvation (if it is meant to come) from other—social—forces; and if it does not come, . . . by God! it will not have been the first time that a powerful civilization has crumbled. Life will know very well how to blossom again from the ruins.

Very affectionately yours,

Romain Rolland

Do you keep up with Bertrand Russell's publications, and are you familiar with the recent American opposition?[3]

–You see, I am perfectly convinced that there shall never be more than a handful of us in the world. In the order of events, we shall always be the conquered, but what of it? The soul is never conquered—except when it consents to it. It is ahead of its times.

377.   To Michele Besso

[Benzingen, Württemberg,] 3 September 1917

Dear Michele,

The watch arrived in good condition and is running properly again. The trip came off well. Ehrat really did come to Schaffhausen.[1] The next morning I passed by the posts,[2] but slowly. At Thaingen a full hour of luggage inspection; it's good that I did not have *more* with me. In Gottmadingen I came almost last in turn and missed the connection.[3] Delay until 5 o'clock. I barely made it to Sigmaringen. Saturday morning I drove to here and surprised my pastor[4] by taking the 1-hour walk along the uphill path to the village.[5] (What does my dear nurse Anna say to that?[6]) But my innards did not rebel in the least, they have not done so at all as yet. Here milk and honey is flowing, literally,[7] and the pastor is a very kind person whose views agree very extensively, if not entirely, with mine, even in things one would hardly suspect.

May I thank you again for the hospitality and loving care? But you both probably do not want to hear anything of the sort. At any rate, I shall recommend the Anna Sanatorium most warmly to the place of "authority," as is its proper due. But otherwise, she must practice her patience and leniency.[8] She should start with a quarter of an hour a day and then increase it (under the motto: it's a hard life). I don't know yet whether I am going away somewhere with Else; that depends on her. My address is thus 5 Haberland St. The move seems to have already been completed.[9]

Warm regards from your

Albert.

378.   To Erwin Freundlich

[Benzingen, Württemberg,] Monday. [3 September 1917]

Dear Freundlich,

Forgive my silence. But I let everything lie in Lucerne in order to abandon myself completely to laziness, with good success. I am coming to Berlin in the course of this month and am living at 5 Haberland St. If my cousin does not come to Thuringia, I shall come around the 10th, otherwise, however, only at the end of the month, which is more likely. Then we can discuss everything. Such things are much better done verbally. I must speak mainly with my closer colleagues.[1]

I am sorry that you were not able to relax satisfactorily.

To a happy reunion! Yours,

A. Einstein.

379.   From Adolf von Harnack[1]

Berlin NW7, Royal Library, 12 September 1917

The Honorable Professor Einstein, Wilmersdorf.

Your Honor,

I am pleased to be able to inform Your Honor that the establishment of the Kaiser Wilhelm Institute of Physical Research, planned jointly with the Leopold Koppel Foundation, is now secured for the period of 1 October 1917 onwards.[2] I have the honor of most humbly enclosing an excerpted transcript of the University Council minutes for your obliging information.[3] The compensation of 5,000 M, provided for Your Honor as director of the Institute,[4] will be paid out to you in advance through the local banking house Mendelssohn & Co. in quarterly installments. Yet I submit to Your Honor to inform me whether payment should be made to your personal address or to a banking institution.

von Harnack
President of the Kaiser Wilhelm Society
for the Advancement of the Sciences.

380.   To Władysław Natanson

[Berlin,] 14 September 1917

Dear Colleague,

The news of the death of our outstanding friend has shaken me deeply.[1] He was one of the most excellent persons I knew. He was endowed besides with great talent and a cheerful nature.[2] May I ask you for his wife's address? I would like to write her, particularly since I know her personally.–[3]

I was away in Switzerland for the entire summer until two days ago in order to recover from an intestinal complaint.[4] I am well again.

One feels increasingly alienated in this hard world. But we appreciate our good-willed brothers of like mind, even if they are located far away in Cracow! I hope fate soon brings us together once again.[5]

With cordial regards to you and yours, I am yours,

Einstein.

381. To Michele Besso

[Berlin,] 22 September [1917]

Dear Michele,

I am still very well. I had only one little attack, which has passed, however.[1] It is touching how worried you both are about me. The food is good, and I am resting a lot. Today I applied the quantum theory to (freely moving) rigid bodies. It is kind of Anna to concern herself about my wife.[2] Tete must return; I cannot come up with the money; I have already written to Miza.[3] I do not believe in the new medical magic with X-rays. I am at the point where I only trust post mortem diagnoses, nothing else.

Dear Michele, now you have to visit me here soon. It is spacious and comfortable enough in the new apartment.[4] Can't you come for a few weeks at Christmas or Easter? Get used to the idea gradually!

Warm regards, yours,

Albert.

382. From Gunnar Nordström

Leyden, 22 September 1917

Dear Prof. Einstein,

Upon taking up my scientific research again now, as a happily married man,[1] one of my first tasks is to finish writing the letter I had drafted already a few months ago.[2] The matter does not apply at all to the big questions in your last gravitation paper[3] but only to some calculations with your formulas from the previous paper (*Berl. Ber.* 1916, p. 1114).[4]

When I began with these matters, I asked myself initially what is meant by a system's mass. The mass must naturally be able to be expressed by an integral that just needs to be extended over the material system but not over those parts of the space where all $\mathfrak{T}_\nu^\mu$'s are zero. Without treating the issue very generally, I soon restricted myself to the case of a reference system in which the gravitational field of the system under consideration of bodies is stationary.[5]

With reference to the mentioned paper of yours, let us initially assume that for a very general field we have[6]

$$-\mathfrak{G}^* = \mathfrak{G} - \operatorname{div}\mathfrak{A},\tag{1}$$

where $\mathfrak{A}$ behaves like a vector under Lorentz transformations and div means the four-dimensional divergence, in the sense of the special theory of relativity. In (1),

$$\mathfrak{G} = \sqrt{-g} \sum g^{il} \left(\left\{ \begin{matrix} ik \\ n \end{matrix} \right\} \left\{ \begin{matrix} nl \\ k \end{matrix} \right\} - \left\{ \begin{matrix} il \\ n \end{matrix} \right\} \left\{ \begin{matrix} nk \\ k \end{matrix} \right\} + \frac{\partial \left\{ \begin{matrix} nk \\ k \end{matrix} \right\}}{\partial x_l} - \frac{\partial \left\{ \begin{matrix} il \\ k \end{matrix} \right\}}{\partial x_k} \right) \qquad (2)$$

$$-\mathfrak{G}^* = \sqrt{-g} \sum g^{il} \left(\left\{ \begin{matrix} ik \\ n \end{matrix} \right\} \left\{ \begin{matrix} nl \\ k \end{matrix} \right\} - \left\{ \begin{matrix} il \\ n \end{matrix} \right\} \left\{ \begin{matrix} nk \\ k \end{matrix} \right\} \right) - \qquad (3)$$

$$-\sum \left\{ \begin{matrix} ik \\ k \end{matrix} \right\} \frac{\partial}{\partial x_l} (\sqrt{-g} g^{il}) + \sum \left\{ \begin{matrix} ik \\ l \end{matrix} \right\} \frac{\partial}{\partial x_l} (\sqrt{-g} g^{ik})$$

$$\mathfrak{A}_k = \sqrt{-g} \sum (g^{ik} g^{al} - g^{il} g^{ak}) \left[\begin{matrix} il \\ a \end{matrix} \right] \qquad (4)$$

The minus sign on the left in (1) and (3) seems to me necessary so that the $\mathfrak{T}_4^4$ calculated from your equations gives the energy density with the correct sign. ($\mathfrak{M}$ should have a sign allowing the variation principle to be written as $\delta \int (\mathfrak{G}^* + \mathfrak{M}) d\tau$.)[7] It appears to me that Droste also has the same sign error in equation (17) of his dissertation,[8] and I am going to investigate the matter properly, together with him. But because it is entirely possible that I am completely wrong, I don't want to go into further detail about the sign question yet. The matter does not affect the results derived in the following.

From the expression in (3) for $-\mathfrak{G}^*$ and from equations (29) and (32) in your *Annalen* paper,[9] it is evident that $\mathfrak{G}^*$ consists of nothing but terms of the form

$$F(g^{ab}) \cdot g_e^{cd} g_h^{fg}.$$

Hence one has

$$\sum \frac{\partial \mathfrak{G}^*}{\partial g_\alpha^{\mu\nu}} = 2\mathfrak{G}^* \qquad (5)$$

and the last [of your] expressions (20) in the *Berl. Ber.*[10] yields for the diagonal sum of the small t's

$$\sum_\alpha t_\alpha^\alpha = \mathfrak{G}^*. \qquad (6)$$

Because

$$\sum_\alpha \mathfrak{T}_\alpha^\alpha = \mathfrak{G},$$

(1) produces

$$\sum_\alpha (\mathfrak{T}_\alpha^\alpha + t_\alpha^\alpha) = \text{div}\,\mathfrak{A}. \qquad (7)$$

(Here a brief parenthetical remark. An equation quite similar to (7) is obtained from (18) *Berl. Ber.*[11] The vector $\mathfrak{A}$ thus has the same divergence as the vector of which the α-component is

$$-\sum \frac{\partial \mathfrak{G}^*}{\partial g_\alpha^{\mu\nu}} g^{\mu\nu}.$$

Whether these two vectors themselves are identical I do not know.)[12]

Now one should assume that the gravitational field is *stationary*. Then the four-dimensional divergence in (7) reduces to the three-dimensional divergence in the spatial part of $\mathfrak{A}$. Integrated over the entire three-dimensional space, (7) gives furthermore, on the basis of Laue's theorem, the total energy E and hence also the mass of the system:[13]

$$E = \int \sum_\alpha (\mathfrak{T}_\alpha^\alpha + t_\alpha^\alpha) dV. \tag{8}$$

This integral must be extended not only over the material system but also over the entire gravitational field.

E can also be described by an integral over the material system alone. First, we also have for E

$$E = \int (\mathfrak{T}_4^4 + t_4^4) dV \tag{9}$$

integrated over the entire three-dimensional space. However, because for a stationary field $g_4^{\mu\alpha}$ equals zero, your last expression (20) in the *Berl. Ber.*[14] yields

$$t_4^4 = \frac{1}{2} \mathfrak{G}^*. \tag{10}$$

Thus, according to (6),

$$t_4^4 = \frac{1}{2} \sum_\alpha t_\alpha^\alpha \tag{10a}$$

and equation (9) multiplied by 2 and, with (8) subtracted from it, gives the result

$$E = \int (2\mathfrak{T}_4^4 - \sum \mathfrak{T}_\alpha^\alpha) dV = \int (\mathfrak{T}_4^4 - \mathfrak{T}_1^1 - \mathfrak{T}_2^2 - \mathfrak{T}_3^3) dV. \tag{11}$$

Here the energy is expressed by an integral that only needs to be extended over the material system.

The volume integral in (11) can also be transformed into a surface integral over a surface enclosing the material system. This is because your equation (18) in the *Berl. Ber.*[15] shows that $(\mathfrak{T}_4^4 + t_4^4)$ can be expressed as a divergence of a three-dimensional "quasi-vector":

$$(\mathfrak{T}_4^4 + t_4^4 = \operatorname{div}\mathfrak{B}) \qquad \left(\text{where } \mathfrak{B}_\alpha = -\sum_\mu \frac{\partial \mathfrak{G}^*}{\partial g_\alpha^{\mu 4}} g^{\mu 4}\right). \qquad (12)$$

This equation (12) multiplied by 2 and with (7) subtracted from it, taking (10a) into account, yields

$$\mathfrak{T}_4^4 - \mathfrak{T}_1^1 - \mathfrak{T}_2^2 - \mathfrak{T}_3^3 = \operatorname{div}(2\mathfrak{B} - \mathfrak{A}) = \operatorname{div}\mathfrak{C}, \qquad (13)$$

and by applying this and Gauss's law, (11) changes into

$$E = \int \mathfrak{C}_n df, \qquad (14)$$

integrated over surface f enclosing the material system. Unfortunately, however, the quasi-vector $\mathfrak{C}$ is not a four vector, not even in the sense of the special theory of relativity.

I have verified equations (11) and (14) for the case of a spherically symmetric body.[16] When I continued to calculate t_4^4 in the gravitational field outside of the body, however, I obtained a result that does *not* agree with your formula (11) in "Approximative Integration of the Field Equations."[17] [When this had already been written, I discovered a possibility of explaining the inconsistency. See page 5.][18] I would like to present this calculation now.[19]

First I calculate the vector $\mathfrak{A}$ for the field outside the center. In this field one has, e.g., according to Droste (dissertat[ion] equation (28)),[20] if the coordinate system is conveniently chosen,

$$ds^2 = \left(1 - \frac{\alpha}{r}\right) dt^2 - \frac{dr^2}{1 - \frac{\alpha}{r}} - r^2(d\vartheta^2 + \sin^2 \vartheta d\varphi^2). \qquad (15)$$

To avoid difficulties with the integration boundaries, etc., I use Cartesian coordinates, however, and retaining Droste's designations as much as possible, I set[21]

$$g_{11} = -p^2 - \frac{x_1^2}{r^2}(u^2 - p^2), \quad g_{12} = -\frac{x_1 x_2}{r^2}(u^2 - p^2), \text{ etc.} \left.\begin{array}{c} \\ \\ \\ \end{array}\right\} (16)$$

$$g_{14} = g_{24} = g_{34} = 0, \quad g_{44} = w^2,$$

where p, u, w are functions of r. Then[22]

$$g^{11} = -\frac{1}{p^2} - \frac{x_1^2}{r^2}\left(\frac{1}{u^2} - \frac{1}{p^2}\right), \quad g^{12} = -\frac{x_1 x_2}{r^2}\left(\frac{1}{u^2} - \frac{1}{p^2}\right), \text{etc.} \left.\begin{array}{c} \\ \\ \\ \end{array}\right\} (16a)$$

$$g^{14} = g^{24} = g^{34} = 0, \quad g^{44} = \frac{1}{w^2}, \quad \sqrt{-g} = uwp^2.$$

To obtain expression (15) for ds^2 we must set[23]

$$
\left.
\begin{aligned}
&p = 1 \text{ (corresponding to } v = r \text{ for Droste)} \\
&u^2 = \frac{1}{1 - \frac{\alpha}{r}}, \quad w^2 = 1 - \frac{\alpha}{r}.
\end{aligned}
\right\} \tag{17}
$$

Hence this holds for a very specific coordinate system.

In a point of space where $x_1 = r$, $x_2 = x_3 = 0$, you obtain

$$
\left.
\begin{aligned}
&g^{11} = \frac{-1}{u^2}, \quad g^{22} = g^{33} = -\frac{1}{p^2}, \quad g^{12} = g^{13} = 0, \\
&\frac{\partial g_{12}}{\partial x_2} = -\frac{1}{r}(u^2 - p^2), \quad \frac{\partial g_{22}}{\partial x_1} = -2pp' = \frac{\partial g_{33}}{\partial x_1} \quad \left(p' = \frac{dp}{dr}\right) \\
&\quad = \frac{\partial g_{13}}{\partial x_3} \qquad \frac{\partial g_{44}}{\partial x_1} = 2ww'
\end{aligned}
\right\} \tag{18}
$$

Now $\mathfrak{A}_1$ must be calculated at a point where $x_2 = x_3 = 0$. Thus

$$
\mathfrak{A}_1 = \frac{1}{2}\sqrt{-g}\sum(g^{i1}g^{al} - g^{il}g^{a1})\left(\frac{\partial g_{ia}}{\partial x_l} + \frac{\partial g_{1a}}{\partial x_i} - \frac{\partial g_{il}}{\partial x_a}\right). \tag{19}
$$

You will notice, first of all, that the terms with $a = i$ contribute zero. To obtain a term differing from zero, moreover, either i or a must have the value 1, and the remaining two of the three indices a, i, l must be equal.

$i = 1$, $a = l \neq 1$ contributes

$$
\frac{1}{2}\sqrt{-g} \cdot g^{11}\left(g^{22}\frac{\partial g_{22}}{\partial x_1} + g^{33}\frac{\partial g_{33}}{\partial x_1} + g^{44}\frac{\partial g_{44}}{\partial x_1}\right),
$$

and $a = 1$, $i = l \neq 1$ gives

$$
-\frac{1}{2}\sqrt{-g} \cdot g^{11}\left\{g^{22}\left(2\frac{\partial g_{12}}{\partial x_2} - \frac{\partial g_{22}}{\partial x_1}\right) + g^{33}\left(2\frac{\partial g_{13}}{\partial x_3} - \frac{\partial g_{33}}{\partial x_1}\right) - g^{44}\frac{\partial g_{44}}{\partial x_1}\right\}.
$$

$\mathfrak{A}_1$ consists of these two terms. Introducing the expressions (18), one obtains

$$
\mathfrak{A}_1 = -4\frac{wpp'}{u} + \frac{2}{r}uw\left(1 - \frac{p^2}{u^2}\right) - 2\frac{p^2 w'}{u}. \tag{20}
$$

For the special coordinate system, where equations (17) apply, one finds

$$
\mathfrak{A}_1 = \frac{\alpha}{r^2}. \tag{21}
$$

For reasons of symmetry, we have

$$\mathfrak{A}_2 = \mathfrak{A}_3 = 0. \tag{21a}$$

At an arbitrary point (21) and (21a) give $\mathfrak{A}_\alpha = \dfrac{\alpha}{r^2}\dfrac{x_\alpha}{r}$ $(\alpha = 1, 2, 3)$. From this it

follows

$$\operatorname{div}\mathfrak{A} = 0, \tag{22}$$

and because outside of the body $\mathfrak{G} = 0$ as well, (1) yields $\mathfrak{G}^* = 0$ and finally, (10) yields

$$t_4^4 = 0. \tag{23}$$

The gravitational field therefore would have no energy outside of the body, and this does not agree with your equation (11) in "Approximative Integration of the Field Equations."

Continuation [*Fortsetzung*], 28 Sept.

I had just written all this down when this idea shot through my head: The entire inconsistency could lie in the fact that we are using different coordinate systems. I also think that this is so, even though I don't obtain exactly the same result as you for your coordinate system. As you know, your coordinate system is characterized by the velocity of light being equal in all directions,[24] and this requirement agrees with $p = u$ in (16) and (16a). The gravitational field in such a coordinate system is also examined by Droste (dissertation, p. 20). Droste's formula (31) gives

$$p = u = \left(1 + \frac{\alpha}{4r}\right)^2 \qquad w^2 = 1 - \frac{\alpha}{r\left(1 + \frac{\alpha}{4r}\right)^2}. \tag{24}$$

So this takes the place of (17). For $\mathfrak{A}$, at a point where $x_2 = x_3 = 0$, because $p = u$, (20) initially produces

$$\mathfrak{A}_1 = -4wp' - 2w'p. \tag{25}$$

In order to achieve the same degree of accuracy as in your paper,

$$p = 1 + \frac{\alpha}{2r}, \qquad w = 1 - \frac{\alpha}{2r}$$

must here be inserted for p and w. For p' and w', the expressions obtained from this through differentiation may not be used, however, but rather the following more precise ones:

$$\rho' = -\frac{\alpha}{2r^2}\left(1 + \frac{\alpha}{4r}\right) \qquad w' = \frac{\alpha}{2r^2}\left(1 - \frac{\alpha}{2r}\right).$$

In this way the approximate expression for $\mathfrak{A}_1$ is obtained

$$\mathfrak{A}_1 = 2\frac{\alpha}{r^2}\left(1 - \frac{\alpha}{4r}\right) - \frac{\alpha}{r^2} = \frac{\alpha}{r^2} - \frac{1}{2}\frac{\alpha^2}{r^3}.$$

This clearly is valid at a point with $x_2 = x_3 = 0$. For an arbitrary point, we find the β-components

$$\mathfrak{A}_\beta = \frac{\alpha x_\beta}{r^3} - \frac{1}{2}\frac{\alpha^2 x_\beta}{r^4}. \qquad (\beta = 1, 2, 3) \qquad (26)$$

The divergence of this is

$$\mathrm{div}\mathfrak{A} = \frac{\alpha^2}{2r^4} \qquad (27)$$

and equations (1) and (10) yield

$$t_4^4 = \frac{1}{4}\frac{\alpha^2}{r^4}. \qquad (28)$$

This is indeed an energy density of the order of magnitude one would expect in a theory of gravitation that excludes action at a distance. What is most peculiar, though, is that—assuming my calculations above are correct—this energy can be transformed away to zero in the entire field outside of the body by means of a very small change in the coordinate system.[25] The reason for this is obviously that the t_μ^ν's are not tensor components. I must say, though, that the interpretation of t_4^4 as the energy density does not seem as useful to me now as it did earlier.[26]

Now what remains to be examined is whether expression (28) agrees with your expression (11) in "Approxim. Integr." When your values

$$\gamma'_{44} = -\frac{\kappa}{2\pi}\frac{M}{r} \qquad \text{(remaining } \gamma'_{\mu\nu} = 0)$$

are inserted in this expression, one obtains, as far as I see, for t_4^4 a value twice that of (28) with an opposite sign.[27] This contradiction may possibly stem from an incorrect approximation.[28]

Well, please excuse me for conveying to you all these fairly boring computations. Perhaps you will find something of interest in it nevertheless, and at all events you can see that I am occupying myself continually with your theory.– It has now just been a year since you were here,[29] and we are all hoping keenly that there will soon be a repetition. There is very, very much to discuss of a scientific nature, and science aside, your visit would bring great joy. So consider the matter.– Here courses begin next week. Lorentz will be lecturing on quantum theory this year as well.

Cordial regards from my wife and me, yours,

Gunnar Nordström.

383. To Edouard Guillaume

Berlin, 5 Haberland St., 24 September 1917

Dear Guillaume,[1]

I am just in receipt of the report by the Swiss Society with a renewed reprinting of your interpretation of the Lorentz transformation.[2] This interpretation is impossible, however, because the three equations

$$u' = \beta(u - \alpha x)$$

$$\frac{du}{dt} = c$$

$$\frac{du'}{dt} = c'$$

cannot be reconciled with one another. The t you introduce does not exist if the Lorentz transformation is retained. I shall not return to the matter publicly if you do not force me to by emphasizing it constantly.–

This summer I could not come to Berne because of an ailment that forced me to rest;[3] next year again, I hope. Politically, a more salutary wind is beginning to blow;[4] all shall be well yet. Best regards from your

A. Einstein.

384. To Walter Schottky

[Berlin,] 26 September 1917

Dear Mr. Schottky,

Your grave misfortune goes very much to my heart.[1] I certainly noticed how happy you were and can imagine what this means to you. Now it is important, though, that you do not abandon yourself aimlessly to this grief. Don't stay alone too much when you are not actually working or reading, but interact with your colleagues, so that you don't lapse into dark brooding. I would invite you on walks if I did not have to refrain from such things owing to my frailness.[2]

It would certainly be good if you sought me out more often,[3] so that we can discuss science and other matters. I now live at 5 Haberland St. (top floor).

Cordial regards, yours,

Einstein.

385.   From Edouard Guillaume

Berne, 3 October 1917

Dear Einstein,

Finally I get something from you![1] First I wrote you, then sent my papers (admittedly to your former address)[2] without getting any sign of life from you.

I think that a misunderstanding exists. Namely, I just say the following: The transformation expressed in homogeneous coordinates:

$$x' = \beta(x - \alpha u), \qquad y' = y, \qquad z' = z, \qquad u' = \beta(u - \alpha x)$$

can be regarded, in purely formal terms, as a normal projective relationship between 2 point systems. If the coordinates are then derived with reference to a parameter t, rules are obtained for the addition of the "velocities" $\dot{x}$, $\dot{y}$, $\dot{z}$, $\dot{u}$, $\dot{x}'$, $\dot{y}'$, $\dot{z}'$, $\dot{u}'$, which permit the classical optical experiments, etc., without admitting the "contraction," and this without any inner inconsistency.

That is all.

I would be very grateful if you would reply to me at length, and it would be my pleasure to have your refutation appear in the *Archives*. For I believe that it is worthwhile. There has been much discussion about "relative" time and the true "existence" of contraction. H.-A. Lorentz, for ex., writes (*Das "Relativitätsprinzip*," Teubner 1914,[3] pp. 22 & 23): "What importance must we ascribe to the principle of relativity? This is a question into which the physicist, strictly speaking, does not need to delve all that deeply." A discussion of relative motion and the importance of simultaneity follows. He even takes the view that absolute simultaneity may well be retained under the assumption that it corresponds to numerically different values for t and t'.[4] He then concludes thus: "Einstein says, in short, that all the questions just mentioned have no meaning." Nonetheless, Lorentz seems not to be quite satisfied with that, since he adds: "The evaluation of these concepts (relativity, time) belong mainly within epistemology, and the verdict can also be left to this field, trusting that it will consider the issues under discussion with the necessary thoroughness." I am convinced, though, that epistemology *alone* will never be in a position to throw much light on the question. You know, of course, what the Greek did to show that motion is possible: He walked! That is why I examined whether it was possible *mathematically* to arrive at the *known final equations* when "universal" time is simply *postulated.* On the other hand, Minkowski has emphasized that the great merit of your 1905 paper lies in the fact that it explains the Lorentz contraction by the relativity of simultaneity.[5] The reciprocal to it has been lacking until now, though: if absolute simultaneity is introduced, the contraction must disappear. Now, if you can prove that the path I have taken is impossible for any reason, then, in my opinion, a very significant part of the discussion of general importance will be clarified. The question of whether the Lorentz transformation would have to suffer alteration in some form as a result appears insignificant to me against the possibility of describing the natural processes from a new angle. It was precisely in order to bring clarity to all these issues that I produced my publication, and hence I can only welcome it being responded to publicly as well.

I hope that you have found the recovery you sought in Switzerland.[6]

Your optimistic hopes for the future[7] pleased me very much.

In looking forward to hearing from you again soon, I send you my kind regards, as yours,

Ed. Guillaume.

386.   To Adolf von Harnack

[Berlin,] 5 Haberland St., 6 October 1917

Highly esteemed Colleague,

At the kind advice of our colleague Planck, I request herewith that the Board of Directors of the Kaiser W[ilhelm] Institute of Physics be invited to a meeting to confer particularly on the following subjects:

a) Statutes.[1]

b) Announcements about the Institute in the press.

Planck advised me to add that it would be very desirable if a man versed in business affairs (he named you personally or Mr. Trendelenburg[2]) were to chair the meeting.–

This is just supposed to be a tentative suggestion. If you consider it expedient to choose another agenda or another path to set about these preparational tasks, everyone will certainly be in agreement. I also am willing to meet you in the library[3] at a time specified by you, if you deem this useful.

Very respectfully,

A. Einstein
Telephone: Nollend[orf] 2807.

Recipient's note: "9 October 1917. N.B.: To be answered that I shall convene the meeting as soon as the Koppel Foundation has named its representative, von Harnack."

387.   To Edouard Guillaume

[Berlin, 9 October 1917]

Dear Guillaume,

That parameter t just does not exist.[1] The equation

$$u' = \beta(u - \alpha x)$$

excludes both of the equations

$$\left.\begin{array}{l} u = ct \\ u' = c't, \end{array}\right\}$$

from which, rather than and in contradiction to the above equation, the equation

$$u' = \text{const} \cdot u$$

would indeed result.–

Should I really let this be printed in the *Archives*? I unfortunately did not receive your earlier letter. My move was being attended to precisely during my absence from Berlin.[2] If my health improves by next year, I shall come again to Berne. Then we can discuss these matters. For the moment, however, you will see from the simple statement above that your method is not feasible. I also recommend geometric illustration to you.

With best regards also to your wife,[3] I am yours,

A. Einstein.

388. To Walter Schottky

[Berlin, 10 October 1917]

Dear Colleague,

I believe you are right with your interpretation. A singular case really is all that is involved, which thus cannot claim principal significance.

I have similar curiosities as well.[1] E.g.:

Let a little rod S be freely rotatable on a plane (1 degree of freedom). Specific possible motions exist there, consistent with quantum conditions. If the little rod is suspended on an arbitrarily thin torsion thread, then the whole thing changes fundamentally.

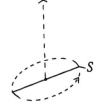

The case is similar, incidentally, to that of a mass-point moving freely to and fro between two elastic walls. In order to know what a system may do and allow, it has to have a memory that, depending on the case, must cover immense periods of time, whereas we are used to the assumption that the state of a system, which causally determines the subsequent states, is an *instantaneous* state.

Best regards, yours,

A. Einstein.

389. From Adolf von Harnack

Berlin, 10 October 1917

The Honorable Professor A. Einstein, locally.

Highly esteemed Colleague,

To the obliging message of the 6th of this mo.[1] I humbly reply that I have agreed with the Trustees of the Leopold Koppel Foundation on convening the Board of Trustees and the Board of Directors of the Kaiser Wilhelm Institute of Physical Research shortly for a joint constituent meeting.[2] The date cannot be scheduled at the moment, as the Board of Trustees has not yet been fully assembled. Mr. Franz Stock, whom the Kaiser Wilhelm Society had elected as member of the Board of Trustees, did not accept the nomination, and the Council must therefore conduct new elections.[3] The Board of Trustees of the Leopold Koppel Foundation has also not yet made the selection of the member it is to designate.[4]

As soon as the composition of the Board of Trustees has been concluded, I shall take the liberty of convening the planned joint meeting.

In great respect for Your Honor, very sincerely,

sig. von Harnack
President of the Kaiser Wilhelm Society
for the Advancement of the Sciences.

Added note: "s[ee] inquiry by Pr[ivy] C[ouncillor] Planck of 2 November 1917 in II 14."

390.   To Hans Albert Einstein

[Berlin,] 15 October [1917][1]

My dear Albert,

You have certainly inherited your letter-writing diligence from your father. What are you up to all the time, you little rascal? Do write me sometime about what you have to do at school,[2] and how you like the individual subjects. What do you do in your free time? Do you also play the piano? I am quite busy this semester and since leaving Zurich have already written a nice paper.[3] What is the food like at home, and the heating?[4] Have you possibly gotten thinner already? If anything troubles you or something is wrong, write me about it. I still remember exactly how I felt at your age as a schoolboy in Munich,[5] so I'll understand everything you say. Do you still dislike reading? Do you have friends with whom you go on walks, or do you go alone? Do you write to Tete?[6]

When you reply to me, take this letter in hand and answer my questions. Then writing will be easier for you as well. How is Mama doing? Ask her if she has received the 1,000 marks for Tete,[7] and write me again about that.

Affectionate greetings from your

Papa.

Regards also to Mama and Aunt.[8]

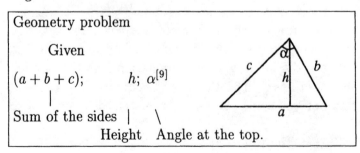

Theorem

In a triangle, the corners
of which are the height and
base points of a triangle, the
heights halve the angles.

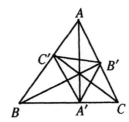

The heights $A\,A'$, etc., halve the angles $\angle B'\,A'\,C'$, etc.
I now live at 5 Haberland St.

391.   To Werner Weisbach

[Berlin,] 15 October 1917

Esteemed Colleague,

Quite belatedly I declare my approval of the decisions reached by the "Association of the Like-Minded" on 30 May 1917.[1]

I was pleased that the danger of fragmentation, which all associations face that are based on their members' ideals, not on realistic aims, has been eliminated.[2] I intend to make it my business to win over colleagues to this organization. The experience of the last few weeks has unfortunately shown again how deeply anchored the religion of might is among German academics.[3] Hard experience certainly will work incomparably more in our direction than our conscious advocacy can; yet this little must occur nonetheless, if only to spare the enlightened of the discouraging feeling of isolation.

With cordial regards, I am very truly yours,

A. Einstein.

392.   From Edouard Guillaume

Berne, 17 October 1917

Dear Einstein,

When integrating, the *integration constants* do not have to be omitted![1] The equations:

$$\frac{du}{dt} = c, \qquad \frac{du'}{dt} = c'$$

upon integration result in:

$$u = ct + r \qquad u' = c't + r',$$

where, although r and r' are independent of t, they are nevertheless dependent on x or x'.[2] *Example*: If in the equation

$$\dot{x} = \frac{\dot{x}' + \alpha}{1 + \alpha \dot{x}'} \qquad \dot{x} = \frac{dx}{du}; \quad \dot{x}' = \frac{dx'}{du'}$$

one sets $\dot{x}' = 0$, one obtains the following equation,

$$\dot{x} = \alpha,$$

which upon integration can be in agreement with

$$u' = \beta(u - \alpha x)$$

only when one sets the integration constant $= \dfrac{x'}{\beta}$:

$$x = \alpha u + \frac{x'}{\beta}, \text{ from which } x' = \beta(x - \alpha u).$$

Incidentally, it is important to note that, in general, c and c' are *not* constants, but are themselves functions of t; they should be interpreted as the "fourth homogeneous velocities," analogous to the fourth homogeneous coordinates. Later, I shall give you a complete report about this new "kinematic theory of relations," since my current papers do not contain any sufficient explanation of it.

   With best regards, I remain yours.

393.   From Gunnar Nordström

Leyden, 44 Nannie Street, 23 October 1917

Dear Prof. Einstein,

   Cordial thanks for your kind letter. It has strengthened my conviction that you really will be coming to Leyden. You can imagine how pleased Ehrenfest and Lorentz were when I told them that your arrival is not improbable. We all hope keenly that your health will allow the trip. Naturally, this time you must have a complete rest here and not be taxed as much as a year ago.[1] Ehrenfest and Lorentz share my opinion, though, that this can certainly be attended to. You will have such a calm and nice time here that your sojourn will certainly be favorable to your health; and you know that you are welcome.

   Now to science and the little t_μ^ν's. The calculation I made, based on formula (50) of your *Annalen* paper,[2] using the approximate values (70), again yields

$t_4^4 = 0$ for $g_{\mu\nu}$ and not the expression you obtain.[3] Since I do not know where the inconsistency lies, I would like to present my calculations to you quite comprehensively. Formula (50) initially produces to the approximation indicated by you

$$\kappa t_4^4 = \frac{1}{2} \sum_{\alpha\beta\mu} \left[\begin{matrix} \mu\beta \\ \alpha \end{matrix} \right] \left[\begin{matrix} \mu\alpha \\ \beta \end{matrix} \right] - \sum_{\mu\beta} \left[\begin{matrix} \mu\beta \\ 4 \end{matrix} \right] \left[\begin{matrix} \mu 4 \\ \beta \end{matrix} \right] =$$

$$= \frac{1}{8} \sum_{\alpha\beta\mu} \left(\frac{\partial g_{\mu\alpha}}{\partial x_\beta} + \frac{\partial g_{\beta\alpha}}{\partial x_\mu} - \frac{\partial g_{\mu\beta}}{\partial x_\alpha} \right) \left(\frac{\partial g_{\mu\beta}}{\partial x_\alpha} + \frac{\partial g_{\alpha\beta}}{\partial x_\mu} - \frac{\partial g_{\mu\alpha}}{\partial x_\beta} \right) -$$

$$- \sum_{\mu\beta} \left(\frac{\partial g_{\mu 4}}{\partial x_\beta} + \frac{\partial g_{\beta 4}}{\partial x_\mu} \right) \left(\frac{\partial g_{\beta 4}}{\partial x_\mu} - \frac{\partial g_{\mu 4}}{\partial x_\beta} \right)$$

The last sum is equal to zero, as can be seen by exchanging μ with β. By reformulating the first sum, we find

$$\kappa t_4^4 = \frac{1}{8} \sum_{\alpha\beta\mu} \left\{ \left(\frac{\partial g_{\alpha\beta}}{\partial x_\mu} \right)^2 - \left(\frac{\partial g_{\mu\alpha}}{\partial x_\beta} \right)^2 - \left(\frac{\partial g_{\mu\beta}}{\partial x_\alpha} \right)^2 + 2 \frac{\partial g_{\mu\alpha}}{\partial x_\beta} \frac{\partial g_{\mu\beta}}{\partial x_\alpha} \right\} =$$

$$\kappa t_4^4 = \frac{1}{8} \sum_{\alpha\beta\mu} \left\{ 2 \frac{\partial g_{\mu\alpha}}{\partial x_\beta} \frac{\partial g_{\mu\beta}}{\partial x_\alpha} - \left(\frac{\partial g_{\alpha\beta}}{\partial x_\mu} \right)^2 \right\}. \tag{1}$$

For a point *on the x_1-axis*, expressions (70) of your *Annalen* paper[4] yield

$$\frac{\partial g_{12}}{\partial x_2} = \frac{\partial g_{13}}{\partial x_3} = -\frac{\alpha}{r^2}, \qquad \frac{\partial g_{11}}{\partial x_1} = +\frac{\alpha}{r^2}, \qquad \frac{\partial g_{44}}{\partial x_1} = +\frac{\alpha}{r^2}. \tag{2}$$

All the other $\dfrac{\partial g_{\mu\nu}}{\partial x_\alpha}$'s are zero at such a point.

For a set of values, α, β, μ, inserted after the summation sign in (1) to contribute to κt_4^4 a component differing from zero, at least one of the indices α, β, μ, must be equal to 1; if all three are not equal to 1, the remaining two must equal each other. Hence the following components to κt_4^4 result:

$$\alpha = \beta = \mu = 1 \quad \text{yields} \quad +\frac{1}{8}\frac{\alpha^2}{r^4}$$

$$\begin{cases} \mu = 1,\ \alpha = \beta = 2 \quad '' \quad +\frac{1}{4}\frac{\alpha^2}{r^4} \\[2ex] \mu = 1,\ \alpha = \beta = 3 \quad '' \quad +\frac{1}{4}\frac{\alpha^2}{r^4} \end{cases}$$

$$\begin{cases} \alpha = 1,\ \beta = \mu = 2 \quad \text{yields} \quad -\frac{1}{8}\frac{\alpha^2}{r^4} \\[2ex] \alpha = 1,\ \beta = \mu = 3 \quad '' \quad -\frac{1}{8}\frac{\alpha^2}{r^4} \\[2ex] \beta = 1,\ \alpha = \mu = 2 \quad '' \quad -\frac{1}{8}\frac{\alpha^2}{r^4} \\[2ex] \beta = 1,\ \alpha = \mu = 3 \quad '' \quad -\frac{1}{8}\frac{\alpha^2}{r^4} \end{cases} \quad \begin{array}{l} \alpha = 1,\ \mu = \beta = 4 \text{ and} \\[2ex] \beta = 1,\ \mu = \alpha = 4 \\[2ex] \text{yield zero.} \end{array}$$

$$\mu = 1,\ \alpha = \beta = 4 \quad '' \quad -\frac{1}{8}\frac{\alpha^2}{r^4}$$

κt_4^4 is made up of these components and the resulting sum is zero. Thus I find exactly the same thing I had reported to you earlier, and the inconsistency between our computations remains. I have calculated out the matter in a third manner as well, namely, with formula (52) of your *Annalen* paper.[5] Owing to the differentiation on the left, more precise expressions than (70) must be used for the $g_{\mu\nu}$'s. The calculation is very simple, and one obtains

$$t_1^1 = t_2^2 = t_3^3 = t_4^4 = 0.$$

With the best of intentions, I cannot get anything else, and I do not see where the inconsistency lies either.[6]

   Because I have so much space left on the page, I would also like to say a few words about a topic I discussed with Fokker. But it is purely a matter of taste, namely, whether it is generally useful to regard the $g_{\mu\nu}$'s as pure numbers ($\sqrt{-g}$ established as not equal to 1). Is it not more practicable to have different units for distances and coordinate lengths in natural measure? If these units are denoted as σ and ξ, then $g_{\mu\nu}$ would be of the dimension, $\sigma^2\xi^{-2}$. The centimeter would probably be the unit of natural distances; the coordinate lengths could even be considered pure numbers.

Enough science for this time. In a week it's gravitation evening again at Ehrenfest's colloquium. De Sitter will be speaking. It would be nice if you could be there, but you are unlikely to be able to come so soon. In any event, we all hope that the waiting period will be extremely short. Ehrenfest will probably also write about this. Cordial regards from my wife[7] and me and

Auf Wiedersehen! Yours,

Gunnar Nordström.

394.   To Edouard Guillaume

[Berlin, 24 October 1917]

Dear Guillaume,

In my opinion, this new point of view[1] is also untenable. If t is supposed to be a function of u and x, it must *be possible to specify* this function *explicitly.* Upon calm consideration, you will yourself become convinced of the nonexistence of a t to which the role of universal time could be ascribed. If it existed, there would also have to be a preferred family of surfaces in Euclidean geometry, since the former is *also* obviously characterized by linear, orthogonal coordinate transformations.

With best regards, yours,

A. Einstein.

395.   From Franz Selety

Vienna I, 11 Zedlitz Alley, 29 October 1917

Esteemed Professor,

I thank you most cordially for the lengthy letter you were so kind as to send me.[1] I had naturally not expected you to devote so much time to me and had not demanded it. Although I had hoped to receive an answer to my questions, I certainly had not intended to cause you so much trouble. In my present reply I am not going to pose any more questions; nonetheless I do take the liberty of answering you in some detail again since, obviously, reading requires less time than writing. I also prefer that you not read the philosophical sections of this letter right away, which constitute the greater part of it, just because you have received it, but that you do so *in no hurry* when you have time for it; for I know how burdensome something can become, even if it is intrinsically of interest, when it is forced upon one when something of urgency is waiting to be done, and in this case, one is not in the mood to comprehend it well. I shall also have this letter transcribed again so that perusal proceeds as painlessly as possible.

Even though I have made an effort to become somewhat versed in physics, my knowledge is nevertheless quite limited, even more so my ability to work independently with the subject matter. For a long time it has been a great wish of mine to become not only a philosopher but also a trained theoretical physicist; but I see increasingly that although I am able to understand the mathematics, I cannot master it to the point that I can work mathematically. Thus I have become acquainted with the general theory of relativity not only through popular descriptions but also through studying your original papers. I may be able to understand covariant theory, etc., yet I do not have an overview of it. That is how it came about that in my letters to you I overlooked some things I had learned and which are self-evident, of course, such as, that according to the general principle of relativity, the law of the conservation of momentum must be valid for every coordinate system. Besides, from your letter I saw clearly what is involved in the selection of a coordinate system and in what way the laws of momentum conservation do not work. From it I also learned, in particular, what I had heard many times already but what a philosopher repeatedly forgets, namely, that an idea lacking mathematical formulation has no value in theoretical physics as yet, and that expressions like "most closely Euclidean,"[2] etc., merely state the problem.

As concerns electron theory and Ehrenhaft, certain critical considerations by Ehrenhaft, quite aside from the experiments, do seem to me very cogent. Ehrenhaft thinks virtually nothing of all theories and, even though he does go too far in this, he certainly shares this fault with the great Mach.[3]

Ehrenhaft points out legitimately that α and β rays merely measure the relation $\frac{e}{m}$. He considers it unproven that α particles were helium atoms with two elementary charges.[4] The fact that helium emerges from it is not adequate proof for him. Even so, this certainly does seem to be a very plausible hypothesis. On the other hand, his criticism concerning β rays really is very convincing. Indeed, it is completely unproven, but nonetheless obviously not necessarily false, that β particles carry the elementary charge and that their mass is $\frac{1}{2000}$ of the hydrogen atom. But since it is acknowledged, at least, that matter sometimes holds several units of elementary charge, the already so firmly rooted belief that we knew of masses that were $\frac{1}{2000}$th of the hydrogen atom is quite unfounded. I believe this very simple critical objection is certainly justified if it can be assumed that the mass of β particles is smaller than that of the hydrogen atom. It certainly is a mistake that, for years now, all popular books have been treating as dogma that we knew of bodies whose mass was $\dfrac{1}{2000}$ of the hydrogen atom, with the result that all laymen (and all physicists as well) regard this as secure knowledge. I think anyone who considers this must find that the hypothesis is not based on experience.–

I come now to your reply to my philosophical theory.[5] You write, first of all, that a consequence of my theory would be that the only real facts of pure experience are those of a present state of consciousness. Indeed, I myself have always emphasized this to be correct in a specific sense, namely, that a momentary state would notice nothing if, apart from itself, the entire remaining world did not exist and that hence instantaneous solipsism is logically possible. Thus I also now find the expression, facts of pure *experience*, to be not very well chosen, giving rise to this misunderstanding. I actually intend something else and do not seek in any way to grapple with solipsismal and ultrasolipsismal doubts.– I initially assume that, by its quality and quite generally speaking, consciousness is an actual and absolute reality and is not merely fictitious, such as are physical objects perhaps. What types of consciousness actually do exist concretely is another question. Whether another absolute reality also exists in addition to conscious reality, I regard at the very least as uncertain, and even assume that the only reality type known to us is also the only one that really exists. Now, with this notion I approach our normal positive worldview and attempt to establish what of it belongs within this reality of consciousness and what does not. Since this reality of consciousness is the only one we really know, this isolation of it from our worldview seems to me to be one of the most important philosophical missions. In doing so, I arrive at the frequently mentioned result of the orderless set of states of consciousness. In consequence of starting out from the positive worldview, I have nothing directly to do with the epistemological problem of (instantaneous) solipsism and merely use as an auxiliary consideration the possibility of assuming that for each state nothing outside of itself exists, in order to show that the order of the states does not belong directly within consciousness. It would therefore be better if, instead of facts of pure experience, I say *facts of pure consciousness.*

My method of determining whether something is a part of pure consciousness is, as you know, that I negate the matter in question and see whether something is changed in the world of consciousness.[6] This method is, purely logically, a simple tautology, and one can establish what belongs within pure consciousness and what does not without this auxiliary consideration; it merely serves as a more forceful illustration. I hope these distinctions have become clear and that when I say pure consciousness rather than pure experience, this objection no longer applies to me. I am not concerned with how we can experience something and what we know for certain, but rather with isolating out of the given worldview a specific quality that seems to me the genuinely real basis of it, whereas the rest are auxiliary constructions and fictions. Mach and Avenarius have also posed this problem, of course, but they arrived at different results.[7]

On this basis I assume furthermore, on the one hand, that our human consciousness must be regarded as the thing-in-itself [*Ding an sich*], as the basically real for the cerebrum, *so that we know exactly the reality underlying this one physical object* and, on the other hand, that other physical objects are based on something more or less analogous to our consciousness: resembling it to the extent that other living bodies with our brain share general physical characteristics, and differing from it to the extent that they differ. In this way I believe a metaphysics is possible which, although not provable, is nevertheless admissible in principle, since it construes the unknown by analogy with the really known. The idea of constructing metaphysics on an empirical basis in this way, in that what really underlies things is conceived as analogous to our psyche, seems to me to be Leibniz's great philosophical achievement. My metaphysics differs from that of Leibniz in that for him the entire soul (with the passage of time as content) is regarded as a monad; for me, however, it is the state, and I therefore imagine the monads unknown to us as analogous not to the soul but to the state of consciousness. In any event, it seems to me the only metaphysics possibly admissible would be one that builds upon the foundation of the only reality of consciousness truly known to us. Perhaps no metaphysical approach will appeal to you; nevertheless, I invite you to consider whether the above chains of reasoning are not legitimate nonetheless.

As far as your other remarks on my philosophy are concerned, I unfortunately found that you did not understand me correctly. This was troublesome to me for various reasons and muted my delight in your letter. I was sorry, first and foremost, that I was not understood properly by *you* in particular, since then only bleaker prospects must await me for comprehension by the public at large; for if *you* do not immediately understand the matter completely, it is very much to be feared that most persons will not be capable of doing so at all. In these points I do still hope, though, to be able to convince you that you had not grasped the matter clearly enough and that this is the reason for your subsequent comments. I naturally assume that you will not take offense at my frank rebuttal.

To start with, you write that the more intently one concentrates on pinpointing the instantaneous present, the more it shrivels up into something that quite resembles a Machian element. Moreover, a bit further down you use the term "content-deficient individual state."[8] But this is a completely incorrect interpretation, as can evidently be seen upon adequately strict reflection, especially if one considers that if nothing other than the momentary state existed, this state could notice nothing, as you yourself had emphasized at the beginning of your letter. Therefore, all the manifolds that we really experience, the whole abundance of our discrete experiences, must find expression in the momentary state. First of all, it is certain that in spatio-optical sensation, when we see an object, we have a strictly simultaneous manifold in our consciousness, since otherwise we clearly could not speak of seeing an object but would merely have a point of color in our consciousness.[9] If the other points were in our consciousness even only a millionth of a second earlier or later, we would not have the total optical impression. The past perception would, of course, be in the other world of the past state. Nevertheless, we do see complete, extended spatial images. At the same time, however, we certainly see distinctly areas that are larger than the one that immediately falls on the eye in one moment of focussed looking, which is probably explained physiologically in that we move our eye very rapidly and combine the individual partial images, be it as a result of the physiological aftereffect in the eye or in the brain. It is not good, though, and is just confusing, to allow oneself to be influenced at all by such physiological considerations in judging what is experienced directly. In any case, one thing is "evident," though, and this in the literal sense, namely, *that we see a manifold simultaneously.*

But by no means does this exhaust the immense variety of what must necessarily be found within a momentary state of consciousness, although these subtle things are so difficult to catch. It is a matter of the immeasurable manifold of grades of coloration featured in the state of consciousness mentioned in the foregoing. It is clearest and simplest if we deal with the considerations of these aftereffects in connection with musical interpretation, even though this also applies to the interpretation of the spoken sentence; yet in this case there is the added complication that, in the first place, one must agree that the sense of the words corresponds to something within immediate consciousness. Thus as concerns music, you will also find briefly discussed on page 2 of my paper that we would have no notion of a complete melody if, with each received tone, a very definite lingering effect of the foregoing tones were not *simultaneously* in the consciousness.[10] If this were not the case, we would then just have a perception of isolated tones or harmonies. This aftereffect naturally does not consist in the graphic concept of the preceding tones, otherwise we would obviously have a terribly dissonant impression; rather, it is something very specific. When you consider these simple circumstances a little, you must understand what an immeasurable manifold of states of consciousness there is of the subtlest differences in hue and how much such a manifold differs from a Mach element. Every musical sequence we hear and take in, however great it is in length, is unequivocally assigned to a very discrete state of consciousness; that is, this state corresponds in our temporal order to each moment of the totality of what has been heard until then, for if this were not the case, we could not understand the musical sequence, and what has gone before would have been lost. These circumstances are illustrated very well in the following diagram, which can be found in the "Psychology" article of the *Encyclopaedia Britannica*.[11]

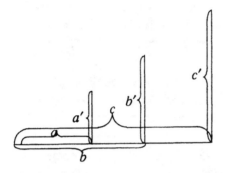

You see how each section experienced projects the momentary final state, otherwise we could not characterize the section in question as experienced by us at all and thus it would be as little present to us as to any other person unknown to us. This is because, for each state as such, any others are not there; what it experiences is merely the projection of the other.

Hence there are as many different states of this kind as there are different sections of the various musical works. For ex., the state we experience during the xth quarter note of the yth measure of the zth movement of the uth symphony by Beethoven is a very specific one and different to every other one of the sort, just as the entirety of the portion of the symphony played up to then differs from any other section of this and any other symphony. The manifold of these states is rich enough for someone who is just capable of grasping short melodies; how much richer and more finely graded are these states for the musically inclined! It is true that within a state we comprehend more clearly the part having elapsed only a short time before, and those further away increasingly less; yet for the musically minded this continuity must also extend beyond the individual movement to the entire piece, since otherwise it would not matter to us during the second movement whether the first movement had been played beforehand, or instead nothing at all, or another movement altogether. I believe, when you think about these things, you must realize the necessity for these subtlest of gradations of the projections in each state and how rich they are in the conception of a great musical work of art. You must just keep clearly in mind that each momentary state possesses nothing and knows about nothing other than what it itself contains, as if the whole rest of the world did not exist.

The same considerations as those about music also apply to the apprehension of the spoken sentence; yet in order to realize that something quite different from word melody apprehension is involved here in addition, one must keep clearly in mind that our abstract thought also is composed of characteristic colorations of our consciousness and not merely of words or of graphic notions. Even if one forms no graphic notions, when hearing the word "triangle" one has a different state of mind from someone who does not understand German and a different one from when one hears the word "quadrangle," and likewise with every other word. A very distinctive coloration of our consciousness is even attached to such words as: but, yet, although, etc.; otherwise these words would be little more than Chinese vocables to us. And hence, upon hearing a spoken sentence in each moment, we have a very characteristic state that is determined by what has been heard up till then, that is, not by the word melody perhaps, but by the meaning of what has been heard. Also added to all of this are the peculiar colorations of anticipations both in music as well as in speech, since we also expect a very specific continuation that makes sense in the musical context, of course, just as in the verbal one. In addition to these various infinitely graduated colorations of consciousness, aftereffects, the sense of words, and anticipation, there are others that do not concern us here, though. Thus, for ex., when we try in vain to recall into our consciousness something forgotten, we are in a different state of mind, depending on what the forgotten thing is, etc.

The great American psychologist James was the first to deal thoroughly with all these psychological subtleties and to present them in the vividest and clearest form in his work.[12] He calls the various aftereffect colorations *fringes* (*Fransen*); his German advocate Cornelius calls them relation colorations [*Relationsfärbungen*].[13] Any psychology that does not take these things into account, and this unfortunately includes the majority, lends a rough, wooden impression to those familiar with them, and certainly also to any open-minded person not familiar with them. If it is the mark of a genius to be the first to describe things that everyone can observe and yet that no one has noticed, then James has earned this title to the fullest degree. He was the first to succeed in comprehending the directly given facts of consciousness in all their vibrancy instead of having them pressed into a herbarium, whereby all the fine details are lost. Thus my reply to your comment that the momentary state resembled a Mach element and was deficient in content has become a little descriptive psychology. It would also be of particular satisfaction to me if I should have succeeded in showing you how intriguing this normally so boring discipline actually is. I hope to have succeeded in convincing you that the present state of consciousness must be immeasurably rich, a microcosm, since all the manifolds we experience must, of course, somehow be expressed in the momentary state. I hope to have succeeded in presenting clearly the matter of multifarious aftereffects and of musical apprehension through projection in the instantaneous state, whereupon I refer particularly to the diagram.

With reference to this presentation, I am so bold as to quote the charming sentence from your popular essay: "Before you have conceded this to me with conviction, dear reader, do not read on!"[14] Once you have recognized, however, that the current state is rich in content, you will agree with me more readily also in the following.

You write verbatim: "In any case, once you have broken down the world of facts of consciousness into a vast quantity of such states deficient in content, surely you must paste them together again with some sort of suitable *glue*, so that this world is formed again out of them." (With glue you apparently do not mean much more than temporal order, since it is scarcely likely that you mean the stronger glue of the substantial identity of successive states in the sense of the old theory of mental substance.) I saw with dismay from your statement that you had not grasped the gist of my entire argument after all. For I demonstrate that it is impossible for a glue between the states, an order between them, to be able to belong within pure consciousness. I was all the more astonished that you did not see this, since it follows easily from a fact you yourself had acknowledged earlier, namely, that all things lying beyond the current state are as if they were not present for it, and that it would not notice if they did not exist. This does not exclude that beyond this state, 1, 2 . . . n, infinitely many other states analogous to it and each likewise with a world of its own also existed, which we all firmly believe, of course, but it is impossible that anything between them—a glue or an order—can belong to the total consciousness existing in the world, to that quality so well-known to us all. With this it is naturally not proven that such a glue or such an order does not exist as something unconscious and metaphysical; Hume cannot prove with his criticism either that some metaphysical causal relation really does exist apart from the facts of consciousness demonstrated by him; but that this relation does not belong within pure consciousness can be shown, and whoever believes that nothing exists other than what is more or less analogous to the consciousness familiar to us, will not believe in an unconscious metaphysical causal relation and in a metaphysical order in time.[15] As paradoxical as the notion of temporally unordered states of consciousness is, it is nonetheless evident that we must arrive at this concept if we eliminate everything that is not an immediate fact of consciousness to any one of them. It is probably clear from my description at the beginning that I do not construe immediate facts of consciousness as merely what we can know very certainly to exist. For the only thing we can know certainly is the existence of our momentary state; but with an immediate fact of consciousness I mean something that has a very distinctive quality (of being intrinsically and directly given, of the type of perception and sensation) if it exists, whether we know about it or not. That is why the totality of all immediate facts of consciousness can very well be a multiplicity of elements of consciousness, but each is immediate on its own. For whom, though, is an order of elements of consciousness an *immediate* fact of consciousness? The firm *conviction* that such an order exists (this is projection along the vertical line), which admittedly is an immediate fact of consciousness, in most human states, nonetheless is far from making *this order itself* (the *horizontal* line in the diagram) into an immediate fact of consciousness. Neglecting this distinction is the error most others commit.

From the fact that each state knows nothing directly aside from itself and is a closed world, it directly follows, as I have emphasized, that in the case where a multiplicity of such states is present in the world, it is *impossible* for any order between them to be a direct fact of consciousness, although, as it seems, this consequence cannot be accepted so simply. It is best we have another look at the diagram. There we see that only the vertical lines are facts of consciousness, but their arrangement in the horizontal ones are not. Let us imagine the individual vertical lines *exchanged randomly* on the horizontal ones but each in itself is left entirely unchanged, then nothing would be changed in the immediate world of consciousness. In the state b', for ex., the conception would be exactly thus, that the states followed the sequence in the way projected within it, even if this is not the case. I hope I have now succeeded in showing you that the order or any other glue between the states is not a fact of consciousness. *Only the projections are facts of consciousness, the individual vertical lines, each in its own world, not their arrangement along the horizontal lines.* Also with reference to this fact I dare to quote your statement: "Before you have conceded this to me with conviction, dear reader, do not read on."

From the above it follows that the order normally assumed by us is determined only by the inner quality of the individual states. For even though an arrangement of the states does not in itself exist, if we deny all unconscious elements, through their quality (in our diagram through the differing lengths of the vertical lines) an arrangement is calculated nonetheless (in which the individual vertical lines are ordered according to size). This can also be illustrated with the following image. We imagine the continuous spectrum broken down into all of its lines and these are mixed together randomly; then the old order would surely be determined and discerned by quality, since it is the only one with which there is continuity in color quality. Thus the temporal arrangement is also the only one in which the continuity of the qualities of consciousness can be found.

As concerns your comparison of a painting as being composed of daubs of paint which nevertheless do not make up the painting, precisely this can be used against you, since a painting involves specifically the *simultaneous* concept of all the daubs of paint and only the simultaneous concept can cause the order of the daubs to be a true fact of consciousness. Your example shows that the momentary state of consciousness differs from the Mach element not only insofar as a whole set of these elements are found assembled within it in an undefinable manner, but also that the state is not correctly described as a mere quantity of such elements, because in that way the so-called gestalt quality is ignored, which cannot be construed as an element at all, since it is not conceivable independently of the set of elements. (The gestalt of the arrangement of n elements cannot be regarded as a type of $(n+1)$th element, such as perhaps the feeling evoked by it, but the gestalt quality is also a fact of consciousness differing from this element.)– As concerns the ensuing relations analogous to the gestalt qualities within the musical sequences, the result here is, as I have shown, that to the extent that they are facts of consciousness, these relations must be found in the vertical projections within the individual states of consciousness. This is as if the painting's gestalt were projected within each point of color and, apart from this projection and the illusion of the, in principle, unordered points of color, did not exist at all. From this it is apparent that the psychological relations of spatial gestalten and musical experiences are very different and are not exactly analogous. In the first case, a real order within the consciousness is involved, in the other case, merely real projections, again within the state of consciousness and an imagined, or at least in no way conscious, order. Compare with the last page of my paper.[16] At least two spatial dimensions are directly given as a concept; by contrast, there is no direct concept of succession, rather only a concept *of* succession in the *momentary* consciousness. Thus, on the one hand, space and time differ in the consciousness; on the other hand, however, according to my system there is an exact analogy in that the states of consciousness are as little ordered in time as they (or "souls") are ordered amongst themselves in external spatial ordering relations, which but few assume nowadays anymore. Indeed, I am surprised that, as a consequence of this analogy, this philosophical theory did not arise long ago out of your and Minkowski's physical theory of time.

When it becomes clear to you that the state of consciousness is anything but deficient in content, you will no longer feel such a compelling need for a glue between the individual states, since everything that one could desire and that is possible is found within this small world, and you will rather understand that you were gripped so strongly by this entirely unfulfillable wish simply as a consequence of a false conception of the individual state. If my representation has not already done so, I hope sincerely that you will come to agree with the correctness of my arguments, when you have found the time to think about it. Should this be the case, I would naturally be very pleased if you would inform me about it; this can easily be done without much effort by postcard, of course. I mean the idea that pure consciousness in the world is none other than an *orderless set of unitary states of consciousness*; by no means more than that, scarcely less.

You wrote, furthermore, that you do not believe there is actually anything new to be said about these things. The idea that a momentary state would be completely ignorant about whether nothing else apart from itself existed has long been known, as well as that the concept of time must occur in the present state. As far as I know, the novelty in my approach is the virtual integration of this idea with the conception of the entire existing consciousness and ultimately all being on this basis. However, the problem of working out the totality of what exists directly (as consciousness), according to its quality, from the naive world of experience has already been dealt with by many under various labels.   In my opinion, Mach and Avenarius were justified in considering it the fundamental problem of philosophy. This principle, however, has not been proposed until now, as far as I know. Although the individual elements of my idea were known and these were also recognized by you in your letter, your objections do prove that some misunderstandings are still possible before one can concede the correctness of my summary. In projecting oneself into my ideas, it becomes apparent that, for all its abstract rigidity, this summarizing consideration does not lack a certain attraction. One recognizes above all that the *formal* relationship of two states of the same person is no different to that of two differing states and that the difference is based merely on the *inner qualities* of the states. From the idea of instantaneous solipsism, only *one* common trait has been acknowledged, namely, that we cannot know absolutely certainly about the existence of a preceding state of ours, just as about that of a foreign consciousness; however, that the temporal distinction is absolutely *identical* to the distinction between the consciousness of different persons seems to me to be a very important simplification of our worldview.

In thanking you once again for your long letter and hoping that sooner or later I shall receive word from you, be it only a short note, I am yours truly,

Dr. Franz Selety.

396.   To Edgar Meyer

[Berlin,] 30 October 1917

Dear Colleague,[1]

Even before receipt of your letter I had taken steps on Epstein's behalf, turning to Nernst in particular. In general, though, I hear that in such a case only with great difficulty can permission to leave the country be obtained,[2] quite contrary to my expectations. Nernst stated bluntly that he could do nothing in this affair. Only people who know E. personally very well and who have the necessary connections could possibly achieve anything, provided they came here and took up the matter *in person.*–

With great interest I hear from Abraham[3] that you are behind having Ehrenhaft's experiments tested.[4] This really would be very worthwhile. For this matter cannot be settled by declaring "sources of error must be there, although it is unknown which ones." The fantastic matter of "negative light pressure" also cries out for elucidation.[5] I am very curious what you will achieve.

With best regards and best wishes for these pending projects, I am yours very truly,

A. Einstein.

397.   From Zofija Smoluchowska-Baraniecka

Cracow, 27 Studencka, 8 November 1917

Highly esteemed Professor,

I ought to have answered your letter of September 23rd long ago, which conveyed such warm words and profound feelings.[1] But it was too hard for me to reply to those letters that moved me.

I thank you very much for your note, which I value especially highly because it comes from someone for whom my husband has always had sincere admiration and the greatest respect and true sympathy.

Although there can be no question here of consolation, it is nonetheless bittersweet to see how everyone admired my husband's work, how respected and loved he was:—indeed—he knew how to win the hearts of all.

A few days ago I was informed in a letter by Dr. Berliner that you are kindly honoring the memory of my husband by intending to write an obituary of his so tragically and abruptly shattered young life.[2]

I immediately sent Dr. Berliner the biographical information that might possibly be needed for it, which he had requested. I would like to thank you especially for your intention.– I remember very well—how pleased my husband was with your postcard last year, in which you, Professor, informed him that you were discussing his papers in your seminar;[3] he always wanted to answer you—but was so overburdened with his responsibilities as dean that I do not know if he managed to do so.–

I am really pleased at the honor of having my husband's eulogy flow from your pen, in particular, which is so highly esteemed throughout the entire scholarly world.–

It is a pity that we cannot discuss in person so many things—especially what he often said about himself, how he counted himself among the romanticists in science (according to Ostwald's classification)[4]—and what a profusion of started papers, of research topics he left behind!–

And his personal life!– His passion for serious, classical music—then primarily for Wagner and Bruckner!–

You knew him personally, of course, in all his youthful vigor and zest for life! and his interest in everything that can be called culture and intellectual creativity on the one hand, and his feel for nature—and the need for a closer association with it, a related passion for traveling, skiing, and rowing—on the other, not to mention a sense for all that is sublime, good and ethical!–

Yes, that was a consummate individual—no fault, no dark side.–

–I do not know whether you possess a photograph of him—if not, I shall be glad to send you one—we had one here of you from Prague–[5]

Those were fine days back then!–[6] I am enclosing a German death notice.[7] Owing to the war conditions, they were printed late, and I also did not get around to sending them out immediately.

If you would like to have a complete listing of all the publications by my husband, it would be a pleasure for me to send it to you—in case you need it for the obituary.– Best regards, and in expression of my utmost respect, from

Sophie von Smoluchowska

P. S. This letter was already completely finished and was supposed to be sent off when I learned that a new edition of the complete works of my husband is being planned and that the wish had been expressed that you, Professor, be asked to write the introduction.–[8]

I agree that this would lend much greater importance and special value to the entire work, and that is why, when I gather that the editors of the new edition will turn to you about it, I also shall take the liberty of soliciting your consent, since I truly believe it to be best and most desirable.–

Once again, my compliments and respect,

Sophie Smoluchowska.

398. From Rudolf Förster[1]

Essen, Krupp Factory, A. K. Laboratory, 11 November 1917

Esteemed Professor,

Immensely inspired by your bold ideas on gravitational theory, I take the liberty of robbing you of some of your time with these lines. Allow me to make some comments. In doing so, please note that the mathematical literature is accessible to me only to a very limited degree.

1) In the treatment of specialized problems, a theorem by Schläfli, which I found in Pascal, *Repertorio*,[2] seems to me to be very beneficial: For a given line element $ds^2 = \sum_1^0 g_{ik} dx_i dx_k$, a system of at most 10 functions $y_i(x_1, x_2, x_3, x_4)$ can always be found, such that $ds^2 = \sum_1^{10} dy_i^2$. (The world thus lies in a linear Euclidean R_{10}.) The number of y_i's, reduced to 4, is called a class of R_4.

Upon insertion of the y_i's, the Christoffel and Riemann tensors were simplified considerably: As $g_{i\kappa} = \sum_j \dfrac{\partial y_j}{\partial x_i} \dfrac{\partial y_j}{\partial x_k}$, hence

$$\left[\begin{matrix}\mu\nu\\\sigma\end{matrix}\right] = \sum_j \frac{\partial^2 y_j}{\partial x_\mu \partial x_\nu} \cdot \frac{\partial y_j}{\partial x_\sigma}; \qquad \left\{\begin{matrix}\mu\nu\\\tau\end{matrix}\right\} = \sum_{\alpha,j} g^{\tau\alpha} \frac{\partial^2 y_j}{\partial x_\mu \partial x_\nu} \frac{\partial y_j}{\partial x_\alpha},$$

furthermore, on applying the formula $dg^{\mu\nu} = -\sum_\alpha \sum_\beta g^{\mu\alpha} g^{\nu\beta} dg_{\alpha\beta}$:

$$B^\rho_{\mu\sigma\tau} = \sum_{\alpha,\beta,\gamma,j,k} g^{\rho\beta} g^{\alpha\gamma} \frac{\partial y_j}{\partial x_\alpha} \frac{\partial y_k}{\partial x_\gamma} \left(\frac{\partial^2 y_j}{\partial x_\mu \partial x_\sigma} \cdot \frac{\partial^2 y_k}{\partial x_\tau \partial x_\beta} - \frac{\partial^2 y_j}{\partial x_\mu \partial x_\tau} \cdot \frac{\partial^2 y_k}{\partial x_\sigma \partial x_\beta} \right) - $$

$$- \sum_{\beta,j} g^{\rho\beta} \cdot \text{(The same parentheses but with } k = j)$$

$$B_{\mu\nu} = \sum_{\alpha\beta\gamma\delta jk} g^{\alpha\beta} g^{\gamma\delta} \frac{\partial y_j}{\partial x_\alpha} \frac{\partial y_k}{\partial x_\beta} \cdot \left(\frac{\partial^2 y_j}{\partial x_\mu \partial x_\nu} \frac{\partial^2 y_k}{\partial x_\gamma \partial x_\delta} - \frac{\partial^2 y_j}{\partial x_\mu \partial x_\delta} \cdot \frac{\partial^2 y_k}{\partial x_\nu \partial x_\gamma} \right) - $$

$$- \sum_{\gamma\delta j} g^{\gamma\delta} \cdot \text{(The same parentheses but with } k = j).$$

If $y_1 = x_1, y_2 = x_2, y_3 = x_3, y_4 = x_4$ are set generally, it is possible to limit oneself in the j and k summations to the interval $5 \rightarrow 10$, since the 2nd derivatives disappear.

Example 1: Gravitational field of the 1st class:[3] $s_1 = x_1$; $s_2 = x_2$; $s_3 = x_3$; $s_4 = x_4$; $s_5 = f(x_1, x_2, x_3, x_4)$. If the derivatives for f are marked with indices, then

$$g_{ii} = 1 + f_i^2; \quad g_{ik} = f_i f_k (i \neq k); \quad g = 1 + \sum f_i^2;$$

$$g^{ii} = 1 - \frac{f_i^2}{g}; \quad g^{ik} = -\frac{f_i f_k}{f} (i \neq k)$$

$$\sum_{\alpha\beta} g^{\alpha\beta} f_\alpha f_\beta = \frac{\sum f_i^2}{1 + \sum f_i^2}; \quad B^\rho_{\mu\sigma\tau} = \frac{-1}{g} \sum_\beta g^{\rho\beta} (f_{\mu\sigma} f_{\tau\beta} - f_{\mu\tau} f_{\sigma\beta});$$

$$B_{\mu\nu} = \frac{-1}{g} \sum_{\alpha\beta} g^{\alpha\beta} (f_{\mu\nu} f_{\alpha\beta} - f_{\mu\alpha} f_{\nu\beta}).$$

If furthermore one forms $B_\sigma^\alpha \equiv \sum_\tau g^{\alpha\tau} B_{\tau\sigma}$ then

$$B_\sigma^k = \frac{-1}{g} \sum_{\alpha\beta\gamma} g^{\alpha\beta} g^{\gamma k} (f_{\alpha\beta} f_{\gamma\sigma} - f_{\alpha\gamma} f_{\beta\sigma}), \text{ and}$$

$$B = \sum_k B_k^k = \frac{-1}{g} \sum_{\alpha\beta\gamma\delta} g^{\alpha\beta} g^{g\delta} (f_{\alpha\beta} f_{\gamma\delta} - f_{\alpha\gamma} f_{\beta\delta})$$

result. Thus the mixed energy tensor of matter according to the field equations of 1916[4] becomes

$$T_k^i = \frac{1}{\kappa g} \left[\sum_{\alpha\beta\gamma} g^{\alpha\beta} g^{\gamma i} (f_{\alpha\beta} f_{\gamma k} - f_{\alpha\gamma} f_{\beta k}) - \frac{1}{2} \delta_k^i \cdot \sum_{\alpha\beta\gamma\delta} g^{\alpha\beta} g^{\gamma\delta} (f_{\alpha\beta} f_{\gamma\delta} - f_{\alpha\gamma} f_{\beta\delta}) \right].$$

Example 2: Arbitrary class. $s_1 = x_1$, $s_2 + x_2$; $\dfrac{\partial g_{ik}}{\partial x_1} = \dfrac{\partial g_{ik}}{\partial x_2} = 0$: (Special class of two-dimensional static or one-dimensional dynamic fields.) The above portrayal shows immediately that all B_{ik}'s except for B_{33}, B_{34}, B_{44} vanish. The line element becomes

$$ds^2 = dx_1^2 + dx_2^2 + edu^2 + 2f\,dudv + gdv^2$$

$B_{33} : B_{34} : B_{44} : 1 = e : f : g+ : \dfrac{1}{k}$, where $\underline{k}$ is the Gauss curvature of the line element $edu^2 + 2f dudv + gdv^2$. Furthermore, $B_3^3 = B_4^4 = k$; $B_4^3 = 0$.
But unfortunately $T_\sigma^\alpha = 0$ for all the indices.

2) Hence the space can be curved even without the presence of matter! Something is evidently wrong here. For from $T_\sigma^\alpha = 0$, although not ds^2, the curvature ought to be established, if I have understood the matter correctly. At the same time, I venture to request being informed about how the second term on the right-hand side of your field equations $-\kappa(T_{\mu\nu} - \frac{1}{2} g_{\mu\nu} \cdot T)$ is justified.

3) Regarding your field equations of 1917 (February 8):[5] $G_{\mu\nu} - \lambda g_{\mu\nu} = -\kappa(T_{\mu\nu} - \frac{1}{2} g_{\mu\nu} \cdot T)$. (a) Surely you assume λ as the universal constant only because ρ is supposed to be constant? (b) You have disposed of the boundary conditions at infinity, but in exchange the periodicity conditions result.

4) Re your gravitational wave theory.[6] One still tends to say gravitation is propagated with the velocity of light. Atomic physicists still tend to attribute

all finer processes to electrical oscillations, because the Maxwell equations are flexible enough to be adapted to all sorts of processes for which conventional mechanics is inadequate. I have a new premonition, though. In a few years one will say: Light and electrical waves are propagated with the velocity of gravitation, the dielectric constant will be derived from the energy tensor of matter, and a new theory of electromagnetism will be created in which it is not the process of induction, perhaps, which forms the underlying phenomenon but electrification through friction. Maybe a covariant six tensor will be found that explains the occurrence of electricity and results easily from the $g_{\mu\nu}$'s rather than being taken over as a foreign element.

5) An idea which admittedly stems only in part from me and has arisen from discussions in which I attempted to acquaint an average engineer with your ideas: You limit yourself to the mathematical development of the theory. For the purpose of disseminating the new views on gravitation more broadly, a graphic interpretation would be of great use, though. On discussing Michelson's experiment, this engineer countered that the negative result seemed obvious to him, as the light ether surely moves along with it. After resisting a long time, pointing to Fizeau's water experiment as well as the difficulty encountered when the light source and the observer move in opposite directions relative to each other so that the poor ether does not know with which of the two it ought to go, I did warm to this view, and now in anticipation of your esteemed reply would like to state the following hypothesis:

"The field equations of gravitation are to be interpreted as: differential equations of the motion of the ether." (Instead of ether, one could also say space.) The ether is the medium of the gravitational field and of gravitational waves, its state at one place is described by the $g_{\mu\nu}$'s. To start with, I would also like to burden the same with the light phenomena as well as with radiating heat, and finally, preferably also with the electromagnetic field.

My occupation unfortunately just allows me to devote a few hours of leisure to these matters and I therefore can see no other way than turning to you personally, for which I would like to beg your pardon again herewith.

With greatest respect,

Dr. Rudolf Förster, Engineer.

399. To Paul Ehrenfest

[Berlin,] 12 November 1917

Dear Ehrenfest,

I really was heartily pleased about your invitation,[1] and you can believe me that nothing is more appealing to me than a trip to my dear Dutch friends, with whom I share such close and kindred feelings in everything. Yet, with the prevailing travel conditions[2] and my weak health, I may not think of it. I am bound to a strict diet and am never really safe from a painful attack.[3] In the last half year it has admittedly been improving steadily; however, I must follow a clinically regular bourgeois lifestyle.

Scientifically there is nothing of note to report. Your objection to my quantum paper of 1914 is thoroughly justified;[4] I became aware of the same recently upon studying your paper of 1916.[5] I do believe, though, that the matter can very probably be corrected in this way: The equation $S = \lg Z$ is initially only proven for purely thermal changes. It was wrong, now, to conclude the invariability of Z according to the adiabatic hypothesis.[6] Instead I avail myself of the circumstance that I can choose the system's external conditions and that, depending on this choice, other processes are "purely thermal." Hence, for ex., a rise in temperature with a constant volume is a "purely thermal" process or a rise in temperature at constant pressure, depending on the choice of these conditions. Therefore any state of the system is attainable through "purely thermal" changes. That is why the equation $S = \lg Z$ applies generally if it is valid for purely therm. processes.

Cordial regards, yours,

Einstein.

400. To Rudolf Förster

Berlin, 5 Haberland St., 16 November 1917

Esteemed Sir,

From your letter[1] I see that I am dealing with a man of extraordinary theoretical talent. It would be a pity if you did not have enough leisure to think about these fine problems. Now I would like to reply immediately to your individual points in the order you followed.

1) The introduction of y_i in place of $g_{\mu\nu}$ is, without a doubt, possible and is also suitable for reducing the number of functions sought, from 10 to 6. But it is questionable whether the expressions for $B_{\mu\nu}$ obtained in this way really are

more convenient in the calculation. In any event, this matter is worth publication, especially if you work out a special case according to your method, that of a point of mass, for ex.

We do not have to be surprised at the fact that the curvature does not necessarily become infinitesimal with the vanishing of the matter's energy components, as long as no boundary conditions have been set. It is analogous for Newton; as long as the behavior of φ at ∞ has not been laid down, $\varphi \neq 0$ is possible, even if throughout finitude $\Delta\varphi = 0$. (As a solution we have $\varphi = xy$, for ex.). It is just this situation which forces me to consider boundary conditions, if I do not want to admit a disturbing ambiguity.[2]

2) The second term $-\kappa\left(T_{\mu\nu} - \dfrac{1}{2}g_{\mu\nu}T\right)$ on the right-hand side of the field equation is necessary so that, through the field equations, the energy-momentum law (equation (56) in the booklet)[3]

$$\sum_{\sigma} \frac{\partial(t_{\mu}^{\sigma} + T_{\mu}^{\sigma})}{\partial x_{\sigma}} = 0$$

applies. A result in agreement with this is obtained from Hamilton's principle, whereby one is led directly to the form of the field equations where $-\kappa T_{\mu\nu}$ appears by itself on the right-hand side. You will draw this from the paper I am sending you by the same post.[4]

3) In the equations

$$G_{\mu\nu} - \lambda g_{\mu\nu} = -\kappa\left(T_{\mu\nu} - \frac{1}{2}g_{\mu\nu}T\right)$$

under no condition is λ taken to be constant so that a solution in which ρ is constant can exist. If λ were an invariant function of the coordinates, it would need another additional differential equation that λ would have to satisfy as a function of $x_1 \ldots x_4$. The gravitational field would then be described by the $g_{\mu\nu}$'s *and* λ. So if one wants to maintain that the $g_{\mu\nu}$ quantities alone determine the gravitational field, then λ must be a universal constant.

It is correct that the periodicity conditions (condition of spatial closure) take the place of the boundary conditions at infinity. What speaks for the former and against the latter possibility is the fact *that boundary conditions satisfying the postulate of relativity cannot be postulated for infinity.* On the other hand, the closure condition is relative. Added to this is the physical argument elucidated already by Seeliger, which is also presented in my Acad. paper of 8 February 1917.[5]

4) The goal of unifying gravitation and electromagnetism by tracing both groups of phenomena back to the $g_{\mu\nu}$'s has already cost me much frustrated effort. Maybe

you will be luckier at the search. *I am firmly convinced that, in the end, all field quantities will emerge as essentially identical [wesensgleich].* But premonition is easier than discovery.

5) Whether the $g_{\mu\nu}$'s are ascribed to an ether or not is totally unimportant. At all events, one should not think of ether as anything analogous to normal matter insofar as a uniform translatory motion relative to it seems to have absolutely no physical reality. That is why it does seem better to do away with this concession to handed-down conventions.

In great respect,

A. Einstein.

401.   From Hans Thirring

[Vienna, 3 December 1917]

After an interlude of several months, which was filled with intensive practical work,[1] I now finally come back to scientific pursuits and am just busy writing up my little relativity paper of the summer and publishing it in two articles in *Phys. Zeitschr.*[2] In the process, some doubts arose which I want to hold back from the publication, for the time being, but which I would like to discuss with you directly.

According to the gen. theory of relativity there is a spatio-temporal distribution of mass, the gravitational field of which is equivalent to a "centrifugal field." One could also say from observations of fixed stars that this spatio-temporal distribution is such that, on the whole, the masses rotate like a rigid body around an axis. Let us say, at angular velocity ω around the Z axis. If the gravitational field of these masses is to be equivalent to a "centrifugal field," then the following field components[3] must exist at a point of origin on the Z–X plane ($x_4 = t$):

$$\Gamma^1_{44} = C\omega^2 x \qquad + 2\Gamma^1_{24} = C'\omega$$

where C and C' are constants.[4] (The calculation of the field for a rotating hollow sphere actually does yield such Γ's as well, of course.) Now, naturally one will demand that no force phenomena occur on a mass-point that is likewise orbiting around the Z axis at angular velocity ω, since from its point of view everything is at rest. The x component of the motion equation now is[5]

$$\frac{d^2x}{ds^2} = \{\Gamma^1_{44} + 2(\Gamma^1_{14}\dot{x} + \Gamma^1_{24}\dot{y} + \Gamma^1_{34}\dot{z})\}\left(\frac{dx_4}{ds}\right)^2 + [\dot{x}, \dot{y}, \dot{z}]_2.$$

For the point of mass (which is just located in the XZ plane) rotating along with it, $\dot{x} = \dot{z} = 0 \qquad \dot{y} = \omega x$. Hence

$$\frac{d^2x}{ds^2} = (C\omega^2 x + C'\omega^2 x)\left(\frac{\partial x_4}{\partial s}\right)^2 + [\,.\,.\,\dot{x}, \dot{y}, \dot{z}]^2.$$

In order for this expression to disappear also in the limiting case of tiny velocities, $C = -C'$ must apply. In other words: the factors $\omega^2 x$ in the centrifugal force and ωv in the coriolis force must be numerically equal to each other. But if the Γ's are supposed to represent our centrifugal field in conventional mechanics, then because of the formulas $m\omega^2 r$ and $2m\omega v$:

$$2C = -C' \text{ would have to be true.}^{[6]}$$

How can these differences be reconciled?

    Another disturbing two occurs elsewhere as well. In a stationary gravitational field in which all $g_{\mu\nu}$'s are independent of x_4, the equations apply in first approximation[7]

$$-\Gamma^1_{44} = +\Gamma^4_{14} = \frac{1}{2}\frac{\partial g_{44}}{\partial x_1} \text{ etc. } \Gamma^4_{44} = 0.$$

The motion equations of a mass-point in this field are[8]

$$\frac{d^2x}{ds^2} = \{\Gamma^1_{14} + 2(\Gamma^1_{14}\dot{x} + \Gamma^1_{24}\dot{y} + \Gamma^1_{34}\dot{z})\}\left(\frac{dx_4}{ds}\right)^2 + [\ldots \dot{x}, \dot{y}, \dot{z}]^2$$

$$\cdot \quad \cdot \quad \cdot \quad \cdot \quad \cdot \quad \cdot \quad \cdot \quad \cdot \quad \cdot \quad \cdot \quad \cdot \quad \cdot$$

$$\cdot \quad \cdot \quad \cdot \quad \cdot \quad \cdot \quad \cdot \quad \cdot \quad \cdot \quad \cdot \quad \cdot \quad \cdot \quad \cdot$$

$$\frac{d^2x_4}{ds^2} = \langle\Gamma^4_{44}\rangle + 2(\Gamma^4_{14}\dot{x} + \Gamma^4_{24}\dot{y} + \Gamma^4_{3}\dot{y})\left(\frac{dx_4}{ds}\right)^2 + [\ldots]$$

Now $\dfrac{d^2x_4}{ds^2} = \dfrac{d}{dt}\left(\dfrac{v^2}{2}\right)$, applies for slow velocities,[9] whereas the highest ranking terms on the right-hand side of the 4th equation represent *twice* the power.[10] The energy balance thus is not correct. I do not doubt that this will be resolvable. Schrödinger[11] thought the source of the error was that the field lost its stationary nature through the motion of the mass-point, thus that in the approximation used, Γ^4_{44} no longer can be set equal to 0. I replied to this that in electrodynamics the motion of an electron also disturbs the field, but that there $\dfrac{d^2x_4}{d\tau^2} = 1 \cdot (\mathfrak{Ev})$ nonetheless applies.[12]

    Maybe it is only I who am now too far afield from the theory with my range of thought to grasp the matter, but *item*, there is nobody in Vienna who would be able to solve these paradoxes for me in a satisfactory manner.

With best regards, yours truly,

Hans Thirring.

402.   From Erwin Freundlich

Babelsberg, Berlin, 4 December 1917

Dear Mr. Einstein,

My research agenda for the immediate time to come will be the following:[1]

1) For verification of *light deflection in the gravitational field*: Development of a method of photographic imaging of stars in the proximity of the Sun; furthermore, preparatory researches for a solar eclipse in 1919 and development of Kapteyn's parallax method for stellar occultations by Jupiter. First of all, I am going to acquire at Potsdam the requisite practical skills in astrophotography.

2) For verification of a *red shift of spectral lines* for fixed stars: Development of the new measurement technique with the photoelectric measuring instrument presently under construction[2] and development of both the statistical method and the direct one with spectroscopical binary stars, both components of which are visible. I shall in part try to use existing data, and in part have to obtain new data myself. Here also, first of all, accumulation of practical knowledge in the production of spectral images from fixed stars.

With best regards,

E. F. Freundlich.

403.   To Heinrich Zangger

[Berlin,] 6 December 1917

Dear friend Zangger,

It is shameful how long it has been since I wrote you; I seem to have inherited my sloth in writing from my Albert, for he is even lazier.[1] I had trouble reading your postcard. Were you in Lausanne with *Perrin*? I know him very well from the Brussels conferences and from my stay in Paris[2] and do like him. Do you know what my wife's illness is exactly? Is it an ulcer on the spinal cord, or multiple sclerosis perhaps? Is there, in your opinion, any prospect of recovery?[3] Please let me know, if possible, by return post, how much you have disbursed for me in total (Tete, packages, etc.).[4] It seems to be auspicious for money transfers at the moment.[5]

Healthwise, I am feeling very well. I have gained ca. 4 pounds since the summer, thanks to Elsa's good care. She cooks everything for me herself, since this has proved to be necessary. This is possible because I live in the apartment next to hers (for the interim).[6]

I am working hard, lecturing on statistical mechanics and quantum theory[7] and am also writing a thing or two, but nothing of particular importance. Truly novel inventions only come during one's youth; later one becomes ever more experienced, renowned and—blockheaded. Debye has done something very fine on the X-ray spectrum, elaborating on Bohr and Sommerfeld.[8] I shall tell you about it when I come to Switzerland again. Do you know that Smoluchowski died of dysentery?[9] What a pity. Will there be peace soon? No one knows. About a week ago I sent a lot of papers to Weiss[10] but received no notice of receipt. He had asked me for the dispatch many times; but it had been delayed owing to the impediments. Please kindly forward the papers to Weiss, in case he should be absent from Zurich.

The general theory of relativity is being received with downright enthusiasm among my colleagues.[11] How can it ever be that this culture-loving time be so dreadfully amoral? I am becoming increasingly inclined to rate everything as trivial against charity and philanthropy. Would it not be good for the world if degenerate Europe were to wreck itself totally? When I hear the new repugnant term "toughening up" [Ertüchtigung], my stomach turns. All of our exalted technological progress, civilization for that matter, is comparable to an axe in the hand of a pathological criminal. I believe, in all seriousness, that the Chinese ought to be held in higher esteem than we and hope that their healthy proliferation will outlast the extinction of our "toughened" comrades.[12]

My financial situation has improved, so I don't need to worry for the time being.[13] But I definitely want Tete to be brought back down again from up there in the New Year.[14] I am opposed to having such a child spend his youth in such a ⟨distillery⟩ disinfector. Our life is corrupted not just technologically but also medically—which is actually only a kind of technological pollution.

Dear Zangger, do not be angry at my wild ranting; people must vent themselves in order not to burst. Do write again sometime, very legibly, to your old

A. Einstein.

Do please put in a good word for me when you are in Berne again, so that I continue to receive the "family packages."[15] For otherwise, because I am living here as a single Swiss, they just grant me "bachelor's packages," in which there is virtually nothing that I can eat without retribution.[16] But otherwise I do not want to have anything out of the ordinary but cut my coat fairly and squarely according to my cloth.

404.   From Erwin Freundlich

Babelsberg, Berlin, 6 December 1917

Dear Mr. Einstein,

With this letter I am sending you the draft contract, as I discussed it yesterday with Dr. Oppenheim, and once again, very briefly, my research plans.[1]

1) The day before yesterday I immediately spoke with Struve.[2] He fully approves of my leaving my current position on 1 Jan. 1918 and also instantly agreed to my staying nevertheless at my current residence until April 1. He intends to inform the Ministry of this right away. We also discussed all that is necessary for the smooth continuation of my current researches.

2) As concerns the draft contract, I hope that we have considered all that is required.

As far as the second paragraph of §1 is concerned, I must still inquire at the Ministry how these generally applicable provisions read, especially concerning my pension entitlement. I have directed a petition to Undersecretary Naumann asking that I be placed on leave of absence for the duration of the coming 5 years and that these years be credited to me later. *Would it not be advisable that the Board of Directors of the K. W. I. endorsed this petition?* (I am enclosing a transcription of this petition.)[3]

Regarding §2, 7, 8, you might have to discuss this with Planck sometime. For each small purchase, for phot[ographic] plates, for ex., or for a new cell for a measuring instrument, I cannot first submit a long-winded written petition. What leeway per month (see §7) am I going to be granted? Based on what I have been told in Potsdam, for ex., I count on being granted smaller purchases without complications, e.g., for plates, out of the budget of that institute.[4] But I certainly would have to have some kind of recourse.[5]

I have added §6 just for the sake of completeness, because the Jagor Foundation,[6] for ex., always emphasizes explicitly that it be mentioned in the publications, in a clearly visible spot, that it was the donor of funds for the relevant research.

§8. For §8 it must be considered if I can ask to have the right to keep things that I had procured out of K. W. I. funds until the conclusion of the researches or only until the termination of the contract. The former seems more appropriate to me for the good of the research. Well, one cannot think of everything. A good part of such a contract must be left to trust and faith.

I hope that, essentially, everything in the contract appears to you carefully thought-out.

With best regards,

Erwin Freundlich.

405. To Hans Thirring

[Berlin,] 7 December 1917

Esteemed Colleague,

First to your second question.[1] Your equations of motion are correct. They read, if negligible terms are omitted:[2]

$$\frac{d^2x_1}{ds^2} = -\frac{1}{2}\frac{\partial g_{44}}{\partial x_1}$$

$$- - - - - - -$$

$$\frac{d^2x_4}{ds^2} = -\left(\frac{\partial g_{44}}{\partial x_1}\dot{x}_1 + \cdot\cdot +\right)\frac{dx_4}{ds} = -\left(\frac{\partial g_{44}}{\partial x_1}\dot{x}_1 + \cdot\cdot +\right).$$

However, one should not set $\dfrac{d^2x}{ds^2}$ equal to the time derivative of the kinetic energy.[3] A point's negative momentum and energy, of mass 1 in natural measure, with a coordinate choice according to $g = -1$, form a covariant four vector of the components[4]

$$\sum_\alpha g_{\mu\alpha}\frac{dx_\alpha}{ds}\ \left(\text{not }\frac{dx_\mu}{ds}\right).$$

The energy is thus

$$\sum_\alpha g_{4\alpha}\frac{dx_\alpha}{ds}$$

or accurately enough for us,

$$g_{44}\frac{dx_4}{ds}.$$

If this is derived for s, then

$$\frac{dg_{44}}{ds}\frac{dx_4}{ds} + g_{44}\frac{d^2x_4}{ds^2}$$

is obtained, or approximately enough

$$\left(\frac{\partial g_{44}}{\partial x}\dot{x} + \frac{\partial g_{44}}{\partial y}\dot{y} + \cdot\right) + \frac{d^2 x_4}{ds^2}.$$

It is this expression which indicates the vanishing of the last equation of motion.[5]

For $\dfrac{dx_4}{ds}$, one obtains, if one sets for the $g_{\mu\nu}$'s (incorrectly, by the way)

$$
\begin{array}{cccc}
-1 & 0 & 0 & 0 \\
0 & -1 & 0 & 0 \\
0 & 0 & -1 & 0 \\
0 & 0 & 0 & g_{44}
\end{array}
$$

$$\frac{dx_4}{ds} = \frac{1}{\sqrt{g_{44} - v^2}}.$$

The root of your error lies in that $\dfrac{1}{\sqrt{1 - v^2}}$ cannot be inserted in its place,[6] since $\dfrac{d}{dt}(g_{44})$ and $\dfrac{d}{dt}(v^2)$ are of the same order of magnitude.–

I probably do not understand your first difficulty.[7] You say entirely correctly: If I were within the interior of a rotating, hollow rotational body, mechanically I would have to find myself to be in the state of a rotational system when I am "at rest." A point of mass must be able to move in a circle *free of forces* when it is moving at a suitable angular velocity around the Z-axis (in the rotational body's sense of motion).

Now, you require, however, that $\dfrac{d^2 x}{ds^2}$ vanish for such a point when it is on the X-Z plane. This does not apply to rotational motion, though. What must rather apply is

$$\frac{d^2 x}{ds^2} = -\omega'^2 r = -\omega'^2 x,$$

whereby ω' is the point's rot. velocity, which ought to relate to your ω as

$$\omega'^2 = C\omega^2.$$

Your equation then yields, as it should,[8]

$$2C = -C'.$$

The fact that this cannot be otherwise is, incidentally, guaranteed by the general covariance of all equation systems in the theory.–[9]

In the hope that I have thus eliminated your difficulties, I am with best regards to you and your sister,[10] who visited me last summer at my sister's home in Lucerne, yours truly,

A. Einstein.

406.   To Hans Albert Einstein

[Berlin,] 9 December 1917

Dear Albert,

Your last little letter pleased me very much. The money for the insurance, which I immediately instructed to have sent, has arrived, I hope. In your letter the postscript was the most amusing, though; it was not from you at all but from the censor. He had provided the entirely correct solution to the problem I had posed for you[1] in place of yours and had written in pencil under your letter: It goes like this:[2]

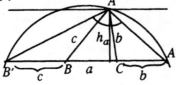

given $\angle A$, height h_a and $(a+b+c)$, you noticed correctly that it involves the construction of the $\triangle AB'C'$, but then you ran out of steam. You realized that the $\angle$ at the tip is equal to $\alpha + \dfrac{\beta+\gamma}{2}$ or equal to $\dfrac{\alpha}{2}+90°$. Therefore, in the triangle $AB'C'$, the base line, the $\angle$ at the tip, and the height are given. The $\triangle$ can be constructed from it, though. For there are two geometrical locations for the tip A

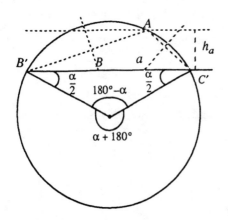

1) the parallel to the base line at distance h_a
2) the circle describing $(a+b+c)$ with the peripheral angle

$$\frac{\alpha}{2}+90°.$$

When you consider that the peripheral angle is equal to half the central angle, you easily find the center by the method indicated in the figure. The apothems of AB' and AC' intersect the points B and C.

In this way two triangles result, which you should also prove are congruent to each other.

I have sent you both 5 *very fine and interesting* books for Christmas.[3] This occupied me for an entire day, because the selection was difficult and, in addition, I needed an export permit from the military authority. They belong to you and Tete together. I did not divide them between you. Reading and understanding is what is important, not *possession.* Write me a postcard as soon as the books have arrived. Everything is so uncertain nowadays! Well, Tete is going to be back home with you soon, as you know;[4] you must be glad! As soon as he's back, write me how he is and how he looks. Now he must start going to school as well and, if possible, not together with the littlest ones, because it would surely be far too boring for him. Write me about how you all consider it. Also write me how Mama is and whether everything is all right over there. Michele Besso is away now, of course,[5] and Zangger knows only a little about you all, so I really am dependent on your reports.

My health is in good order. But my stomach has become so sensitive and weak[6] that I have to follow a diet like that of a small child. If this persists, that's the end of our traveling; because someone must always cook something separately for me.

Tell me a little about school and about your life otherwise. What do you do in your free time? Are you overworked?

Greetings and kisses from your

Papa.

Greetings to Mama and Aunt.

407. To Gustav Mie

Schöneberg, Berlin, 5 Haberland St., 14 December 1917

Highly esteemed Colleague,

You would be doing me a great favor if you sent me a reprint of your interesting lectures held in Göttingen.[1] Studying from journals is very inconvenient for me now, because I am not supposed to go out much owing to illness.

With kind regards, yours truly,

A. Einstein.

408.   To Felix Klein

Esteemed Colleague,

I am returning to you your second lecture notebook, which I obtained from our colleague Sommerfeld a few days ago, with many thanks.[1] It does seem to me that you are very much overestimating the value of purely formal approaches.[2] The latter are certainly valuable when it is a question of formulating definitively an *already established* truth, but they almost always fail as a heuristic tool. Thus I am convinced that the covariance of Maxwell's formulas under transformation according to reciprocal radii can have no deeper significance; although this transformation retains the form of the equations, it does not uphold the correlation between coordinates and the measurement results from measuring rods and clocks.[3]

In great respect, yours truly,

A. Einstein.

409.   To Wilhelm von Siemens

Berlin, 5 Haberland St. [before 16 December 1917][1]

Highly esteemed Privy Councillor![2]

At the meeting of the Board of Directors and Board of Trustees[3] we resolved to make an announcement about the K. W. Institute through public notice. I drafted the notice with Mr. Planck and Mr. Harnack and have submitted it for printing to the newspapers and periodicals we had selected.[4]

Meanwhile, I have paid and recorded the costs that arose out of this and other miscellaneous minor purchases. I think that, toward achieving the simplest possible accounting, it would be beneficial if small sums were put regularly at my disposal for the defrayal of running disbursements. It would probably be best for me to have an Institute cashbox at my home containing a few hundred marks and to render an account of it at specified intervals.[5]

Furthermore, I would like to request that a secretary be granted me for three half days per week. As compensation I am thinking of 50 M a month.

Very respectfully,

A. Einstein.

P. S. Should I be able to turn elsewhere for such small matters, I kindly request that you inform me accordingly.

410.   From Gustav Mie

Halle-on-S[aale], 17 December 1917

Highly esteemed Colleague,

I would have already sent you a separatum of my paper, if I had received the offprints.[1] As soon as I have them, I shall send you one. By the way, I had asked the editorial office of the *Phys. Z[ei]tschr[ift]* to have the correction proofs forwarded to you, since I had no duplicate of my manuscript; I hope that it has done so. I very much wish I could discuss these matters with you once, but bad as the traveling conditions are now, I shall probably not be coming to Berlin anytime soon. Above all, I wish that you will recover soon from your illness!

With best regards, yours truly,

G. Mie.

411.   From Heinrich Zangger

[Zurich, 17 December 1917][1]

Dear friend Einstein,

All told, things are in order here. When I travel through Chur, I shall possibly visit your little Tete in Arosa, in order to give you a report.[2] I just don't know when I can travel, since now Trudy[3] has a fever (probably measles, I expect). There is no mention of you, even by the children—that is always suspicious. We have serious issues here which, under normal psychological circumstances, would have been solvable without much trouble. I had two sleepless nights. I received this letter from the addressee very confidentially as a medical doctor for an opinion on the question of what illness could possibly be looming in Nicolai's case.[4] After the report, I was alarmed: A breakdown is imminent from which the man can recover only in a matter of years, if at all. Monakow was, upon confid. consultation, of the same opinion.[5] At the outbreak of the war, we saw even severe conditions that recovered completely. Is recuperation in Switzerland not possible, with suspension of sentence to after the end of the war? Persons who earnestly clear the air ought not to perish now in conflicts that would otherwise not exist. We also have people here like Fritz von Unruh,[6] you know, who are very thankful.

Best regards from Enriques.– Besso[7] and

Zangger.

412.   From Heinrich Zangger

[Zurich, 17 December 1917][1]

Dear friend Einstein,

The arrival of your letter[2] today was a great joy among many oppressive things—with your profession of philanthropy and charity as the foremost principles in life—there was a time when you used almost to pity this penchant of mine; I could not do otherwise. You see from the letter, overdue since 3 months, which was sent out a 1/2 hour before yours,[3] that I had thought of many of the items—I just cannot carry out everything now for a while, above all, going to Arosa in order to see what Tete can be expected to go through—the beastly February and March weather![4] His mother is doing tolerably well—the spinal cord has certainly remained completely healthy,[5] the inflammation process has spread, as is very rarely the case, to the bones, also according to the X-ray plate.   The disks (which supply the elasticity) have narrowed, thus now any dislocation pinches more than otherwise for relatively small dislocations.   But there has been no fever for a long while; flexibility is not worse but *any strain* is dangerous, owing to buckling of the vertebral column and bending of the spinal cord canal[6]—repose;

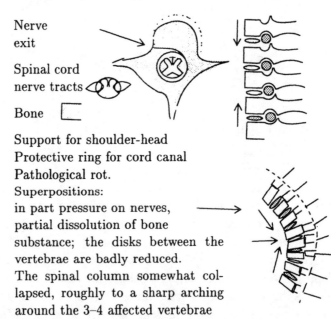

Nerve exit

Spinal cord nerve tracts

Bone

Support for shoulder-head
Protective ring for cord canal
Pathological rot.
Superpositions:
in part pressure on nerves,
partial dissolution of bone substance; the disks between the vertebrae are badly reduced.
The spinal column somewhat collapsed, roughly to a sharp arching around the 3–4 affected vertebrae

as in so many cases, nature compels the formation of a kind of boney sleeve which can bear again, although a bit stiffly.

Before your letter arrived, the day had saddled me with several things already: Trudy is developing a high fever,[7] I ought to leave on a business trip abroad, F[ederal] C[ouncillor] Forrer was here,[8] had difficult patients, became dean of a society that is not always of one mind, and had to accept,[9] and am annoyed that I cannot do any of the physics work. Your mathemat. bridges, λ, spanning to the end of the world are too lofty for me,[10] but many of your experiences during the discovery, which you retraced for me, are present in me and often work so creatively, as though I possessed all the bases for this thinking.[11] Thus relativity in this general, overall form finds its way into our heads; the problem of human decency, of trust in loyalty and faith, must therefore be much more difficult, or—Europe must be particularly badly grounded.[12] Your scolding is always like an elevating liberation for me. You see from this letter that I have been through other burdensome things today as well. It is difficult to find out whether the letter arrived for the deliberations and at the right time.[13] We also have atrocious conflicts that are not even allowed to be divested of their tragedy, if anything is to be gained from them.

Besso has appeared like a bold from the blue.[14]

Best wishes for a steady recovery,

Zangger.

413.   To Hendrik A. Lorentz

[Berlin,] 18 December 1917

Dear and esteemed Colleague,

It has been such a long time since I had the pleasure of having personal dealings with you. And now, a somewhat boring matter is the reason for my letter. On 26 May of this y[ear], you submitted a short communication by Mr. de Donder to your Academy which is based on an error.[1] If this matter involved nothing more than a priority claim, I would not have to come back to it; but since it contains a misleading assertion, a correction must follow in some manner in the interest of science. I leave entirely to you the form in which this correction is best carried out. Mr. de Donder might best clear up the facts himself in a second note.[2]

The following is at issue:

The field equations set up by me in my paper "Cosmological Considerations"[3] (Equations (9) in de Donder's note) are supposed to be equivalent to equations (1) of de Donder's note, the latter of which Mr. de Donder had already specified previously.[4]

The statement of this equivalency is based on error. In order not to bother you superfluously at length with this affair, I shall come straight to the false conclusion. De Donder concludes unjustifiably that the vanishing of the curvature scalar C (equation 16) follows from his equation (15).[5] This false conclusion is based on the illegitimate inference that (15) were an *identity* whereas, in reality, it is an *equation of condition* for the space functions λ, ρ and the $g_{\mu\nu}$'s.

In consequence of this manner of deduction, one could just as well conclude as follows: No curve exists according to the equation

$$x^2 + y^2 = R^2;$$

because since the right-hand side of the equation is independent of x, the left-hand side also must be!

De Donder unfortunately already erroneously asserted the vanishing of C in an earlier paper, which assertion can only be the case with vanishing energy scalar T.[6] The energy scalar does not vanish for actual matter, however.–

I am constantly very depressed about the endless tragedy we must witness.[7] Even the habitual flight into physics does not always help. Surely it is the same for you as well. My health is not the best either.

Cordial regards to you and your wife as well as to Mr. and Mrs. de Haas, yours,

A. Einstein.

414.   From Max von Laue

Würzburg, 19 December 1917

Dear Einstein,

The event, which Planck had announced to me in October as forthcoming and which you now report to me as having come to pass is, of course, extremely welcome.[1] I just don't know what use I can draw from it before my transfer to Berlin has been secured.[2] And for that reason I am also unclear about what more I should write you except that I would very much enjoy conducting experimental investigations with X rays on crystalline structures and would like to apply the support of this new foundation toward this end.[3] All specific details would probably have to be delayed until my future has otherwise become clear.

Otherwise, I am, with most cordial regards, yours,

M. von Laue.

415.   To Otto Marx[1]

Zurich [Berlin], 22 December 1912. [1917][2]

Marx had requested Einstein's assistance in solving a problem with an airplane.

. . . I was delighted with your gorgeous and witty gift. When looking through the fine album, I shall think back often with amusement about your . . . and my escapade into the realm of praxis . . .[3]

With best regards and good wishes for the holidays, yours truly,

A. Einstein.

416.   To Gustav Mie

[Berlin,] 22 December 1917

Highly esteemed Colleague,

Cordial thanks for sending me your very interesting lectures, which I have studied more than once already.[1] Our remaining differences of opinion are relatively insignificant. The main distinction consists in that I do not believe in the "flat basic shape" of the world continuum. If it were to hold, your suggested coordinate choice through orthogonal projection on R_4 in a 10-dimensional Euclidean space would be very reasonable (provided that the hyperbolic character did not add difficulties).[2] I do not believe, though, that the universe is essentially empty and flat and, while extending into infinity, is filled with matter only within a finite region. That smacks of a geocentric interpretation. Many already, Seeliger among them, wanted to have nothing to do with that.[3] I cannot believe that the $g_{\mu\nu}$'s (hence also inertia) are determined not only mainly by boundary conditions at infinity but also a bit by the remaining matter, in addition. In my opinion, the $g_{\mu\nu}$'s are determined entirely by the matter, insofar as they do not have to remain undetermined because of the arbitrariness of the choice of coordinates.[4] I do not agree with the reflections on the necessity of the existence of preferred coordinate systems. They can easily cause confusion.

We must discuss all these things one day at our leisure and in private, not in the presence of Philistines (Lecher).[5] I may be a democrat, but in these subtle matters I think little of the opinion of the majority. I am sending you a paper which I would like you to read in the meantime so that you also become acquainted with my conception of certain problems you have touched upon.[6] I agree with your views on matter.[7] Cordial regards, yours,

A. Einstein.

On page 555, the field equations are quoted incorrectly. There is a term missing.[8]

417.   To Hans Albert Einstein

[Berlin, 24 December 1917]

My dear Albert,

Your postcard about the merry ski trip thrilled me. Today is Christmas (Eve). I am staying in bed for 4–6 weeks because of my stomach ulcer;[1] but I feel all right. My books have arrived, I hope; I sent them to you all in time for the Christmas celebration.[2] I hope that Tete joins you again at New Years.[3] If not, then send the *1001 Nights* stories up to him. Write me whether Tete is returning home.

Greetings to you, Mama, Tete, and Aunt, yours,

Papa.

418.   From Walther Nernst

Berlin, 25 December 1917

Dear Einstein,

I am just in receipt of the draft contract from Haber[1] and receive your kind postcard at the same time. I am naturally in agreement with everything, only, I would advocate striking §2. Despite not really being able to forbid Mr. Freundlich from publishing, we do simultaneously assume a certain responsibility by examining his results and approving publication, which is scarcely our intent.[2] I would like to see the last sentence in §7: "In the case . . . ." struck, because surely we cannot possibly be permitted to bind ourselves, even to the slightest degree, in any way for longer than three years.[3]

With best regards, yours,

W. Nernst.

[3]Einstein's emendation: "retracted by Nernst by telephone." Struck passage in draft contract: "In the case that the provisions of this contract prove appropriate, the Board of Directors of the K. W. Institute will consider extending it by an additional two years."

419.   From Michele Besso

Zurich, 27 December 1917

To Prof. Einstein, Berlin

Dear Albert,

From your most recent letter, which our friend Zangger has shown me, I gather with great dismay that now you are virtually bedridden.[1] Maja already told me that you are not feeling well and that she has urged you very strongly to take in *mountain air* again, after all, which evidently agreed with you so well last summer.[2] I can only support this idea vigorously: unfortunately in a quite selfless manner, because I should gain nothing from it if you were to come here soon, since I must return again to the site of my new and, as I imagine, permanent sphere of activity in the middle of next month already.—[3] But it should anyway be a *very* extended stay, so that I may perhaps still have the pleasure of getting together with you often in the months of May, June, and July (which you ought to devote to the thorough fortification of your health in the mountain air, and I to a semester lecture course, which I would still like to hold here[4]).

Our friend has relented now, after all, in preparing a bill for you,[5] and a very substantial one at that, after careful discussion with your family about the most urgent necessities they are expected to face, in order to create a bit of reserves under the current favorable conditions.[6] Also I, or shall we say Anna, whose conscientiousness also in financial matters is known to you, will vouch for a *thoroughly* proper use of it.

Warm regards and best wishes for 1918, also to your loyal and devoted nurse, yours,

Michele

Handwriting and thoughts a bit wobbly: I am freezing a little for a change.

Our descendants will probably also have to make do with little or no coal;[7] but then they must also become a bit more hardy than my generation. In this— and probably in other respects as well!

420.   From Rudolf Förster

Essen, Krupp Factory, AK Laboratory, 28 December 1917

Esteemed Professor,

I would like to express my sincerest thanks for your kind answer[1] to my communication as well as for sending me an offprint of your paper on Hamilton's principle in gravitation theory.[2] The reason I did not do so long ago is that I have been busily making calculations using the theory, in search of something new. Some little things have come out of it, yet it seems that I am not making any more progress for the time being and would like to take the liberty of reporting to you about my efforts. Although it seems probable to me that you have long been acquainted with these sidetracks, among the lot connected to your theory, and have come farther along them than I, it could nevertheless be possible, and it would please me greatly, if you were to benefit from one or the other in some way.

First of all, I would like to correct an error in my earlier letter.[3] I gave an example there of a space of nonvanishing curvature in which the energy tensor of matter was supposed to vanish. Subsequently I noticed nonvanishing components in a hidden corner of the tensor, though.

1) My main efforts since receipt of your letter have been devoted toward unifying the electromagnetic field with the gravitational field.[4] I have not yet found a solution. I performed larger calculations in one vein by proceeding from an asymmetric fundamental tensor $g_{\mu\nu}$. I would like to present the following approaches:

a) Reduction into a symmetric and an antisymmetric part $g_{\mu\nu} = s_{\mu\nu} + a_{\mu\nu}$. Determinant: g, or subdeterminant s, divided by g (or s) itself: $g^{\mu\nu}$, or $s^{\mu\nu}$. $\sum g_{\mu\alpha} g^{\nu\alpha} = \varepsilon_\mu^\nu = 0$ or 1, likewise $g_{\alpha\mu} g^{\alpha\nu} = \varepsilon_\mu^\nu$, by contrast $g_{\alpha\mu} g^{\nu\alpha} = a_\mu^\nu (\neq 1)$; $g_{\mu\alpha} g^{\alpha\nu} = i_\mu^\nu$; hence $g_{\mu\nu} = a_\nu^\alpha g_{\alpha\mu} = i_\mu^\alpha g_{\nu\alpha}$; $g^{\mu\nu} = a_\alpha^\mu g^{\nu\alpha} = i_\alpha^\nu g^{\alpha\mu}$; $i_\alpha^\nu \cdot a_\mu^\alpha = i_\mu^\alpha \cdot a_\alpha^\nu = \varepsilon_\mu^\nu$.

$dg^{\mu\nu} = -g^{\alpha\nu} g^{\mu\beta} dg_{\alpha\beta}$; $dg_{\mu\nu} = -g_{\alpha\nu} g_{\mu\beta} dg^{\alpha\beta}$; $\dfrac{\partial \log g}{\partial t} = g^{\mu\nu} \dfrac{\partial g_{\mu\nu}}{\partial t} = -g_{\mu\nu} \dfrac{\partial g^{\mu\nu}}{\partial t}$

(: t one parameter :).

b) If A^ν is a vector, then

$$f_{\mu\nu} \equiv g_{\mu\alpha} \frac{\partial A^\alpha}{\partial x^\nu} + g_{\alpha\nu} \frac{\partial A^\alpha}{\partial x^\mu} + A^\alpha \cdot \frac{\partial g_{\mu\nu}}{\partial x^\alpha} \text{ is a tensor, likewise}$$

$$g^{\mu\alpha} \frac{\partial A^\nu}{\partial x^\alpha} + g^{\alpha\nu} \frac{\partial A^\mu}{\partial x^\alpha} - A^\alpha \frac{\partial g^{\mu\nu}}{\partial x^\alpha}.$$

Others in addition to this through multiplication by $g^{\mu\sigma}$, or $g_{\sigma\nu}$, etc.

c) From the 64 derivatives of $g_{\mu\nu}$ the 40 second derivatives of the x^σ's can plainly be calculated if 24 relations exist between the $g_{\mu\nu}$'s and $g'_{\mu'\nu'}$'s. These relations imply that the following system of quantities is a tensor:

$$A_{\mu\nu,\sigma} \equiv \frac{\partial g_{\mu\nu}}{\partial x^\sigma} - g_{\mu\alpha} \cdot \left\{ \begin{matrix} \nu\sigma \\ \alpha \end{matrix} \right\}_s - g_{\alpha\nu} \cdot \left\{ \begin{matrix} \mu\sigma \\ \alpha \end{matrix} \right\}_s \qquad \text{(: Expansion of the fundamental tensor :)}$$

where the attached s indicates that the symbols for the symmetric tensor $s_{\mu\nu}$ must be formed (: They are defined below for the asymmetric ones :)

$A_{\mu\nu,\sigma} = -A_{\nu\mu,\sigma} \cdot A_{\mu\nu,\sigma}$ can be expressed using $a_{\mu\nu}$ and $\left\{ \begin{matrix} ik \\ l \end{matrix} \right\}_s$ as:

$$A_{\mu\nu,\sigma} = \frac{\partial a_{\mu\nu}}{\partial x^\sigma} - a_{\mu\alpha} \left\{ \begin{matrix} \nu\sigma \\ \alpha \end{matrix} \right\}_s - a_{\alpha\nu} \left\{ \begin{matrix} \mu\sigma \\ \alpha \end{matrix} \right\}_s .$$

The combination appearing in the integral $\int\int\int A_{\mu\nu\sigma} dx^\mu dx^\nu dx^\sigma$:[5]

$$f_{\mu\nu\sigma} \equiv A_{\mu\nu,\sigma} + A_{\nu\sigma,\mu} + A_{\sigma\mu,\nu} = \frac{\partial a_{\mu\nu}}{\partial x^\sigma} + \frac{\partial a_{\nu\sigma}}{\partial x^\mu} + \frac{\partial a_{\sigma\mu}}{\partial x^\nu}$$

is the antisymmetric expansion of the six tensor $a_{\mu\nu}$.

From $A_{\mu\nu,\sigma}$ another tensor is derived through multiplication by $g^{\mu\alpha}g^{\beta\nu}$, which is likewise antisymmetric in μ, ν:

$$A^{\mu\nu}_\sigma = \frac{\partial g^{\mu\nu}}{\partial x^\sigma} + g^{\mu\alpha} \left\{ \begin{matrix} \alpha\sigma \\ \nu \end{matrix} \right\}_s + g^{\alpha\nu} \left\{ \begin{matrix} \alpha\sigma \\ \mu \end{matrix} \right\}_s .$$

The reduction of this tensor according to the first (or last) upper index yields

$$1)\ I^\mu = \frac{\partial g^{\mu\alpha}}{\partial x^\alpha} + g^{\mu\beta} \left\{ \begin{matrix} \alpha\beta \\ \alpha \end{matrix} \right\}_s + g^{\beta\alpha} \left\{ \begin{matrix} \alpha\beta \\ \mu \end{matrix} \right\}_s$$

d) Asymmetric triple index symbols: $\left[\begin{matrix} \mu\nu \\ \sigma \end{matrix} \right] = \frac{1}{2} \left(-\frac{\partial g_{\mu\nu}}{\partial x^\sigma} + \frac{\partial g_{\sigma\nu}}{\partial x^\mu} + \frac{\partial g_{\mu\sigma}}{\partial x^\nu} \right).$

$$\left[\begin{matrix} \mu\nu \\ \sigma \end{matrix} \right] - \left[\begin{matrix} \nu\mu \\ \sigma \end{matrix} \right] = -f_{\mu\nu\sigma} .$$

A generalization of Riemann's tensor (: not the only one :).

$$B_{\mu\nu,\sigma\tau} = \frac{\partial}{\partial\sigma} \left[\begin{matrix} \nu\tau \\ \mu \end{matrix} \right] - \frac{\partial}{\partial\tau} \left[\begin{matrix} \nu\sigma \\ \mu \end{matrix} \right] + s^{\alpha\beta} \left(\left[\begin{matrix} \nu\sigma \\ \alpha \end{matrix} \right] \left[\begin{matrix} \mu\tau \\ \beta \end{matrix} \right] - \left[\begin{matrix} \nu\tau \\ \alpha \end{matrix} \right] \left[\begin{matrix} \mu\sigma \\ \beta \end{matrix} \right] \right).$$

$-B_{\nu\mu,\sigma\tau} = -B_{\mu\nu,\tau\sigma} = +B_{\nu\mu,\tau\sigma} = +B_{\mu\nu,\sigma\tau} \neq B_{\sigma\tau,\mu\nu}$. This tensor thus has 36 components. The 2nd-order terms read: $\dfrac{1}{2} \left(\dfrac{g_{\mu\tau}}{\nu\sigma} + \dfrac{g_{\nu\sigma}}{\mu\tau} - \dfrac{g_{\nu\tau}}{\mu\sigma} - \dfrac{g_{\mu\sigma}}{\nu\tau} \right)$, where the fractions signify second differential quotients. If in the last term $s^{\alpha\beta} \cdot (\cdots)$, the $[\ ]$ are replaced by $[\ ]_s$, $B_{\mu\nu,\sigma\tau}$ changes by $-\frac{1}{4}s^{\alpha\beta} \cdot (f_{\sigma\nu\alpha}f_{\tau\mu\beta} - f_{\tau\nu\alpha}f_{\sigma\mu\beta})$. The term resulting after subtraction of this tensor differs from the fully symmetric one in the terms:

$$\frac{1}{2}\left(\frac{a_{\mu\tau}}{\nu\sigma} + \frac{a_{\nu\sigma}}{\mu\tau} - \frac{a_{\nu\tau}}{\mu\sigma} - \frac{a_{\mu\sigma}}{\nu\tau}\right).$$

One can see that although the fundamental tensor is asymmetric, $B_{\mu\nu,\sigma\tau}$ becomes symmetric if the 4 equations are satisfied: (2) $f_{\mu\nu\sigma} = 0$. I would like to treat equations (1) and (2) as Maxwellian, but there is a hitch in the energy components.

e) Out of the expansion of the fundamental tensor, three scalars (for Hamilton principles, or the like) can be formed:

$$A_{\mu\nu,\sigma} \cdot A_{\alpha\beta,\gamma} \cdot \begin{cases} \cdot s^{\mu\sigma} \cdot s^{\alpha\gamma} \cdot s^{\nu\beta} & \text{or} \\ \cdot s^{\mu\sigma} \cdot s^{\nu\beta} \cdot s^{\sigma\gamma} & \text{or} \\ \cdot s^{\mu\alpha} \cdot s^{\nu\gamma} \cdot s^{\sigma\beta} & . \end{cases}$$

2) Gravitational fields derived from a "potential"

a) Assumption: $C \cdot g_{\mu\nu} = g_{\mu\alpha}\dfrac{\partial A^\alpha}{\partial x^\nu} + g_{\alpha\nu}\dfrac{\partial A^\alpha}{\partial x^\mu} + A^\alpha\dfrac{\partial g_{\mu\nu}}{\partial x^\alpha}$. If special coordinates are introduced, so that $A^1 = A^2 = A^3 = 0$; $A^4 = 1$, then results $C \cdot g_{\mu\nu} = \dfrac{\partial g_{\mu\nu}}{\partial x^4}$;

$g_{\mu\nu} = e^{Cx^{(4)}} \cdot h_{\mu\nu}(x_1, x_2, x_3)$. For $C = 0$, the $g_{\mu\nu}$'s would thus depend on only three variables. Such an approach appeals to me very much, because obviously, in reality also, only the $16 \cdot \infty^3$ initial values can be given for the $g_{\mu\nu}$'s, out of which the subsequent ones develop by natural processes. The integral curves of the equations $dx^1 : dx^2 : dx^3 : dx^4 = A^1 : A^2 : A^3 : A^4$ must mean something like the individual local time. (World lines? Or are they connected to a kind of four potential?) At any rate, my conceptions of this are still very vague.

b) Assumption. A vector A_ρ exists such that $\dfrac{\partial A_\rho}{\partial x^\nu} - A_\tau \cdot s^{\alpha\tau} \cdot \left[\begin{array}{c}\rho\nu \\ \alpha\end{array}\right]_s = 0.$

Through a special choice of coordinates $A_\rho = s_{\rho 4}$ can be obtained. Then results (1) $\dfrac{\partial s_{\rho\nu}}{\partial 4} = 0$; (2) $\dfrac{\partial s_{\rho 4}}{\partial \nu} = \dfrac{\partial s_{\nu 4}}{\partial \rho}$, $A_\rho = \operatorname{grad}\varphi$; $\operatorname{rot}\mathfrak{A}_\rho = 0$; A_ρ thus cannot be the four potential.

3) A formal series for which I have not yet found any proper meaning:

$$A_\mu dx^\mu + \left(s_{\mu\nu} + \frac{\partial A_\mu}{\partial x^\nu} + \Gamma^\tau_{\mu\nu}|_s A_\tau \right) dx^\mu \delta x^\nu +$$

$$+ \left(h_{\mu\nu\sigma} + \frac{\partial^2 A_\mu}{\partial\nu\partial\sigma} + \Gamma^h_{\mu\nu}\frac{\partial A_h}{\partial\sigma} + \Gamma^h_{\mu\sigma}\frac{\partial A_h}{\partial\nu} + \Gamma^h_{\sigma\nu}\frac{\partial A_\mu}{\partial x_h} + \right.$$

$$\left. + A_h \left(\frac{\partial\Gamma^h_{\mu\nu}}{\partial\sigma} + \Gamma^\tau_{\mu\sigma}\Gamma^h_{\tau\nu} + \Gamma^\tau_{\sigma\nu}\Gamma^h_{\mu\tau} \right) \right) dx^\mu \delta_1 x^\nu \delta_2 x^\sigma + \dots$$

Addendum to (2a): If one sets $g_{\mu\alpha}\dfrac{\partial A^\alpha}{\partial\nu} + g_{\alpha\nu}\cdot\dfrac{\partial A^\alpha}{\partial\mu} + A^\alpha\dfrac{\partial g_{\mu\nu}}{\partial\alpha} = 0$, then $\dfrac{\partial g_{\mu\nu}}{\partial x^4} = 0$ results as above (with the special coordinate choice $A^\alpha = 0, 0, 0, 1$). If a four potential is introduced: $\varphi_\alpha = g_{\beta\alpha}A^\beta = g_{4\alpha}$, then the electromagnet. six tensor becomes: $F_{i\alpha} = \dfrac{g_{4i}}{\alpha} - \dfrac{g_{4\alpha}}{i}$ $(i, \alpha \neq 4)$; $F_{i\alpha} = -\dfrac{g_{44}}{i}\cdot 2\cdot\begin{bmatrix} 4i \\ \alpha \end{bmatrix} = F_{\alpha i}(i, \alpha \neq 4)$. Here it is advisable to limit oneself to the symmetric tensor $s_{\mu\nu}$. Then $2\cdot\begin{bmatrix} i4 \\ \alpha \end{bmatrix} = F_{\alpha i}$ also results; furthermore, $2\cdot\begin{bmatrix} 44 \\ i \end{bmatrix} = F_{i4} = -F_{4i}$ $(i \neq 4)$. In the gravitational energy tensor[6] t^α_σ (or rather in its part) $-s^{\mu\nu}\Gamma^\alpha_{\mu\beta}\Gamma^\beta_{\nu\sigma}$, terms then appear: $-\dfrac{1}{4}s^{44}\sum_{\beta\epsilon\rho}s^{\alpha\epsilon}F_{\beta\epsilon}\cdot s^{\beta\rho}F_{\sigma\rho}$ that suspiciously resemble the Maxwell tensor. Other terms also appear, however: $-\dfrac{1}{2}\sum_{\nu\epsilon\beta}s^{4\nu}\left\{\begin{matrix} \nu\sigma \\ \beta \end{matrix}\right\}s^{\alpha\epsilon}F_{\epsilon\beta}$ that must correspond to a reciprocal energy of matter and electricity.

If, for instance, one now attempts to develop Schwarzschild's solution[7] $g_{11} = f_1(x_1)$; $g_{22} = -\dfrac{f_2(x_1)}{1 - x_2^2}$; $g_{33} = -f_2(x_1)\cdot(1 - x_2^2)$; $g_{44} = f_4(x_1)$; $f_1 f_2^2 f_4 = 1$ on this point a bit further along his lines, then $A^1 = A^2 = A^3 = 0$; $\dfrac{\partial A^4}{\partial x^4} = 0$ results; thus $\varphi_\alpha = (0, 0, 0, f_4\cdot A^4)$; $F_{12} = F_{23} = F_{13} = F_{24} = F_{34} = 0$; $F_{14} = -\dfrac{\partial\varphi_4}{\partial x^1}$, and the entire fine correlation is lost because f_4 merges completely with A^4.

4) Normal forms of the line element. (a) At variance with spaces with low numbers of dimensions, in general, the orthogonal form $ds^2 = \sum_i g_{ii}(dx_i)^2$ cannot be obtained, only when $B_{12,34} = B_{13,24} = B_{14,23} = 0$ (:2 equations:). This is probably known to some, but could be emphasized once.[8] (b) The geodesic form that can be found in the *old* Bianchi:[9] $g_{i4} = 0$, $g_{44} = 1$. (c) "Light ray coordinates." The equation $g^{i\kappa}\dfrac{\partial\varphi}{\partial x^i}\dfrac{\partial\varphi}{\partial x^\kappa} = 0$ $(g^{i\kappa} = g^{\kappa i})$ [: in the case of the special

theory of relativity: $\left(\dfrac{\partial\varphi}{\partial t}\right)^2 = \left(\dfrac{\partial\varphi}{\partial x}\right)^2 + \left(\dfrac{\partial\varphi}{\partial y}\right)^2 + \left(\dfrac{\partial\varphi}{\partial z}\right)^2$:][10] apparently has

a complete integral which contains, apart from an additive and a multiplicative integration constant, 2 other essential ones. Knowing one such integral, $\dfrac{\partial\varphi}{\partial C_1} = C_3$;

$\dfrac{\partial\varphi}{\partial C_2} = C_4$ must be the equations for the total "light rays," or better, "gravitation rays." This would be the characteristics of the total diff[erential] eq. $ds^2 = 0$.

If $x'^i = \varphi(x^1,\ x^2,\ x^3,\ x^4,\ a_i,\ b_i)$ is chosen for four arbitrary value systems and a_i, b_i as new coordinates, then all the g'^{ii}'s disappear. If one can, which is probable, also choose the a_i, b_i's such that $\nabla(\varphi(a_1,\ b_1);\ \varphi(a_2,\ b_2)) = 0$ and $\nabla(\varphi(a_3,\ b_3);\ \varphi(a_4,\ b_4)) = 0$, then g^{12}, g^{34} also vanish; $g_{11},\ g_{22},\ g_{33},\ g_{44},\ g_{12},\ g_{34}$; $\sqrt{g}$ becomes rational $=$ $\begin{vmatrix} g_{13}g_{14} \\ g_{23}g_{24} \end{vmatrix}$; the expressions for $B_{\mu\nu,\sigma\tau}$ become only minimally simpler, though.

For special problems, however, light-ray coordinates offer great advantages.

(d) Example. For the sake of variety, cylindrical symmetry now. The processes allow the group: $\begin{cases} x' = x\cos\varphi + y\sin\varphi \\ y' = -x\sin\varphi + y\cos\varphi \\ z' = z + c \end{cases}$. x and y may occur only in the re-

lations $x^2 + y^2$; $xdx + ydy$; $dx^2 + dy^2$; z only in the form: dz.

From this with $x^2 + y^2 = r^2$; $ds^2 = f(r,\ t)dt^2 - g\cdot(dr^2 + r^2 d\varphi^2) - h(xdx + ydy)^2 - ldzdt - mdz^2 - n(xdx + ydy)dz - kdrdt$ or expressed somewhat differently:

$ds^2 = fdt^2 - \underline{hdr^2} - gd\varphi^2 - mdz^2 - \underline{kdrdt} - ldzdt - ndrdz$. If here in the r, t plane light-ray coordinates are inserted, then the underlined terms are dropped.

The tensor becomes $\begin{vmatrix} 0 & -k & -n & 0 \\ -k & 0 & -l & 0 \\ -n & -l & -m & 0 \\ 0 & 0 & 0 & -g \end{vmatrix}$.

The equations for the contravariant four vector introduced under (2a) above are simplified considerably as a result, since two of the coordinates are parameters of the characteristics of the occurring partial diff. eq. In any case, I did not want to omit turning your attention to these coordinates.

In the pleasant hope of receiving a reply from you soon, I remain very respectfully yours truly,

Dr. Rud. Förster
Engineer.

P. S. Incidentally, I would also like to note that my employment contract prohibits me from any writing activities.[11] (The lot of the industrial slave.)

421.   To Gustav Mie

[Berlin, 29 December 1917]

Dear Colleague,

Cordial thanks for the gala copy of the lectures.[1] I am sending you a few more recent papers. Unfortunately I no longer have the *Annalen* article; however, it was published by Ambr[osius] Barth and is available at any bookstore.[2] Among these papers I recommend the one on "Hamilton's Principle . . ."[3] and the one on quantum theory in the *Physikalische Zeitschrift*.[4] The art[icle] "Approximative Integration . . ." contains calculational errors.[5] I intend to submit a cleaned-up version of it once again to the Academy;[6] it's a pity when an idea is presented so badly.

Best regards from yours truly,

A. Einstein.

Happy New Year.

422.   From Mercur Aircraft Company

Berlin S.E. 36, 29 December 1917

To Professor Einstein, Schöneberg, Berlin, 5 Haberland St.

With respectf[ul] reference to the last conference, held at the house of our Mr. Wankmüller,[1] Dr. Zehden offered us his additional opinion regarding the lifting airship.[2] In order to give you, Professor, the opportunity to comment now on Dr. Zehden's point of view for the conference, possib[ly] to be held imminently, we are conveying to you with the enclosed a transcription of Dr. Zehden's letter for your kind notice.

Respectfully,

Mercur Aircraft Design Co., Ltd.[3]

Enclosure.

423.   From Max Planck

Grunewald, 29 December 1917

Dear Colleague,

Of course I agree entirely with your intention of omitting §2 of Freundlich's contract. I had to laugh heartily about your sensitive perception of Nernst's reasoning; it thoroughly suits your good nose for science.[1] However, I would like to emphasize, as a matter of principle, that in general I do set some store in a provision corresponding to §2 in other contracts of this sort. For I could very well imagine the case where someone who has been supported by the Kaiser Wilhelm Institute does not come up with anything worthwhile in the end and then, on his own initiative, publishes his worthless results on the authority of his relation to the K. W. Institute in the hope of promoting himself this way. This could cause us some embarrassment, and that is why, in general, we ought to try to protect ourselves from such incidents. But in Freundlich's case, there is no danger, of course, to the extent that you have him in hand, so to speak. I generously forgive your omission of my name in the "announcement." It is fortunate, though, that it wasn't some other name, by chance.[2]

May the New Year bring you full recovery from your health complaints![3] I hope that during the same year your sympathies for the German Party grow as well; it is ever ready for peace, despite our being in such a favorable position militarily now as never before.[4] I am also pleased with the rapid rise in our exchange rate.[5]

So, on the 2nd we have colloquium. Will the room be heated then, I wonder?[6]
With cordial greetings, yours,

Planck.

424.   From Heinrich Zangger

Zurich 7 (Fluntern), 25 Berg Street, 31 December 1917

Prof. Einstein, Berlin

You desire a rendering of accounts concerning your boy in Pedolin Sanatorium in Arosa, whom I just visited.[1]

We have paid for 3 months
@ Fr 300 –	Fr 900
for miscellaneous expenses ca.	Fr 300
	Fr 1200

We shall show you the bills later. Our friend Besso is in charge of checking.[2]

We wish you happiness and good health for 1918, yours,

Zangger.

425.   To Wilhelm von Siemens

[Berlin,] 4 January 1918

To Privy Councillor von Siemens

Esteemed Privy Councillor,

Regarding the draft contract, which we sent you on 31 November 1917,[1] we note the following. It is the intention of the Board of Directors to arrange that the astronomer Dr. Freundlich receive a kind of stipend for a number of years. This way, he should be placed in the position to devote himself for some years exclusively to practical astronomical research with the purpose of testing the general theory of relativity.[2] As Dr. Freundlich is married and without means, and because living quarters in the proximity of the New Babelsberg Observatory are very expensive, we believe that the planned amount is not excessive.[3]

The draft contract was examined and approved by all members of the Board of Directors. However, Mr. Nernst desires that §2 of the draft be omitted;[4] Mr. Planck and I consent to the striking of the §,[5] the other gentlemen presumably also,[6] who could not yet be consulted. We would be grateful to you and the other Trustees for a prompt decision on this matter, especially since the solution to Dr. Freundlich's professional relationship with the Roy. Observatory in New Babelsberg is linked to it.[7] In the event that the draft contract is approved by the Board of Trustees, I shall consider myself authorized to conclude the contract with Dr. Freundlich. On this occasion I ask the Board of Trustees to inform me whose signatures are required in such cases.

Very respectfully,

A. Einstein.

426.   From Wilhelm Schweydar[1]

Potsdam, 21 Leipziger St., 4 January 1918

Dear Mr. Einstein,

If your kind letter of Jan. 2nd, for which I thank you very much, had not contained the news that you are seriously ill, it would have sparked in me perfect joy. I am very sorry that you must keep to your bed,[2] and I wish with all my heart that you will be completely healthy again very soon. I am glad that it is possible for you to do something on behalf of the matter of such importance to us, despite your sickness, and I expect the best to come of it.[3] Thank you cordially for having decided to do so at all.

I hope that with the enclosed separate letter[4] I have complied with your wishes.

I consider it very commendable to ask Eötvös, since he enjoys great repute among domestic and foreign geodesists and certainly is interested in the filling of the position.[5] His address is

Baron Roland Eötvös, Roy. University, Budapest.
I can think of no one else whom you could contact now.

It would be desirable that a pure geodete be avoided for the Institute; we just cannot function without geophysics here, no matter how tenaciously pure geodesists from the polytechnics cling to the designation "*Geodet.* Inst."[6]

Your kindness emboldens me to reveal to you a secret wish. With the change in directorship now, I would like to become department head at the Institute, in order to secure for the future as well the greater independence I had already had as an observer under Helmert,[7] and in order to ensure that my area of research cannot be taken from me. As a matter of fact, I already direct a department; I direct the Seismic Service, and all practical and theoretical research connected with the study of the Earth's elasticity devolve on me.[8] Added to this are the practical researches using the torsion balance. I have been engaged in other projects at the Institute (rotation of the Earth),[9] but the ones mentioned are entrusted to me officially. Since I am not department head *formally*, all this can change sometime.

A department head position will certainly become vacant with the change in directorship, because Krüger will either become director, or he will go into retirement.[10] Since he is currently department head, such a position will be available in any case. The Ministry assigns the position, usually on the petition of the director; but with the change, the Ministry could also take the initiative itself.

I would be very grateful to you if you would assist me in this at some suitable opportunity.

Thank you also for the kind news that my researches are being valued.

With best wishes for your health and cordial regards, I am yours very truly,

Schweydar.

427.   From Wilhelm Schweydar

Potsdam, 21 Leipziger St., 4 January 1918

Dear Mr. Einstein,

The main purpose of the Geodet. Inst.[1] is to determine the Earth's gravitational equipotential surfaces, a problem which cannot be separated from the question of the Earth's constitution. Connected with this is the study of the temporal alteration in shape of these equipotential surfaces as they are determined by the fluctuations of the Earth's rotational axis and elastic pliability towards deforming forces. The largest project of the International Geodetic since 1900 concerns researching the variation in geogr. latitudes and the procession of the Earth's axis. Modern geodesy can no longer be segregated from geophysics, either in theoretical or methodological respects, and there is certainly room for development mainly in the direction of geophysics.

Advances and problems in geodesy have been moving in this direction as well for a long time, and the Geodet. Inst. can be referred to equally well as a geophysical institute.

For the direction of the Institute, it would be desirable that an individual be chosen who is equipped with ample qualifications in mathematics and physics and who has great interest in geophysics and related problems of a geodesic and astronomical nature. Wiechert would be very suitable and also acceptable, in consequence of his international reputation as director of the Central Bureau of the Intern. Geodetic[2] which is connected to the Institute. With his now famous paper "Über die Massenverteilung im Innern der Erde" ["On the Distribution of Mass in the Interior of the Earth"] (Göttinger Nachrichten)[3] he has made a significant advance in the theory of the Earth's shape and the assessment of the Earth's interior, which was fruitful for many geophysical and geodesic problems. He was also the first to indicate successfully ways to utilize earthquake observations for knowledge of the composition of the Earth.[4] With his talents in mathematics and physics, it is to be expected that he will promote the projects at the Institute, above all, from the theoretical point of view, and will act as a stimulus and helpful adviser to his co-workers at the Institute. His lectures on geodesy at the University of Göttingen also do not draw him far afield from the responsibilities of pure surveying; a series of his lectures on geodesy is published as "Über angewandte Mathem. u. Physik in ihrer Bedeutung für den Unterricht an den höh. Schulen" ["On Applied Mathem. and Physics and Their Importance for Instruction at Schools of High[er] Education"]. Lectures, etc., published by F. Klein and Riecke (Teubner 1900) as Einführung in die Geodäsie [Introduction to Geodesy].[5]

Schumann, who was previously employed at the Geod. Inst. and is now instructor of geodesy at the Polytechnic in Vienna, would also have to be mentioned.[6] Through a series of more extensive calculations for the Internat. Geodetic and smaller geodetic researches, he has demonstrated that he has a good command of geodesy. He is known mainly for his discussion of the observations of fluctuations by the Earth's axis, in which he developed new ideas.[7] Although these ideas have not met with general acceptance and must certainly be characterized as misguided, they nevertheless served as a stimulus and show that he does address problems, which cannot be said of some candidates for Helmert's post.[8] He is inferior to Wiechert insofar as he lacks scope in the area of theory.

His achievements in investigating sources of error in measurements of gravitation by means of pendula must also be mentioned.

With cordial regards, your very truly,

Schweydar.

428.   To Michele Besso

[Berlin,] 5 January 1918

Dear Michele,

I thank you heartily for both your letters and am glad that your new occupation is taking on a permanent nature.[1] Ever since taking to bed, I feel quite well.[2] [I have a gastric ulcer at the pylorus or duodenum. No blood at bowel movement, no growths (X-ray exposure). Sensitivity toward shaking. Stomach acid.] The local sensitivity has disappeared,[3] and I also look well. I am going to rest in bed for another 4 weeks or longer, which is also particularly desirable because of the poor heating.[4] For now, I don't plan to travel to Switzerland this year, possibly never. My longterm care is provided for, in fact, exactly according to doctor's orders.

It does not concern me that Zangger is sending me a large bill. I don't have the power to remove Tete from Arosa,[5] but apart from what my wife can reasonably claim, I shall send Fr 6,000 annually, no more.[6] Last year I sent 12,000 M to Zurich. I shall give to Zangger what he spent on my behalf last year. But I refuse to tolerate constantly being used like a gullible schoolboy. I have no intention of opening an account in Zurich over which I would grant you both access; it would probably not be possible, even if I wanted to do so.[7] We are faced here with a vicious circle unless I put an abrupt end to it. This I am doing, though, here and now, and no one can persuade me otherwise.–

Scientifically I am just working on minor things currently. For it, I am studying and reading copiously, which is no small achievement either. The K. W.

Institute involves quite a large amount of correspondence;[8] even so, my correspondence is steadily on the rise. I am having very pleasant scientific exchanges; in short, I am doing fine.[9] In the fall, Carathéodory, a mathematician from Göttingen, is coming to the local university.[10] I am looking forward to him very much.

Anna's[11] letters are a delight to me; for it is almost the only way I hear about the boys. You see, Albert is no letter writer before the Lord.[12] I am not saying this to complain about him, though. The little that he does write is altogether nice. I'll stop now, because writing in bed is so uncomfortable.

Affectionate regards to all of you, yours,

Albert.

429.   To Roland von Eötvös

Schöneberg, Berlin, 5 Haberland St., 5 January 1918

Highly esteemed Colleague,

With the death of Prof. Helmert,[1] the director's position at the Potsdam Geodetic Institute has become vacant. The Academy, the University, and Ministry are thus left with the responsible task of looking for a successor. Various colleagues have now asked me to collect opinions from impartial experts in the field.[2] It seems to me that you, highly esteemed colleague, are the only one on whose opinion we ought to assign weight in this matter; I therefore ask that you send us your counsel.[3]

Without wanting to anticipate your comments in the least, I request that in your letter you please also devote some words to Messrs.

Schumann (Vienna)[4]
Wiechert (Göttingen)[5]
Krüger[6]
Kohlschütter[7] } (Potsdam)
Schweydar[8]

because these men have already been brought to the attention of the authorities.

Comments about the importance to science of each of the last three gentlemen listed would be desirable also in the event that they do not come into consideration for the filling of the position, when compared to the first two.

Looking forward to your highly valued response with great interest, I am respectfully, yours very truly,

A. Einstein.

430.   From Karl Scheel

Dahlem, Berlin, 28 Werder St., 5 January 1918

Esteemed Colleague,

I heard yesterday to my dismay from Mr. Jahnke[1] that you must keep to bed for an extended period,[2] which I regret very much. On the other hand, Mr. Jahnke told me that, despite your confinement to bed, you are able to occupy yourself with things other than your illness. This gives me the courage to turn to you with two confidential matters.[3]

1) A Mr. Vrkljan,[4] who has applied for membership in the Society in a special-delivery letter, sent me the enclosed paper for publication in the *Verhandlungen*. I would not like to bear the responsibility entirely on my own, and that is why I ask you for a brief statement of your opinion. We cannot be overly severe in these bad times nowadays, of course. I would naturally still have to clean up the paper stylistically.

2) As you probably know, Mr. Planck is celebrating his 60th birthday on April 23rd. It has been suggested that the event be observed by arranging the meeting of the Society on April 26 in his honor,[5] to which invitations are extended widely, at which an address is delivered (possibly Warburg)[6] on Planck, then 3 or 4 summarizing talks about Planck's fields of research (e.g., Einstein, Sommerfeld, Born, Laue). What would you think of this?

If you consider this plan worthy of discussion, namely, in the sense that it would please Planck, I am very willing to devote myself to the realization of this plan. But then it would be advisable that we discuss the details sometime. Perhaps your state of health could tolerate my visiting you once. It does not matter, of course, that you would be lying in bed.—Maybe you can leave me a message sometime by telephone at the Reich Institute (Wilhelm 5515).

With best wishes for your recovery and kind regards, yours very truly,

Scheel.

431.   From Hugo A. Krüss

[Berlin,] Ministry of Culture, 6 January 1918

Esteemed Professor,

Many thanks for the interesting statement by Schweydar.[1]

Concerning Dr. Freundlich, I spoke with P[rivy] C[ouncillor] Müller of the
Astrophys. Obs. in Potsdam.[2] He would like to consider the matter for a few
more days. Perhaps you could provide him with a few more details about the
matter in a few lines and persuade him a little.

With best regards, yours very truly,

Krüss.

432.   From Pieter Zeeman

Amsterdam, 158 Stadhouderskade, 8 January 1918

Highly esteemed Colleague,

With the same post I permit myself to send you as printed matter a short
Dutch paper about some experiments regarding gravitation as well as on the
inertial and ponderous mass of crystals and radioactive substances.[1] I hope that
you will allow me to indicate very briefly (also because the English translation of
the paper is not yet ready) the results of the communication.

1° Weight determinations of quartz spheres with horizontally, then vertically positioned axes prove that at the new orientation, the weight does not change by 1
to 26 000 000.—These experiments were conducted a while ago.

2° Experiments with the torsion balance, performed according to the method by
Eötvös, on *quartz* cylinders prove that an influence of the orientation of a *quartz*
crystal on the proportion of inertial and ponderous mass must be smaller than 1
to 30 000 000.

3° Similar result for *calcspar.*

4° As regards radioactive substances, it is shown that for *uranyl nitrate*, the
proportion mentioned deviates less than 1 to 5,000,000 from the proportion of
masses for quartz.[2]

In the case of uranium oxide, initially a larger discrepancy was found, which
is probably attributable to admixtures of traces of iron.

I hope to be able to make the method roughly ten times more sensitive, about which I hope to report later.

With kind regards and very respectfully, yours most sincerely,

P. Zeeman.

433.   From Hugo A. Krüss

[Berlin,] Ministry of Culture, 9 January 1918

Esteemed Professor,

I am sending you the enclosed letter with the request that you kindly *return* it upon perusal. P[rivy] C[ouncillor] Müller's position on the matter is not very encouraging.[1] If in your opinion special and close relations exist between the problems to be addressed by Dr. Freundlich and astrophysics, I would be obliged to you for information about it. Also, I would appreciate learning about how Schwarzschild stood on this question.[2]

It would also be of interest to me to speak with Dr. Freundlich himself sometime.[3] Perhaps you can arrange to have him visit me in the next few days, although he should agree upon the time with me beforehand by telephone, so that he does not come in vain.

With best regards, yours very truly,

Krüss.

434.   From Gustav Müller

Potsdam, 9 January 1918

Highly esteemed Professor,

In reply to your inquiry regarding the possible engagement of Dr. Freundlich at our Observatory,[1] I inform you that currently only the position of research assistant is available which, however, I have already promised to another applicant.[2] The hiring of Dr. Freundlich could be possible only if the Ministry were inclined to support a second research assistant position at our Observatory in the next state budget. I already indicated this solution in a letter to Prof. Krüss yesterday[3] and would consider it helpful if you were to recommend my suggestion to Prof. Krüss when the occasion arises.

In any event, I am prepared to support the planned investigations by Dr. Freundlich toward testing the general theory of relativity,[4] in that I am granting him permission to acquaint himself with the various observation methods employed at our Observatory, and in that I am willing to take care of his instruction and, as far as is possible, to place the Observatory's equipment at his disposal.

I would be pleased if, when your health has been restored,[5] you could come to Potsdam sometime and discuss this matter further with me.

Respectfully and most sincerely yours,

G. Müller.

435.   To Hugo A. Krüss

[Berlin,] 10 January 1918

Highly esteemed Professor,

Many thanks for forwarding the letter by Director Müller, which I return in the enclosed.[1] I can understand his point of view entirely; Mr. Freundlich does not have any experience in astro*physical* observation yet, only in *metric* astronomy (star-position and meridian angular measurement). It is correct that Mr. Freundlich must still learn certain observational methods before he can set about certain ones for the planned investigations. Under these circumstances I consider it necessary to explain why it seems desirable to me that Mr. Freundlich be employed at the astrophysical institute.[2] I beg your pardon in advance for the unavoidable lengthiness of my account.

The advances in astronomy in recent decades have been based far more on the perfection of precision in observational methods than on fundamental, theoretically based innovations. That is how it came about that purely practical observational skill was valued more highly than theoretical knowledge and expertise. The employees were chosen in keeping with this approach and eventually acceded to the leading positions at the observatories. The consequence of this is that collection of the most precise data possible became an end in itself and that, in general, leading astronomers have quite poor theoretical training and insight.[3] Thus we see that, among astronomers, the more modern efforts in the area of gravitation theory is generally met with a complete lack of understanding and interest, even though the empirical investigation of gravitation must be seen as astronomy's most important task.

The only striking exception mentionable is the regrettably deceased Schwarzschild,[4] who combined a brilliant theoretical talent with sufficient practical skill in the observational methods to bring far-reaching viewpoints to bear in the research at his institute.

Only by again having staff members with not only practical but also theoretical training at the large observatories at New Babelsberg and Potsdam and by also granting them the necessary freedom of action, can an improvement in the current conditions be brought about. Among the younger astronomers at New Babelsberg and Potsdam, to my knowledge, Mr. Freundlich is the only one with a solid knowledge of mathematics, celestial mechanics, and gravitation theory. Although considerably less talented than Schwarzschild, he nevertheless recognized the importance of modern gravitation theories to astronomy many years before the latter did and has worked fervently toward verifying the theory along astronomical or astrophysical lines.[5] His director's negative stance, however, prevented him seven years long from carrying out his research plans to test the theory. Instead, year after year, Freundlich was forced to perform a task that any normally gifted 18 year-old boy could be assigned to do after brief training.[6] In the interest of science and equity it would be very desirable if he could be employed at one of the two large institutes in such a way that he could, within certain limits, freely choose the topic of his research. The fact that Mr. Freundlich has not yet had the opportunity to acquire practical skills in certain areas of astrophysics is infinitely less significant compared to the favorable qualifications mentioned. This deficiency can be offset within a short time. It is quite inconsequential to the *issue* whether Mr. Freundlich ought to be employed at New Babelsberg or at Potsdam; formally speaking, Potsdam would be most suitable, because testing gravitational theory must be seen as an astrophysical task.

Very respectfully,

A. Einstein.

436.   From Rudolf Humm[1]

Göttingen, 14 Planck St., 15 January 1918

Esteemed Professor,

While preparing a talk on your and de Sitter's boundary conditions at infinity,[2] I noticed the following, which relates to your new gravitation equations.

These equations are:

$$K_{\mu\nu} - \lambda g_{\mu\nu} = -\kappa \left(T_{\mu\nu} - \frac{1}{2} g_{\mu\nu} T \right)$$

$$T^{\mu\nu} = \rho \frac{dx_\mu}{ds} \frac{dx_\nu}{ds}$$

(1)

By presupposing these equations, we have Euclidean geometry not when all matter disappears, as was earlier the case, but when a very specific value is prescribed for the material tensor, a value which results from the equations

$$-\lambda\delta_{\mu\nu} = -\kappa\left(T_{\mu\nu} - \frac{1}{2}\delta_{\mu\nu}T\right). \tag{2}$$

The $T_{\mu\nu}$'s are determined from it such that only the diagonal terms occur, that is, all equivalent to one another. (λ cancels out.)

Equations (2) are valid upon presupposing a coordinate system S in which $g_{\mu\nu} = \delta_{\mu\nu}$ throughout; in such a system,

$$\rho\left(\frac{dx_1}{ds}\right)^2 = \ldots = \rho\left(\frac{dx_4}{ds}\right)^2$$

should apply,[3] that is, the matter ought to appear to be moving with reference to S at the velocity of light.

However, this contradicts experience, according to which an approximate Euclidean geometry can be observed in a coordinate system against which matter is at rest.—One could take the pressure into account in the material tensor, but this is no substantial improvement; it seems most likely to work with internal tensions. Indeed, $T_{\mu\nu}$ is interpreted as the el[ectro]m[a]g[netic] energy tensor; hence the equality of the diagonal terms probably means that we have a matter consisting in pure el.mg. pressure energy.

In any case, your consideration is tailored to ponderable matter, and the material tensor has always been $\rho\dfrac{dx_\mu}{ds}\dfrac{dz_\mu}{ds}$. This tensor originates from the generalization of Euler's equations of motion and has nothing to do with tensions; inviscid fluids exist. It indicates the density and velocity distributions, it has more similarity to kinetic energy than to stress energy.

If *this* tensor is retained then, precisely because density and velocity occur in it, the possibility immediately arises of constructing your static world, by inserting $T^{44} = T = \rho$ in (1). Yet, the difficulty with Euclidean geometry presents itself, although I do not know for sure whether it really is a difficulty.

If this tensor is *not* retained and is replaced with one taken from the elasticity, shall we say, the latter difficulty can perhaps be avoided, but I do not know at present how it would then have to be inserted specifically in order to obtain your or de Sitter's world ($T^{44} = $ const ?, $T^{\mu\nu} = 0$).[4]

I would be very obliged to you if you would kindly clarify these questions for me. With best regards, yours truly,

R. Humm, stud[ent].

437.   To Pieter Zeeman

Berlin, 16 January 1918

Esteemed Colleague,

Cordial thanks for your articles and for the commentary thereon in your letter,[1] as well as for your wonderful earlier papers on the ether-drag coefficient.[2] Among your recent investigations, those about the inertial and gravitational mass of uranyl nitrate interest me most.[3] Regarding the investigations on the gravitation and inertia of crystalline substances, the theoretical background to the problem is unfamiliar to me;[4] even so, I understand that the issues can be addressed without such theoretical points of view, particularly at this time of revision of the general foundations of gravitation theory.

Cordial regards, yours very truly,

A. Einstein

(This letter was dictated from bed, because I am ill.)[5]

438.   To Erwin Freundlich

[Berlin, before 17 January 1918][1]

Dear Freundlich,

Perhaps it is better if I use my personal influence concerning that instrument, since no bellicose odium weighs on me.[2] I am on the Council of the Anti Orlog Raad, you know.[3] I am very pleased that you are being received so amicably at Potsdam.[4] The ice seems finally to have broken.

The notice by de Donder is scandalously superficial. I am amazed that Lorentz was taken in by it, or accepted it without closer consideration.[5] The error in the chain of reasoning is, as you yourself indicated in your letter, that (16) does not follow from (15). (15) is not an *identity* but an *equation of condition*. The deduction is approximately thus.

Assertion:

No circle can exist on a plane.

Proof:

The circle equation

$$x^2 + y^2 = R^2$$

is intrinsically inconsistent; for since the right-hand side is independent of x, the left-hand side must also be. The equation is therefore inadmissible!

The nonsense $C = 0$ already haunts an earlier paper of de Donder's. The factor $\frac{1}{4}$ in (1), incidentally, is closely connected with it.[6] Actually, $\frac{1}{2}$ should appear there. In short, we have here a genuine counsel of confusion.

To whom should I write about the instrument? I shall do so as soon as I have the address. I am returning to you, with best regards, your report as well as the famous notice by de Donder. Yours,

A. Einstein.

I shall see that we put forward your contract for resolution on Thursday.[7] I think you should say in your report that the author's equations are not only not equivalent with mine but not even possible on their own. For it ought to be known that you discovered de Donder's error. I was very pleased that you spotted the crux of the problem.

I have left to Lorentz the exact manner of correction in the Amsterd. Academy Reports. It is best that the author submit a correction himself.[8]

439.   To Rudolf Förster

[Berlin,] 17 January 1918

Esteemed Colleague,

First of all, I sincerely apologize for my long silence. But I have been lying in bed for almost 4 weeks already, owing to a refractory stomach ulcer and am very much hampered in everything. Your letter[1] was extraordinarily interesting to me and confirmed my opinion that, with you, we are dealing with a born theoretician. You must not take your industrial slavery too much to heart. I also was one of those during my best years, as an employee of the Swiss Patent Office. My genuinely original ideas all stem from that time, for all too soon does one become an avuncular old boy, and the little fount of imagination dries up.[2] Permission to publish purely scientific ideas will surely be granted you despite your employment contract, that I guarantee you, especially considering that Krupp is very positively disposed toward all scientific endeavors.[3]

Now to your various affairs. First to what I did not grasp. I do not understand why the condition

$$B_{12,34} = B_{13,24} = B_{14,23} = 0$$

suffices for the feasibility of the line element

$$\sum g_{ii}dx_1^2,$$

because I do not see why the above condition is invariant (provided that the other $B_{ik,lm}$'s have some value or other). For that, a tensor would obviously have to exist that has no components other than those mentioned above.

I also was occupied for a long time with proceeding from a nonsymmetrical $g_{\mu\nu}$; but I have almost given up hope of getting to the bottom of this (gravitation–electromagnetism) unification mystery in this way. Various reasons arouse serious doubts:

1) Electricity has the two constants, the electron's charge ε and mass μ, which cannot be invented or drawn out of a purely mathematical consideration.

2) The mathematical investigations show me, time and again, that the symmetrical and antisymmetrical components of $g_{\mu\nu}$ ($= s_{\mu\nu} + a_{\mu\nu}$) always separate like oil and water. I draw my examples from your own interesting reports:

Let us consider the two equation systems you propose to be electromagnetic[4]

$$\frac{\partial g^{\mu\langle\nu\rangle\alpha}}{\partial x_{\langle\sigma\rangle\alpha}} + g^{\mu\langle\alpha\rangle\beta} \left\{ \begin{matrix} \alpha\langle\sigma\rangle\beta \\ \alpha \end{matrix} \right\}_s + g^{\langle\alpha-\rangle\beta\alpha} \left\{ \begin{matrix} \alpha\beta \\ \mu \end{matrix} \right\}_s = \langle 0 \rangle I^{\mu} \tag{1}$$

$$-f_{\mu\nu\sigma} = \left[\begin{matrix} \mu\nu \\ \sigma \end{matrix} \right] - \left[\begin{matrix} \nu\mu \\ \sigma \end{matrix} \right] = 0.$$

If in the first system you insert $g^{\mu\alpha}$, $g^{\alpha\nu}$, $s^{\mu\nu} + a^{\mu\nu}$, etc., for $g^{\mu\nu}$, then the portion stemming from $s^{\mu\nu}$ disappears, so that in fact only the divergence of $a^{\mu\nu}$ remains (in the sense of absolute differential calculus).

If the same is done analogously in the second system, there remains

$$\frac{\partial a_{\mu\sigma}}{\partial x_{\nu}} + \frac{\partial a_{\sigma\nu}}{\partial x_{\mu}} + \frac{\partial a_{\nu\mu}}{\partial x_{\sigma}} = 0. \tag{2}$$

The fact that $a_{\mu\nu}$ is the antisymmetrical portion of $g_{\mu\nu}$ does not play the slightest role in (1) or (2). The hypothesis of interest to us at the outset therefore truly plays no role in the result; rather, the latter agrees exactly with my earlier electr. equations for the vacuum.–

Your generalization of the Riemann tensor would be interesting if a transparent geometrical or analytical way to its derivation could be given that does not make use of the resolution $g_{\mu\nu} = s_{\mu\nu} + a_{\mu\nu}$. But it does seem suspicious that $s^{\alpha\beta}$ appears as a factor of the second term.

3) A study of the gravitational waves has shown me that something formally different to light waves is involved. The $g_{\mu\nu}$'s themselves do not occur in the energy but essentially their *first derivatives*;[5] this I cannot reconcile either with the fact that $a_{\mu\nu}$ signifies the electromagnetic field.

Your other comments are just as interesting *per se* and new to me. I am not so bold as to want to decide whether they will lead to an advance. In any event, I wish you luck in your efforts!

With best regards, I am yours,

A. Einstein.

440.   To Rudolf Humm

[Berlin,] 5 Haberland St., 18 January 1918

Esteemed Colleague,

You are, in fact, entirely correct.[1] The equations

$$(+)\lambda\delta_{\mu\nu} = -\kappa\left(T_{\mu\nu} - \frac{1}{2}\delta_{\mu\nu}T\right)$$

$$T_{\mu\nu} = \rho\frac{dx_\mu}{ds}\frac{dx_\nu}{ds}$$

cannot be satisfied in any way. The equations

$$\rho\left(\frac{dx_1}{ds}\right)^2 = \ldots = \rho\left(\frac{dx_4}{ds}\right)^2$$

read more completely as (since $x_4 = it$ must be set so that $g_{\mu\nu} = -\delta_{\mu\nu}$ also for $m = n = 4$):

$$\frac{\rho\left(\frac{dx}{dt}\right)^2}{c^2 - v^2} = \frac{\rho\left(\frac{dy}{dt}\right)^2}{c^2 - v^2} = \cdot = \frac{-\rho}{c^2 - v^2},$$

which cannot be satisified.

Changing anything in the energy tensor is out of the question.–

But this finding presents absolutely no difficulty, according to my conception of the boundary condition problem. The supplementary term serves specifically to allow the quasi-spherical (or quasi-elliptic) world to take the place of the quasi-Euclidean one. Thus it is desirable that the equations just preclude any quasi-Euclidean world.

I presented in detail the reasons that led me to this interpretation in my article of last year.[2] To repeat, I only note that the boundary conditions for a quasi-Euclidean world ($g_{\mu\nu}$ = const. at spatial infinity) are not *relative*, i.e., not valid for any system that remains infinite at infinity. This difficulty can be avoided by replacing the boundary condition at infinity with the condition of closure. But this condition cannot be realized without the supplementary term, with the given structure for the energy tensor of matter.[3]

De Sitter's matter-free solution is not admissible, in my view, because at infinity it is not free of singularities (with the static and spatio-spherical coordinate choice, $g_{44} = 0$ on one surface). This actually means the localization of all matter on one surface.[4] De Sitter's consideration, in particular, strengthens my opinion that, on the basis of the additional term, no $g_{\mu\nu}$-world exists without matter and without singularities. This conduct by the equations is required from the standpoint of the general theory of relativity (complete dependence of the $g_{\mu\nu}$-system on the matter).[5]

With best regards, yours truly,

A. Einstein.

441. From Wilhelm von Siemens

Berlin W. 9, 4 Voss Street, 21 January 1918

Board of Trustees of the Kaiser Wilhelm Institute of Physics.

To His Honor the Director of the Kaiser Wilhelm Institute of Physics, Professor Einstein, locally.

The members of the Board of Trustees of the Kaiser Wilhelm Institute of Physics have declared their approval of the enclosed draft contract with Dr. Erwin F. Freundlich, subject to the omission of §2 as suggested by Privy Councillor Professor Nernst.[1] At the same time, the Board of Trustees has authorized Your Honor to close the contract with Dr. Freundlich on your part.

I humbly look forward to kind information about the final execution as well as about where the funds placed at Dr. Freundlich's disposal should be remitted.

Chairman of the Board of Trustees
von Siemens.

Recipient's note: "Answered on 1 February. E[instein]."

442.   To Hans Albert Einstein

[Berlin,] 25 January 1918

My dear Albert,

Your letter and your postcard delighted me. I see that not only are you my own son, but also that really affectionate relations are developing between us. Your concern about my illness was especially gratifying to me. So, let it be said with pleasure: I am content with you and am happy when I think of you. I feel that the maturer and more independent you become, the better friends we become.

I approve entirely of your going to *Realgymnasium*. For a person with your kind of aptitude, it is not good when too many languages are crammed into you.[1] You seem to feel this yourself as well.

It has been just a month now that I have been lying in bed;[2] and I probably will have to stay lying down for sometime longer. My ailment involves a stubborn ulcer at my stomach exit, which is healing only very reluctantly. There is no doubt that I have had this illness longer than you have been in this world.[3] The pain I always used to have was caused by the same source. I am probably never going to become truly healthy again but am going to have to eat a kind of infant food for the rest of my life and avoid moving about much. I am not worried about that but am happy and content. I can attend to my work very well in bed, you know, and my cousin does a splendid job in seeing that I have enough food to "peck" at [*Vogelfutter*][4] so that I am well cared for in all respects.

I don't yet know how to arrange it so that we can meet this summer. But if there is a will there is a way, and we'll want it, won't we? Give Tete my affectionate regards. Is he back at home with you already?[5] I am firmly convinced that it's wrong to pamper him up there for so long. The longer he has been up there, the surer it is that he will not be able to tolerate the Zurich climate anymore when he comes down again. Added to this is the misfortune that an enormous amount of money is being spent so that all my savings are being used up.[6] One fine day I am going to die and nothing will be left for you all. My friends in Zurich, who are irresponsible in this regard, are to blame for this calamity.[7] And ultimately it is you boys who will be affected if you have not been provided for adequately. But my hope lies in that you are gradually getting older and I am going to be able to arrange everything with you without our needing the help of strangers.

Greetings and kisses from your

Papa.

Kind regards to Mama and Aunt. Kisses to Tete!

443.   From Roland von Eötvös

Budapest, 7 Eszterházy Alley, 27 January 1918

To Prof. A. Einstein, Berlin

Highly esteemed Colleague,

It is a great honor and also a great pleasure for me to have received a letter from you, even more so to be invited to express my opinion on such an important question as the filling of the position of director at the Potsdam Roy. Pr[ussian] Geodetic Institute.[1]

A response is not easy for me, though. I have spent my life, including my scientific life, as if on a deserted island, secluded from general contact. After reliable provisioning at Heidelberg and Königsberg, I retired to this island of mine, erected my hut there, and settled down.[2] Following my inclination to solitude, only rarely did I leave this working place of mine in order to see with my own eyes what was happening on the neighboring islands and continents. I have received word of all the great and fine things that have happened there mostly only in writing.

Thus I have remained very deficient in the personal relations which are of considerable significance in deliberating the question at hand.

It is possible, however, that this impersonal judgment of mine is just what seems of advantage to you, so upon mature consideration I shall offer my opinion very candidly, as best as I am able.

The responsibility of geodesy, in the strictest sense of the word, and thus also of a geodetic institute, is no more than the most precise determination possible of the equipotential surfaces of terrestrial gravity. As such, it renders valuable services to practical surveying, geography and astronomy. In accordance with this abstract responsibility, this science strives to achieve the greatest possible precision in the observations and flawless calculation of its results, free of all hypotheses unattached to this goal. It is an exact science in the strictest sense and should always remain so as well.

Yet I scarcely believe that a geodesist is ever content with his series of latitude, longitude, and altitude determinations and that his scientific imagination has not carried him away to seek the causes for why that portion of the equipotential surface that he has established has precisely that form and none other. The mind's eye thus plumbs the inaccessible depths of the Earth's interior and questions are posed that belong within the scope of geophysics and geology. Here hypotheses and analogies accompany the theorist who now follows a clue that may seem alluring and promising but that does not yet nearly attain the rigor and reliability of classical geodesy today and perhaps never will.

Should the geodesist therefore not banish such efforts as annoying reveries? With his theodolite and gauging rods, should he not rather build up the foundation walls upon which many a lofty castle in our knowledge of the cosmos could sometime later find a firm basis?

Such questions do not require long discussion, for who doubts that the splintering of our natural sciences into so many specialties finds justification only in the fact that each of these strives with its own energy and means toward the common goal of knowledge, thus also that geodesy is prepared to offer all assistance allotted it toward the construction of the cosmos? For this reason, a geodetic institute should not and may not become a geophysical institute, yet it should direct its activities also to those geodesic researches that promote what today are only nascent fields of geophysics and geology.

And as the institute, so also its director.

Now, who should this director be? Who should it be at Potsdam, at the institute that the International Geodetic has legimately chosen as its leader?

Happy Germany! The choice is difficult when one must select the best suited among the many who are equal to the task.

When I reflect upon the matter, my attention is directed first and foremost to *Prof. L. Krüger*, the great geodesist, who through his investigations, encompassing the entire Earth, on the form of the geoid and the best manner in which to depict it,[3] has advanced the main mission of geodesy in a genuinely Gaussian sense, and has thus contributed a good portion of the fame indisputably attributable to the Roy. Pruss. Geodetic Institute.

Another prominent geodesist is *Prof. R. Schumann*, and I discern with pleasure the inclination in his work also to address problems that are more closely related to physics. His researches on the fluctuations in pole height and on isostasy are splendid proof of this.[4] Prof. *J. E. Wiechert*, also stands out among many others as a geodesist, and even more so as a physicist, who is apparently interested in the Earth mainly because it is the site of earthquakes and, as one of the founders of earthquake theory, is engaged in all tasks related to the shapes of the Earth also within its interior.[5]

Last but not least, I would also like to name one of the younger candidates, Prof. *W. Schweydar*. Under the direction of the unforgettable Helmert,[6] he has gone through magnificent training. As the reports and publications of the R. Pr. Geodetic Institute demonstrate, there is virtually no type of geodesic research in which he has not participated and to which he has not contributed as assistant and also as independent researcher. Aside from these researches at the institute, he has also endeavored to broaden his own sphere of research, which borders on the frontiers of physics. In him I see a man who, as a devoted student of Helmert, while constantly bearing in mind the imposing responsibilities of geodesy, also has a command of the allied field of physics and who is equal to the most difficult tasks both in observational method and in their theoretical evaluation. His papers on the elasticity of the Earth are sufficient confirmation of this assertion of mine.[7]

Listing more names and then proposing a single one amongst all of them would be presumptuous of me. For such types of appointments, personal motives as well as order of precedence must usually also be drawn into the consideration, of course, and judging this is not my business, yet may I be permitted to close with a word of warning. In the appointment, youth should not be regarded as a shortcoming but as an advantage, "for art is long, life short." This old saying rings true most convincingly in research like that in geodesy, which requires a lifetime and more in order to come to a certain close.

More I cannot and should not say regarding the question posed to me; it was a pleasure for me, though, and an honor to be allowed to be so expansive already. Yet I must ask you to make allowances for the delay in my response. After months of constant illness and, at the end, a serious operation,[8] I had a large heap of items in arrears that had to be attended to in the last few weeks.

Accept, esteemed Colleague, the expression of my deepest respect, yours truly,

Roland Eötvös.

444.   From Heinrich Zangger

[Arosa, 28 January 1918]

Dear friend Einstein,

With Tedi in Arosa.[1] It has all been arranged until spring; you don't need to worry about anything unusual anymore. Tedi is feeling very well. D O. will take another X-ray. He can probably come down in 1 month—February—and go to school in the spring.

Regards from both of us,

Zangger
Tedi.[2]

445.   From Fritz Haber

[Berlin, before 29 January 1918][1]

Dear Einstein,

It appears to me that you are deeply shocked by the tactic I have chosen. That is why I am writing to clarify to you that the matter meant nothing more to me than amusement at your lack of familiarity with military formalities. However, since I can just assume that Mr. Nordström has made a petition as unconventional as is your letter, the only way left open to me in finding support for you was to attempt to awaken personal interest at the highest level.[2] For, with or without my recommendation, the departments directly involved could not have responded otherwise than demanding that a petition with the required supporting documents be submitted, & then you would have gotten only as far as before. There would be no alternative to going there, seeking out the official in charge, inquiring about the requirements, passing them on to Nordström, etc. You cannot do this since you are bedridden,[3] and I can do it only with tremendous difficulty, as my time is taxed to the limit, especially during working hours.[4] Consequently, I tried to interest the head of the act[ing] General Staff[5] personally on the reasoning that he might issue an order on the basis of which his office would relieve him of these steps, would itself attend to the formal procedures for you, or the like. One word from him would suffice for this, but this he offers only when the case strikes as unusual, very unusual, and I turned *his* attention to the fact that this is the case here in the form I considered feasible.

Life, not only in the field of mathematical physics, is based on knowledge of and adherence to certain formal correlations; and not only in the field in which you perform wonders is success often not attainable without the formal laws. As for myself, I merely ask for your faith in that I am happy and pleased to be of use to you, that I have too much respect for your character & achievements to ridicule anything you do, & that I am fond of you personally.

It is not worth the effort to reply to me. Take care of yourself and recover, this is my special wish, and give Mrs. Einstein my regards, your friend

Fritz Haber.

446. To Fritz Haber

[Berlin,] 29 January 1918

Dear Haber,

Although you have already ruled out receiving an answer to your nice letter,[1] I now have occasion to write you nonetheless. Your commending lines have accelerated the petition's effectiveness to such an extent that today already the news arrived that no more obstacles stand in the way of Nordström's trip.[2] Once again, thank you very much indeed, also in the name of Nordström, who will probably write you himself. My little letter was written in a brash rather than sensitive tone;[3] but the inanimate characters do not convey this. At any rate, the delicate consideration you have shown for my caprices, with such an overwhelming workload as yours, has filled me with a mixture of joy and shame which, thank heavens, I don't have to analyze.

Best regards, yours,

Einstein.

447. From Max von Laue

Würzburg, 40 Mergentheimer St., 30 January 1918

Dear Einstein,

I thank you and Planck most cordially for the kind letter of Jan. 25[1] and ask you not to hesitate to make this answer of mine accessible to him, since I could not think of the slightest reason for not showing equal trust in him as in you.

You both were entirely right in seeing that the question of my relocation is not clear, above all, to what extent I should draw advantage from the foundation of the K. W. Institute for my move there.[2] To be more precise, this is unclear to me particularly because I know that the now available funds are earmarked for purely material disbursements. One or two years ago it overwhelmingly seemed as though Dr. Fleischer in Wiesbaden intended to endow a post to be designated to me personally;[3] and when I met Sommerfeld in August 1917 in Berchtesgaden[4] and he explicitly mentioned, without any actual prompting, that he had just happened to be corresponding with Dr. Fleischer and about very inconsequential matters, I was led at that time already—August 1917—to assume that my suspicion was correct. But this may, of course, have changed in the meantime. At all events, it is a matter of fact that Dr. Fleischer has not concerned himself with me for many months, in contrast to the period from the end of 1915 to the close of 1916.

I do not want to deny that this would have been most preferable to me. If it is no longer obtainable now, however, I am also prepared to switch with Born,[5] provided the Ministry does not reduce his salary below the offer of August 1914.[6]

Immediate election to the Academy would also be a nice recompense, of course, although not in the pecuniary sense.[7]

This is all that I can think of in answer now. If Planck and you still have a question about this, though, please write me. In the next few days I shall probably not be reachable by mail, however; from Feb. 2 until Feb. 10 I am going to be prowling around in the lines of communication area behind the western front. From Feb. 10 until Feb. 20 I am going to be on the Feldberg in the Black Forest (Feldbergerhof). Thereafter, here again.

Incidentally, to broach another entirely different subject, I would like to ask you if you know of any more suitable place, that is currently accessible to us on the Earth's surface, than Berlin for the treatment of an ailing stomach?[8] I do not want to meddle in your doctors' business, but I would think that now in wartime about *any* other place is better for this purpose. You must not consider this comment presumptuous; we physicists all have a shared interest in your health.

With cordial regards to Planck and you, yours,

M. von Laue.

448.   To Max Planck

[Berlin, after 30 January 1918][1]

Dear Colleague,

Now the case is clearly before us. Laue thus had a prospect unknown to us until now.[2] His expression of willingness is unambiguous; I am in the dark about his motives, though, likewise about the conversation with Sommerfeld[3] regarding the position. If the roguish biblical passage "Man's will is his paradise" is not deceptive, heaven on Earth may indeed fall to his share.

With cordial regards, yours,

Einstein.

449.   To Mileva Einstein-Marić

[Berlin,] 31 January 1918

Dear Mileva,

The endeavor finally to put my private affairs in some state of order prompts me to suggest the divorce to you for the second time.[1] I am firmly resolved to do everything to make this step possible. In the case of a divorce, I would grant you significant pecuniary advantages through particularly generous concessions.

1) 9 000 M instead of 6 000 M,[2] with the provision that 2 000 of it be deposited annually for the benefit of the children.

2) The Nobel Prize—in the event of the divorce and in the event that it is bestowed upon me—would be ceded to you in full *a priori*.[3] Disposal of the interest would be left entirely to your discretion. The capital would be deposited in Switzerland and placed in safe-keeping for the children. My payments named under (1) would then fall away and be replaced by an annual payment which together with that interest totals 8 000 M. In this case you would have 8 000 M at your free disposition.

3) The widow's pension would be promised to you in the case of a divorce.[4]

Naturally, I would make such huge sacrifices only[5] in the case of a voluntary divorce. If you do not consent to the divorce, from now on, not a cent above 6 000 M per year will be sent to Switzerland. Now I request being informed whether you agree and are prepared to file a divorce claim against me. I would take care of everything here, so you would have neither trouble nor any inconveniences whatsoever.[6]

My friends report to me regularly about your and the children's well-being. I'm glad that you don't have to endure fever and attacks anymore.[7] Albert's letters delight me exceedingly; from them I see how well the boy is developing intellectually and in character. I hope that through the long stay in the mountain air[8] Tete has not become too sensitive to the impure town air and will come home soon reinvigorated.

With amicable regards also to your sister,

Albert.

Kisses to the children.

Reply please by registered mail.

[5]Deleted text in draft: "only, for the sake of my liberation, . . ."

[6]Draft version: "I would take care of & pay for everything here, so you would have neither trouble nor any difficulties whatsoever, and the entire proceedings would take place here."

450.   To Roland von Eötvös

[Berlin,] 31 January 1918

Highly esteemed Colleague,

I thank you from the bottom of my heart for the extremely illuminating and valuable response to my inquiry.[1] I shall do my utmost to have it taken to heart. We all know that we have no one here upon whose judgment we can rely so much in the important question at hand as on yours. Based on your letter, the most auspicious solution appears to me to be: Krüger as director and Schweydar alongside him, with as much independence as possible, the latter so that geophysics does not come off badly.[2]

I do not want to let this exchange of letters pass, though, without expressing to you my gratitude for the advancement that our knowledge of the identity of gravitational and inertial mass has made through your investigations.[3] In the last few years I have been concerning myself with the theoretical side of this problem and would like to take the liberty of sending you a little booklet from which the principle aspects can be drawn without taking up too much time.[4] The mathematics can be found in an *Annalen* article of 1916[5] and in some papers of the *Berliner Sitz. Berichte.*[6]

Wishing that you may soon recover fully from your illness,[7] and in thanking you once again for your valuable information, I am with great respect, yours very truly,

A. Einstein.

451.   To Hugo A. Krüss

[Berlin,] 31 January 1918

Highly esteemed Prof. Krüss,

I just received Mr. Eötvös's reply[1] and hasten to forward it to you. In my opinion, the value of this report by a man of such unquestionable objectivity and expertise cannot be overestimated.

It is evident that Eötvös would entrust the directorship of the Institute to Prof. Krüger. I must add the remark that in my inquiry I had named Wiechert and Schumann as first choices.[2] The recommendation of Krüger against Wiechert and Schumann is underscored even more in that, at over 80 years of age, Eötvös recommends assigning the management to a younger employee.[3]

Furthermore, it is very conspicuous that he does *not* name Mr. Kohlschüt-ter,[4] despite my having asked him to take him into consideration.

Finally, the praise he lavishes on Mr. Schweydar's achievements is notable. I can add that I also have a very favorable impression of this man's work. He is an authority mainly on geophysical issues. It is unmistakable that, of all of them, Eötvös discusses him with the greatest *warmth.* Certainly it would be advantageous if Schweydar could be made independent in some way so that his initiative could make itself felt; otherwise there is a danger that geophysics would be stifled by a "strictly geodete" director.[5] Not every director has such broad horizons as Helmert![6]

It must be pointed out that there is no one at the Academy comparable to Eötvös to whom we could turn for the relevant expertise.

With all due respect,

A. Einstein.

452.   From Cornelia and Gunnar Nordström

Leyden, 31 January 1918

Dear Prof. Einstein,

Heartfelt thanks for your kind postcard. How sorry we are that your health is still not good.[1]

It is very kind of you to want to take the trouble to make our trip possible.[2] Unfortunately, I must now tell you that it is very questionable when we can travel. The situation in Finland has deteriorated dramatically very recently.[3] The day before yesterday we spoke with the Finnish delegate (who had announced the independence of the country to the Dutch government), and he strongly advised us not to travel now. It is therefore very possible that we can leave only in the summer, perhaps not even before August.—Thank you very much indeed, though, for your readiness to help.

Warm greetings from both of us and from Ehrenfest. Yours,

Nel and Gunnar Nordström.

453.   To Arnold Sommerfeld

[Berlin,] 1 February 1918

Dear Sommerfeld,

Your warm-hearted letter pleased me very much, even though my illness is not at all so troublesome that I should deserve so much sympathy. It is an antiquated gastric ulcer, thanks to my neglect, which is now supposed to be combatted with prolonged bed rest.[1] I thank you just as much for the kind package consisting of that now rare material.

The endeavor to master dispersion makes sense to me for slow waves, the indications about the rest were not clear to me.[2] It is striking how much can be done satisfactorily with classical mechanics. If only it were possible to clarify somewhat one day the principal ideas behind quanta! My hope of witnessing this is diminishing steadily, though.

At the end of April we are holding a meeting in honor of Planck's 60th birthday. You would do us all a great favor if you were willing to present a talk at this occasion on the development of radiation and quantum theory during the special session of the German Phys. Society.[3] Nothing more specific has been arranged yet; but I am telling you now already so that you hear of it as early as possible. No one here could do it nearly as well as you; and I think that anything that would please Planck ought to be done.[4]

Nothing new can be found from general relativity anymore: identity of inertia and gravity. The metric ⟨laws⟩ behavior of matter (geometry & kinematics) ⟨are⟩ is determined by the interaction of the bodies; independent properties of "space" do not exist. With this, in principle, all is said. I am also convinced that a coherent theory without the hypothesis of spatial closure is not possible.[5] How the world radius is determined by the world mass can be seen superficially in that[6]

$$\left(\frac{\kappa}{c^2}\right) \cdot \frac{M}{R} = \text{number (of the order of 1)}$$

$$\sim 10^{-23}$$

κ nor[mal] gravitational constant
c light velocity
M world mass
R world radius

Schlick's exposition is masterful.[7] This must surely be an inspired *teacher* there, as well. It is to be hoped that he does not remain stranded too long in Rostock.–

I am very much looking forward to your coming. It is admirable how many fine things come to your mind. Perhaps I can go out again by then. If not, I then request a little *privatissimum*.

In thanking you cordially once again for your moving kindness and concern, I am, with best regards, yours,

A. Einstein.

454.   To Arnold Sommerfeld

[Berlin,] 5 Haberland St. [after 1 February 1918][1]

Dear Sommerfeld,

I am writing you now once again and, to a certain extent, officially about the little celebrative meeting that we would like to hold in Planck's honor on Friday the 26th of April.[2] On behalf of the G[erman] Ph[ysical] Society I request that you give a speech on Planck's importance in the development of radiation and quantum theory. We have conceived the matter as follows:

1) Introductory address (by Warburg), ca. 20 min.

2) Sommerfeld (radiation & quantum theory), 25–30 min.

3) Laue (Planck's thermodynamic achievements), 20–25 min.

4) Planck's scientif. personality, 20 min.[3]

Suggested changes are gladly accepted. I beg you not to deny us your help; Planck would certainly be very pleased if you came. The trip would be paid for by the Society. I am looking forward to the evening now already, despite the gods having completely withheld from me the gift of oration, because I am very fond of Planck and he will certainly be pleased when he sees how much we all like him and how highly everyone regards his lifework.[4]

Cordial regards, yours,

A. Einstein.

455.   From Ernst Troeltsch[1]

Charlottenburg, Berlin, 4 I, Reichskanzler Place, 4 February 1918

Esteemed Colleague,

On the first day of vacation[2] I finally find the time to thank you for your very welcome and amicable letter occasioned by my Volksbund speech.[3] I do such things only out of a sense of duty, you know, and actually against my inclination. But something must be done about the terrible illusions dominating the public and about the Fatherland Party's[4] treacherous incitement of all the passions. The Volksbund was the only thing that we could manage to do.[5] The current strike is a serious test of endurance for it.[6] It is thus all the more important that our intellectuals profess their support and join it. I should ask you to do so as well. Now each and every person is needed for the cause of reason. Unfortunately, the newspapers, which have been bought up directly and indirectly by the opposing party, are very hard for us to obtain. This makes private action even more important. If the Social Democrats are eventually eliminated, a league of intellectuals could come out of it. The great delusions will certainly dissipate quickly, and then our program will gain importance and influence.

In utmost respect, most devotedly yours,

E. Troeltsch.

456.   From Gustav Mie

Halle-on-S[aale], 47 I Magdeburger St., 5 February 1918

Highly esteemed Colleague,

I do still extend my very cordial thanks to you, albeit somewhat tardily, for sending me the offprints of your latest papers![1] After the Christmas holidays, I was unfortunately very preoccupied with teaching and therefore had to postpone until now my plan of writing you. I would now like to add some comments to your essay, "Cosmological Considerations,"[2] which I read with particular interest. I see as the most important result of this study the statement: "The parallel axiom is independent of the axiom of the general transformability of the fundamental equations in physics."[3] I find this result quite extraordinarily interesting. However, I must admit that even so I personally feel no inclination toward abandoning the parallel axiom, which contributes essentially to the wonderful simplicity of the general mathematical foundations of our world, which is so complex in its finer details. What is going on at infinity in space and time actually does not matter to me at all,[4] and although I see from your writing that this is not so for all scientists, I do know that there are some other people of my persuasion. As your paper demonstrates, pondering about these things can certainly provide an impetus for very interesting reflections, but one will probably not be able to arrive at the definitive conclusions on the basis of them.[5]

Now I would like to venture a remark on the statement expressed in eq. (15) of this paper: $\rho = \dfrac{2}{\kappa \cdot R^2}$.[6] This relation seems to state that the radius of curvature R of the space is determined by the density of the total mass distributed approximately evenly within it. This claim would be wrong, though.[7] If, as is commonly done, we define the unit of mass as the mass of a specific atom (1 gr $= \dfrac{N}{16}$ oxygen atoms, where N is Avogadro's number) and define the units of length and time through a particular atomic process (1 cm $=$ 15 531.6403 $\times$ wavelengths of the red Cd line; 1 sec $= \dfrac{3 \cdot 10^{10}}{15\ 531.6403} \times$ period of this light), it is then easy to see that, for any given value of R cm for the radius of curvature of the space, the density of the world's mass distributed evenly throughout the space can have an entirely arbitrary value $\rho \dfrac{\text{gr}}{\text{cm}^3}$. If I call the value that would result from equation (15) specifically for the density, ρ_1, hence $\rho_1 = \dfrac{2}{\kappa \cdot R^2} \dfrac{\text{gr}}{\text{cm}^3}$ then, for $\rho = \rho_1$, the gravitational potential in the regions without matter between the uniformly distributed individual celestial bodies in the space would be equal to -1, $g_{11}^\infty = g_{22}^\infty = g_{33}^\infty = -g_{44}^\infty = -1$; but if ρ were to have some other constant value, the result would then simply be: $g_{11}^\infty = g_{22}^\infty = g_{33}^\infty = -g_{44}^\infty = -\dfrac{\rho}{\rho_1}$; this would be the only difference.[8] The gravitational potential in free space depends on the average density of the celestial bodies filling the space, otherwise this density has no influence on the events in the world. If desired, the units of mass, length, and time can, of course, be selected for any contents of matter in the world such that $g^\infty = -1$ results exactly; in this mass system, eq. (15) then also applies unchanged.–

I did want to write this, at the risk of imparting things long since familiar to you, to make you aware that eq. (15) can possibly be misunderstood. But I was naturally particularly interested in what you say on p. 148 about the approximation of the "real" spatio-temporal continuum with a mathematically simple space-time of constant curvature, because there you express the same idea I had presented—for a flat space-time, of course—in my third lecture.[9] I find your comparison to the geodesists' method of approximating the Earth's surface with an ellipsoid superb.[10] As indispensable as this ellipsoid is for physical geography, which must yield definite figures for the heights of mountains and the depths of the oceans, so also an "absolute" space-time scheme, in which the real world must be fitted, is equally indispensable for physics.[11] In some respects this arranging must be based on conventions, just as the selection for the Earth's ellipsoid is obviously a convention. Mathematicians tend to talk freely of "arbitrary" conventions whereas, from the standpoint of logic and epistemology, these conventions must frequently be assessed differently. As an example I would like to mention a very important convention that is always implicitly made about space and time and that is also a part of the axioms of geometry, namely: that the space-time scheme in which we fit the world must be thoroughly uniform in and of itself. I remember having once read a very insightful illustration of non-Euclidean geometry by Poincaré, in which he tentatively abandons this convention by assuming that the velocity of light in vacuous space is not the same everywhere but a simple function of position[12] and that at the same time the natural laws, atomic dimensions, atomic weights, elementary quanta, and so forth, are also altered according to laws such that the world is again described correctly. Space, which on assuming a constant velocity of light would be Euclidean, would, on assuming the new conventions, thereupon become non-Euclidean. Thus one can arrive at a description of the natural phenomena which, although more complicated, is nonetheless mathematically precisely just as correct as the usual description. Mathematical consistency—I would like to add—is by no means the same as being logically sound, though. From the logical point of view, the world system Poincaré constructs to illustrate non-Euclidean geometry is not equivalent to the world system accepted by science, for in Poincaré's system one must assume the deviations at various places in the world as actually existing, without being able to provide any reason other than Poincaré's mathematical rule. Mathematically, the requirement that the fundamental equations of physics must satisfy can be expressed in this way: They must not explicitly include the coordinates and time. This requirement is contained in the principle that Kant calls the principle of causality; it is not satisfied in Poincaré's system. However, it would be equally contradictory to the logical basic tenets of physics if one wanted to assume arbitrarily construed gravitational fields, although, as you have shown, this fiction could be followed through mathematically and even in such a way that coordinates and time do not appear explicitly in the basic equations you have formulated. Nevertheless, exactly as in Poincaré's system, deviations arising only from the mathematical rules would exist at various places in the world and thus contradict the causality principle. To demonstrate this quite drastically I have used the example of the writhing rod,[13] and I consider it also very essential that through such a frightful exemplum it be shown whither one can be led when the principle of general transformability is taken without closer reflection; for I believe that it actually has already caused quite some confusion. It even seems as though, at one place in your *Annalen* essay, in the introduction, where you discuss two spheres rotating relative to each other,[14] you yourself lost sight of home-baked logic, blinded by the mathematical spell of the general transformability of the fundamental physical equations. For, as a matter of fact, it is possible to declare both coordinate systems S_1 and S_2 as completely equivalent only when a world permeated by an arbitrarily construed gravitational field in accordance with physical laws not caused by matter is considered as equally admissible as a world whose space-time scheme is inherently homogeneous and in which only gravitational fields caused by matter occur. At least, I cannot see any other possibility and am incapable of regarding both these schemes as equivalent. If one wants to declare S_1 and S_2 as equivalent, then one must also admit that the snaking rod in my example is logically equivalent to a rigid straight rod. The same error in logic is made, in principle, except that in the example I have chosen it manifests itself a bit more drastically.

For the time being, incidentally, I do not yet want to assert definitely whether the projection of the so-called "Hilbert world"[15] into the smooth space-time scheme (which I assume as flat, whereas you prefer to attribute a curvature to it) is simply feasible by the method described in my third lecture. For it seems to me that fitting the world points into the space-time scheme would have to occur according to very specific logical and methodological rules or at the very least that, among several possibly logically admissible methods, a best method must exist. I consider it necessary first to consider according to which principles experimental physics would actually proceed in measuring lengths and times in gravitational fields, and in making the necessary corrections to the directly obtained measurement results in accordance with your theory of gravitation. It is only then that one can see which projection must be used to assign the points of "Hilbert's world" to the "physical world." I named the simple vertical projection as an example of just one possibility, without intending to propose that it is the right one.[16]

Once the railways are better accessible to civilians again,[17] I hope to be able to speak with you someday. But perhaps with this letter I shall have succeeded in clarifying the situation a bit in advance. I would be very pleased if you found the time to reply to it, so that I could attempt to familiarize myself a little with your mode of thought (which really is quite different from mine). With best regards, yours truly,

G. Mie.

[2]Deleted text in draft: "I already read with very great interest your essay "Cosmological Considerations" during the holidays."

[7]Draft formulations: "is, however, in some way only apparent"; "can be easily misunderstood"; and "its content is somewhat less notable than what one might think at first glance."

[16]Deleted text in draft: "I have been told, Felix Klein has discovered that, upon assumption of general transformability, the energy principle becomes an identity? I did not allow this story to spoil my joy in your marvelous theory, for this version of Klein's discovery seems improbable to me. I have sometimes considered Hamilton's principle as not being an independent axiom but rather derivable mathematically from the axiom of general transformability. Could this connection perhaps have played a role in Klein's discovery? We shall surely soon learn more, maybe you already know what?"

457. From Mileva Einstein-Marić

[Zurich, after 6 February 1918][1]

You will understand that with my current illness[2] it is difficult for me to come to a decision.[3] I do not have an overview of the situation—I must first accustom myself to the idea, also for the children's sake. I understand that you want an unhampered future; I don't know whether it is necessary for you and your work, but I don't want to stand in your way and obstruct your happiness. But all of it, communications, everything, would seem to me easier after the war. I've asked Dr. Zürcher to inquire into the procedure.[4] It doesn't seem so simple; I am so worried about agitations and must do everything so that we are secure and so that nothing can take this security from me, for the children's sake. Have your lawyer write to Dr. Zürcher about how he envisions it all, how the contract should be. I must leave the upsetting things to objective persons and experts in the practical matters. You see, I do not want to stand in the way of your happiness, if you are resolved, but with the illness it is harder than you think—especially nowadays—to do anything decisive.

458. From Ernst Troeltsch

Charlottenburg, Berlin, 4/1 Reichskanzler Place, 7 February 1918

Highly esteemed Colleague,

Cordial thanks for your kind information. I knew nothing about these circumstances, as you yourself presumably are safely assuming. They do make it entirely self-evident, though, that political activity, particularly for you, is out of the question.[1]

Otherwise, I am pleased to hear about the organization established by Mr. Weisbach.[2] I am all the more pleased about it since my former colleague in Heidelberg, Curtius,[3] whom I esteem very highly, is involved in it. All reason has thus not quite abandoned German academia as completely as it may initially appear, after all.[4] If only one could hear a bit about it and take note of it! Historians and jurists, at least a good proportion of the more liberal minds among the former,[5] gradually are drawing attention to themselves. It is too formidable a task to combine the essential realistic political requirements, the factual possibilities of a war situation admittedly unknown to us, and the basic political and moral standards while remaining sufficiently up-to-date in the changing situation. But it must be tried, and what is primarily lacking in Germany is the moral courage of its intellectuals.[6]

It is a pleasure to encounter any like-minded person, and so allow me to express herewith my thanks and appreciation. Yours most truly,

E. Troeltsch.

459. To Hedwig Born

[Berlin, 8 February 1918][1]

Dear Mrs. Born,

Your long letter with the consoling expression of sympathy and confidence pleased me very much. In exchange, I also would like to answer as though I were talking to myself, eradicating completely the hideous I-you gulf.

Laue wants to come here. When some time ago he had prospects of receiving a kind of privately endowed researcher position here without teaching obligations,[2] he based his efforts to come to Berlin on his distaste for teaching. Now that this plan is apparently not materializing, he is thinking of exchanging positions with your husband.[3] Therefore, primary wish: "To Berlin." Motive: ambition (of his wife?). Planck knows about it; the Ministry, hardly likely. I did not discuss this with Planck yet.[4] I imagine that his efforts are aimed at becoming Planck's successor. The poor man! Nervous vacillation. Striving after a goal that conflicts starkly with his natural desire for a calm life without complicated human relations.[5] On this, please read Andersen's nice little tale about the snails.[6] The objective likelihood of Laue's plan being realized depends on two conditions:

1) Compensation for your position sufficient for Laue,[7]

2) your husband's inclination to exchange positions.

Let us assume (1) were fulfilled, then the question arises whether you ought to consent; this obviously is the question you are agonizing over today already. My opinion is:

Definitely accept.

I probably do not need to assure you both how fond I am of you and how happy I am to have you as friends and mind-mates in this—desert. But such an ideal position in which one is entirely independent should not be declined. The professional scope there[8] is larger and freer than here, offering a better opportunity for your husband to develop his talents. But the main point is: living next to Planck is a joy. When Planck cedes his post one day, however, you are then not certain, even if you stay here, whether your husband will take his place. If it was someone else, though, it could possibly become less pleasant. All cases must be looked at face-on. You should not subject yourselves unnecessarily to that one.

Take good care of yourselves, and take me as a warning example. For me, the jump upwards is no longer attainable. Warm regards to you, your children, and your master,[9] who will be returning soon, I hope, yours,

Einstein.

460.   To Gustav Mie

[Berlin,] 8 February 1918

Dear Colleague,

I hasten to respond to your long letter,[1] from which I see that our views on the fundamentals still diverge very much.[2] First to the simplest issue. The relation[3]

$$\rho = \frac{2}{\kappa R^2}$$

certainly ought to express a relation between density and radius of curvature that is independent of any coordinate choice. This is not refuted by what you say. In the relation,

ρ means the density of the stellar matter (conceived as evenly distributed) in the world, compared to the density of liquid water.

(2)πR, the length in natural measure of the geodesically measured circumference of the world[4] (i.e., measured with the cm scale ($= 15\,531.-$ wavel[ength] of the Cd line))

That this is the meaning of ρ and R according to my papers is easily drawn from my considerations. The crux of the matter is that R must not be conceived as a "coordinate length" but as a length in "natural measure," and thus is defined, just as ρ, independently of the coordinate choice.–[5]

What divides us most of all is ultimately the circumstance that you do not take the standpoint I call relativistic: the behavior (continuation of the temp[oral] course of states) of every single physical body is such that it is uniquely determined by its own state and by those of all the other bodies (this statement must, of course, be interpreted without instantaneous action at a distance). Only the observable objects of our physical world are to be understood here as "natural bodies." The varying behaviors of S_1 and S_2 (in my pamphlet)[6] must be explained in that the state of motion of the remaining world of bodies relative to S_1 is different from the one relative to S_2. Any other interpretation is not relativistic. (Mach has already recognized this very clearly.)

Now, if you do not share this relativistic point of view, then *I do not understand in the least why you assign any significance to the general covariance requirement.*[7] Either the hypothesis that in physics only relative motions exist is taken seriously, i.e., that only *relative* motion appears as efficacious in laws of nature, or absolute space is introduced as an independent element in the causal construction; a compromise point of view between these two does not exist.

In justification of the "Cosmological Considerations," the following:[8] What happens at infinity is all the same to me as well. Nevertheless, I must know whether what I interpret as the relativity principle can be thought through without encountering contradictions.[9] For this, it is necessary that the gravitational field be determined entirely by matter[10] which, however, is not true for the quasi-Euclidean world. The latter rather has the following properties:

1) The mean density (in natural measure) of matter inside a spherical surface must go to zero with the spherical surface's radius. (That is, the world is essentially empty and has a center.)

2) Its boundary conditions at infinity are not covariant under arbitrary transformations, i.e., not relativistic. The inertial phenomena and metric properties
of the space-time continuum are basically determined by these nonrelativistic
boundary conditions, not by the *bodies*.

These deficiencies are eliminated with the λ-hypothesis. According to it, there can
be no singularity-free $g_{\mu\nu}$-world without matter.[11] The $g_{\mu\nu}$'s rather seem entirely
defined by matter, obviously up to the 4 arbitrary functions which correspond to
the *physically* ⟨meaningless⟩ vacuous choice of coordinates. You should not say
that I prefer the curved world (over the flat one) but that I prefer the matter-filled
world over the *essentially empty* one. Matter necessarily causes curvature.

I do *not* agree *at all* with your reflection about the bent (flapping) rod.[12]
All physical descriptions that yield the same observable relations (coincidences)
are equivalent in principle,[13] provided that both descriptions are also based on
the same laws of nature. The choice of coordinates can have great practical
importance from the point of view of clarity of description; in principle, though,
it is entirely insignificant. It means nothing that "arbitrary gravitational fields"
occur, depending on the coordinate choice; it is not the fields themselves that
lay claim to reality. They are merely analytical auxilliaries in the description
of realities; in principle, one can actually only learn something about the latter
by eliminating the coordinates. The ghost of absolute space haunts your rod
example; the argument works "*ad hominem*" but does not hit the mark at all
as I interpret it. It is not a question of a violation of logic. That a rod must
either be straight or bent but cannot be both corresponds exactly to the objection
advanced by philosophers against the special theory of rel. that the same body
(at the same ⟨instant⟩ space-time point) cannot be simultaneously at rest and in
motion.

Are my sketchy remarks clear enough, I wonder? Once the world has wo-
ken up again from its miserable nightmare, once the railways can again be used
with pleasure,[14] and provided we both are still alive, let us make our planned
conversation a reality.

With best regards, I am yours truly,

A. Einstein

Read Schlick's booklet on space & time.[15] Dr. Berliner is sending it to you![16]

461.   From Emil Warburg

Charlottenburg, 25b March Street, 8 February 1918

Highly esteemed Colleague,

Many thanks for your amiable letter of the 5th this mo. I have arrived at somewhat different interpretations, though.

I. In the + pile, just as many anions (electrons) wander inside to the cathode side as cations to the anode side. For some reason, about which I could only speculate, these ions do not recombine with one another.

In the following I consider only the case of an entirely "elastic" collision between electrons and gas molecules.

II. The electric current of the unit of volume and [sc.] is denoted as

$$i \cdot \Delta$$

and is expended in two ways:

1) Also for an entirely elastic collision, the electrons transmit energy (as heat) to the gas molecules.

2) By ionizing a gas molecule, an electron loses its field energy, one of the two electrons recombines again immediately, whereby the electron's pot. energy (field energy) is converted into radiation and heat.

Thus one can set

$$\Delta = \Delta_1 + \Delta_2;$$

for Δ_1, I find $\Delta_1 < \dfrac{3RT}{2,9650} \cdot \dfrac{1}{l} 4 \cdot \left(\dfrac{q_2}{q_1}\right)^2 \cdot \left(\dfrac{m_2}{m_1}\right)^2$

R gas const., T abs. temp., l av. length of path of the electrons, q, m velocity and mass 1 relating to the gas molecule, 2 to the electron.[1]

For nitrogen of 1 cm pressure, I find $\Delta_1 < 0.002$ Volt/cm, whereas the observation $\Delta = 27$ Volt/cm resulting in Δ_1 is thus negligible.

I am engaged—as far as my time permits[2] [I have not been able to occupy myself much with this matter up to now.]—in calculating Δ_2 out of the ionization gap, pressure, etc. The equation obtained must contain the solution to the problem, or resp., must show whether the assumptions made are admissible. More specific justification of [this] I reserve for oral discussion. Unfortunately, I am probably not going to be able to visit you on Sunday.

With cordial regards and wishing a speedy recovery,[3] yours most truly,

E. Warburg.

462.   From Max Planck

Grunewald, 13 February 1918

Dear Colleague,

It is how I had suspected. If for a quasi-elastic central motion the Lagrangian function is $L = T - U$ (T kinetic, U potential energy) and m is the mass of the oscillating point, then $-\dfrac{\partial L}{\partial m}$ is the "force" that must be exerted externally so that m remains constant during the motion.

If, furthermore, a is the constant for the quasi-elastic force, then $-\dfrac{\partial L}{\partial a}$ is the force that must be exerted externally so that a remains constant.

(With natural motion, m and a obviously are constant because they have *prescribed* values, in the sense of firm conditional equations linking the coordinates.)

If, now, m and a are simultaneously altered such that the relation $\dfrac{m}{a}$ remains unchanged, and the period along with it, then

$$\frac{\delta m}{\delta a} = \frac{m}{a}.$$

Then the total work exerted from outside upon the system is:

$$A = -\frac{\partial L}{\partial m}\delta m - \frac{\partial L}{\partial a}\delta a = -\frac{\delta a}{a}\left(\frac{\partial L}{\partial m}m + \frac{\partial L}{\partial a}a\right) = -\frac{\delta a}{a}\cdot L,$$

because $\dfrac{\partial L}{\partial m} = \dfrac{T}{m}$ and $\dfrac{\partial L}{\partial a} = -\dfrac{U}{a}$. Consequently, the temporal mean value $\overline{A}$ for one period $= 0$ (because $\overline{T} = \overline{U}$).

From this follows that, for the reversible adiabatic change in state under consideration, the energy of the system $(T + U)$ does not change, hence neither does its half: $\overline{T}$. And since the period also does not change, the time integral of T does in fact remain invariant throughout a period. I thus think that the adiabatic hypothesis can very well be applied to changes in *mass* without encountering an inconsistency.[1]

Good-bye until Saturday the 23rd.

Cordial regards, yours,

Planck.

463.   From Rudolf Förster

Essen, 6 Kunigunda St., 16 February 1918

Esteemed Professor,

I thank you kindly for your nice letter of 17 January.[1] Again I could not make up my mind to answer it immediately, since very many questions I hoped to clarify were passing through my head. Now that the full extent of the mathematical difficulties has become evident to me, however, I cannot postpone it any longer.

Your objections to the expositions in my earlier letter[2] apparently are based on misunderstandings for which, owing to my so frugal choice of words, I myself am at fault.

1) $B_{12,34} = B_{13,24} = B_{14,23} = 0$ should not be a satisfactory condition but only a necessary one which allows ds^2 to be changed into the orthogonal form.

2) I derived directly from the transformation formulas my generalization of the Riemann tensor $B_{\mu\nu,\rho\sigma}$ to the case of an asymmetric fundamental tensor, without resorting to splitting it into a symmetric and an antisymmetric component. The occurrence of the connected *symmetric* contravariant tensor $s^{\mu\nu}$ automatically results as a requirement.

3) If $g_{\mu\nu}$ is reduced in the specified manner to $s_{\mu\nu} + a_{\mu\nu}$ and forms the $g^{\mu\nu}$'s (or $s^{\mu\nu}$'s or $a^{\mu\nu}$'s) from the determinant of the $g_{\mu\nu}$'s (or $s_{\mu\nu}$'s, or $a_{\mu\nu}$)'s, then *under no circumstance* is $g^{\mu\nu} = s^{\mu\nu} = a^{\mu\nu}$. Rather, $g^{\mu\nu} - s^{\mu\nu}$ depends not only on the six tensor of the $a_{\mu\nu}$'s, but also on all the $s_{\mu\nu}$'s besides in a quite complicated way. It is thus not correct that in my tentatively suggested electromagnetic field equations the symmetric and antisymmetric components separate like oil and water. This separation occurs, in the form I have used, only in the 2nd Maxwell equation; in the first one, the $g^{\mu\nu}$'s enter completely. I had unfortunately omitted to emphasize this explicitly.

4) My considerations on "gravitational fields that are derived from a potential" have of themselves nothing to do with those on the asymmetric fundamental tensor; indeed, they are in conflict with them to a certain extent. Whereas the introduction of the asymmetric tensor is a generalization, that of the potential is a specialization. Your objection that the field energy depends essentially on the first *derivatives* of the $g_{\mu\nu}$'s does admittedly apply to my $a_{\mu\nu}$'s but not to my potential.[3]

I have abandoned asymmetric $g_{\mu\nu}$ again for another reason besides, which I want to go into now.

In a note to appear shortly in the *Astronomischen Nachrichten* under the pseudonym *Rudolf Bach*,[4] I conduct probability considerations on the attractive force of an infinite stellar system and find that the *physical* objection to the infinity of the universe and the mass contained within it, based on Newton's law implying an infinite force, is invalid. On the other hand, I have also not been able to convince myself of the necessity of assuming a finite world on *mathematical* grounds.[5] You think that the boundary conditions cause problems at infinity. In the general theory of relativity, however, since it interprets the world with differential equations, thus with function theory, infinity ($x^1 = \infty$, $x^2 = \infty$, $x^3 = \infty$, $x^4 = \infty$) is not a limiting case for a large sphere, perhaps, but for a *point* just like any other point and can be transformed into any other point.– Thereupon I reflected on which boundary conditions would then be at all suitable to define clearly a solution to the system of field equations. For this purpose [: in the case of geodesic coordinates[6] $g_{i4} = 0$; $g_{44} = 1$, which is not a limitation of the generality, in connection with the equally admissible assumption that the coordinate system is "*quasi-orthogonal*" in the vicinity of the observed point $x^i = x_0^i$ ($i = 1, 2, 3, 4$) (: i.e.: the g_{ii}'s differ from 1 just by infinitesimal amounts of the 2nd order, the other g_{ik}'s are themselves in the 2nd order :) :],[7] I tried to apply the system of field equations to *Tresse's* canonical form of a "passive system" (*Encyclop. d. math. Wiss.* II A 5, No. 2),[8] and in doing so found that, in order to single out a solution, the following can be prescribed arbitrarily at the origin: (a) All derivatives of g_{12}, g_{13}, g_{23} with the exception of each of the following four fourth derivatives:

$$\frac{\partial^4}{1144}, \frac{\partial^4}{2244}, \frac{\partial^4}{3344}, \frac{\partial^4}{4444}$$ and all of their derivatives, in addition, (b) the derivatives of g_{33} with the exception of six specified ones (of 2nd–4th order) and their derivatives, in addition, (c) those of g_{22} with the exception of 5 determined ones (of 2nd and 3rd order) along with their der[ivations], and finally, (d) those of g_{11} with the exception of those of the 2nd order according to 22, 33, 44, 23, 24, 34 and their derivatives. From this, the arbitrary functions appearing in the general solution can also be deduced (: The ranking used here: g_{12}, g_{13}, g_{23}, g_{33}, g_{22}, g_{11} is obviously arbitrarily chosen :). Certainly a very complicated boundary value problem.

   With this the matter is far from finished, though. The system of your field equations conceals many more surprises. I would just like to point to one. Through adept elimination, 20 equations of the 4th order can be derived from your general equations, which (except for lower-order derivatives of all the $g_{\mu\nu}$'s) contain 4th-order derivatives just for each of *4* (or 3) of the unknowns ($g_{\mu\nu}$), namely for each only a second derivative of the first differential parameter. In the above-mentioned geodesical, quasi-orthogonal coordinates, in the main, 6 of these equations remain of interest, for which the highest-order terms read:

$$\frac{\partial^2}{44}\left(\frac{\partial^2 g_{ik}}{11} + \frac{\partial^2 g_{ik}}{22} + \frac{\partial^2 g_{ik}}{33} + \frac{\partial^2 g_{ik}}{44}\right) \quad (i, k = 1, 2, 3) \text{ (These terms thus contain}$$

only *one* unknown each of the 4th order.). Unfortunately, they then also contain some nasty terms of the 2nd order that are quadratic to the other unknowns.

On this occasion, I also discovered some differential relations between the $B_{\mu\nu}$'s, or B_σ^α, or $B_{\alpha\beta,\gamma\sigma}$. Those between the B_σ^α's lead to your energy-momentum law and read (without the constraining condition⟨s⟩ $\sqrt{-g} = 1$):[9]
$\frac{\partial B_\sigma^\alpha}{\partial x^\alpha} - \frac{1}{2}\frac{\partial B}{\partial x^\alpha} = B_\sigma^\alpha \cdot \Gamma_{\alpha\beta}^\beta - B_\beta^\alpha \Gamma_{\alpha\sigma}^\beta.$ These relations are the direct consequence of more general, apparently still entirely unknown relations:[10]
$\frac{\partial B_{ik,lm}}{\partial x^n} + \frac{\partial B_{ik,mn}}{l} + \frac{\partial B_{ik,nl}}{m} = \Gamma_{kl}^\alpha B_{\alpha i,mn} + \Gamma_{km}^\alpha B_{\alpha i,nl} + \Gamma_{kn}^\alpha B_{\alpha i,lm} -$ (3 analogous terms with i and k exchanged.). I am probably going to take the opportunity to publish this, after all.

My calculations showed me now that the field equations are inadequate for determining the $g_{\mu\nu}$'s, that is, not even to the extent of allowing general arbitrariness for the coordinates. Yet the *space-time geometry* in the vicinity of a point should be determined entirely by the energy state—etc.—prevailing within it. Hence *conditions* must necessarily *be added* to your field equations. For this, it seems to me, my condition of potential is very suitable, which states that a four vector A^μ must exist such that $g_{\mu\alpha}\frac{\partial A^\alpha}{\nu} + g_{\alpha\nu}\frac{\partial A^\alpha}{\mu} + A^\alpha\frac{\partial g_{\mu\nu}}{\alpha} = 0$ (: or, as far as I am concerned, equal to another 2nd-order tensor :).[11] For it seems to me that, on the basis of this assumption, the $g_{\mu\nu}$'s are definitely determined essentially by the energy tensor and the electromagnet. field (: if the electromagnet. field is derived from A^μ through the formulas $\varphi_\sigma = g_{\mu\sigma}A^\mu$; $F_{\rho\sigma} = \frac{\partial \varphi_\rho}{\sigma} - \frac{\partial \varphi_\sigma}{\rho}$, as I have done in my last letter :).

All these reflections are terribly drawn out and tricky, and I am almost inclined to give up hope of arriving at a positive result here with the current state of mathematics, with my lack of access to the literature and with my limited time. I cannot spend my 17-day vacation on a thorough study of mathematics either, disregarding that nothing would come out of it in such a short time. I therefore would like to ask you to direct your attention to these problems yourself or, if your time also is taxed too much otherwise, to refer some mathematicians to your equations. My esteemed teacher Hilbert[12] will certainly find some results here. The author of the Encyclopedia article cited above, Mr. *Ed. von Weber*,[13] might possibly be in the position to shake some theorems effortlessly out of his rich store of special mathematical skills, which might illuminate the applied boundary value problem.

In the hope that your health is fully reinstated again,[14] I am very respectfully yours truly,

Dr. Rud. Förster.

464.   From Arnold Sommerfeld

Munich, 16 February 1918

Dear Einstein,

Such an amiably reiterated invitation calls for a resounding yes![1] I shall thus do my best: Even though you naturally would have much more to say about quanta and radiation than I, and even though your eloquence is much superior to mine. But I assume that you are supposed to spare yourself and perhaps will say the closing personal note. I am faced with a certain difficulty in that I have already written a laudatory article on Planck for the *Naturwiss.* which will have been printed by then;[2] so I myself have already drained some of the water from my torrent of words.

You write nothing about your state of health; your previous letter just said that you deserved no sympathy. I hope you are soon over the worst and then are healthier and more cheerfully industrious than before![3]

So until we meet again, latest in April, yours,

A. Sommerfeld.

465.   From Gustav Mie

Halle, 47 I Magdeburger St., 17 February 1918

Dear Colleague,

The statement that you denote as the basis of the "relativistic" stance, "the behavior of every single physical body is such that it is uniquely determined by its own state and by those of all the other bodies,"[1] still seems to me to allow various kinds of interpretations. If I did not happen to know that you mean something else by it, I would believe that this statement only contains the causality principle, just as I interpret it and use it as a basis for my theory of matter.[2] Thus the difference between our views certainly cannot be formulated in this way. If you would like to become acquainted with my standpoint, please just read through the last section: "VIII. Relations to Philosophy," in the booklet by Dr. Schlick sometime, for the mailing of which I thank you heartily. This section was not printed in the journal *Die Naturwissenschaften*, as Dr. Berliner writes me.[3] It coincides with my views so thoroughly that I am almost astonished that someone else has written it. I concur thoroughly with the author especially in his criticism of Mach which, despite its very friendly tone, is really quite sharp in content.[4] Above all, I emphasize the following statements (on p. 62 of the booklet): "Now, it is possible, however, to describe *the same* factual findings by *various* judgment systems, consequently, various theories can exist for which the criterium of truth applies in the same way, which thus all can take care of the observations to the same degree and lead to the same predictions. Various notation systems are simply assigned to the same objective reality, and various forms of expression convey the same factual findings. *Among all the possible approaches that contain the same core of truth in such a way, one of them has to be the simplest; and that we give preference to this one in particular is not based merely on practical economy, on a kind of mental convenience (as has sometimes been thought), but is logically grounded in that the simplest theory contains a minimum of arbitrary factors.*"[5] It is exactly in this sense that I spoke of "home-baked logic" in my letter to you, which could be lacking in a mathematical theory because mathematics is, as you know, in the position to make arbitrary definitions to suit itself.[6] I could not possibly think you capable of a gross logical error, as some philosophers evidently have done in assessing your old theory of relativity;[7] it would not be in keeping with my opinion of you. When one says, "the Earth rotates," a large group of phenomena—Foucault's experiment, the trade winds, the deflection of a falling stone, etc., etc.—are summarized most briefly in this judgment.[8] Naturally, according to your principle of the transformability of the fundamental equations, it is possible to adopt a coordinate system in which the Earth is at rest, but then the scheme of ⟨spatial⟩ space-time order in which the objects are placed must be assigned certain characteristic $g^{\mu\nu}$'s which, to use Schlick's expression, contain "arbitrary elements" [*willkürliche Momente*] since they are *not determined by the presence of real bodies*; thus, although the representation describes the phenomena correctly, it now becomes unnecessarily complicated and one must therefore reject it.[9] Hence the expression "the Earth rotates" will always retain its reasonable content even despite general transformability. I would like to demonstrate this from another angle as well. Take a field of an electron at rest. According to the principle of relativity of linear motions, through a linear transformation I can immediately obtain from it a new integral of fundamental equations which represents the "uniformly moving electron." We all know of what significance this has been for experimental physics in particular, which certainly deals with real factual findings. Now, is it possible also to derive the integral that represents the rotating electron from the general transformability principle? I believe it would be of great importance for science to obtain this integral, but it simply cannot be done. There you see quite clearly that the principle of general transformability is not a principle of general relativity.[10] It is exactly this that I considered necessary to express once, because some actually have gained the impression from your presentation that with the new principle the fields of arbitrarily moving electrons could now be obtained. I heard this view mentioned even in Göttingen at the discussion.[11] If this were so, if there really were a principle of general relativity, then this principle would absolutely have to read as it was recently formulated in the *Annalen* by Dr. Kretschmann, and I consider it of great merit that he has demonstrated for once in full detail that *such* a general principle of relativity absolutely cannot exist.[12]

Nevertheless, I am a very enthusiastic supporter of your theory of gravitation. I can only explain your mystification at this by the probably somewhat incomplete presentation of my ideas. Most importantly, in my Göttingen lectures I expressed merely in words the principle that, in my opinion, is the core of your theory of gravitation;[13] I did not yet express it in mathematical formulas, though. These formulas can also be formed, however, without the least difficulty in principle, and I intend to do so as soon as I have free time for these investigations. I would also be very pleased if my ideas could offer you some inspiration, just as you have stimulated me so much already.[14]

19 February 1918[15]

In the meantime I have been thinking a bit more about your curved space. You are entirely right with your assertion $\rho = \dfrac{2}{\kappa \cdot R^2}$; I was mistaken.[16] On the other hand, I now have severe doubts about whether your interpretation of this equation really is correct. Surely the field equations $G_{\mu\nu} - \lambda g_{\mu\nu} = -\kappa(T_{\mu\nu} - \frac{1}{2}g_{\mu\nu}T)$ should not just have statistical validity, so to speak? For then they would not be fundamental equations. But if they are valid in every space element, the integral you have found would surely have to be interpreted such that the entire space, wherever there are no ordinary accumulations of matter, must be permeated evenly by a *continuous*, extremely fine material haze of density ρ. For in regions that are at an "infinitely remote" distance from any matter, thus in the expansive empty spaces between the fixed stars, the gravitational potential must satisfy the condition that the space exhibits the same geometry throughout. It can be flat, for this the condition is $\lambda = 0$, then $\rho = 0$; or on the other hand, it can have a constant radius of curvature R, then λ must differ from zero and then, as you have proven, $\kappa \cdot \rho = 2\lambda$. Hence the point is simply this, that in the flat space the atmosphere of highly rarefied energy, which surrounds the mass particles, converges very rapidly toward $\rho = 0$, while in the curved space, on the other hand, it converges very rapidly toward $\rho = 2\frac{\lambda}{\kappa}$. This seems to me to be the only difference. Nothing can be stated, however, as it seems, about the amount of discrete material particles in the world, also not in the case of curved space. In the latter case there would be conceivable a world in which there were only 1 electron or 1 atom, while the entire remaining region would be permeated continuously by an extremely thin material haze of density ρ; but there could also be an entirely arbitrary number of atoms in the world.[17] If you want to interpret ρ in your equation $\kappa \cdot \rho = 2\lambda$ as the average density of the celestial bodies' distribution of mass in space, then in my opinion you must first show that a value for the gravitational potential can exist in the vast empty intermediate spaces, where one actually has to set $\rho = 0$, which satisfies the condition of the continuous uniformity of these spaces. If no such integral solution for the equations $G_{\mu\nu} - \lambda g_{\mu\nu} = -\kappa(T_{\mu\nu} - \frac{1}{2}g_{\mu\nu}T)$ can be found,[18] then the interpretation that I just described must be the only correct one. Then, however, the assumption of curvature could offer no apparent advantage. What is immediately conspicuous about this theory and speaks little for it is, of course, that the principle of relativity of uniform motions becomes invalid within it. For with the cylindrical curvature assumed by you for the four-dimensional space-time region, the time axis as a cylindrical axis is preferred over the three spatial axes ⟨in the cylinder cross-section⟩, which the three cylinder circles form. If the theory of cylindrically curved space-time were correct, it would have to be possible to conduct experiments that did not yield a purely negative result, as did the Michelson experiment, provided, of course, that we could attain the necessary precision. This seems to me to be a very important difference of fundamental principle between the theory of flat and of curved space.

  With best regards, yours truly,

Gustav Mie.

[7]Deleted line in draft: "the same body cannot be simultaneously at rest and in motion."

  [14]Deleted text in draft: "I hope . . . I can win you over to this interpretation as well."

  [15]Draft passage: "I am quite shocked to note now that in the new theory absolutely none of the principles that I had seen as the nicest in your theory of gravitation are valid anymore. The principle of the relativity of motion certainly is no longer valid, because within the universe itself the time axis plays a different role from the three spatial axes; hence there must be experiments that do *not* produce a negative result like Michelson's

experiment, provided that adequate precision can be attained. But above all, as far
as I can see, even the principle of the relativity of the gravitational potential becomes
invalid, because a particle is capable of existing but at one place, where ρ and thus also
the gravitational potential has a specifically prescribed value. In places where the mass
in the world does not have the exactly prescribed density, it must be in a constant
state of internal change, which has the purpose of correcting the mass distribution,
quite irrespective of the effects of the forces, as I would see it, solely as a consequence
of the mismatched gravitational potential. This is quite different from diffusion as a
result of irregular motions, through which the uniform distribution of the molecules of
a gas is brought about in space. I must admit that now after these considerations I
have completely lost all confidence in the new theory's possibility and strongly doubt
whether any satisfactory theory can be found in which the parallel axiom is not valid."

Draft postscript: "In order to give you the opportunity to test whether my last
mentioned assumption might just rest on fallacies, I would like to describe to you
briefly how I arrived at it:

Let: $\kappa \cdot \rho = 2 \cdot a \cdot \lambda$, where $a \neq 1$ is naturally positive, real.

I now set $\alpha = \frac{1-a}{1+a}$, then results as an integral of the field equations: $G_{\mu\nu} =$
$-\kappa \cdot \left(T_{\mu\nu} - \frac{1}{2} g_{\mu\nu} \cdot T \right)$ in the vicinity of the point $x_1 = x_2 = x_3 = x_4 = 0$ the
following:

$g_{\mu\nu} = - \left(\delta_{\mu\nu} + \frac{x_\mu \cdot x_\nu}{R^2 - (x_1^2 + x_2^2 + x_3^2 - \alpha \cdot x_4^2)} \right) \quad \mu, \nu = 1, 2, 3.$

$g_{\mu 4} = - \frac{\alpha \cdot x_\mu x_4}{R^2 - (x_1^2 + x_2^2 + x_3^2 - \alpha x_4^2)} \quad \mu = 1, 2, 3.$

$g_{44} = \left(1 - \frac{\alpha^2 x_4^2}{R^2 - (x_1^2 + x_2^2 + x_3^2 - \alpha x_4^2)} \right)$, where $R^2 = \frac{(1+\alpha) \cdot (2-\alpha)}{2\lambda}$

g_{44} thus changes at point $x_1 = x_2 = x_3 = 0$ with the time; specifically, it always gets
smaller, whether $a > 1$ or $a < 1$ is true."

[18]Deleted passage in draft: "In any case your explanation, that ρ is the mean
density of the distribution of the celestial bodies, appears to me to be very disputable
and in need of a more definite proof. What does the integral look like at places where
ρ really is supposed to be zero? It seems to me impossible to be able to claim there the
existence of a field at equilibrium that obeys the equations $G_{\mu\nu} - \lambda g_{\mu\nu} = 0$ and that
at the same time exhibits the properties of the field at a point located very far away
from all matter, as your integral does. You yourself prove that a gravitational potential
satisfying the conditions that must be set for very remote distances from all matter is
consistent only with the condition $\kappa \cdot \rho = 2\lambda$, don't you? Besides, I must admit that I
also find curved space unappealing."

466.   From Max von Laue

Feldberg (Black Forest), 18 February 1918

Dear Einstein,

I have received here your forwarded postcard of the beginning of this month and your letter of the 13th of this mo. Many thanks for the extremely welcome news.[1] As soon as I am home again, and this I am going to be on the 24th of this mo., I shall write to Planck. I would like to wait until then in order to discuss these things with my wife[2] and because there I have access to the typewriter again, which is of advantage to *everyone* involved. Should the letter to Planck be urgent, please telegraph me here as soon as possible; the address: "Feldberg (Black Forest)" suffices.

It seems to me of more urgency to write you about the Planck celebration.[3] The task of portraying Planck's meritorious contributions to thermodynamics is very honorable, but quite difficult for me. For it, one must have actually experienced the development of thermodynamics, shall we say, from 1875–1895. Since this is not the case for me, I would certainly have to conduct *thorough* studies, especially since I heard earlier that personal disagreements arose at that time over thermodynamical problems which, from what one hears, have not been overcome in all quarters to the present day.[4] Now, please do not take this as though I wanted to doubt Planck's merits in the field of thermodynamics in any way; it just seems to me necessary to weigh this very carefully. By no means do I want to decline the talk either. However, I attach a condition to it. If I do not make any proper headway with it, then you Berliners must allow me to come to Berlin and ask you there for advice; I also count on Warburg's[5] help a bit in this; for he obviously knows the mentioned time span from personal experience.

Finally, I would like to inform you *confidentially* that the Natural Sciences Faculty at Frankfurt has decided to award Planck an honorary doctorate of science [*Dr. phil. nat.*] for his 60th birthday and wants additionally to have an address formally presented to him.[6] Now, if I am to give a speech in Berlin on the 26th, I would like to come over on the 23rd already and would deliver the diploma and the address myself. Can you write me whether occasion for this presents itself? If, for example, you and Rubens[7] went to visit Planck in his apartment on the 23rd, could I join you?[8] But please do not let Planck know about this in advance. You can discuss this with Rubens; he is anyway already informed.

With cordial greetings, yours,

M. von Laue.

467.   To Rudolf Förster

[Berlin,] 19 February 1918

Esteemed Colleague,

I was very pleased with your new letter,[1] from which I see how actively you are involved with the problems of relativity. I am sorry that you must devote a major part of your energy to such pathetic endeavors! If you follow your ideas through and publish them, though, we can hope that you will succeed in exchanging your profession with that of a teacher, so that you no longer have to do what is becoming the curse of mankind.[2] I shall reply in sequence again:

1) It seems to me that $B_{12,34} = B_{13,24} = B_{14,23} = 0$ is valid *for an arbitrary coordinate system* only when $B_{ik,lm} = 0$ for any choice of the indices.

2) and 3). Here you are entirely right. I had overlooked that the relation

$$g_{\mu\nu} = s_{\mu\nu} + a_{\mu\nu}$$

does *not* correspond to the relation

$$g^{\mu\nu} = a^{\mu\nu} + s^{\mu\nu}.$$

Your generalization of the Riemann tensor thus seems not to be trivial, no less than your attempt at a novel formulation of electromagnetics. Otherwise, we both seem to be abandoning the nonsymmetrical $g_{\mu\nu}$'s for the reasons mentioned.

Obviously, I can only form a judgment about the article you are publishing under the pseudonym[3] when I have it in front of me. There is *one* thing I would just like to point out, though. The infinitely remote can be regarded as a point like any other[4] only when the world is not infinitely large, when measured with a cm-gauge, that means, when the $g_{\mu\nu}$'s degenerate at ∞ in such a way that[5]

$$\int_{-\infty}^{+\infty} \sqrt{-g}\, d\tau = \text{finite}$$

$$\int_{-\infty}^{+\infty} \sqrt{g_{11}}\, dx_1 = \text{finite for any } x_2, \, x_3, \, x_4,$$

etc., so that, for ex.,

$$\lim_{x_2 = +\infty} \left(\int_{-\infty}^{+\infty} \sqrt{g_{11}}\, dx_1 \right) = 0.$$

Then an arbitrarily small body can fill the "infinite realm" and the world is *only apparently* infinite (just owing to the choice of coordinates). If, however, the $g_{\mu\nu}$'s do not reduce to zero in the indicated way for infinitely large x_μ's

($\mu = 1, 2, 3$) then, metrically, the infinite realm does not have the nature of a point and cannot be transformed into finitude, without the appearance of an inadmissible singularity. But if such a transformation is possible, *then I simply name the world finite* (i.e., finite as measured in standard measure). Then it can be represented as a continuum closed in upon itself, and special boundary conditions are eliminated. I claim that, from a consistently relativistic point of view, this is the only possibility.

Your remark on the continuation of the $g_{\mu\nu}$'s from one position seems to me to be supported by a physically illegitimate postulate. The legitimate requirement reads as follows: It must be possible to continue the universe beyond a spatially complete cross-section by means of the differential equations. This continuation must be uniquely determined, apart from the ambiguity attached to the free choice of coordinates.[6] It is very improbable to me that the φ_ν's can be subjected to additional conditions in the manner you believe. The relation

$$g_{\mu\alpha}\frac{\partial A^\alpha}{\partial x_\nu} + g_{\nu\alpha}\frac{\partial A_\alpha}{\partial x_\mu} + A_\alpha\frac{\partial g_{\mu\nu}}{\partial x_\alpha} = 0$$

changes into an evidently invalid condition for the electrical field at an infinitesimal gravitational field $\left(\dfrac{\partial g_{\mu\nu}}{\partial x_\alpha} = 0\right)$.

Best regards, yours,

A. Einstein.

468.   From Hermann Coenen[1]

Breslau XVI, 21 February 1918

Esteemed Mr. Einstein,

I was delighted with your book, which appeared with Vieweg,[2] and thank you very much. I already had this book of my own accord and have thus passed on the first copy to the Library of the Roy. Surg. Clinic. Your gift copy to me, however, I shall keep with especial pleasure but request the kindness of also writing a little dedication in it. I hope that, with your loaded schedule, this does not inconvenience you too much.

Your handwritten message pleased me even more than the book, though, and in the next few days I shall have the occasion of going over your figures with Prof. Wilkens, the astronomer at the local observatory,[3] since I would get stuck alone.

Should you be acquainted in Berlin with my friend Prof. e[xtra]o[rdinarius] and Health Councillor Moritz Borchardt, 6 Dörnberg Street, Director of the Surgery

Department at Virchow Hospital,[4] I would also be glad about that. Borchardt recently performed an operation for an Einstein family. But that probably was another one. Nonetheless I asked him, if it can be arranged, to invite you and me over one evening, for I really would like to make your acquaintance.

Again accept my hearty thanks for everything and do not forget to write the dedication in the book and then return it to me again.

I wish you much success in your continued researches and have great expectations of it, as you yourself also do, I am sure.

With best wishes for your well-being and for unslacked energy in your work, I am yours most truly,

Prof. H. Coenen
Head Doctor of the Roy. Surg. Univ. Clinic.

469.   From Heinrich Zangger

Zurich, 21 February 1918

Dear friend Einstein,

I did not want to bother you with questions. But after your last postcard, a clarification is necessary. Most importantly, you must know what the situation is here. We were hoping that the sister would assume the domestic chores. However, she has become melancholic and had to be attended to in an asylum.[1] In the interim a nurse is here who is taking care of the household. If no suitable individual can be found for a cheaper wage,[2] I would again suggest that your wife be admitted to hospital.[3] Albert would come to us on Berg Street until the summer holidays,[4] as soon as my wife has returned. (Our little daughter is presently ailing from a lung infection at Samaden District Hospital.)[5] We would put your little boy in a children's residence in a healthy location, such as we have in great numbers here in Switzerland. I hope that this would be feasible for Fr 5 and $5\frac{1}{2}$ per day.[6]

With greetings from your respectful colleague,

Zangger.

470. To Gustav Mie

Dear Colleague,

First of all, to what I called the "relativistic stance" in my last letter.[1] I maintain that this requires the following interpretation regarding inertia. If L is the actual path of a certain freely moving body and L' is one deviating from it 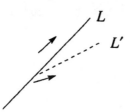 with identical initial conditions, the relativistic point of view requires that the actually described path L be preferred over the, from the logical point of view, equally possible paths L', on the basis of a *real cause*, which has the preference of L over L' as a consequence. According to the theorem you recapitulated, nothing but the (relative) positions and states of motion of all the remaining bodies present in the world can act as such a real cause. These must determine entirely and uniquely the inertial behavior of our mass. Mathematically this means: the $g_{\mu\nu}$'s must be determined *completely* by the $T_{\mu\nu}$'s—up to the 4 arbitrary functions, of course, which correspond to the free choice of coordinates.[2]

This requirement is not satisfied by Newton's theory, but also just as little by mine *as long as the world is conceived as quasi-Euclidean*. For then the $g_{\mu\nu}$'s are predominantly fixed by nonrelativistic boundary conditions at infinity. Then no real cause exists for the preference of path L over certain other L''s (rectilinear ones against non-Galilean rigid coordinate bodies). It is in this sense that I said that Newton's theory violates the causality requirement; but Schlick is right when he finds fault with this form of expression.[3]

I agree fully with the quote from Schlick[4] but not with the use you make of it. I do not deny that the description of the world proves simpler when a reference system is introduced relative to which the Earth rotates in a specific way. But I do contest that the corresponding preference of *one* coordinate system has essential significance. You say, in one case the $g^{\mu\nu}$'s must be assigned particular properties not determined by the matter and that, on the other hand, this was not the case for a "natural coordinate choice." *If you were right in this, though, I would regard my standpoint, indeed, my whole theory altogether, as untenable.* But let us see how this holds up.

When we consider the solar system or, perhaps, the Milky Way, in any case, only a part of the universe, then the differential laws are the same in both cases (for both coordinate choices); only the boundary conditions (for the spatial boundaries of the considered system) are different. Now you will say: In one

case (Galilean coordinate choice), $g_{\mu\nu} = \begin{array}{cccc} -1 & 0 & 0 & 0 \\ 0 & 1 & 0 & 0 \\ 0 & 0 & 1 & 0 \\ 0 & 0 & 0 & 1 \end{array}$ at the boundary, in the

other case, they are certain functions in which not even the equivalence of all the directions in space is observed (a mere axial symmetry); there the first system does indeed appear preferred in principle. I say, though: *I do not believe that the boundary conditions* ($g_{\mu\mu} = 1$, $g_{\mu\nu} = 0$) *apply in principle.* If this were the case, my entire theory would have to be rejected. Since a theory, generally invariant with regard to the differential equations (regarding arbitrary subst.) but *not* generally invariant with regard to the boundary conditions, is a monstrosity.[5] The *usefulness* of the boundary conditions ($g_{\mu\nu} = 1$ or 0) is only based on the circumstance that the parts of the world we are considering are sufficiently small. *A sufficiently small part of a continuously curved manifold can always be treated as flat.* The gravitating individual masses, the fields of which we are examining, then appear as local sources of ⟨perturbation⟩ curvature in the otherwise flat manifold. Belief in the Euclidean nature of the world corresponds fully with the belief of antiquity in the basically flat shape of the Earth's surface; in order to be able to adhere to this belief, autonomous, nonrelativistic boundary conditions for infinity must be introduced, and on top of this, this infinite world must be considered essentially as empty so that it is not curved by its own matter! If the boundary conditions ($g_{\mu\nu} = 1$ or 0) fall away, then all essential grounds for the preference of one specific choice regarding the coordinate system's rotational state also fall away. This becomes completely clear when the world is viewed as (spatially) closed;[6] for then *spatial* boundary conditions are eliminated, so that for any choice of coordinates the events of the world as a whole are determined entirely by the differential equations *alone*.

To your remark about the electron I just comment that the transformation to an electron constantly at rest would indeed be a formal simplification for some problems. The simplification is less great, though, than in the case of pure translation (in the absence of a gravitational field) because of the nonstationary gravitational fields which are generally present. As an example to the contrary, however, I mention the Earth–Moon system, the mechanics of which is represented naturally only when the non-Galilean system is introduced, where the origin constantly occurs in the center of gravity of both masses. The exceptionality of the ordinary th. of rel. lies in that the Euclidean nature is retained through the Lorentz transformation; the importance of special relativity thus extends only to the consideration of regions that are small enough to be regarded with sufficient precision as Euclidean (Galilean).

———————

From the last exposition in your letter I unfortunately see, as in many other instances already, that my Academy paper "Cosmol. Considerations" is being totally misunderstood.[7] Apparently I did not express myself precisely enough. The equations

$$G_{\mu\nu} - \lambda g_{\mu\nu} = -\kappa(T_{\mu\nu} - \frac{1}{2}g_{\mu\nu}T)$$

should be exactly valid everywhere. The T_{44}'s are thus supposed to differ from zero only in the interior of the stars but vanish everywhere else. If I imagine the universe divided into regions of equal size, each of which contains on average 1,000 fixed stars, then each of these regions will contain approximately the same amount of mass. I.e., I make the hypothesis that apart from the local concentration in stars, the matter is uniformly distributed on the large scale. This matter I replace, for the sake of convenience, with homogeneously distributed matter of the same mean density. In this way, the grav. field's local structure is admittedly changed, or disturbed compared to reality. However, the metric character of this field will be preserved on the large scale, so I am correctly informed about the geometric nature of the universe *on the large scale*. Thus, in this consideration I abstract totally from the structure of the field in spaces of the order of magnitude or smaller than the distance from neighboring fixed stars.–

Finally, I note that Michelson's experiment must always come out negatively according to the general theory as well, in the case where the mirror structure is so small and in such a state of motion that the $g_{\mu\nu}$'s, with reference to the structure as a coordinate system, are constant enough in the relevant space.–

Now I believe I have addressed everything you mentioned in your letter. I think we shall come to understand each other's standpoint by this means of "successive approximation." With best regards, yours truly,

A. Einstein.

471.   To Karl Camillo Schneider

[Berlin,] 24 February 1918

Esteemed Sir,[1]

You gave me great pleasure by sending me some of your little tracts. I admire the versatility of your interests and knowledge, as well as your fine and amusing style. I have read almost all of it.[2] The "Academy" with my Mr. Double amused me especially as well; this character is somewhat less precise in his statements than the original,[3] but for it considerably more comical and palatable. Only your strong-arm attitude, sailing under the Germanic flag,[4] goes very much against the grain for me. I prefer to hold with my countryman Jesus Christ, whom you and your mind-mates consider irretrievably obsolete. Suffering really is more preferable to me than exerting force. History may instruct us where this mentality, which you and so many contemporaries in this country extol, is going to lead us; who can know? But moving beyond all designs: *De gustibus non est disputandum*; my taste is otherwise.

Very respectfully,

A. Einstein.

472.   From Hermann Weyl

Zurich, 20 Schmelzberg St., 1 March 1918

Highly esteemed Colleague,

On my instruction, the printer is sending you the printer's proofs of my book *Space, Time, Matter*.[1] Besso had prompted me to publish this lecture held in the previous semester.[2] I hope the book proves suitable for contributing to the dissemination of the great ideas you have introduced into physics!—The proofs are intended for your use, of course; I thought it might interest you to note now already one or two of the things it contains.- Lately I succeeded, I believe, in deriving electricity and gravitation from a single common source.[3] A completely determinate action principle results which, in the electricity-free case, leads to your gravitation equations; in the gravitation-free one, on the other hand, it yields equations that agree in *first approximation* with Maxwell's. In the most general case, the eqs. do become 4th order, though.[4] May I send you the manuscript (about 10 pages) when I have completed it, so that you can perhaps submit it to the Berlin Academy? At the end of March I am coming to Berlin and would be very pleased if I could visit you. Or are you coming to Switzerland again during the holidays? With best regards and utmost respect, yours,

H. Weyl.

473.   From Heinrich Zangger

Zurich, 4 March 1918

Dear friend Einstein,

Mrs. Besso has determined with great diligence all the disbursements, as far as is possible;[1] they are admittedly surprisingly large, but with the nursing change and the sister's falling ill, the entire situation was just affected even more unfortunately.[2] Mrs. Besso is now pressing, as I do, that your wife be placed in a hospital, to be more precise, I am thinking of the Affoltern Civil Hospital, where there is a very competent doctor, Dr. *Grob*,[3] who would certainly do us the favor of admitting your wife. For your little boy we have good accommodations in Appenzellerland for Fr 120 per month + clothing; I expect to take him there shortly so that I can also become convinced of the conditions from a personal impression as well. Why you were angry at me,[4] as you write in introduction, is in actual fact not clear to me, but I also do not desire clarification. So we are taking Albert in again temporarily;[5] Mrs. Besso has found lodging with a secondary-school teacher's family with a boy of about the same age, which would come into consideration in case my wife could not come back owing to our sick little child.[6] We must leave your wife's sister in the asylum for a while longer until we can bring her home.[7] All this would cost something and is ridiculously expensive when looked at from a distance; in emergency situations, not much else can be done, unfortunately, than what we are now presenting to you and what Mrs. Besso has prepared,[8] in order to keep the budget and the ca. Fr 6,000–6,500 in balance.

With greetings from your respectful colleague,

Zangger

Your letters just arrived. Let us look at the facts as physicists should. Albert must take his exams in April;[9] everything must first be arranged.

474.   To Anna Besso-Winteler

[Berlin, after 4 March 1918][1]

Dear Anna,

I'm downright grateful to you for the letter with the complete statement of account.[2] That was the good, trusty Anna again, as I have come to know her these 22 years and respect highly.[3] Now I have *full confidence* again that everything is being done correctly and properly.[4] You'll see that I am very easy to work with. First of all, I agree that under the prevailing conditions a livelihood of Fr 6 000 is unthinkable. Therefore we shall set *Fr 8 000*.[5] From now on, I'm going to send the money to you, because obviously you also send me the accounts. ⟨Fr 900 will be sent to you as well for this quarter⟩. This should be valid from April 1st onwards. Then you will receive Fr 2 000 at the beginning of each quarter. Of this, Maja is to receive the fourth installment if Albert goes to live with her;[6] I'll write you more details when the time comes. After extensive reflection and consultation with my highly competent doctor, my attitude has changed now considerably.[7] But the essential thing for me remains that definitive conditions must be arranged, at last, once and for all, so that these constant agitations finally cease. I imagine the following solution as the best: Albert goes to Maja permanently. Miza, who will remain incapable of heading a household for the rest of her life, is cared for permanently in a sanatorium, in Lucerne if she likes, so that she can see Albert daily. I would give Maja about Fr 2 000 for Albert, so Fr 6 000 would remain for Miza and Tete together. Tete should grow up in a healthy mountain environment, so that his chances of overcoming the illness⟨es⟩ are as favorable as possible. My doctor thinks that he must live in a mild or relatively germ-free climate all his life; otherwise, in the doctor's opinion, as a hereditarily afflicted tubercular,[8] he would be subject to certain early death. Zangger did not want to tell me this directly, in order to spare me; but it's good that I know. I understand that in such circumstances I must attend to the *present*. If I die early and the children are not adequately provided for, this would simply be a blameless misfortune. My health is substantially better, incidentally; this is evident already by the fact that in the last few weeks my weight increased from 127 to 132 pounds. The extended rest and careful diet have helped.[9] In June I'm going to Switzerland and am staying two months at high altitude—probably on Mount Rigi.[10] It's lucky that my position permits me to take such care of myself.

Now I must come back to the divorce again. Take the time to consider how matters stand. Think of the two young girls, whose prospects of getting married are being hampered considerably under the present circumstances, through my fault.[11] Think of the difficulties I have at every turn because, owing to my illness,

I am compelled to live in the same flat as Elsa.[12] Consider furthermore that I am prepared to provide that not the slightest disadvantage arise out of this step for Miza and the children. Is it then astonishing when I sometimes feel bitter and get the impression that all are united in making life unnecessarily hard for me? Do put in a good word for me sometime to Miza and make it clear to her how unkind it is to complicate the life of others pointlessly!

I have already written a mollifying letter to Zangger, in which I informed him of my changed attitude. I don't want to discuss your accounting point by point but only want to comment that I did not use any of the money mentioned in my compilation for myself and Albert. I took this exactly into account in my correction; for clarification, let me mention that I had taken 500 M with me.

Give Michele and Zangger my cordial regards. I must thank you once again for this letter of yours, which pleased and reassured me very much. I hope we'll see each other again in June. Ask Vero to inquire on my behalf about the offprints sent to Weiss.[13] Amicable greetings from your old

Albert.

475.   From Anna Besso-Winteler

[Zurich, after 4 March 1918][1]

Your second letter[2] arrived now as well. I understand the difficult circumstances in which you find yourself, but still, I cannot agree.[3] But do as you think, I just cannot sway Mileva anymore—not after your letter—the responsibility is too great.[4] Debate it out with her. I would only like to add: *1.* Under the current conditions, it is impossible to make anything absolutely *secure*.[5] The finest promises will make no difference. *2.* If Elsa had not intended to make herself so vulnerable, she ought not to have run after you so conspicuously.[6] A mother with children ought to know what she is doing. And if she does so voluntarily anyway out of some noble motivation, then she must also take the consequences with dignity and with a sense of her *inherent* rectitude. *3.* [. . .] *4.* The fact that you are ill is dire fate, but I do not understand how this should be a reason for marriage. So many women now care for *unrelated* men, and added to that, Elsa is your close relative and, after all, it is only a duty that she is fulfilling.[7] Besides, you know that you are very welcome here anytime. *5.* You know that it is my nature to be frank. Although I aired my opinion just as much last summer as now (possibly even more so), you did say: "Oh, Annie, nowhere is it as comfortable as here with you."[8] You also said again and again: "That's true—you are right." Then: "Oh, as far as Elsa is concerned—you know, I really am not going to marry again."–

I'm only telling you this so that you can compare your *present* standpoint with the one *then*.[9]

I would be sorry indeed if the second time you did not find the happiness you seek either. That is what your friends also think, who are both very noble-minded and perspicaciously prudent. It was very misguided of Elsa—also of you—to want to attack these persons, both of whose intellects I esteem very highly (each in his own way).[10] I will not let myself be blindfolded with open eyes.

A few lines from Mich. just arrived.[11] He writes: "Write my poor friend, whose suffering I understand well, that I swear not to do anything to his own detriment. Tell him that—whatever he may say or do—we'll always be fond of him.—Poor pal! What does your health have in store for you?"– I have nothing more to add. I have done what I felt to be a friend's duty—and now I can but keep my peace. I hope that you are recovered soon and send my best regards.

An.

476. To Hermann Weyl

[Berlin,] 8 March 1918

Highly esteemed Colleague,

I am reading with genuine delight the correction proofs of your book, which I am receiving sheet by sheet.[1] It is like a symphonic masterpiece. Every word has its relation to the whole, and the design of the work is grand. What a magnificent method the infinitesimal parallel displacement of vectors is for deriving the Riemann tensor![2] How naturally it all comes out. And now you have even given birth to the child I absolutely could not muster: the construction of the Maxwell equations out of the $g_{\mu\nu}$'s![3] I always thought that the invariant

$$J = R_{ik,\,lm}R^{ik,\,lm}$$

$$(R_{ik,\,lm}) = (ik,\,lm)$$

would have to generate this. But I could not manage to do it. Naturally, I am tremendously eager to see you and your paper.[4] I speak for myself in hoping that you will visit me at the end of March; I still can hardly ever go out, after having been resting for almost a quarter of a year.[5]

Sending you my warm greetings, yours,

Einstein

P. S. The naturalness of the hypothesis of the world's spatial closure can probably best be demonstrated in this way:[6] The boundary conditions must also be expressed covariantly, in keeping with the nature of the thing. The differential equations

$$(ik, \, lm) = 0 \text{ (at infinity)}$$

must take the place of the conditions $g_{\mu\nu}$ = const. at infinity. These are evidently far less natural than the closure condition. It is clear that the latter is not attainable with the original field equations.

I am already looking forward to presenting your paper to the Academy. Do send it to me soon! May I see to the corrections, or is it necessary that they be sent to you? The *Berichte* print the things upon receipt.

477.   From Arnold Sommerfeld

[Munich, 8 March 1918]

Dear Einstein,

I am very glad that, gradually, you are recovering. As I hear, Lenz has also provided a small contribution to your nutrition.[1] Lenz is particularly close to me profession[ally] and personally.

In re the Planck celebration, the following suggestion, which comes from Laue and which I support as plainly serviceable:[2] Take the order Warburg, *Laue, Sommerfeld*, Einstein. Thermodynamics precedes the quanta for Planck temporally and practically.[3] I do not know in the least yet what I am going to say, though.– At the moment I am busy constructing the Lothar Meyer curve for atomic volumes from Bohr's model.[4]

Kindest regards, yours,

A. Sommerfeld.

478.   From Karl Scheel

Berlin, 9 March 1918

Dear Colleague,

Today I finally come back to your last letter after having discussed the Planck issue with a few interested parties, particularly with Mr. Jahnke.[1] The unanimous view is that it is not necessary to convene a meeting of the board; furthermore, your suggestion of fixing the festive meeting earlier for April 19th has been met with approval.[2] *Since the matter is pressing now* and I would like to initiate all the formalities, I request information on the following points:

1) Have you discussed the matter with Planck himself, and does he agree with April 19?

2) What are the titles of the talks: (a) Einstein, (b) Laue, (c) Sommerfeld [I can clarify the matter with Warburg][3] and in what sequence are the speakers supposed to present their talks?[4]

3) Do you have any particular wishes regarding the invitation cards (I think the usual format could be retained) and the persons and associations to be invited (I am considering sending 10 invitation cards with a letter to each of the affiliated organizations: the Electrot[echnical] Assoc., Assoc. of Eng[ineers], Chemical Society, etc.)?

4) What time of day (I am likewise thinking of a $^1/_4$ to 8)?

Please be so kind as to treat this matter as *extremely urgent* now, otherwise our celebration will fall flat. I also especially request that you clarify everything with Laue and Sommerfeld.– I am reachable by telephone in the usual way at the R[eich] I[nstitute][5] (Wilhelm 5515). I am also happy to come and see you, if desired.

But now please, hurry, hurry, hurry!!!

With cordial regards, yours very truly,

Scheel.

479.   From Max Planck

Grunewald, 12 March 1918

Dear Colleague,

It gives me no peace: Before we meet again, I must rid myself of something while also thanking you very much for the suggestion for improving the formulation of my (current) views.

I consider the equation: Entropy $= k \cdot \log$ number of complexions as always valid for an "ergodic" system of a very large, but finite number N of degrees of freedom. I call a system ergodic if in the $2N$th dimensional phase space of the outwardly completely closed system every path in the phase space (line connecting temporally consecutive phase-space points) fills the hypersurface energy = const. densely throughout, that is, it traverses any chosen area of this hypersurface however small it happens to be. I obviously cannot prove at all whether ergodic systems in this sense exist,[1] but I assume it for the moment pending proof to the contrary, and I believe that the statement in the above equation is not at variance with the wave theory of radiation. The equation cannot be applied to black-body radiation, because black-body radiation is not ergodic.

But now I must give you a sharp rebuke, namely, that you did not impart to me that your doctor had prescribed you not to go out for longer than a $\frac{1}{2}$ hour! How much it pains me to think that I had led you to commit a breach of the holiest of your laws. So from now on, I shall think of it a bit more.

I shall keep the surviving beneficiary question in mind and report to you about it later.[2]

Cordial regards, yours,

Planck.

480.   To Felix Klein

[Berlin,] 13 March 1918

Highly esteemed Colleague,

It was with great pleasure that I read your extremely clear and elegant explanations on Hilbert's first note.[1] However, I do not find your remark about my formulation of the conservation laws appropriate. For equation (22) is by no means an identity, no more so than (23); only (24) is an identity. The conditions (23) are the mixed form of the field equations of gravitation. (22) follows from (23) on the basis of the identity (24).[2] The relations here are exactly analogous to those for nonrelativistic theories.

In my view, the formal importance of the $\mathfrak{t}_\sigma^\nu$'s consists precisely in that they occur next to the $\mathfrak{T}_\sigma^\nu$'s in equations (22), which are valid independently of the choice of coordinate system. Their *physical* importance[3] is not only that they give, together with the $\mathfrak{T}_\sigma^\nu$'s, the conservation laws, but also that (23) permits an interpretation that is entirely analogous to Gauss's law

$$\operatorname{div}\mathfrak{E} = \rho \qquad \text{or} \qquad \int \mathfrak{E}_n dS = \int \rho d\tau$$

in electrostatics. In the static case, you see, the number of "lines of force" running from a physical system into infinity is, according to (23), only dependent on the 3-dimensional spatial integrals

$$\int (\mathfrak{T}_\sigma^\nu + \mathfrak{t}_\sigma^\nu) dV,$$

to be taken over the system and the gravitational field belonging to the system. This state of affairs can be expressed in the following way.

As far as its gravitational influence at a great distance is concerned, any (quasi-static) system can be replaced by a point mass. The gravitational mass of this point mass is given by

$$\int (\mathfrak{T}_4^4 + \mathfrak{t}_4^4) dV,$$

i.e., by the total energy (more precisely, total "rest energy") of the system, exactly as the *inertial* mass of the system.

Formally, the possibility of this physically important interpretation is based on the fact that the same quantities, $\mathfrak{T}_\sigma^\nu + \mathfrak{t}_\sigma^\nu$, which enter in the conservation law (22), can, by virtue of the field equations (23), be written as a "divergence" (i.e., in the form $\sum_\rho \dfrac{\partial}{\partial x_\rho} (\ )$) of certain expressions constructed out of the $g_{\mu\nu}$'s & their *first* derivatives.

From (22) it can be concluded that the same integral $\int(\mathfrak{T}_4^4 + \mathfrak{t}_4^4)dV$ also determines the system's *inertial* mass. Without the introduction and interpretation of $\mathfrak{t}$, it is impossible to say that the inertial and gravitational mass of a system coincide.–

I hope that this anything but complete explanation will allow you to guess what I mean. In the first place, though, I hope that you abandon your view that I had regarded an identity, that is, an equation that does not impose any conditions on the quantities occurring in it, as the energy law.

On your comparison of the old theory with the new one, I comment that in the case of the old theory, the $g_{\mu\nu}$'s are not determined adequately by the equations

$$K_{\mu\nu} = 0.$$

Since all

$$(\mu\rho, \sigma\tau) = 0$$

must be satisfied, of course.[4] A substitution by $K_{\mu\nu} = 0$ could only take place if specific boundary conditions were imposed. Thus this problem also leads to the fundamentally important boundary conditions which previously had not been given sufficient notice.

In great respect, yours truly,

A. Einstein.

481.   From Karl Camillo Schneider

Spitz-on-the-Danube, 16 March 1918

Esteemed Professor,

A while ago I took the liberty of forwarding to you an issue of my periodical *Mitteleuropa als Kulturbegriff* (Issue 9/10), in which your special and general theories of relativity were presented in the article "Academy."[1] This was an enormously difficult task for me as a nonphysicist, for which I tried to prepare myself as thoroughly as possible, of course, with the available literature. I would now be extremely obliged to you if you would be so kind as to look through the article and would graciously point out to me the mistakes probably contained in it. It is perhaps an exacting demand, but the importance of the subject—and I am going to have to return many more times yet to your ingenious theory, which is of the greatest significance—will justify it, since it is obviously necessary to present the theory as intelligibly as possible to the general public. That was plainly the intention of my essay. Physicists rarely have an eye for how they could make their theories plausible to the layman; they always refer to topics unfamiliar to him and which he can only incompletely grasp, and above all, they think far too much in mathematical terms to be able to present a world view.

Very respectfully and devotedly,

Univ. Prof. K. C. Schneider.

482.   From Karl Camillo Schneider

[Spitz-on-the-Danube, after 16 March 1918][1]

Esteemed Professor,

I am very much obliged to you for your kind remarks,[2] which encouraged me considerably. It was extremely hard work for me to write the lecture on the theory of relativity, and I am very glad that I did not expose any conspicuous weak spots in doing so.

I would tend to doubt that our views in other respects should diverge so much. I am quite definitely a pacifist and against the current German strong-arm mentality, I just seek to create a foundation for eternal peace based specifically on the *genuine* German character.[3] As far as your statement is concerned, that you prefer to suffer over exerting force, my mind balks at it. A discoverer of your caliber does not quietly endure but exerts the most powerful force conceivable—does he not force our thought on to new paths? This is supposedly not force? Intellectual force, which is completely irresistible, is the most powerful form, and the very title of my periodical evidences that I advocate only this kind: *Mitteleuropa als Kulturbegriff* [*Central Europe as a Cultural Concept*]. It is just that the culture I support is not democratic because the present time teaches us clearly enough, it seems to me, that democracy leads to war and not to peace.

In thanking you very much again, I am very respectfully,

K. C. Schneider.

483.   To Mileva Einstein-Marić

[Berlin, 17 March 1918]

Dear Miza,

I am very willing to comply with your wishes. You shall receive Fr 2,000 around April 1st for this quarter.[1] To my delight, I hear that you are feeling well and that you have prospects of a full recovery. But do be extremely careful regarding Tete. I think perhaps he should be kept permanently far away from the city.[2] Regarding the pension, I still have not received word.[3] Please acknowledge receipt of the money *every time immediately*, so that I know that you are receiving everything regularly; otherwise, under the present circumstances, I can't be sure. If your confirmation notice fails to arrive, I'll complain at the bank.

Kind regards also to the children, yours,

Albert.

484.   To Mileva Einstein-Marić

[Berlin, after 17 March 1918][1]

Dear Miza,

I was very pleased with your letter, because I see from it that you are looking cheerfully into the future again and that you are more conciliatorily disposed toward me as well. Naturally, I am very glad that you can take in the children again. My ideas of having to accommodate the children elsewhere arose only from an incorrect impression of your state of health.[2] Be on the alert about Tete's temperature, so that we can do something right away if the old malady sets in again to the slightest degree.[3] Around the first of April you will receive Fr. 2000.[4] Please send me the acknowledgment of receipt immediately so that I don't have to upset myself about that either. Under these unstable conditions, anything is possible. One of these days I'm going to call on the Ministry about the pension; you'll then immediately receive news of the outcome.[5] Believe me, the security of all of you is just as close to my heart as it is to yours. My health is good enough, but it is still very delicately balanced; I lie down for the greater part of the day and am being fed similarly to a recently weaned infant.[6] Despite this, there are occasional scares, and any agitation is sure to cause a relapse. I'm going to write to Albert right now and shall enclose the letter.

Best regards, yours,

Albert.

P. S. In answer to your postcard to the bank: 1 000 M. were sent off to you three times in October. The first two remittances were for the regular quarterly expenses, the last especially for Tete;[7] it left at the end of October. If you have not received them, let me know. I'll then have it checked at the bank. But I have difficulty imagining that an irregularity has occurred. I sent to Anna the statement sent to me by the bank that the money had been sent off about one month ago as documentation.[8]

Whether the Ministry consents to the transferral of the pension is uncertain at the moment. No decision should be expected soon. If Dr. Zürcher agrees, the following modus could be chosen:

1) A sum (separate from my savings) is deposited notarially, the interest of which should provide you with a pension in the case of my death.

2) The trust account is closed again:
   a) If the Ministry grants the petition.
   b) If the Nobel Prize is allotted to you.[9]

The deposit can be carried out as soon as I have received and approved of your draft.[10]

485.   From Rudolf Förster

Essen, Krupp Factory, AK Laboratory, 19 March 1918

Esteemed Professor,

Many thanks for your valued letter of February 19.[1]

1) First, to come back again to the condition $B_{12,34} = B_{13,24} = B_{14,23} = 0$, it is generally valid for any coordinate system, if all $B_{ik,lm}$'s vanish with the exception of the $B_{ik,ik}$'s, which *Kretschmann* (*A[nn]. der Phys.* 53, 592ff)[2] places in the main diagonal of the $B_{ik,lm}$ scheme.

2) My paper appearing in the *Astronomische Nachrichten* on the forces exerted by an infinite stellar system is firmly based on Newton's mechanics.[3]

3) More in-depth analysis of the essence of infinity has convinced me that the $g_{\mu\nu}$'s for infinite values of the randomly chosen coordinates $x_1 \ldots x_4$ must become infinitesimal in the 4th order. But then the integrals $\int_0^\infty ds$ mentioned in your letter, covering the x_1-axis or one parallel to it, as well as $\int \sqrt{-g}d\tau$, extending over the world, are finite, *provided* that the $g_{\mu\nu}$'s do not become infinite anywhere for finite values of $x_1 \ldots x_4$. However, this *must* occur where the $g_{\mu\nu}$'s are conceived as *analytic* functions; for according to function theory, a uniquely determined function that is regular everywhere in the finite realm and becomes zero at infinity must without exception be $\equiv 0$. But if the $g_{\mu\nu}$'s have poles (The sum of the ordinal numbers must be ≥ 4.), then the $\int ds$ extended over this pole can be finite or infinite, depending on the pole's order. In the latter case, such poles would be physically infinitely far apart. Here various choices plainly still have to be made. (: The Schwarzschild polar field has 2 singular points, $r = 0$ and $r = \infty$, the latter because g_{22}, g_{33}, g_{44} do not become ∞^{-4} at infinity. :)

4) The continuation you desire of the $g_{\mu\nu}$'s from a spatially complete cross-section is probably not uniquely possible. If for $x_4 = 0$, g_{11} g_{12} g_{13} g_{22} g_{23} g_{33} are *arbitrary*, as well as $g_{14} = g_{24} = g_{34} = 0$, $g_{44} = 1$ are given (: which is not a universal limitation :), then through a suitable choice of coordinates, that is, uniquely, $g_{14} = g_{24} = g_{34} = 0$, $g_{44} = 1$ can generally be achieved, so only the continuation of the 6 g_{11} $g_{12} \ldots g_{33}$ remains to be studied. Then your 10 gravitation equations, which I can name in short after the B_{ik}'s occurring in them, are divided into 2 groups. The equations $B_{i,k}$ $(i, k = 1, 2, 3)$ contain the *2nd* derivatives of the relevant g_{ik}'s with respect to x_4. The three equations B_{i4} $(i = 1, 2, 3)$ and the equation $B_{11} - \sum_1^3 \sum_1^3 g^{ik} B_{ik}$ include only *first* derivatives with respect to x_4, namely, linear ones. Thus $\dfrac{\partial g_{ik}}{\partial x_4}$ can be determined from these last 4 of the 6 derivatives. For the 2 remaining ones, the first 6 equations are available, which I do not believe determine the rest.

In any event, with your gravitation equations you have given mathematicians a hard nut to crack.

5) Your objection to my relation

$$g_{\mu\alpha} \cdot \frac{\partial A^\alpha}{\partial x^\nu} + g_{\nu\alpha} \cdot \frac{\partial A^\alpha}{\partial x^\mu} + A^\alpha \cdot \frac{\partial g_{\mu\nu}}{\partial x^\alpha} = 0$$

is not quite clear to me. For one, with an infinitesimal gravitational field ($T_{\mu\nu} = 0$), $\dfrac{\partial g_{\mu\nu}}{\partial x_{[\alpha]}} = 0$ does not need to be true at all,[4] since the gravitational field does not uniquely determine the $g_{\mu\nu}$'s at all, as *Großmann* already admits, of course (*Z[eit]s[chrift] für Math. + Phys.* 62, 226ff.),[5] and then, even if this were so, if, therefore, the case of constant $g_{\mu\nu}$'s existed, one would obtain through transformation to the orthogonal form $\frac{\partial A^\mu}{\partial x^\nu} + \frac{\partial A^\nu}{\partial x^\mu} = 0$. The A^i's would then be linear functions of the form $\sum a_{ik} x_k$, with $a_{ik} = -a_{ki}$, likewise the vector potential $\varphi_\rho = \sum g_{\rho\alpha} A^\alpha$; the field would be constant and homogeneous, the current $I^\mu = 0$. If there are no electromagnetic masses at infinity, then the el.m. field must disappear completely. This condition cannot bring down my assumption, though; for one obviously cannot easily dismiss the fact that the presence of an electrom. field influences the $g_{\mu\nu}$'s *already* by the field energy. The usual requirement that, under the assumption $g_{ii} = 1$, $g_{i \neq k} = 0$, electromagnet. fields can be assumed, is only a first approximation.

6) Furthermore, I would also like to make you aware of a factor which altogether does not suit my purpose, namely, the assumption for your t_σ^α's repeated in your latest article (*Sitzungsberichte* of 14 February [19]18),[6] as well as the quasi-requirement of limiting the coordinate choice by the conditions $\sum \frac{\partial \gamma'_{\mu\alpha}}{\partial x^\alpha} = 0$, in order to arrive at the retarded potentials. Regarding the former, I believe that we shall succeed in putting the right-hand sides of the equations $\frac{\partial T_\sigma^\alpha}{\partial x^\alpha} = -\frac{\partial t_\sigma^\alpha}{[\partial] x^\alpha}$ into a form in which a real tensor takes the place of t_σ^α, but the latter gives me pause. For, as it seems to me, in connection with *Schläfli*'s theorem,[7] which I already introduced in my first letter, *this factor* of your theory, which Kretschmann designates as an *absolute* theory,[8] would thus appear in a manner *extending far beyond previous interpretations* insofar as not only the inner, natural geometry (*g intrinseca*) of the manifold defined by the equation $ds^2 = \sum_1^4 g_{\mu\nu} dx^\mu dx^\nu$ would have *physical* significance, *but also the* coincidental *shape* in which this manifold (: in the sense of the theory of the unraveling [*Abwickelung*] of spatial structures :) would be bent into shape *in its linear 10-dimensional space.*

7) Have you already thought of applying to the theory of gravitation the *integral* equations Hilbert had used with such astonishing success on gas theory?[9] (: I am thinking of this because the $g_{\mu\nu}$'s will not be able to be uniquely determined through differential equations alone. :)

8) Regarding my profession, I can just inform you that I am very satisfied with it and do not wish to exchange it for a teacher's, not even for an academic teacher's, even disregarding the unworthily meager salary. At best, a position at a research institute could attract me. The results of my work here are connected only very remotely to the mass-genocide.[10] I do not design cannons but occupy

myself with electrical measurements, apparatus, calculations, electric. propulsion of mechanical instruments, etc., a very varied and interesting field.

In the last few weeks I worked, on the side, on the formation of electromagnet. waves in a regularly laminate medium (dynamo-metal armature unit, etc.) and am considering publishing these calculations also under a pseudonym.[11] With such concrete problems, a positive result is naturally much more likely.

In the hope that your health is completely reinstated again, I remain with best regards, yours,

R. Förster.

486.   From Max Planck

Grunewald, 19 March 1918

Dear Colleague,

Despite your explicit instructions, I would still like to answer you immediately, to both letters, that is; the business one first. Freundlich should just apply to the Academy with a proposal in which he describes the circumstances and requests that the Academy take steps with the relevant authorities to have the instruments that were confiscated in Odessa returned as soon as possible.[1] I shall then do all I can to support this petition. It would obviously be advisable that F. report this affair to Mr. Struve beforehand and secure his approval of the proposal.[2] For there is no doubt that Struve will be invited in the first place by the Academy to give an evaluation. If it comes out favorably, I then do not doubt that the Academy will gladly take up the matter. That is what it is there for, of course. The other question of the expedition next spring will naturally have to be dealt with quite separately.[3] Something can be achieved here only if a tangible proposal is on hand. It will probably be quite difficult to work out such a proposal. I myself would obviously be pleased to do all I can, provided I could see some feasible way, to contribute toward a positive outcome.

But now to something else:

I. Our opposing views on the relation between entropy and probability can (if I understand yours correctly) perhaps be formulated as follows:[4]

Let us imagine in a 2 Nth-dimensional phase space (N very large) the hypersurface $E = $ const. and designate the quantity (the surface area) of the *total* surface as F, then in *my* opinion (except for terms of no consequence here) the entropy $S = k \log F +$ const. If we consider a "path of phases" [*Phasenbahn*] that develops along this surface, in general, in the course of time it will not cover all the areas of this surface but only certain parts, the total extension of which I denote as $F' \leq F$. Then, according to *you*, $S = k \log F' +$ const. These two equations do not contradict each other only if $F' = F$, i.e., if the system is ergodic or if F' is nearly $= F$.

The latter applies, e.g., to the example you put forward of a system composed of a number of independently ergodic systems. Here the total system is admittedly not ergodic. But the deviation from an ergodic system, or the fraction $\dfrac{F - F'}{F}$, is so small that it is negligible against $\log F$.

II. Your assertion that your oscillation consideration is upheld if a pulverized carbon particle is presupposed to exist within the black-body radiation would be fully valid only if it were demonstrated that, upon introducing a carbon particle, black-body radiation became ergodic.[5] However, I believe such proof cannot be given; for I consider any system whose energy partly consists in black-body radiation as nonergodic. I think this because, in a pure vacuum, processes take place according to Maxwell's equations, hence are entirely uniquely specified, to be more precise, are such that the ones in the interior are determined by the processes at the bordering surface. That is why not all ⟨degrees of freedom⟩ phase regions may be regarded as ⟨statically independent⟩ dynamically possible, and thus an essential precondition of ergodic behavior is removed.

III. Regarding the fluctuations of motion of a diathermanous (or reflective) plate in a radiation space, I would just like to venture to make two comments.

1) Certainly, a plate consisting of a substance that has eigenfrequencies only in the ultraviolet can be considered absolutely homogeneous in long-wave radiation, yet only under the condition that the radiation intensity is absolutely constant. As soon as it is subjected to fluctuations, however, the atomic structure of the plate comes into play, indeed, with greater effect, the more abruptly the fluctuations set in.

2) Strictly speaking, a plate is in statistical equilibrium with the radiation only when it has the very same temperature. But then it may not be regarded as rigid, rather it must take into account the oscillations of its particles, and this amounts to a considerable complication of the processes.

So, now you have probably had enough of this! Therefore only a word about the question of your surviving beneficiary conditions.[6] I also am entirely of your opinion that you attend to the execution of this yourself and for this reason would like to offer you the advice of contacting the official responsible for academic affairs at our Ministry, Privy Councillor Prof. Krüss.[7] The document in which the Ministry informs the Academy that you should be granted widow's and orphan's pensions according to the provisions existing for the survivors of university professors is dated 20 Nov[em]b[e]r 1913 and bears the file number U. I. K. No. 4054.[8] With that you will easily steer Mr. Krüss on to the right track. If in the course of the negotiations you should believe that you may in any way be in need of my involvement, I naturally am always at your disposal. With cordial regards, yours,

Planck.

487.   From Felix Klein

Göttingen, 20 March 1918

Esteemed Colleague,

Here is my long reply now to your v[alued] letter of the 13th.[1]

First of all, my agreement with your closing remark: I had noticed at the t[ime], during proofing, that my statement on p. 13 of my note[2] that the special theory has 10 field equations $K_{\mu\nu} = 0$ could seem misleading (because there are the 20 equations $(\mu\nu, \rho\sigma) = 0$, of course), but had left it because the situation is described clearly at the bottom of p. 5.[3]

Then: on page 9 of my note, l[ines] 3, 4 from above, the $\sqrt{g}$ (which was deleted in the separatum sent to you) must be retained after all[4] and, as I now see, a minus sign must be inserted before the summation signs.–

In other respects, however, I want to stand by the considerations of No. 9 of my note[5] and substantiate them here in that I compare them formula by formula to your paper on Hamilton's principle, etc., in the *Berliner Sitzungsberichte* of 1916:[6]

1) My K's and αQ's, multiplied by $\sqrt{g}$, are directly your G's and M's, insofar as I disregard the special form of Q,[7] to which I restrict myself in my note but which are used only in the definition of δq_ρ in f[ormula] (13)[8] and in the transition from (14) to (14').

2) My Lagrangian derivatives, $K_{\mu\nu}$, $\alpha Q_{\mu\nu}$, again multiplied by $\sqrt{g}$, read for you as:

$$\frac{\partial G^*}{\partial g^{\mu\nu}} - \frac{\partial}{\partial x_\alpha} \left(\frac{\partial G^*}{\partial g^{\mu\nu}_\alpha} \right), \; \frac{\partial M}{\partial g^{\mu\nu}}.$$

For the Lagrangian derivatives, there is indeed no difference in operating with G or with G^*.

3) Instead of $K_{\mu\nu}$, $\alpha Q_{\mu\nu}$, I can now, in order to come closer to your designation, introduce "mixed" components as needed:

$$K^\nu_\sigma = \sqrt{g} \cdot \sum K_{\mu\sigma} g^{\mu\nu} \qquad\qquad \alpha Q^\nu_\sigma = \alpha\sqrt{g} \cdot \sum Q_{\mu\sigma} g^{\mu\nu}.$$

4) For field equations we have my (16a), (16b) and your (7), (8) corresponding to

a) the ten: $K_{\mu\nu} + \alpha Q_{\mu\nu} = 0$, or $K_\sigma^\nu + \alpha Q_\sigma^\nu = 0$
b) the remaining: $Q^\rho = 0$.

5) Between your left-hand sides the identities (17) of my note hold, which I would like to reproduce here in this way:

$$\sum_\nu \frac{\partial([K_\sigma^\nu + \alpha Q_\sigma^\nu] - \alpha\sqrt{g}Q^\nu q_\sigma)}{\partial w^\nu} =$$

$$-\frac{1}{2}\left[\sum_{\mu,\nu}\{\sqrt{g}(K_{\mu\nu} + \alpha Q_{\mu\nu})g_\sigma^{\mu\nu}\} + \sum_\rho \alpha\sqrt{g}Q^\rho q_{\rho\sigma}\right].$$

6) Using field equations (b) I shall cancel out the Q^ρ's on the left and right-hand sides, but only apply field equations (a) on the right-hand side. I then have:

$$\sum_\nu \frac{\partial(K_\sigma^\nu + \alpha Q_\sigma^\nu)}{\partial w^\nu} = 0.$$

7) These equations naturally are physically meaningless, since the $K_\sigma^\nu + \alpha Q_\sigma^\nu$'s are of themselves zero in consequence of (a); the zero on the right-hand side does not result from the differentiation. It is just for this reason that these equations (6) are not analogous to the law of the conservation of energy in classical mechanics $\left(\frac{d(T-U)}{dt} = 0\right).$

8) Now, however, you set, according to (18), (19) of your paper,[9]

$$\alpha Q_\sigma^\nu = -\mathfrak{T}_\sigma^\nu, \qquad K_\sigma^\nu = -\mathfrak{t}_\sigma^\nu + \frac{\partial}{\partial x_\alpha}\left(\frac{\partial G^*}{\partial g_\alpha^{\mu\sigma}}g^{\mu\nu}\right),$$

and you note in (17) that—because G is an invariant built in a particular way against arbitrary transformations of x—

$$\frac{\partial^2}{\partial x_\nu \partial x_\alpha}\left(\frac{\partial G^*}{\partial g_\alpha^{\mu\sigma}}g^{\mu\nu}\right) \equiv 0.$$

9) Your equations

$$\frac{\partial(\mathfrak{T}_\sigma^\nu + \mathfrak{t}_\sigma^\nu)}{\partial x_\nu} = 0$$

hence distinguish themselves from (6) only in that you have inserted under the differentiation symbol a term whose divergence vanishes identically. Besides, this term depends only on the $g^{\mu\nu}$'s and in no way on the q_ρ's.

10) Therefore, by no means am I capable of seeing how any physical statements about the nature of the field $(K, \alpha Q)$ can be obtained from (9).–

So much for criticism. However, I have long since given a positive turn to the matter by presenting to our Soc. of Sci. on 22 February the closely related proposal[10] of avoiding all identities by not attributing to the gravitational field any special energy at all, rather simply choosing

$$Q_{\mu\nu} = -\frac{K_{\mu\nu}}{\alpha}$$

as components of the "energy tensor."[11] As a test, I applied Schwarzschild's formulas for the gravitational field of a homogeneous, incompressible fluid sphere at rest (*Sitzungsberichte* 1916).[12] To simplify the comparison to the statements of classical mechanics, I immediately placed, instead of the energy *tensor*, its simplest *scalar*:[13]

$$-\frac{\sum K_{\mu\nu}g^{\mu\sigma}}{\alpha} = -\frac{\sum K_{\nu}^{\sigma}}{\alpha\sqrt{g}}$$

(which, by the way, converts into $+\dfrac{K}{\alpha}$ according to Hilbert's first note,[14] p. 8) into the foreground and regarded it as analogous to the "Lagrangian" $(T - U)$. This means that I regard

$$\int \frac{K}{\alpha} \cdot d\tilde{w},$$

integrated over any segment of the world, as analogous to Hamilton's integral (and not, for instance, as Lorentz does in his lectures of 1916,[15] $\int \dfrac{K}{\alpha}d\tilde{w} + \int Qd\tilde{w}$).

In Schwarzschild's case, the mentioned scalar, in conformance with your formula of 1914[16] from which Schwarzschild starts, is equal to $\rho_0 - 3p$—or, how I still always prefer to write it in the cm.gr.sec. system $= c^2\rho_0 - 3p$. For p, Schwarzschild indicates here in formula (30) of his paper, if I also add the factor c^2,

$$p = \rho_0 c^2 \cdot \frac{\cos \chi - \cos \chi_a}{3\cos \chi_a - \cos \chi}.$$

Here, to make the transition to classical mechanics, I replace the cos. with their series expansion and have as the lowest term[17]

$$\rho_0 c^2 \cdot \frac{\chi_a^2 - \chi^2}{4},$$

or, with κ as the normal gravitation constant:

$$\frac{2\pi\kappa\rho_0^2(r_a^2 - r^2)}{3}.$$

This now is exactly the expression for the hydrostatic pressure which, in accordance with classical theory, governs the interior of the sphere, and $\int -3p \cdot dv$ is at the same time no more than the potential we attribute to the sphere itself as a consequence of the gravitation.

This agreement, it seems to me, is an important argument for the practicality of my proposal.[18] For the energy components chosen by me, we naturally have the equations which you write thus in (22) of your paper:[19]

$$\frac{\partial T_\sigma^\nu}{\partial x_\nu} - \frac{1}{2} g_\sigma^{\mu\nu} T_{\mu\nu} = 0.$$

To this Runge has remarked in a lecture[20] he gave on 8 March before our Society of Sci. that in each individual case the form you desire,

$$\frac{\partial \mathfrak{T}_\sigma^\nu}{\partial x_\nu} = 0,$$

can be produced by specifying the coordinate system, which is just barely acceptable, so that the sums $g_\sigma^{\mu\nu} \mathfrak{T}_{\mu\nu}$ are eliminated! A true philosopher's stone. (Both communications, mine and Runge's, are just being prepared for publication.[21] In the meantime, I also followed up on your cosmological ideas of 1917,[22] in that I allow χ_a of Schwarzschild's paper $= \frac{\pi}{2}$, in which case Schwarzschild's fluid sphere fills the entire "elliptic" space![23] Thus I come very close to your specifications, only that the numerical factors are not correct yet. Of course—now written in Schwarzschild's first designation—$p = -\rho_0$; the world in the region of constant curvature becomes $\frac{\kappa\rho_0}{3}$, from which $R = \sqrt{\frac{3}{\kappa\rho_0}}$ results, whereas you find $\sqrt{\frac{2}{\kappa\rho_0}}$. The $K_{\mu\nu}$'s become $-\kappa\rho_0 \cdot g_{\mu\nu}$, hence your $\lambda = \kappa\rho_0$, whereas you have $\frac{\kappa\rho_0}{2}$. What could be the cause of these differences?)[24]

Now I am eager to know what you will say to this long letter. I only want to add that, meanwhile, my general lectures on quadratic differential formulas have also been typed.[25] But only in ca. 2 weeks can I send you a copy (which I would then request you pass on to Sommerfeld again).[26]

Very truly yours,

Klein.

[18]Comments in draft: "I find it very remarkable that at the end of his lect[ures] of 1916 Lorentz stated he was satisfied with postulating in addition to field equations (a) a double energy current, the two components of which constantly compensate each other." . . . "It contains precisely the agreement of the grav. mass and inertial mass to which you assign so much weight."

[23]Draft postscript: "P.S. "On Cosmology."

Previously already, and not spherical space but elliptical.

Case of Schwarzschild's sphere for $\chi_a = \frac{\pi}{2}$

4-dimensional space with constant curvature $\frac{\kappa \rho_0}{3}$.

Hence $R = \sqrt{\frac{3}{\kappa \rho_0}}$, whereas you indicate $\sqrt{\frac{2}{\kappa \rho_0}}$.

$p = -\rho_0$ Scheme $\begin{pmatrix} -p & 0 & 0 & 0 \\ 0 & -p & 0 & 0 \\ 0 & 0 & -p & 0 \\ 0 & 0 & 0 & -p \end{pmatrix}$, which does not agree with Schrödin-

ger either. Mass content $\pi^2 R^3 \rho_0$.

The K_ν^ν's all become, according to Runge's letter, $= -\kappa \rho_0$

which agrees with the premise
$K_\nu^\sigma + \kappa \cdot T^\sigma \nu = 0.$

That is why $K_{\mu\nu} = -\kappa \cdot \rho_0 g_{\mu\nu}$.

Thus I find Einstein's $\lambda = \kappa \rho_0$, while E. himself has $\frac{\kappa \rho_0}{2}$. So I am quite close to Einstein, but the numerical coeffs. do not agree. (Schw[arzschild] must have calculated correctly, after all: Fréedericksz and Runge have checked him independently.)"

[25]Draft text: "In particular, another "travel copy" has been made. Prof. Conrad Müller, Hanover, my longtime collaborator at the *Math. Encyclopedia*, has taken it for a while and will send it on to you around Eastertime."

[26]Draft passage: "What I say about Riemann, Beltrami, and Lipschitz will probably be applauded by you right away; it appears to me that at the t[ime] Grossmann had given you too one-sided instructions from the point of view of the more limited Christoffel School. On the other hand, you may find some objection when I consider some of what you aim at with your relativistic conception as long since contained in Lagrange's equations. For this you must not look at me as a one-sided formalistically inclined mathematician but rather as a man who in his life development was led over by chance to the math. side and is now trying to demonstrate the knowledge that he has gained there also with regard to its significance for neighboring disciplines."

488.   From Gustav Mie

Halle-on-S[aale], 47 I Magdeburger St., 21 March 1918

Dear Colleague,

Through personal circumstances, namely, my wife's falling quite seriously ill which, thank God, we have weathered, I was prevented from answering your letter of 22 February sooner.[1] By now, each of us has probably presented his point of view sufficiently, in the main.

I would just like to say one more thing: You think I had totally misunderstood your paper "Cosmological Considerations."[2] This I do not believe, though. I merely make a sharp distinction between what you *wish to attain* with the gravitational equations posed there and what, as far as I can see, really *is attainable* with them.[3] I am convinced that if a reasonable integral solution of your equations really is gained with λ, then ρ in this integral in the vast spaces between the individual fixed stars *must* have a constant value, thus *must* occur as a mass density *continuously* permeating the space. I have considered the matter very thoroughly and do not believe that I am mistaken.[4]

The equation system with λ differing from zero seems to me to be of very great interest for epistemology, though. We all know that the parallel axiom is a physical law;[5] it indicates, to use my terminology for once, a specific characteristic of ether physics and thus is on the same level as the principle of relativity, Hamilton's principle, etc. However, until now it has always been puzzling—to me, at least—at what place exactly the parallel axiom enters into the fundamental physical equations because, given the way in which the fundamental equations had always been formulated before your publication "Cosmological Considerations," the parallel axiom was always silently assumed from the outset. Now this is suddenly different; in your new equations one can finally see clearly in what way the parallel axiom is contained in the fundamental equations of ether physics, which are accessible to experimental verification. It simply involves the experimentally determinable coefficient [λ]; if it is zero the parallel axiom is valid, if it differs from zero then the space has a constant curvature which can be calculated from λ.

I still do not quite understand why you are so unsatisfied with my interpretation of your theory of gravitation of 1915,[6] but on this point we shall probably never be able to agree. To make another attempt, nevertheless, I would like to present my point of view once in an entirely new form that has occurred to me since.

I would like to imagine the Earth as an "Einsteinian coupé,"[7] from which, as a consequence of a perennial thick fog, shall we say, no other celestial bodies at all can be observed. We are then still going to be able to determine the rotation of the Earth around its own axis reliably through purely terrestrial observations (with Foucault's pendulum); we are going to be able to say absolutely: the Earth rotates around this and this axis, at this and this rotation velocity. By contrast, if your theory is correct (which I do not doubt) it could *not* be demonstrated that the Earth has a *curvilinear motion* in space around the Sun. For the effect of the Sun's gravitational field always exactly compensates all the influences of the orbit's curvature. So one would probably say, if only terrestrial observations could be made: the Earth is motionless in space and rotates only around its own axis. It is very interesting to note here, though, that this relativity of the gravitational field[8] is valid only when the Sun's gravitational field within the region accessible to us for performing terrestrial observations is regarded as *constant*. Phenomena originating from a change in the Sun's gravitational field with distance within the range of the Earth's diameter, for example the precession of the Earth's rotational axis, would likewise have to be ascertainable from purely terrestrial observations. Let us assume for a moment that Foucault's pendulum experiment could be conducted with such precision that the precessional motion of the Earth's axis were detected, then one would certainly have a purely terrestrial observation that would prove the inadequacy of the conception of an Earth motionless in space, only rotating around its own axis. The principle of the relativity of the gravitational field is thus attached to the condition that a certain limit in observational precision must be presupposed, as I have also explained in my Göttingen lectures;[9] it really is a very wonderful theorem, though, all the same. If we assume this condition valid, we can then arrive at the assertion: "the Earth describes a curvilinear orbit" only when we can extend the observations over a large enough region in space in which the solar gravitational field is certainly no longer constant, hence through the entire region delimited by the Earth's orbit. If we can peer about as far into the universe from our Einsteinian coupé (it must then, obviously, actually lose its coupé designation!), then—but also only then—we can determine absolutely that the motionless-Earth hypothesis is false.

Generally, this can also be expressed thus: If not the entire universe were accessible to our observations, but rather only a thin "world tube" delimited by geodesic world lines, which is so narrow that in its interior the geodesic lines still form a "coordinate grid" (if I may express myself in this concise manner after my discussion in the Göttingen lectures), then nothing can be concluded about the form of this tube from observations; it can be regarded as straight or also as curved in any manner you like. The construction of an "absolute coordinate system," in which the world-line tube just considered is also assigned a specific shape, is possible only when the region accessible to observation is so large that one can say with certainty whether or not the geodesic lines form a coordinate grid.

Hence, it is only along the path of successive approximation that we can seek to arrive at an "absolute coordinate system." As long as we are just dealing with purely terrestrial observations, we are going to regard the Earth as at rest in the universe (but naturally, rotating around its axis!). The coordinate system acquired in this way suffices entirely to explain the observational data. As soon as the Sun and, indeed, all bodies of our planetary system are drawn into our consideration, however, the Earth *must* be regarded as moving along a curved orbit. One can then assume the planetary system's center of mass as being at rest and thus obtain the familiar coordinate system of Newton's theory of planetary motion. But as soon as the other fixed stars of our Milky Way also affect the phenomena (for which, as far as I know, observational precision certainly does not yet suffice), another coordinate system would be reached in which now the Sun also is in orbit and the common center of mass of the entire fixed-star system is at rest. Whether one ought to stop here we shall never be able to say and, in a certain way, the coordinate system still remains behind the ultimate ideal of an "absolute system." Yet it is very important, nonetheless, to emphasize again and again that, in the last analysis, the reason for this uncertainty lies only in the limited precision of our observations. In principle, though, an "absolute coordinate system" must exist but, in defining it, it is not admissible to limit oneself to observations, which cover only a ⟨very⟩ restricted region of the world.

I may be coming to Berlin in the course of next week. If you are there now, I shall inform you in advance when I am coming and when I might be able to visit you, in the hope that I can talk to you then for an hour or so.

With best regards, yours truly,

Gustav Mie.

489.   From Elisabeth Warburg[1]

Charlottenburg, 21 March 1918

Esteemed Professor,

How kind of you to want to write to my son. I am so very worried and anxious, you know. Would that your letter still reached him in good health. Living hell must have broken loose over there.[2] It is a sin against human beings, an outrage, that these mass murders are being committed on and on. Do we have a right to this?

My husband[3] was always so overjoyed with his son. He once told me in confidence that he was one of the very greatest. These hopes we placed in him, must all of it be for naught? Is he not to fulfill his destiny? I fear he is too much involved over there to extricate himself. Why also must it be that his division, in particular, be deployed right at the very front? This is a boundless misfortune for us. Forgive these desperate pleas; I am not resigned, I still rebel against fate and would like to save what can be saved.   How good that you want to help me with this. I am enclosing two letters for you, written to describe him. With cordial regards, also to Mrs. Einstein, ever yours,

Elisabeth Warburg.

490.   From Georg Helm[1]

Dresden, 22 March 1918

Highly esteemed Professor,

In the negotiations concerning the replacement for the regular professorship in philosophy, made vacant by the death of Elsenhans,[2] which are underway at our Technical University, I turned my colleagues' attention to Prof. Josef Petzoldt who has for a long time been working as private lecturer at Berlin University. From his academic orientation and his manner of representation, I consider him particularly suited to transmit the ideas of philosophy to the disciples of the exact sciences, young technicians, and to be a stimulating and intellectually promotive influence in our city's wider educated circles.

Guided by a remark by Petzoldt, I am turning to you with the request that you write me an evaluation of him[3] that I may use in the continuing negotiations and that will certainly have a very decisive influence on my colleagues. It would probably have to involve primarily eliminating the opposing view that would have Petzoldt, who is already of advanced age,[4] seen as an advocate of an outmoded line.

When about ten years ago the same professorship was filled, Mach himself, the physiologist Hering, and the philosopher Schuppe[5] supported my efforts to draw Petzoldt to Dresden—in vain; the choice fell on Elsenhans. Now that the old voices are silent, I hope that a younger generation will bring success. For this I solicit your kind support!

With obliging thanks for the requested opinion, respectfully and devotedly yours,

Prof. Helm.

491.   To Otto H. Warburg

Berlin W., 5 Haberland St., 23 March 1918

Highly esteemed Colleague,

You are probably surprised at receiving a letter from me, because until now we have only walked past each other, without actually ever getting to know each other. I must even fear rousing some form of indignation in you with this letter; but it *has* to be.

I hear that you are one of Germany's most talented younger biologists with great promise[1] and that presently representatives in your particular field are quite mediocre here. I also hear, though, that you are stationed there at a very vulnerable outpost[2] so your life is constantly hanging by a thread! Now, please step out of your shoes and into those of another discerning creature and ask yourself: Is this not madness? Can this post of yours out there not be filled by an unimaginative average person of the type that come 12 to the dozen? Is it not more important than all that big scuffle out there that valuable people stay alive? You know it yourself very well and must admit that I am right. Yesterday I spoke with Prof. Krauss, who is also entirely of my opinion and is also ready to arrange to have you recalled for another task.[3]

My plea to you, which arises from what has been said above, is therefore that you support our efforts in securing your personal safety. Please, after a few hours of serious reflection, write me a couple of words so that we here know that our efforts will not founder on your attitude.

In the ardent hope that in this matter, for once, reason will exceptionally prevail, I am, with cordial regards, yours truly,

A. Einstein.

492.   To Felix Klein

[Berlin,] 24 March 1918

Highly esteemed Colleague,

I thank you very much for your extensive and lucid letter.[1] I shall do as all do; what I like, I silently accept, what I do not like, however, this I mention . . . I do not concede that the equations

$$\sum_{\nu} \frac{\partial (K_\sigma^\nu + \alpha Q_\sigma^\nu)}{\partial w^\nu} = 0 \ \ldots$$

(1)

are devoid of content. What they contain is *a part* of the content of the field equations[2]

$$K^\nu_\sigma + \alpha Q^\nu_\sigma = 0.$$

This naturally also applies to equations equivalent in substance to (1)

$$\sum_\nu \frac{\partial(\mathfrak{T}^\nu_\sigma + \mathfrak{t}^\nu_\sigma)}{\partial x_\nu} = 0 \dots \qquad (2)$$

The value of (2) against (1) lies in the fact that it provides three integral relations. If a physical system exists, at the spatial boundaries of which the $\mathfrak{T}^\nu_\sigma$'s and $\mathfrak{t}^\nu_\sigma$'s always vanish, then (2) yields

$$\frac{d}{dx_4}\left\{ \int (\mathfrak{T}^4_\sigma + \mathfrak{t}^4_\sigma)dV \right\} = 0.$$

The temporal constancy of these four integrals is a nontrivial consequence of the field equations and can be looked upon as entirely similar and equivalent to the momentum and energy conservation law in the classical mechanics of continua. What distinguishes the $\mathfrak{t}^\nu_\sigma$'s from other quantities, which likewise are capable of giving relations of the form of (2), is the circumstance that the $\mathfrak{t}^\nu_\sigma$'s depend only on the $\left\{ \begin{matrix} \mu\nu \\ \sigma \end{matrix} \right\}$'s *but not on their derivatives.* The $\left\{ \begin{matrix} \mu\nu \\ \sigma \end{matrix} \right\}$'s are the quantities analogous to the field components in electrodynamics.[3] With this letter I am sending you an article[4] in which, with the aid of equation (2), a mechanical system's energy loss through the emission of gravitational waves is calculated.

The idea of making, through a special choice of coordinates, the quantities

$$\sum_{\mu\nu} \frac{1}{2} g^{\mu\nu}_\sigma T_{\mu\nu}$$

equal to zero, in order to be able to uphold the energy law[5]

$$\frac{\sum \mathfrak{T}^\nu_\sigma}{\partial x_\nu} = 0$$

for matter alone,[6] I also had considered sometime ago. This route is not feasible, though, just because according to the theory, energy losses from gravitational waves exist, which cannot be taken into account in this way.[7]

On your comparison of my cosmological consideration to Schwarzschild's case of the sphere, I remark that these two cases differ fundamentally from each other in that, for Schwarzschild, g_{44} is variable and equilibrium is only possible with spatially variable pressure, whereas for me, matter with negligible interior pressure is involved.[8] The numerical discrepancy is connected to the fact that the term with λ, which makes the constancy of g_{44} and the vanishing of the pressure possible in the first place in my case, is missing in the equations in Schwarzschild's case.[9] An interpretation of the world as a whole that can be in any way satisfactory and compatible with the facts does not seem possible without the introduction of the λ term in the field equations.

I am very much looking forward to your new lectures.[10] Lately, I read correction proofs of a book by H. Weyl on the theory of relativity,[11] which made a great impression on me. It is admirable how well he masters the subject. He derives the energy law of matter with the same variation trick you use in your recently published note.[12]

Finally, I heartily thank you and the other colleagues of yours involved in having an immense scientific prize assigned to me. I received word today from Hamburg.[13]

In utmost respect, I am yours very truly,

A. Einstein.

493.   To Gustav Mie

[Berlin,] 24 March 1918

Dear Colleague,

In joyous anticipation of your promised visit, I reply just briefly to your long letter of the 21st.[1] According to my interpretation, the distribution of matter does not need to be such that ρ is constant in reality. It suffices that spaces exist that are large compared to the distance to neighboring fixed stars but that are small enough to be able to look upon the metric deviation from Euclidean behavior as still small. For such a region of space, the field equations can be substituted by *linear* ones, where $T_{\mu\nu}$ and the λ term are on the right-hand side. The field at the boundary of this space then depends just as little on the detailed distribution as in Newton's theory.

Based on my theory of 1915,[2] it is not legitimate to regard an Earth enveloped in clouds as rotating *in reality*. For the gravitational and inertial fields are viewed as an indivisible unit. This total field determines the result of Foucault's pendulum experiment and—with respect to a coordinate system not rotating relative to the Earth—satisfies the differential equations. At the time, I did not set up boundary conditions except to characterize special cases. The differential equations are satisfied everywhere, though, no matter how far the space in which they are to be tested is chosen to extend, and no matter how the coordinate system is chosen. On the basis of my exposition of 1915, therefore, absolutely no preferred system exists.

What remained unresolved then was how the structure of the universe as a whole should possibly be conceived.   *One* reason why the quasi-Euclidean possibility ($\lambda = 0$), ($g_{\mu\nu} = $ const.) does not satisfy me is presented in the enclosed notice.[3] All the rest orally. With best regards, yours truly,

A. Einstein.

494.   From Paul Ehrenfest

[Leyden,] 27 March 1918

Dearest Einstein,

Except for along sl[ow] indirect routes, I have heard nothing of you for a long while now and only little about your health. But I think of you often—my not writing stems from: my chronic discontent with work.– But no more of it!–

In the last few months I became enormously fascinated with very specific problems of politico-economic *theory*, and I have been reading up quite extensively and reflecting on the entire complex of problems.[1] Natural[ly], this is a viol[ation] of my professional duties, but it has taken too strong a hold of me—it can't be helped; it simply fits the structure of my thinking machine so well!

Every time I suddenly advance again to a bit of musical insight or ability, I repeatedly think how nice it would be if I had you here.[2] We just received an article by Sommerfeld in which he describes the impressions he gained on the occasion of physic[al] lectures at Ghent. I feel the urge to write him some details about some of the gentlemen whose blather have fed his impressions. We know them here in Leyden.–[3] Hilbert and Planck have remarkably kindly invited me to the Wolfskehl-week discussions[4]—I wrote Hilbert the superficial and *ideological* reasons why I could not come.[5] Since I requested that he also relay my thanks to Planck, I would appreciate hearing from you whether the letter has arrived. I mean to Hilbert.–

Personally, we are doing very well, and the joy the children give us is beyond description.– Report a bit about yourself.– When are you visiting us here again?!!

Fond regards from all of us, yours,

Ehrenfest.

495.   From Friedrich Kottler

Vienna, 30 March 1918

Esteemed Professor,

I would like to extend my sincere thanks for the mailing of your paper and for your k[ind] letter. Much to my regret did I learn of your illness.[1] Please accept my best wishes for a speedy recovery and do let me know on occasion how your state of health is.

Allow me now to return to the subject of my paper.[2] First of all, I would like to state, in order to avoid misunderstandings: For the case that my exposition should prompt you to a reply, I merely made the request that a discussion by letter precede it. Thus not, for instance, that a public reply not take place at all, insofar as you deem it necessary, of course.

I would like to hope that your describing my presentation as hard to understand this time as well was simply a first impression upon cursory perusal. Permit me to reiterate again the request for patience to the reader made in the introduction.

In addressing the individual objections, I allow myself the following rebuttals:

1) My guiding hypothesis is: $g = -c^2$ in Cartesian coordinates.[3] These are coordinates for which[4]

$$g_{i\kappa} = \delta_{i\kappa} + kh_{i\kappa}$$

($\delta_{11} = \delta_{22} = \delta_{33} = 1$, $\delta_{44} = -c^2$, $\delta_{i\kappa} = 0$ $i \neq \kappa$, $k = 6 \cdot 7 \cdot 10^{-8}\text{gr}^{-1}\text{cm}^3\text{sec}^{-2}$.)

It states clearly:

All solutions that permit only a single description in Cartesian coordinates for which $g \neq -c^2$ are inadmissible.

If there are several descriptions in Cartesian coordinates, then only the coordinate system for which $g = -c^2$ is admissible.

As an example of the latter, I mention the homogeneous field (acceleration γ), s[ee] §37 of my paper:[5]

a) According to the equivalency hypothesis:[6]

$$-ds^2 = dx^2 + dy^2 + dz^2 - c^2 \left(1 + \frac{\gamma z}{c^2}\right)^2 dt^2.$$

b) According to gravitation theory:[7]

$$-ds^2 = dx^2 + dy^2 + \frac{dz_1^2}{1 + \frac{2\gamma z_1}{c^2}} - c^2 \left(1 + \frac{2\gamma z_1}{c^2}\right) dt^2.$$

Both representations are merged in the transformation

$$1 + \frac{\gamma z}{c^2} = \sqrt{1 + \frac{2\gamma z_1}{c^2}}.$$

Both are Cartesian coordinate systems; but only one can give the best possible link with experience. And this then is the justified one. It becomes apparent that this is the case for (b) and that here $g = -c^2$ is satisfied; the equivalency hypothesis, on the contrary, yields an illegitimate metric.

Coordinate systems other than Cartesian ones have no new physical significance. Rather, it must never be disregarded that these are merely mathematical operations without geometrical content. Likewise, a surface does not change its shape if I represent it using other coordinates, except in the case of pliability. It is precisely this case which would occur in the above homogeneous field example if one of the systems had not been excluded. Thus the uniqueness of the designated shape of our curved four-dimensional manifold is established. And the geometric shape is identical here to physical nature.

My condition for these non-Cartesian coordinate systems now undergoes the corresponding alteration

$$g = (g)_{k=0},$$

e.g., for polar coordinates r, ϑ, φ

$$g = -c^2 \cdot r^4 \sin^2 \vartheta \quad \text{etc.}$$

That is why it is by no means possible to confuse it with the constraint $g = -1$ you have been using now and then. I warned against this in a separate footnote to prevent such misunderstandings.

In returning again to the case where $g = -c^2$ in Cartesian coordinates, or generally, $g = (g)_{k=0}$ is not satisfied, it must be mentioned that then the initial conditions (structure of the field, material tension tensor, boundary conditions, etc.) ought to be checked. (Comp. §26 the interior of homogeneous spheres.)[8]

2) I have read your paper on gravitational waves[9] but must confess to being guilty of an omission in my paper. Namely, I apply as gravitational tensors [*Gravitationsspannungen*]

$$G_{i\kappa} = \frac{1}{2} K g_{i\kappa} - K_{i\kappa} = \frac{1}{2} g_{i\kappa} \sum_{r,s,q} g^{(sr)} \{rq, sq\} - \sum_q \{iq, \kappa q\},$$

for which

$$G_{i\kappa} + \kappa T_{i\kappa} = 0$$

applies.[10] (In my paper $G_{i\kappa} = \kappa T_{i\kappa}$ appears because I had calculated the inertial forces as negative.) These $G_{i\kappa}$'s are Hilbert's[11]

$$-\frac{1}{\sqrt{g}}[\sqrt{g}K]_{i\kappa}.$$

They identically fulfill the energy-momentum conservation law:[12]

$$\sum_{r,s} g^{(rs)}G_{ir/s} \equiv 0 \qquad i = 1, 2, 3, 4;$$

they are therefore the complete covariant counterparts to the $T_{i\kappa}$'s. Their adoption as gravitational tensors are thus so plain that, under the influence of Hilbert's papers and the excellent little paper by Herglotz in the Leipzig *Berichte*,[13] I assumed you had already abandoned your $t_{i\kappa}$'s.[14] But now I gather that you still are retaining them and shall therefore point out this difference, which is lacking in my paper, later in the correction stage.[15] I suspect, incidentally, that the $t_{i\kappa}$'s used by Lorentz and Levi-Civita are identical to the $G_{i\kappa}$'s.[16] Since I cannot obtain the papers of these authors here, I would be grateful for your confirmation of my suspicion.

Allow me to add to the arguments against the $t_{i\kappa}$'s and for the $G_{i\kappa}$'s (non-covariance, hence physical insignificance! sic!; arbitrary vanishing upon changes in coordinates,[17] etc.) the following essential points to my interpretation: The gravitational tensors disappear outside of matter (with regard to the electromagnetic energy fields): $G_{i\kappa} = 0$. That is, no gravitational energy exists detached from matter, therefore, also no (pure) gravitational waves, in conformance with experience, which demonstrates a complete lack of propagation effects.

As concerns your argument re. the vanishing of the total amount of energy $G_{i\kappa} + \kappa T_{i\kappa} = 0$,[18] I refer to §6 of my paper, where I say: there is no interaction between material and gravitational energy; gravitation, and only gravitation, is purely conservative. This also follows mathematically from the fact that the $G_{i\kappa}$'s and the $T_{i\kappa}$'s each on their own satisfy the energy-momentum conservation law; thus the $G_{i\kappa}$'s are eliminated from the energy law, etc. I recall here again my emphasis on the "force-free" nature of gravitation; I said in 1916: kinematic versus dynamic conception,[19] which comment you seem to have misunderstood.[20]

3) Hence here we already see that arguing from an analogy between electricity, where free field energy exists, and gravitation, where this energy does not exist, is inadmissible; this therefore applies all the more to your third objection, which seems to me to have been taken over directly from electricity.

For Maxwell's theory treats the boundary between two media ε_1 and ε_2 as follows: An electromagnetic field exists on both sides. Hence, no qualitative differences, only quantitative ones. It then evidently leads to the same thing if, in place of the jump from ε_1 to ε_2, I take a boundary layer in which ε_1 varies constantly up to ε_2; afterwards I simply let this boundary layer become infinitely thin. Then, however, the field equations must apply within the boundary layer; from this follows the existence of the differential quotients, from this again the continuity of the tangential components, etc.

Now, this reasoning is inadmissible here for gravitation. The substitution of the jump with a boundary layer is unfeasible. For on one side is a tensor field: $G_{i\kappa} = -\kappa T_{i\kappa}$, on the other there is none: $G_{i\kappa} = 0$. So qualitative differences do exist. Provided, the $T_{i\kappa}$'s changed continuously from 0 to infinite values. But this is not the case for the mass densities in Schwarzschild's example. That is why (you have probably overlooked this) the differential quotient, according to the normals of

$$g_r = \frac{1}{1 - \frac{r^2}{R^2}} \text{ inside,} \qquad \text{or } g_r = \frac{1}{1 - \frac{a^3}{R} \cdot \frac{1}{r}} \text{ outside,}$$

are not constant for him either! Only g_4 is constant for him, because he assumes $T_1^{(1)} = p$ to be constant. There is a priori no compelling reason for this though, otherwise the mass density (in my depiction: $\dfrac{\varepsilon}{c^2}$) would likewise have to increase continuously a priori from 0 to finite values. Then and only then would you have a link to electrodynamics. But this would apparently not be a significant move for the theory.[21]

As far as the validity of your equations in the boundary surface is concerned (incidentally, a mathematical argument, not a physical one), the Maxwellian boundary layer is eliminated. Therefore we have an abruptly discontinuous surface and here, as is known, the unilateral partial derivatives then occur (hence, in the above example: $\left(\dfrac{\partial}{\partial r}\right)_{r=a+0}$ or $\left(\dfrac{\partial}{\partial r}\right)_{r=a-0}$), by which the continued existence of the equations seems assured.

I thus believe not to have been so "entirely inconsistent." Allow me, on the contrary, to draw your attention to the physical arguments that seem to speak for my solution:[22]

First of all, the isolated sphere's total mass becomes infinitesimal. This, as is generally known, has long been one of your own postulates, which had not been satisfied until then.[23]

Second, the Maxwellian stability problem of the ether (§28 of my paper)[24] requires that gravitation have not just an attractive but also a repulsive effect. This, however, plainly determines the discontinuity of the normal derivatives $\dfrac{\partial g_{i\kappa}}{\partial n}$.

Third, this intramaterial repulsive effect of gravitation, that is, the "impene-trability of matter," seems to be confirmed by some results in radioactive research: deflection [*Knickung*] of α-rays (Rutherford), high velocity of β-rays emitted from the "nucleus," etc. Here I lay as a basis a conception of the atom differing from the common view, however. According to this view, not electrical forces, as for J. J. Thomson and Bohr, but gravitational forces would be the *normal* bonding forces for the atom; the nucleus would normally be at an electrical equilibrium and neutral, etc. This interpretation would, in my opinion, explain Bohr's mechanism (comp. the radiationlessness of the planetary orbits in the cosmos, etc.).

4) As concerns the relativity of acceleration, finally, I request you read the quote from Poincaré (§33), not because I would like to hide behind an authority, for instance, but because I hope that Poincaré is more comprehensible than I.[25]

Although you kindly suggest a verbal discussion, I doubt whether my oral expression could become clearer than the written form. In this connection, do permit me, nevertheless, to take the liberty of stressing that the inner consistency, which based on my intuition is not lacking in my papers, offers a certain guarantee that they would not seem to be entirely worthless. Yet I am convinced that the ideas defended by me would have done better not to have sought out me but another as their advocate who possessed a better talent for presentation, and who possibly might also have had more luck. Then they could certainly have come further, instead of enjoying very extensive disregard today.

But from you, Professor, whom I know to be very objective and to have already adopted other unconventional theories, I hope that, while obviously maintaining the distinct differences in fundamental outlook, you will not ignore my depiction without an attempt at extracting its intrinsic value.

I can say only subjectively, of course: I believe it is worth the effort. Does any objective gauge of value exist at all? You said to me in Vienna after the scientific convention:[26] "Nothing is true, of course!" Allow me to add: "But something must be most probable." This seems to me to be both the goal and limit of all research. Knowing this, there can be no polemics, thus no fight over the truth, which does not exist, rather only a communal effort toward probability.

My letter has become lengthy; for this I must beg your pardon. As concerns my current activities, about which you were so kind as to ask, I am still per-severing in arranging the mentioned study trip to Berlin regarding the Mintrop apparatus,[27] for which I have been temporarily withdrawn from the field and appointed to the Technical Military Committee. Maybe it really will happen; it would then be a particular pleasure for me, assuming your complete recovery, to visit you, not just to discuss my papers with you, but also to receive fruitful advice myself.

In indulging myself in the hope of again receiving a letter from you when the occasion arises, I am, in expressing my most cordial respect, yours most truly,

F. Kottler.

496.   From Romeo Wankmüller

Berlin, 30 March 1918

To Prof. Einstein, Berlin, 5 Haberland Street.

Esteemed Professor,

With the enclosed I convey to you a program for the convention of the regular members of the Scientific Society for Aviation in Hamburg from 16–18 April of the cur[rent year] for your obl[iging] perusal.[1]

Mr. Maschke[2] and I have the intention of attending the lectures there, and I take the liberty of inquiring whether your state of health permits your attending the lectures as well. It goes without saying that my company will assume the costs for the stay in Hamburg.[3]—I think the talk by Prof. Prandtl should be of very particular interest[4] and allow myself to propose that you perhaps order the printed text of the lecture for study which, as the prospectus indicates, is available.

Mr. Maschke is member of the Scientific Society and would like to enroll you and me as members. For this a second sponsor is also needed, whom Mr. Maschke suggests be Dr. von Parseval.[5] In case you are in agreement, may I ask you to inform Mr. Maschke of this and likewise to forward both prospectuses to the same?

I also take this opportunity to wish you a speedy recovery and a happy Eastertide, and to pay my respects as yours very truly,

R. Wankmüller

2 enclosures.

497.   From Hermann Weyl

Elmshorn [near Hamburg], 5 April 1918

Highly esteemed Colleague,

As I see that the calculational execution of the theory, about which I told you,[1] will occupy me for quite a long time still, I would like to publish a report on its foundations beforehand. I would be exceedingly grateful if you would present to the Berlin Academy the note of 16 $1/2$ octavo pages sent out in the same post.[2] It would be preferable to me to be able to read the correction proofs once myself; the resulting printing delay of about 2 weeks is, in any case, not an issue for me, and I am then sparing you this effort as well.[3] At the end of next week I shall be back in Zurich. I hope it is possible for you to come to Switzerland in the summer for a longer period and convalesce there thoroughly! With many thanks in advance and best regards, reverentially and devotedly yours,

H. Weyl

P.S. I have accepted the call to Breslau.[4]

498.   To Hermann Weyl

[Charlottenburg,] 6 April 1917[18][1]

Dear Colleague,

I shall be pleased to submit your paper[2] to the Academy next Thursday[3] and see to it that you receive the correction proofs.[4] I am very curious about the content. I heartily wish you luck with your new post.[5] For Zurich it is unfortunately a great loss!

With best regards, yours very truly,

A. Einstein.

Your paper has arrived. *It is a first-class stroke of genius.* However, I have not been able to settle my measuring-rod objection yet.[6] More details on this another time.

499.   To Hermann Weyl

[Berlin,] Monday [8 April 1918]

Dear Colleague,

I beg you to send me *immediately* a brief abstract of your paper for the *Sitzungsberichte.*[1] (Length 5–6 printed lines.) This really should definitely come from *your* pen. In any event, I am submitting the paper on Thursday. If your summary is not here by then and I cannot delay the printing anymore, I shall write the abstract myself.–

Your chain of reasoning is so wonderfully self-contained.[2] The deduction of the dimension number 4 impressed me very much as well.[3] The decomposition of your invariant of "weight" zero also is very striking.[4] Except for the agreement with reality, it is in any case a grand intellectual achievement.

With cordial regards, yours,

Einstein.

500.   To Felix Klein

[Berlin, 10 April 1918]

Esteemed Colleague,

I have received your wonderful expositions[1] and am studying them diligently. When I am finished with them, I shall give them to Sommerfeld. Levi-Civita and Weyl have offered a new, very intuitive interpretation of the Riemann curvature tensor,[2] Weyl additionally a very ingenious theory through which he attempts to merge gravitation and electromagnetism into a unified whole.[3] I am presenting his paper[4] tomorrow to the Academy.

With respectful greetings, yours truly,

A. Einstein.

501.   From Willem de Sitter

Leyden, 10 April 1918

Dear Einstein,

I just received your paper "Critical Remarks on a Solution Offered by de S[itter]."[1] You demand "that equations (1) be valid for *all* points within the *finite realm*."[2] Formulated this way, it is a *philosophical* requirement. To make it into a *physical* one, one must say "all *physically accessible points*." The surface $r = \frac{1}{2}\pi R$, however, is physically *inaccessible*, as I have shown in *M. N.* LXXVIII, pages 17–18,[3] and my solution therefore satisfies the *physical* requirement, but not the *philosophical* one.[4] You naturally have the right to lay down the philosophical condition and thus to reject my solution. I, on the other hand, also have the right to reject the philosophical requirement, but not the physical one. Whether my solution really can be interpreted as a world in which all matter is concentrated on the surface $r = \frac{1}{2}\pi R$,[5] would have to be calculated out—I do not know whether it is so or not.

With cordial regards, yours,

W. de Sitter.

502.   To Hugo A. Krüss

Berlin, 5 Haberland St. [before 11 April 1918][1]

Highly esteemed Professor,

With reference to the interview you recently granted my cousin relating to the request concerning my terms of employment, I take the liberty of presenting the situation to you again in writing.[2] The following is involved.

Linked with my employment is the right to a widow's pension. I now intend to be divorced in order to remarry. Through this, my present wife would lose her pension entitlement to the advantage of the second wife. This situation would entail a hardship for my present wife, however, and is at odds with my sense of justice, especially since my financial circumstances would not make it possible for me to compensate for this loss sufficiently otherwise.

My request now consists in having my employment contract modified so that in the case of my death my current wife becomes the sole beneficiary of the pension, even though she be divorced from me.[3]

I entreat you not to look upon my request as presumptuous, and I add in its favor that the granting of it would not create any troublesome precedence case; for my engagement is of a thoroughly unique nature and does not fall under any government-position category.[4] Furthermore, this concession would result in absolutely no disadvantage to the State Treasury.

I would be greatly obliged to you for your support in this matter.

In utmost respect, I am yours truly,

A. Einstein.

503.   To David Hilbert

[Berlin,] Friday [12 April 1918][1]

Highly esteemed Colleague,

That was terribly bad luck I had, not being able to see you while you were traveling through, despite your having been so nice as to look me up. Yesterday was the second time, after a 3-$\frac{1}{2}$-month confinement to my room, that I could attend a meeting of the Academy,[2] and I used the occasion to eat dinner afterwards at a friend's, without having to go out especially for it. Through the maid's negligence, I received your kind little card so late that now I cannot seek you out anymore. I am very disappointed at having thus been deprived of the pleasure of seeing you.

I take the opportunity to thank you warmly for having arranged that I receive that scientific prize, the substantialness of which put me in a state of happy astonishment. I can imagine that you were a critical factor in the decision.[3]

You can imagine how much I would like to be present at Planck's lectures in Göttingen.[4] But I believe it unlikely that I could afford myself this pleasure, because my mobility has been diminished very much by the gastric ulcer. Recently, for inst., I had a nasty attack that had apparently been brought on by my having played the fiddle for just an hour or so.

My friend Ehrenfest wrote me a postcard, from which I gather that he has written you a quite foolish letter but which he doubts you received. It is a reply to a friendly invitation he had received from you and Planck in which he presents in detail why he feels unable to come to Germany now.[5] His understandable bitterness over politics has swollen to such a degree that now he must give vent to his heart on such an entirely unsuitable occasion. I am convinced that you see the decent fellow behind this clumsy, almost childish conduct and do not hold this blunder against him so much. Such a character is surely a thousand times more welcome than the many miserable fawners who populate our faculties in the main. Please do inform me whether you have received his letter, so that I can answer him and—give him a good dressing down. I am going to write to him as follows:

Dearest little Paul, imagine you had been appointed to Göttingen a few years ago instead of to Leyden, and in the spring of 1918 you had amicably invited our colleague Keesom and you received from him a knock on the head![6] I believe you would be less than enchanted by it and would rather proclaim that Keesom was a lout.—(Application example for the relativity principle.)

I have been exchanging a bit of correspondence with Prof. Klein regarding the energy principle.[7] My t_σ^ν's are being rejected by everyone as unkosher.[8] But *one* thing I must note, without wanting to beautify my skewed holinesses in any way. It is not a matter of *making do*[9] with an energy law of the form

$$\sum_\nu \frac{\partial \mathfrak{T}_\sigma^\nu}{\partial x_\nu} \cdot \frac{1}{2} \sum_{\mu\nu} \frac{\partial g^{\mu\nu}}{\partial x_\sigma} \mathfrak{T}_{\mu\nu} = 0.$$

For it can never be concluded from it, for ex., that the mechanical energy of the planetary system (e.g., also the average kinetic energy) cannot extend into infinity.[10] But this is one of our most certain general observations, and it is clear that this must be expressed.

H. Weyl submitted through me a highly interesting paper to the Academy here, in which he seeks to comprehend gravitation and electromagnetism as a unifying geometrical system of concepts.[11] Mathematically, the thing is wonderful. But *physically* it does not seem acceptable to me.[12] I am curious what you will say to it.

With cordial regards, I am yours very truly,

A. Einstein.

504.   From Wilhelm Schweydar

Targsorul vechi [Romania] 14 April 1918

Dear Colleague,

Your amicable letter of 6 April delighted me; here in surroundings foreign to me[1] I very particularly appreciate any cause for joy. Thank you for the fine idea of sending me to Odessa to retrieve the instruments.[2] I shall be glad to do it, if Odessa is accessible to me and I am not forced to exchange correspondence with military authorities about it (they grate on my nerves here, there is nothing worse). I could possibly get there by ship from Constanţa.[3] In mid-June I intend to interrupt my research here,[4] because the temperature conditions will have become too unfavorable, so I could travel to Odessa in the second half of June, if the Academy approves. I imagine the trip there to be very beautiful and interesting.

Your report that Wiechert will be the director of our institute pleased me as well.[5] I also consider it the best solution and believe that Helmert would certainly have approved of this solution as well, if he could have been asked;[6] he thought very highly of Wiechert. Thank you for the information that Eötvös had written so appreciatively of me and had suggested me for such an honorable post;[7] but I do feel a bit dispirited by it, because I fear that the gentlemen are overrating me.

I made Eötvös's acquaintance in Hamburg at a conference in 1912 and found him very congenial.[8] He is strikingly modest and benevolent. Helmert also was similar.   How drawn one feels toward a person who is not only intelligent but also has a fine character! Only rarely does one come across such people, though. You are right, intelligence without good character turns people into beasts. This simple truth was clear to me long before the war, and I did not understand why persons whom I had respected before the war sought means suitable for destroying millions of people in a single stroke. If I had not had Ad. Schmidt[9] with me at Potsdam, I would have felt completely isolated. Here in the occupied region one also sees more clearly the other side of what had been commended to me so loudly at home as patriotism and what I had once in agitation called the most dangerous of all virtues.

I hope Wiechert accepts the appointment.[10] A Göttingen man has certainly already been designated as his successor. In the winter I sent Hilbert and Wiechert some of my papers, but the gentlemen will probably barely remember that by now, and my prospects of coming to Göttingen are likely as slim as the chance of an imminent peace.

The arrival of the crate so quickly and in such good condition pleased me very much. If you have a craving for beans and peas, I can procure them for you very easily; this is the only thing that can still be gotten hold of easily.

Hoping that you are feeling better now than you did in the winter, I am with best regards, yours truly,

Schweydar

P. S. In accordance with your wish, I also inform you that my disbursements amount to about 4 M.

505.   To Mileva Einstein-Marić

[Berlin, before 15 April 1918][1]

Dear Mileva,

I have received the draft contract and send my cordial thanks to Dr. Zürcher for his help.[2] Now I, on my part, have emended suggested alterations, which I am going to comment on right away. I have entered them on *my* copy as well. You can then send this copy back to me with your suggested changes, in case you would like to add any. Then, when we are in agreement, I'll have two duplicates made, *one* of which you'll receive signed.

My changes concern the following points.

1) I have set my annual payment at Fr 8 000.[3] This is certainly plenty. I cannot go further without having to fear that I myself fall into difficulties.

2) The deposit of the money *in Switzerland* will not be possible now. I already have problems, you know, in getting the regular remittances for your upkeep sent over. The sum can be secured just as well if it is placed in trust *here* with the necessary safeguards. Moreover, I'll see that the affair at the Ministry is settled soon. If you were to receive the pension, that would be the best solution.[4]

3) It would be all the same to me, for easily guessed reasons, if the Nobel Prize eventually came into *your possession*. We could then arrange it so that in this case your income was raised to Fr. 9 000.

4) Surely you don't want to demand that I always come to Switzerland (in peace-time) to see the children. No fair person knowing the circumstances would condone such an imposition.

Do you insist on having the legal process conducted in Switzerland? It seems to me that it would go more quickly here; if you all think that it would go smoothly in Switzerland as well, I am satisfied if it's done there. The reservations I had expressed against it[5] seem to have been unfounded.

I am curious what will last longer, the world war, or our divorce proceedings. Both began essentially at the same time.[6] In comparison, this little matter of ours is still much the more pleasant.

Amiable greetings to you and kisses to the boys from your

Albert.

P. S. As long as you are alive, the children should have no right of disposal to the available money (Nobel Prize) except, of course, in case you get married. The money seems best secured if it is in Switzerland and is your property.

506. To Willem de Sitter

[Berlin, 15 April 1918]

Dear Colleague,

I understand your point of view, since although the metric distance to the "equatorial surface" is finite, the time that a mass-point needs to move there is infinite.[1] Now H. Weyl has actually shown, though, in a forthcoming book[2] that your continuum can be understood as the limiting case of a fluid distributed around an "equator."[3] The calculation is very simple. It thus really does involve a surfacelike singularity which is completely analogous to that of a mass-point.[4]

With best regards, yours,

A. Einstein.

Best regards to Ehrenfest, to whom I shall reply shortly.

507.   To Hermann Weyl

[Berlin,] 15 April 1918

Dear Colleague,

I presented your paper on Thursday[1] but did not yet submit it for printing, because I wanted to wait still for your short abstract (6–8 lines). But you apparently did not receive the postcard I sent you requesting it in time.[2] I am definitely handing in the paper on Thursday and shall write a short abstract to it myself. You can then replace it with another one more to your liking at the correction stage.

As pretty as your idea is, I must frankly say that in my opinion it is out of the question that the theory corresponded to nature.[3] For, the ds itself has real meaning.[4] Imagine two clocks running equally fast at rest relative to each other. If they are separated from each other, moved in any way you like and then brought together again, they will again run equally (fast), i.e., their relative rates do not depend on their prehistories.

I imagine two points P_1 & P_2 that can be connected by a timelike line. The timelike elements ds_1 and ds_2 linked to P_1 & P_2 can then be connected by a *number* of timelike lines upon which they are lying.[5] Clocks travelling along these lines give a fixed relation $ds_1 : ds_2$ independent of which connecting line is chosen.—If the relation between ds and the measuring-rod and clock measurements is dropped, the theory of rel. loses its empirical basis altogether.

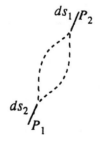

Best regards, yours,

Einstein.

508.   From Hugo A. Krüss

Berlin W. 8, 15 April 1918

To the Hon. Prof. Einstein, Berlin, 5 Haberland Street.

Esteemed Professor,

I am deeply sorry not to be able to give you a more favorable response to your inquiry.[1] The regulation on widow's pensions is grounded on general legal foundations[2] from which exceptions cannot be made in individual cases by special agreement. Pursuant to it, the widow's pension can be paid out for disbursement only to the current lawful widow; a transferral is explicitly ruled out pursuant to the law. Accordingly, I can think of a settlement only in the form of your taking out a life insurance policy to the benefit of your first spouse, in the case of your death,[3] and in the amount of the lawful widow's pension. In this way, your first spouse would be provided with the full equivalent of a state widow's pension.

What annual premium payments will be required for this I cannot estimate, yet it does not seem out of the question to me that the Minister[4] would be prepared to help you, in case you were to regard it as too great a burden on your annual outlay.

Perhaps you can familiarize yourself more closely about the options of such an insurance and then let me know more about it. I shall be glad to assist you in this matter, as far as it lies in my power.

With all due respect, yours very truly,

Krüss.

509.   From Hermann Weyl

Zurich (20 Schmelzberg St.), 15 April 1918

Esteemed Colleague,

Both your postcards,[1] for which I thank you most warmly, likewise for the presentation of my note[2] before the Berlin Academy, unfortunately only reached me here upon my return to Zurich via Breslau.[3] I am sorry that now you probably have had to draw up the abstract.[4] I had great reservations about sending you the note in its current state, which I had already had lying around for 3 weeks, and would have liked to have carried the physical consequences a bit further beforehand. But I was so occupied at that moment and the calculations to be performed are so extensive that I decided upon such a provisional communication after all. The idea is, at all events, worth being considered and carried through to completion. With cordial thanks and regards, yours very truly,

H. Weyl.

510.   From Margarete Hamburger[1]

Berlin, 16 April 1918

Esteemed, dear Professor,

Accept my very sincere thanks for your dear little letter. Your so many generous words in it about a couple of small favors I was allowed to render you touched my heart, and now I am quite embarrassed. It has always done me so much good to be allowed to do something for you, at least in this form, and you would please me very much also to want to see behind the little material gifts a mentality full of respect and gratitude for the wonderful person Einstein.

Your inspired drawing, which gives me a glimpse of your psychophysis and is surely applicable to all other human beings,[2] amused me very much—after some reflection. I am full of admiration for the simplicity and grace with which you know how to make physics useful in an intuitive way to very personal experience and how to draw soul and nature under a single heading. My son, who explained to me the physical apparatus, also enjoyed this original symbolism very much.

With warmest regards, which my son joins me in extending, also to the best of all women, I am yours truly,

Margarete Hamburger.

511.   To Hermann Weyl

[Berlin, 18 April 1918][1]

Esteemed Colleague,

Busily involved in studying the details of your book,[2] I constantly admire anew the beauty and elegance of your derivations. Now, though, in the last § I come upon a conclusion that seems to me to be wrong.[3] Namely, you find that static, spherically symmetric solutions correspond to the elliptic type, since they are all symmetric with respect to an "equatorial surface."[4] However, the latter does not apply according to your own solution. For you examine the case of an equatorially distributed fluid. But your calculation[5] immediately yields the following asymmetrical case, which I show with a figure:[6]

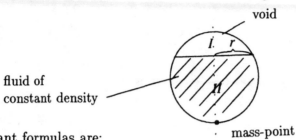

The relevant formulas are:

for I: $\dfrac{1}{h^2} = 1 - \dfrac{\lambda}{6}r^2$

for II: $\dfrac{1}{h^2} = 1 + \dfrac{2M}{r} - \dfrac{2\mu_0 + \lambda}{6}r^2.$

The boundary condition

$$M = \frac{\mu_0}{6}r_0^3$$

is exactly the same as in your case, whereas M is the mass of a real point mass.[7] Evidently, for this M, one can also substitute an extended homogeneous fluid.

It thus seems that no grounds are provided for space possessing the connectivity properties of elliptic geometry.

With cordial regards, I am yours very truly,

A. Einstein.

I hope you have received my postcard,[8] in which I reported to you more exactly the objection bothering me with regard to your new theory. (Objective meaning for ds, not just for the ratios of different ds's originating from one point.)

512.   To Hermann Weyl

[Berlin,] 19 April 1918

Dear Colleague,

Another letter from that Einstein! This time I must give a complete report on the submission of your article;[1] for a difficulty has arisen which I have unfortunately not been able to master until now.

A week ago yesterday I presented the paper at the class meeting.[2] I sketched the train of thought, first from a purely geometrical point of view, and then its application to the theory of relativity. At the end I also described my objection to it, which you already know. (In my view, ds itself has physical meaning.)[3]

Then Nernst stood up and protested against acceptance of the paper without further comment; he demanded that I at least attach a note in which I describe my different standpoint. Planck then suggested I consider the matter for a week and then submit the paper again, with or without comment, as I consider appropriate.

This I did yesterday at the general Academy meeting. There I did not present the paper factually once more but just characterized briefly the existing problem. Since I considered it tasteless to attach a kind of "protest" to the paper, I submitted it, declaring that I would express my differing viewpoint at some other time. Nernst then maintained his view again and backed it with a "precedent." The Secretary (Diels)[4] took on his (Nernst's) position. According to him, the article can be accepted without objection if at the end of your paper (e.g., in a postscript) you address my objection. The latter you could formulate, for instance, as follows:[5]

If light rays were the only means of establishing empirically the metric conditions in the vicinity of a space-time point,[6] a factor would indeed remain undefined in the distance ds (as well as in the $g_{\mu\nu}$'s). This indefiniteness would not exist, however, if the measurement results gained from (infinitesimal) rigid bodies (measuring rods) and clocks are used in the definition of ds. A timelike ds can then be measured directly through a standard clock whose world line contains ds.

Such a definition for the elementary distance ds would only become illusory if the concepts "standard measuring rod" and "standard clock" were based on a principally false assumption; this would be the case if the length of a standard measuring rod (or the rate of a standard clock) depended on its prehistory.[7] If this really were the case in nature, then no chemical elements with spectral lines of a specific frequency could exist, but rather the relative frequencies of two (spatially adjacent) atoms of the same sort would, in general, have to differ.[8] As this is not the case, the fundamental hypothesis of the theory unfortunately seems to me not acceptable, the profundity and boldness of which must nevertheless instill admiration in every reader.

I am terribly sorry that the affair has turned out to be so problematic. But I beg you not to be angry with me on this account. I *could* not withhold my differing view. I had absolutely not expected that I would encounter difficulties in the paper's acceptance for this reason. I am letting the paper lie now until I have received word from you about what I should do. Perhaps I ought to have waited with the submission until you had reported to me your position on the objection; but I did not consider myself entitled to hold back your manuscript for so long.

With cordial regards, yours very truly,

A. Einstein.

513.   To Hermann Weyl

[Berlin, 19 April 1918]

Dear Colleague,

My third message since yesterday![1] There is an error in my letter of yesterday. The indicated solution was physically meaningless, because the (positive) constant M implies a negative mass.[2] Another solution that is not symmetric around the equator,[3] without a point mass and without negative densities, is obtained as follows:

The solution is in three parts according to the following model:[4]

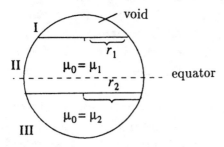

I: $$\frac{1}{h^2} = 1 - \frac{\lambda}{6}r^2$$

II: $$\frac{1}{h^2} = 1 + \frac{2M}{r} - \frac{2\mu_1 + \lambda}{6}r^2$$

III: $$\frac{1}{h^2} = 1 - \frac{2\mu_2 + \lambda}{6}r^2$$

The two boundary conditions are

$$\frac{M}{r_1^3} = \frac{1}{6}\mu_1.$$

$$\frac{M}{r_2^3} = \frac{1}{6}(\mu_1 - \mu_2)$$

Choose r_1 and μ_1. Then M is determined by the first equation. If one now chooses $r_2 > r_1$, μ_2 can always be chosen (with a positive sign) in such a way that the second boundary condition is also satisfied. In this way we have a non-"elliptic" type of solution that uses no negative masses.

Cordial regards, yours,

A. Einstein.

514.   To Heinrich Zangger

[Berlin,] 22 April 1918

Dear friend Zangger,

I was truly dismayed at your having been pestered behind my back by Else with the milk affair. I had strictly forbidden her to do so and ask you never to listen to her again.[1] It was not right to trouble you with such things after your misfortune,[2] and it is anyway very questionable whether it is objectively justifiable to deprive Switzerland of this nutritive foodstuff now, little though it may be, especially since I am otherwise perfectly well supplied.[3] Since the thing is done now, I thank you warmly; I would also thank Mr. Forrer,[4] but I daren't, considering that it would look like a request for the future. When you see him, though, do thank him modestly on my behalf.

Your misfortune, combined with the painful thought of having been unfair to you as well in what I had said, wrings my heart.–[5]

My wife is behaving very nicely towards me now. We are negotiating everything (even the divorce matter) directly with each other, and a thoroughly friendly and good tone has developed between us. I see that I did much damage earlier with my vehemence and severity. But all will be well and everyone will be satisfied.

The horrifying goings on in the wide world seem to want to continue forever, and all superior and finer things, to go into hibernation. The attitude of my personal contacts is becoming increasingly inflexible and unpleasant.

Scientifically there is little news. I have managed to explain a light edge on an X-ray image very probably as total reflection. Thus there is the prospect of finding out the refractive index of the substances for Roentgen rays, which is very interesting.–[6] Presently I am reading Rousseau's *Confessions*, a wonderful book and—the Bible.

Heartfelt greetings, also to your sorely afflicted wife, yours,

Einstein.

515.   To Mileva Einstein-Marić

[Berlin,] 23 April 1918

Dear Mileva,

Thank you very much indeed for the friendly postcard about the children's good state of health. I hope Albert will write me soon also about how Tete behaved on beginning school.[1] I received a negative reply from the Culture Ministry, which is nonetheless favorable to the extent that they are otherwise willing to be of assistance, although along a different route.[2]

I have come to the conclusion that, with the planned course, you all are not sufficiently provided for if I do not receive the Nobel Prize. I've found ways and means to make the following possible:

1) About 40 000 M in securities will be deposited at a Swiss bank.[3] In the case of a divorce, they will become *your property* with right of disposal of the interest, but without authority over the disposal of the *principal.*

2) In addition, 20 000 M will be held in deposit at a bank here with the instruction that you receive the interest after my death, in case I do not receive the Nobel Prize.[4]

Based on my inquiries, it will be possible after all, with the help of the Ministry of Culture, to transfer the 40 000 M to Switzerland.[5]

---

3) In the event I receive the Nobel Prize, it belongs to you (without authority over the *principal*), less 40 000 M.[6]

4) During my lifetime, I'll send you, quarterly, enough so that the remittances, together with the interest indicated under (1) of the capital deposited in Switzerland, amount to Fr. 8 000.

Thus it will be guaranteed that you all will not become destitute, even if I die in the immediate future.

I hope that this plan will be final at last. Inform Dr. Zürcher[7] of this letter, but request that he wait still with the definitive formulation until its executability is secured. You then just need to send me *one* copy, because we can transcribe it here.

I'm coming to Switzerland in July, if possible, to take the boys with me into the mountains; I'll visit you all then in Zurich.

Best regards to the three of you, yours,

Albert.

I've been so occupied now with the conditions following my death that it seems quite odd to me that I am still alive.

[6]Recipient's emendation: "and will be depos[ited] in Switz[erland]."

516.   To Auguste Hochberger

[Berlin, before 24 April 1918][1]

Dear Guste,[2]

From Mama's letter I learned with great dismay that you have lost your brother.[3] I can imagine how heavy a blow this is for you. Rest assured of my deep sympathy. Also please convey my condolences to Mrs. Victor and the children.

In big things as in small, life is hard, and destiny is blind and without mercy. But there are honest persons who go through life steadfastly and honestly nonetheless and are able to keep a sound mind while doing so. You are such a creature; that is why even after this severe blow you will regain your balance. May your family and friends, my mother among them, support you in this!

My mother's Oppenheimer-fatigue makes an imminent change of air desirable.[4] As soon as the ice has melted in the railway cars, she should be coming up north to us or to Uncle Jacob to the south or both.[5] It is a pity that you old girlfriends will thus be separated once again.

With best regards, yours,

Albert.

[. . .][6]

517.   To Auguste Hochberger

[Berlin, before 24 April 1918][1]

Dear Guste,

You sent me a very splendid thing indeed with these apples, fit to provoke the envy of all of Berlin. Every day I eat one with such intense contemplation as though it involved the execution of a religious ⟨rite⟩ function. Thank you from the bottom of my heart.

You would not believe what a cozy situation one gets into, having such an illness. One appears to others in almost as rosy a light as if one were already dead, and yet life is very enjoyable. Everything, even laziness, is gilt by sympathy, everyone is so concerned, and no one is malicious or sly. In short, being sick is nice, but it is naturally not allowed to get out of control . . . .

Now Mama's Heilbronn days are numbered. That will be a leave-taking between the two of you to melt a heart of stone. Long, long ago, in Heilbronn I also had the pleasure of sharing solumn farewells of this kind.[2] But this time it is in earnest, for Mama's love for Mr. O. seems to belong so entirely to the past[3] that what remains could not even serve to blacken a fingernail. We are very eager to have her here with us.[4] Now is also the best time for walks, you know; but in order not to starve along the way, one must bring along one's slice of bread [*Stulle*].[5]

With warm regards and once again many thanks, yours,

Albert.

[. . .][6]

[5]Berlin dialect for open-faced sandwiches.

518.   From Felix Klein

Göttingen, 25 April 1918

Esteemed Colleague,

You will be seeing Sommerfeld, of course, in the next few days on the occasion of the Planck festivities;[1] I have already written him that he best discuss my elaboration with you personally. Besides, I would appreciate it if Dr. Freundlich also were willing to review my elaboration.[2]

A factual remark (about which I wrote to de Sitter lately in more detail):[3] In your "Cosmological Considerations" of 1917, the spherical space cannot be substituted by an elliptic one:[4] Namely, because the "elliptic" plane is a one-sided surface (or as I said earlier: a double surface) where the direction of rotation of an indicatrix ↻ reverses again when sent over the surface, a specific arrow cannot be assigned to the world lines of its points—in other words, no distinction can be made between past and future.[5]

Very truly yours,

Klein.

519.   To Mileva Einstein-Marić

[Berlin,] 26 April 1918

Dear Mileva,

I'm giving in about the children because I have now come to the conviction that you want to handle matters in a conciliatory manner. Of course, I would absolutely not have the children travel here alone in these times. Maybe you yourself will later also take the view that you can give the boys to me here without reservations. For the time being, I'll see them in Switzerland. The contract could perhaps state "outside of Prof. Einstein's town of residence" or something similar, instead of "in Switzerland";[1] I do not insist on it, though. I hope to be able to satisfy you enough that you'll have the motivation to be accommodating toward me as well. I hope that Albert will write me soon.

Best regards also to the boys, yours,

Albert.

520.   To Hans Albert Einstein

[Berlin, after 26 April 1918][1]

My dear Albert,

I was delighted with your letter. It's a pity that your projects are suffering so under the wartime hardships, but slowly, very slowly you'll manage. The main thing is, though, that there is still enough to fill your hungry mouths, even if an awful lot of time must be spent in getting everything.[2] I'm particularly pleased with your cheerful mood and with the fact that you enjoy music so much.[3] That's something that remains staunchly with you, your whole life long. I am feeling reasonably well, I just shouldn't walk much and am on a special diet.[4] But I'm in high spirits and work just as well as in the earlier days. Also, I am lecturing again at the university.[5] Recently I gave a speech on Prof. Planck for his 60th birthday, which pleased him and many others.[6] I would have liked to have seen Tete on his first day at school.[7] Tell me a bit about it. He should also write me a few words of greeting, as best he can, at the bottom of your next letter. Is his health good, his friend the little Zürcher boy's also?[8] Does he still spend much time together with him?

In 6 weeks I'll be coming to see you all and am looking forward to it very much. We are going to go somewhere up in the mountains. I am still not allowed to walk much, but that doesn't matter. We might be going to the Engstlen Alp.[9] Get information about other places as well that are at an altitude of at least 1,500 meters.

Kisses to both of you from your

Papa.

521.   To David Hilbert

[Berlin, before 27 April 1918][1]

Esteemed Colleague,

Countless times in these desolate years of general nationalistic delusion, men of science and the arts issued statements to the public that have already inflicted incalculable damage to the feeling of solidarity that had been developing with such promise before the war among those who devote themselves to higher and freer purposes.[2] The hue and cry raised by strait-laced preachers and servants of the bleak principle of power is becoming so loud and public opinion is being misled to such a degree by methodical silencing of the press that those with better intentions, feeling wretchedly isolated, do not dare to raise their voices. Day by day the danger is growing that even those who have been clinging with all their might to the ethical ideals of a happier phase in human development will eventually despair and will also fall victim intellectually to the general derangement. This serious situation places those, who through fortunate intellectual achievements have gained an elevated position among scholars throughout the entire civilized world,[3] before a mission they must not evade: They must make a public declaration that could serve as support and consolation for those who in their solitude have not yet lost their belief in moral progress.

I am thinking of the following. Each of us provides an avowal, in the form of a short essay[4] of up to about 10 printed pages, designed to make an impact in the above-indicated sense. These essays appear, assembled together in a little volume, on the book market as evidence for international sentiment, that is, probably issued initially in neutral countries abroad. In order to stress the international character, we could try to acquire contributions by men from lands currently in a state of war as well as from neutral nations.

This letter is being sent not only to you but also to a very few reputable men.[5] Please tell me frankly whether you approve of this undertaking and whether you would be inclined to make a contribution.

In great respect,

A. Einstein.

522.   To David Hilbert

[Berlin, before 27 April 1918][1]

Esteemed Colleague,

I wrote you a letter[2] a while ago in order to excuse myself for not being able to get in touch with you during your last visit here. A chain of unfortunate circumstances deprived me of the pleasure of seeing you. At that time I also asked you for information about whether a letter directed to you by Ehrenfest had arrived; Ehrenfest apparently has doubts and therefore requests of me to ask you.[3] Did you receive my letter? Please do clear up both of these issues for me.

The attached little communication concerns an endeavor that was inspired by[4] some like-minded persons.[5] The Rascher publishers in Zurich have taken on the printing.[6] I would be very pleased if an imposing little volume of welcome statements of opinion were compiled. For the moment this letter is going out, in addition to you, also to your colleague Prof. Lehmann (Göttingen), Prof. Troeltsch (theologist, Berlin), Prof. Schücking (Marburg), Brentano (Munich), and Weber (Heidelberg).[7] No one whose name has been tainted by chauvinistic declarations of any sort is being invited. Please tell me *very candidly* your opinion of this effort.

With cordial regards, yours very truly,

A. Einstein.

Could you perhaps name a few suitable persons to whom I could also turn concerning this matter? Among mathematicians and physicists in Germany you are unfortunately the only one to whom I can write about this affair.

[4]Deleted text: "initiated by the Kassirer publishers. This man is in Switzerland and has placed his renowned publishing house completely in the service of international causes. Upon his inspiration I wrote this letter. For a start, it will [. . .]" Marginal note: "The crossed-out text was based on a mistake."

523.   To Felix Klein

[Berlin,] 5 Haberland St., 27 April 1918

Esteemed Colleague,

I already passed on your lectures to Mr. Sommerfeld the day before yesterday. Now I do not know whether it is your intention that I ask him to hand it over to Mr. Freundlich first.[1] I think that Mr. Sommerfeld will soon have picked out the choicest plums. Then, if you wish, he could possibly send it on to Mr. Freundlich.

It seems to me that no argument can be raised against your objection to the conception of the world as quasi-elliptic with respect to the *spatial* relations.[2] Your objection seems correct to me, upon elementary examination, just for *even* (spatial) dimensions.

I imagine an nth dimensional spherical manifold R_n (in an $n+1$th dimensional Euclidean space R_{n+1})

$$x_1^2 + \ldots + x_{n+1}^2 = \langle R^2 \rangle 1.$$

An elementary orthogonal "n-bein" is situated within an element of R_n. It can be moved along R_n.

I call the point $0 = x_1 = x_2 = \ldots x_n$, $x_{n+1} = \pm 1$ the "south pole" or "north pole" of R_n, respectively. The n-bein lies initially at the south pole in such a way that its legs

$$b_1 \ldots b_n$$

are parallel to the positive axis orientations

$$x_1 \ldots x_n, \text{ respectively.}$$

Now I move the n-bein on the sphere R_n from the south pole to the north pole in the following way:

1) The motion is restricted to motion along the meridian $x_2 = x_3 \ldots = x_n = 0$.
2) The legs $b_2, b_3 \ldots b_n$ constantly remain perpendicular to this meridian while in motion (determined in space R_{n+1}).

Upon arrival of the n-bein at the north pole, observed from within R_{n+1}, the orientation is hence given by the scheme

$$
\begin{array}{cc}
b_1 & b_2 \ldots b_n \\
- & + \ldots +,
\end{array}
$$

i.e., the orientation is the same as initially, except that the first leg has reversed its direction.–

Now if R_n represents an elliptic space, then centrically symmetrical points are identical. Our n-bein situated on the north pole is then identical to one to be found at the south pole with legs pointed in opposite directions (seen from R_{n+1}). This corresponds to the scheme

$$
\begin{array}{cc}
b_1 & b_1 \ldots b_n \\
+ & - \ldots - \, .
\end{array}
$$

It now comes down to whether during this round trip in the elliptic world the n-bein changes into a congruent or symmetrical one (viewed from R_n). It is easily

seen that through a rotation around b_1 (in R_n) the *n-bein* can be brought back
to its initial position

$$b_1 \qquad b_2 \ldots b_n$$
$$+ \qquad +\ldots+,$$

if n is odd, otherwise not, however. It thus applies in the three-dimensional case,
just as in the immediately clear 1-dimensional case. That the mirror image results
in the two-dimensional case is likewise immediately clear.—This, as it seems to
me, is evidently the crux of the matter. Am I not right with this consideration?–

By the way, in a wonderful book on general relativity, which is appearing in
the near future,[3] Weyl arrives at the result that the elliptic type *must* apply,
because the static solutions to the equations automatically reveal that spherical
[*zentrische*] symmetry. I find, though, that he is mistaken.[4] It seems the question
must remain open.

In great respect, yours truly,

A. Einstein.

524.   From David Hilbert

[Göttingen, 27 April 1918]

Dear Colleague,

Your earlier kind letter[1] pleased me very much; I wanted to reply to you
more completely and thus did not get to it at all until now. Ehrenfest's negative
response did arrive properly and was read immediately by my wife while I was
still in Bucharest. Neither she nor I took E's honesty badly, though.[2] I shall
answer your last letter, which I just received, in the next day or so, because I
would like to discuss the matter with some of my friends here beforehand. It
goes without saying that the ideas in your letter are a most appealing stimulus
to me.–[3] The most welcome news is that your health is so much better. Even I,
many years your senior,[4] still wish or hope to live to see sensible times, and for
this health above all is necessary, of course; one must therefore think of guarding
it more than before.

With best regards, yours,

Hilbert.

525.   From Hermann Weyl

Zurich, 27 April 1918

Esteemed Colleague,

Thank you sincerely for your hail of messages;[1] I hope nonetheless that it will not flatten my young seedlings entirely! You are undoubtedly right with your objections to the last pages of my book; indeed, I even see that the solution corresponding to a point mass demands not only a "mass horizon" but a permeation of mass covering a complete cap that is larger than half of the world.[2] I am making changes as best as I still can; you will see them on the printer's proofs. I am exceedingly grateful to you for having pointed out this error to me.– On the other hand, I do not grant you your objection to my theory of electricity and gravitation.[3] I am formulating my reply to it in a separate communication so that you can poss[ibly] present it to the Academy. Should objections again be raised, I would like to refrain from publication in the Academy's *Berichte*. I regret very much having caused you such trouble with the paper;[4] please do not hold it against me! (Through illness I was prevented from replying for a couple of days.)

With thanks and best regards, ever yours,

H. Weyl.

526.   From Hermann Weyl

Zurich, 28 April 1918

Esteemed Colleague,

Here is my reply now to your objection that ds^2 *absolute* has a real meaning.[1] I hope the fact that I am not receptive to your objection is not based solely on egoistical and mathematical infatuation with my theory. Various things might still be added, but the reply has anyway become somewhat lengthy already, and the essentials have surely been said. If you are convinced by my response (which I hardly dare to hope, though), I naturally also am agreed not to have this "addendum" printed along with it if you consider it better and the Academy consents. If the Academy is prepared to accept the paper *with* this addendum, I am then likewise satisfied.[2] Should you consider it called for, finally, to clarify the controversial question some more through written discussion before publication (which is unfortunately considerably hampered by the censorship delays), I am quite prepared to do that as well. But if you are entirely agreed to present the paper once again and it then again encountered rejection, I would request that you then state my desire to refrain from publication in the *Sitzungsber.* I do not find the Academy's position justified; but I must concede to Nernst:[3] if your objection is correct, which would mean that my theory consequently has nothing to do with reality, then it is worthless, despite all the "profundity and boldness"[4] you extol in it.   Thus I still do have faith. But these difficulties you have with the paper's submission are quite exasperating for me. Once again, cordial thanks! With best regards, yours,

H. Weyl

(You know what my attitude is toward you; I may surely dispense with repeatedly couching them in conventional formalities like "most devotedly," etc.)

527.   From Marga Planck[1]

[Berlin,] 30 April 1918

Esteemed Professor,

Thank you very much indeed for sending us your "sermon," which we read to each other once more yesterday.[2] We delighted in your thoughts again and now appreciate possessing them as a souvenir of the wonderful evening.[3] Also—this I would like to express to you today—I personally am quite especially pleased that my husband has found such a warm friend in you!

With best regards, yours,

Marga Planck.

528.   To Paul Ehrenfest

[Berlin, 1 May 1918]

Dear Ehrenfest,

Hilbert writes me that he received your letter and that he understands your attitude. I think it is not right to blow off steam at such men.[1] Imagine if you had been appointed to Göttingen a few years ago instead of to Leyden and were now held to account for the great reprehensible machinery! I have said nothing about the matter to Planck. He is like a child in public matters and would understand you as little as a cat understands the Our Father.–

Healthwise I am doing decently well, but I am going to have to abstain from a trip to see you for the time being; there is no lack of inclination, though! I convey my sincere thanks to Nordström for his interesting papers.[2]

Best regards to all of you from your,

Einstein.

529.   To Hermann Weyl

[Berlin, 1 May 1918]

Dearest Colleague,

Thank God your letter with the addendum arrived.[1] I am delighted about your conduct in this affair.[2] The paper is being submitted tomorrow (with addendum) and it naturally will be accepted. Planck is displeased that the Academy took this position; its members ought not to presume to form a kind of higher authority in scientific questions. I was very afraid you could take your paper away from us in protest of our policy. Planck's view is, in fact, rather closer to yours than to mine. However, your reply has not yet convinced me. In any event, I do recognize the naturalness and beauty of your chain of reasoning and do agree that the results must be compared against experience.

I am *very glad* that the affair has taken such a pleasant course through your exceptional conduct.

Cordial regards, yours,

Einstein.

Brief characterization of your theory.

Riemann: El[ementary] length (measuring rod) independent of transportation curve.

(or shape and size of the ri[gid] body)

Weyl: Shape of the el. rigid body [independent of transportation curve].

(but not size).[3]

530. From David Hilbert

Göttingen, 1 May 1918

Dear Colleague,

After in-depth discussion with all of my friends, incl. Max Lehmann, I must unfortunately advise against your well-meaning and appealing undertaking.[1] Such declarations, if they are to be even only passably outspoken and are not to appear entirely weak and insipid, would be tantamount to self-denunciations, which all our enemies in the faculties would be extremely glad to cite. Even your name would not shield you—since the very word "international" is like a red cloth to a bull for our colleagues, when they feel *in corpore* among themselves. But we would also damage the cause. Just as the *Deutsche Zeitung* by its attack solidified Kühlmann's shaken position,[2] just as Count Spee is working at this moment toward saving equal voting rights,[3] thus we also would achieve the contrary of what we intend right now. What is most important, though: we would fire off our gunpowder at the wrong time and possibly also at the wrong persons. I can understand perfectly the urge, which I feel just as you do, to do something; but I would like to recommend waiting until the mad hurricane has spent itself and reason has the opportunity of returning—and this time is sure to come.[4] We would have to restrict ourselves to German profs., since they alone are thoroughly known to us here and also have most to do with it. Other peoples must wash their own dirty laundry. I would then suggest our writing a joint open letter to Germany's professors and scholars, every word of which would have to be unassailable and act like a resounding blow, where we say what science is and what the responsibilities of science are. And that is why I make you this suggestion first of all: as soon as your health permits, come here for some time. G[öttingen] now has its prettiest season, as regards nature; you would feel very comfortable here among our circle. Here you would have simple but good care, could also be on your own in your room as much as you like, and—we could discuss and prepare all of these plans for the future. Finally, third, does the list of names you yourself have drawn up not appall you?[5] Max Lehmann is essentially good; but he cannot possibly fulfill your wish. Troeltsch has, as you know, gone over to the annexationists.[6] Schücking is a politician and is very wellknown as such;[7] he can say nothing new, and what he says is immediately stamped as pure politics and filed away. Weber (Heidelberg)[8] is generally not appealing to me—at least not altogether.

Now let me give you my cordial regards, and write me that you are kindly taking into consideration my suggestion of coming here: You would delight all of us here with it, but especially my wife and me. Yours,

D. Hilbert

P. S. My wife sends you word that she has a great many good preserves left over from the piglet. By the way, I have not yet received separately from you your last notices from the *Physik. Zeitschrift* and the *Berliner Berichte*.[9]

D. H.

531. From Ernst Troeltsch

Charlottenburg, Berlin, 1 May 1918

Highly esteemed Colleague,

I would like to respond very candidly to your first and, to me, very moving letter.[1] Grave problems are connected with the matter. If it is intended to console and unite the few who are not shattered by the war and who believe in the community of the intelligentsia, thus dispensing with any political intervention and influence, it is possible. The war has grown beyond the dimensions of other wars and threatens the very roots of European civilization. Anyone who has his beloved fatherland in mind will be entitled and inclined to maintain and profess his loyalty to this fatherland. But one must then know that it is a platonic profession and that the facts are not changed. These facts are running their course and cannot be restrained and overcome by reason anymore. As things have developed now, for us Germans it is really a matter of not allowing ourselves simply to be crushed by the Anglo-Saxons and of doing all we can to save ourselves from this grave danger.[2] An appeal to reason does not save us, because everywhere reason is powerless; we must unfortunately fight the matter out, allow the havoc to take its course, and save our own skins in the midst of the horrendous perils. For I still see the war as extremely dangerous for us.

A declaration, if it were possible for me, would have to express both of these ideas. However, I doubt whether this conforms with your intention and purpose. Then I doubt whether you could win anyone over from the enemy camp. The plan's effectiveness is just as questionable to me, though. As you know, I have made resolute statements so frequently, at one time in an article in the *Frankfurter Zeitung* specifically attacking hatred among nations,[3] that my position is known. But on the other hand, I am too much of a realist not to sense the imprecision of the whirlwind and the predicament resulting for us out of it. That is why simple flight into the realm of academia and into the society of the few believers is not possible for me, because I am too much concerned, after all, with the outcome of the matter for our real existence.   I would therefore necessarily produce an unequivocal and yet two-sided declaration, and the question is whether it would then not be better that I abstained. Should your idea be realized, I could participate under the said condition, but would have to leave it to your discretion whether you can make use of such a thing.

With many thanks for your confidence, yours very truly,

E. Troeltsch.

532.   From Gustav Mie

Halle-on-S[aale], 47 I Magdeburger St., 6 May 1918

Dear Colleague,

It has been a long time since I wrote you, even though I have been thinking a great deal about our conversation in Berlin in the meantime.[1] My long silence is explained in that I have a very strong suspicion that our written exchange of ideas in fact consists mostly of a mere squabble over words, from which we can probably emerge only after repeated occasional discussion in person, although we may not be completely of one mind. In any event, I wish from my heart to have the pleasure of seeing and talking to you many more times in my life. You may possibly have misunderstood my parting comment when you spoke of wanting to return my visit. You could give me no greater pleasure than to want to meet me here once, but your traveling here in your ailing condition seemed so impossible to me that I did not want any promises from you. You yourself will have felt, I hope, how much I would otherwise have liked to accept your promise to visit me. In any case, I may surely always look you up when I come to Berlin again.

Meanwhile, I would like to try once again to explain myself in writing.[2] I ask, though, that you permit me to talk only about the "pseudo-Euclidean" theory.[3] I still cannot acquire a taste for your gravitation theory with curved space.[4] The more time goes by, the sharper my old objections become for me, which through our conversation in Berlin had perhaps been dulled a bit for half an hour, but the legitimacy of which I believe to have recognized very quickly again.[5] The questions I would like to address in this letter are of such a general nature that it is maybe not even of importance whether we wish to think in a "pseudo-Euclidean" manner or in a "nonpseudo-Euclidean" one. Still, I would like to speak from a firm basis.

Recently I took a closer look at Schwarzschild's papers on the gravitational field of a sphere[6] and have also read your essay on "Approximative Integration" of 1916.[7] There at the end I find the remark: "Therefore, although in this analysis it has proved convenient from the outset not to subject the coordinate system choice to any constraints, . . . our last result does nevertheless show that there is profound physical justification for a coordinate choice according to the condition $\sqrt{-g} = 1$."[8] This is entirely the same point of view that I take in my Göttingen lectures.[9] To be more precise, the reason leading you to make this remark is also almost exactly the same one I have explained with the slithering rod.[10] In your analyses, you came upon a coordinate system in which such a snaking would occur, and you reject this "undulatorily oscillating coordinate system" as nonphysical.[11] From this I see what I have always suspected, namely, that our views are basically in far-reaching agreement; I should like to say: we both simply think essentially in physical terms. The mathematical manner of thinking can be of a completely different kind; I find that mathematical talent and physical talent are more different from one another anyway than what is usually thought. If one just wants to solve the problems mathematically, the most paradoxical assumptions may be employed in order to facilitate the manipulation of the calculations. Mathematics is riddled with such paradoxes, of course. The scientific logic of a physicist is much stricter and more constrained. To give a very elementary example as an illustration of what I mean, I point out that in analytical mechanics, time-dependent forces are easily accepted. The physicist must take exception to such a force; he must first seek to explain it: he encounters a problem where for the purely computational mathematician everything appears nicely settled. Thus the mathematician can also be satisfied when Foucault's pendulum experiment and the phenomena related to it can be mastered mathematically with the aid of some curious coordinate system in accordance with the "general theory of relativity." The physicist, on the other hand, can never ascribe the same legitimacy to a coordinate system, in which what you call "apparent" or I call "fictitious" gravitational phenomena occur, as he can to one in which only "real" gravitational fields exist. As soon as you say, now, that all coordinate systems are equivalent "in principle," you are assuming the mathematician's standpoint, who can define everything as he wants "on his sovereign authority."[12] By contrast, if one says, as I always do, that they are not equivalent in principle, then one is speaking as a physicist who is bound by stricter principles of scientific logic. Thus in the sentence quoted above, you also speak as a true physicist. Allow me to sketch my opinion on this in somewhat greater detail; I hope I do not bore you with it. It is precisely these principles of scientific logic, alien to the mathematician, that are of most importance in theoretical physics. For obviously, theoretical physics does not merely aim at solving mathematical problems; its highest goal always is, as your researches in particular have demonstrated so brilliantly, to point out new paths for research. The purely mathematical, manner of thinking without following principles can never lead to this; the guiding stars for this are precisely these principles of scientific logic, principles whose ultimate basis is the postulate of a real, objectively existing world, without which there could be no empirical science at all. These principles guide the physicist, be it consciously or unconsciously.[13] And you discovered all the wonderful consequences of your theory, which point I also particularly emphasized in my Göttingen lectures,[14] precisely because you did not adhere consistently to the purely mathematical point of view of "Hilbert's world," but rather thought as a physicist.[15] Imagine for a moment that one wanted to support very consistently the view that there were no gravitational fields, just non-Minkowskian spatial domains! Perhaps one can attack many problems mathematically in this way, but one would never arrive at the predictions of the bending of light rays, the alteration of atomic frequencies within a gravitational field, and so on, because for it thinking in physical terms is necessary.

A problem, now, which initially will not affect the mathematician at all but which seems to me to be of the greatest importance to the theoretical physicist is the question of a natural coordinate system ⟨altogether⟩ determinable according to generally valid rules and permitting the description of the phenomena with the greatest possible simplicity, in which, for example, "apparent" waves and similar monstrosities are excluded in principle, in which the field of a sphere has spherical symmetry, and in which the principle of relativity for constant velocities applies. I believe we do share the same opinion in that the question of this preferred natural coordinate system has both sense and purpose. But in this preferred coordinate system, like it or not, the Earth will rotate around its axis. From Schwarzschild's paper[16] I noticed with very particular interest that also in the case of spherically shaped symmetry the additionally inserted condition $\sqrt{-g} = 1$ is not quite sufficient. In his solution Schwarzschild obtains two other arbitrary integration constants, one of which, α, is determined by the central body's mass; the other, however, which he calls ρ, remains entirely arbitrary. Schwarzschild did determine ρ, though, by the condition that the discontinuity in the solution occurs at the origin of the coordinate system. Nevertheless, he was apparently led to this just out of reverence for the old Newtonian potential $\dfrac{m}{r}$ since it is altogether incomprehensible why the discontinuity cannot lie anywhere else within the interior of the ⟨gravitational⟩ central body. Thus despite $\rho = \alpha^3$ being laid down quite arbitrarily, Schwarzschild says on p. 195 at the bottom that, in order to obtain the *strict* solution to Mercury's orbit, $(r^3 + \alpha^3)^{1/3}$ must be inserted into the solution indicated by you in place of R. To what extent this solution is supposed to be stricter than the one in which, let's say, $\alpha = 0$ is set, hence $R = r$, or any other value, is not at all clear, though. Any other value for α simply means a different coordinate system. But one sees from this yet again how convinced Schwarzschild very obviously was that *one* preferred coordinate system had to exist. It would indeed be a quite intolerable situation, of course, if such arbitrary quantities always had to be dragged along within the solutions, without knowing what it actually means if a certain value is inserted for these arbitrary variables, if one does not even know whether some nonsensical consequences, such as apparent waves or the like, are thereby introduced into the solution. Therefore I believe it can be said that, as beautiful as the general transformability of the fundamental equations is seen to be from the mathematical point of view, physically it does signify a weakness in the theory which must first be eliminated, and I see from a statement in a letter of yours that you also are in complete agreement with me about the legitimacy of the recipe offered in my Göttingen lectures for how, by means of a general rule, one can always obtain a coordinate system that is unobjectionable and that, as far as I can see, contains the minimum of arbitrary elements.[17] I am quite convinced that this coordinate system indicated by me is *the* natural coordinate system and that one would be hard pressed to find another as good, let alone a better one.[18] Well now, I have attempted to examine what my suggested coordinate system yields in the case of the gravitating sphere. This problem is soluble without the least difficulty and one arrives exactly at Schwarzschild's solution, but with $\rho = 0$. Outside of the sphere, that which Schwarzschild denotes as R (where $R = (r^3 + \alpha^3)^{1/3}$), measured in the "natural" coordinate system, is the radius vector itself; within the interior of the sphere filled uniformly with mass, the radius vector $R = \eta^{1/3} = \sqrt{\dfrac{3}{\kappa \cdot \rho_0}}$

$\cdot \sin \chi$ must be inserted.[19] It goes without saying that, in the natural coordinate system, the solution is fully unique. In addition, it is evident that the coordinate system chosen by Schwarzschild, in which $R = (r^3 + \alpha^3)^{1/3}$ is inserted, depends on α, i.e., on the gravitational mass of the central body. Hence, for each central body a differently defined coordinate system is taken, a procedure which would, extremely probably, lead occasionally to all sorts of inconsistencies if one wanted to move from the sphere's solution to some other more general solutions. This is avoided with the "natural" coordinate system. It is especially interesting that in my "natural" coordinate system, $\sqrt{-g} = 1$ does in fact result in the field outside of the sphere, though not in the interior of the sphere. Whether some general principle lies behind this or whether it is just a coincidence I do not know, but one can see clearly from this exactly why, as you say, "there is profound physical justification for a coordinate choice according to the condition $\sqrt{-g} = 1$." This condition plainly leads to the natural coordinate system, in the case of spherical symmetry, if one also chooses $\rho = 0$ in addition.

But now comes the question[20] I would have liked most to have discussed more with you: What sense does the whole wonderful theory have if the general transformability of the fundamental equations really is removed again in the end by other additions, namely, the choice of a preferred natural coordinate system? This question was actually the topic of my Göttingen lectures, and I must admit that it has often saddened me that you have not yet uttered a word in reaction to the positive answer I gave to this question, with which I hoped on my part to make a modest contribution to your wonderful theory.[21] My criticism of "general relativity" was only supposed to clear the way for new positive results, and I am convinced that I have arrived at some. (The most important result of my considerations is, in my judgment, that the assertion of the transformability of the equations can be carried further along the same lines: (that the fundamental ⟨equations⟩ physical laws essentially retain their form unchanged, also in non-Minkowskian regions, thus where a transformation is no longer at issue). It is precisely this very general validity of the fundamental laws, which they can have, however, only if general transformability also exists, which actually seems to me to be the very core of the theory. I have been thinking a great deal recently about attacking a specific problem that attracts me very much with the aid of this "extended principle of the relativity of gravitational effects." I shall give it a try, even though it will not be easy mathematically. Should my attempt succeed, I shall be able to show you more clearly with it than with words what I mean. But maybe this letter already will result in some possibilities for us to discuss this further as well. This much I would still like to remark, that the plain coordinate transformations, as studied first by Mr. Born in his doctoral thesis[22] and later in more detail by Mr. Kottler,[23] can certainly be very amusing mathematically, but the problems themselves can never be probed, one is only beating around the bush. Mr. Kretschmann, I find, has given a very nice explanation for this.[24] Perhaps you will see from this what I am driving at now.

With kind regards, yours truly,

Gustav Mie.

[2]Deleted text in draft: "individual points actually do exist, where a clarification of the views can promote our knowledge. I am generally of the opinion that a physicist does not need to be very familiar with epistemology; when a scientist is too critically epistemological, it can even be an obstacle to free and original research."

[13]Deleted passage in draft: "If experimental physics were to adopt the mathematical standpoint of complete freedom in making definitions, then it would be thoroughly unproductive. The nicest test case is Faraday, who liberated us from the exclusively

mathematically meaningful concept of action at a distance and hence took the greatest step forward."

[20]Draft text: "About the negative part of my Göttingen lecture, we are, as it appears to me, actually in complete agreement factually; I just express myself somewhat differently than you do."

533.   To Mileva Einstein-Marić

[Berlin, before 8 May 1918][1]

Dear Mileva,

Only about death can we be *secure*, not about possessions of any kind. There's no changing this. The certificates that I want to send you aside from money are as secure as German currency.[2] The main advantage to you is that the things become your property and come to Switzerland. With a devaluation of German money, these papers would also be devalued correspondingly (and vice versa), because the *interest* would be devalued with the money. Therefore, whether you have these certificates or cash in hand is identical in this regard; in the latter case, though, you would lose the interest.

It is another question whether the money or the certificates be exchanged for Swiss financial instruments. This is a gambling question, through and through, and not a matter of careful thought. I can only say to you that the persons of money here who have loans to pay back in Switzerland prefer to borrow in francs in Switzerland and to allow the obligations to grow through interest than to exchange German currency into Swiss now. Whether they are right, I do not know. Nowadays everything is uncertain; but I do tend toward the opinion that it is more appropriate to deposit German financial instruments as such. You can believe me that I have deliberated on all of this very well and have not allowed any other concern to govern than the best possible security for you and the children. You must also bear in mind *that the 2 000 M pension that you are surrendering*[3] *involves something just as unstable as what you are receiving in this way.* The children naturally keep their 1200 (jointly), which they receive by their 18th year of age. At all events, the best that I can bequeath to my boys, or that they can inherit from me, is not money but a good mind, a positive outlook, and an unblemished ⟨and good⟩ name that is known everywhere on Earth where science-loving people live. If they become well-rounded persons, they will have a less difficult time than I had 20 years ago.[4]

So be satisified, and do not speculate too much about the inextricably con-voluted future fate of your money, and be glad if the Reichsbank is at all willing to give us permission for the transfer. That already entails a significant improvement of your situation. As soon as I receive permission,[5] you'll receive word. Ask Mr. Zürcher[6] to draft with you the necessary changes to our contract soon.

So once again. A liquidation of the certificates would be senseless; for they are the safest thing to be had, exactly as safe as currency. Besides, you are receiving a considerable portion in money.–

Talk to Michele. He'll certainly confirm to you the correctness of what I'm writing you here.

It must be stated in the contract that during my lifetime you are not allowed to wield authority over the principal capital or over its investment without my consent. It is your property, though, and you receive the interest. It is all that I own, a large part of it consists of two large scientific prizes that I received ⟨last year⟩ recently.[7]

With best regards, yours,

Albert.

534.   From Paul Ehrenfest

[Leyden,] 8 May 1918

Dear Einstein,

I have been lying in bed with jaundice for 10 days—now things are slowly looking up again.– Received your postcard today with many thanks.–[1] I did not "blow off steam" at Hilbert—I wrote to him in a way that merely expressed how I felt, in loyal attachment to the atmosphere pervading my Göttingen student days.[2]—I am firmly convinced that you would have approved of the letter— besides, I did what I had to do. [Neither was there any big show about it!][3]

— · —

Of almost all your last papers I have no reprints. Particularly not of your quantum whatnots in the German Phys. Society *Berichte*.[4] From the Academy I only have the "Cosmolog. Consider.,"[5] nothing else after that.

— · —

When am I going to see you again?!– Now I am miles away from all physics, sitting over problems of politico-economic theory (with Papa Lorentz's shoulder-shrugging blessing).[6] When am I going to return to physics again with qualms of conscience and remorse? At the moment it is unspeakably odious to me.

I've been reading: Musaeus's *Folk Tales*[7] in bed, with boundless pleasure.— You know what? That's a fine thing for low moments.

Warm regards. Yours,

P. Ehrenfest.

535.   To Hermann Weyl

[Berlin, 10 May 1918]

Dear Colleague,

On May 2nd your paper with your addendum was accepted by the Academy for the *Sitzungsberichte*.[1] The brief summary also arrived in time to appear. So it all went as desired, thank heavens. I must tell you again, though, that I am firmly convinced that, as interesting as it is, your line of reasoning does not correspond to reality;[2] your rejoinder[3] did not persuade me, but lengthy written disputations would perhaps not be very fruitful. I think you will discard this interpretation again of your own accord.

Your corrected treatment of the zone problem is still not clear to me. I absolutely do not understand why an equatorial fluid distribution should be impossible.[4] I do not understand at all the reason provided at the top of p. 226.[5] The possibility of a world filled only equatorially with matter seems beyond dispute; the boundary condition $p = 0$ is fulfillable for both spheres.

With best regards, yours,

Einstein.

I called Springer to have them wait for your instructions in the matter but do not know whether it is still possible.[6]

536.   To Ilse Einstein

[Berlin, 12 May 1918]

Dear Secretary,[1]

With me, action only inhabits my head. Rarely does it stray down into my fingertips. But my joy at your letter and the pretty avian card brings it about. The fowl appealed to me inversely proportionally to its size. The little nest and its surroundings seem to be very nice, veritable nature, almost unspoilt by humans.[2] Some things have in fact arrived for the Institute, although nothing worth mentioning.[3] I'll save it all up until you come, so that you feel right in your element. Then we'll piece together the shocking balance *viribus unitis.* Nicolai's zeal in *mathematicis* is truly funny; the yearning for objective things is making itself evident, far away from human concerns; I know all about that!

Enjoy yourself and come soon to your

Principal.

537.   To Georg Nicolai

[Berlin, 12 May 1918]

Dear Nicolai,

Lehmann is a genuine mind-mate, this I know from Hilbert.[1] He is esteemed very highly here as a historian, despite the liberty with which he has treated pa-triot[ic] history.[2] I do not know whether he made any form of public appearance in the last 4 years.[3] In any event, he is a person of character and not a careerist. I do not know him personally. It is a thoroughly welcome idea to be named with him.[4] It is *I* who does not deserve being named in this connection; because I have done nothing to restore public opinion, and because I am Swiss. At best, I would have to be *rebuked* if I, as a Swiss, took a different stance. *If* I am to be rebuked, then it is only because I am sitting here idly. But I myself do not know whether I should blame myself for my passivity. I could not see quite clearly in this regard.[5]

Best regards to you all, yours,

Einstein.

538.   From Max Wien[1]

Jena, 12 May 1918

Highly esteemed Colleague,

From Mr. Born[2] I hear that at the last meeting of the German Physical Society I was chosen as Chairman.[3] I convey my most cordial thanks for the honor bestowed on me by this and for the trust given me but, at the same time, also my great regret that I am unable to accept the nomination. My military position occupies me so completely that I have no time left for any other unofficial functions. For this reason, last year, despite my strong urge to inform myself of the latest advances in physics, I was able to attend the meetings of the Physical Society only 2 times, unfortunately. With the best of intentions, coming more often is not going to be possible for me in the future either, no more so taking part in the Chairman's tasks.

Under these circumstances I could just be Chairman on a purely formal basis, and in my view this would be neither in the interest of the Society nor in my own interest: it goes against my innermost feelings to take on an office without being able to assume its duties along with it; and the painful feeling of inadequacy and dissatisfaction I have developed through years of neglecting my responsibilities toward my science, my institute, and my family, would only be aggravated by this.

In thanking you cordially once again, I therefore request being relieved from acceptance of this nomination.

Our colleague Born also spoke now of the paper allotment for the *Berichte* turning out to be very small.[4] I would naturally be very pleased to back, as far as is in my power, the application for an increase in the appropriation.

With best regards, I remain ever yours truly,

M. Wien.

539.   From Charlotte Weigert[1]

Copenhagen, 43 Nansensgade II, 15 May 1918

Esteemed, dear Professor,

You did not consider me ungrateful, I hope, for not letting you hear from me for so long! Only "planetary" life is to blame for my having lacked initiative for the deed! How often have I spoken in raptures about you to persons interested in science and also to our mind-mates in world organization! I hope that your condition is *such* that you are able to work; that is equivalent, of course, to an ascending motion of your life line! How much I wish to hear a bit about your splendid ideas again! If I think back on the evil war years, the conversations with you were one of the few positive things of value to me, which I still remember with gratitude!—Oh, dear Professor, where is the fulfillment of our hopes and plans, which on our recent evening walks we believed to be so near at hand? The world is more sinister than ever, and it is a mercy for those who have the ability and energy to retreat into science and art!– Thus I am also especially thankful not only that I have an intimate audience here on foreign soil,[2] but also that there were particularly enthusiastic persons among the refined artists and writers. It was especially nice that so many Danes participated since, as such, there is no particular interest here in hearing lectures in the German language. You see, though, that I have succeeded in spinning a little "web of sympathizers," as I had planned. Of most importance to me (if I may say so to you, considering the kind interest you have always shown me), is an infinitely rich friendship with the writer Sophus Michaëlis whose books had always been the most precious to me of what Europe's literature has to offer![3] An inspired, ingenious poet who writes books of classical beauty, reaching beyond the current trends and national boundaries and yet full of musically rhythmic life! Mrs. Elsa must read *Giovanna* to you sometime, which little Lena Katz owns.[4] Now, farewell, esteemed Professor.

Unfortunately, one cannot expect the censor to read lengthier letters more frequently! But a sign of life is sufficient, you know. These postcards are naturally intended for Mrs. Elsa as well, to whom I extend my *very* cordial regards. *Please* send me a *short* note sometime.

With amiable regards, your gratefully devoted

Charlotte Weigert.

540.   From Felix Klein

Göttingen, 18 May 1918

Esteemed Colleague,

I have been hearing from various quarters about a note on gravitation theory that you presented on 7 March to the Berlin Academy;[1] I would be very grateful if you could procure me an offprint of it.

Based on our recent correspondence, Runge and I have deferred our planned publications and intend to return to them, should the occasion arise, once we have a complete overview of the currently available literature.[2]

The Planck week was very stimulating.[3] Pl[anck] will give you the best report about it himself, and wanted in particular to make the request that you join the German Mathematicians Association as a member (annual contribution 2 marks, registration with Privy Councillor Prof. Krazer in Karlsruhe in B[aden],[4] 57 Westend Street; I can also arrange the matter myself, of course).

Very truly yours,

Klein.

541.   From Georg Nicolai

Eilenburg, 54 Torgauer Street, 18 May 1918

To Prof. Albert Einstein, Berlin W. 30, 5 Haberland Street

Dear and esteemed Professor,

Many thanks for your lines; I did want to say to you, though, that you very much underrate yourself in thinking you "ought not be named in this connection";[1] because you evidently *have* spoken out, and nowhere is the Appeal to Europeans,[2] coauthored by you, forgotten. Indeed, it would probably never have come to light if you had not participated in it. I believe, at least—such an "if" can never be exactly determined, of course—that I would have done nothing on my own.

You are being unfair to yourself in something else as well, I believe. You are absolutely not to be "rebuked" for sitting in Berlin and working; if anyone has the right to act as a second Archimedes toward the mercenaries, crying out *"noli tangere circulos meos,"*[3] then it is *surely you.*

But finally, you are also not quite right—in my opinion, at least—on a third point either: namely, when you put forward your Swiss citizenship; for it is only in the second place that citizenship is involved here. *You are* a German (and moreover you represent a piece of German culture entirely in its own right), and then you are also a European, and it is this which must be emphasized now above all. I, at least, am much more strongly convinced today than at the time we were drawing up the Manifesto to the Europeans that deliverance from the coming cultural collapse will only be possible if the European idea—pure and simple—can be carried through. To be precise, this idea of a worldwide cultural organization must inspire the masses at least as much as the idea of Christianity had 2,000 years ago. A few ideologues standing up for such ideas will not cure the world.

Mass suggestion must be added. But just because the danger is so great at the moment, just for this reason, I believe that humanity's common sense will instinctively understand and not tolerate that this fair Earth be so desecrated. I believe that what we are advocating today as a very isolated few, will tomorrow be in the possession of all. Some occasion—perhaps a very banal one—some person—perhaps a very foolish one—will come along, and recovery will be upon us.

You might be able to say now, whoever is so firmly convinced in something does not need to fight for it to come about.—You are entirely right, I also find it quite superfluous; but then again! It may be specifically my act that provides the minimal impetus necessary to tip the unsteady balance and supply this organism with a new position of equilibrium, which would then be able to persist again for another few thousand years.

You see, esteemed Professor, this foolish war and its outcome do not interest me at all anymore, but for times to come, for the future of Europe, something ought to be done, and perhaps you will reconsider whether you might want to speak once after all *as a European to Europeans* and help in collecting together all the separate things that are being shoved into corners and evaded, to allow them to be heard.[4]

Today would bring more success than four years ago, even in Germany.[5]

One just does not have to turn to the *party bosses* exactly!

With cordial and very respectful regards, sincerely and devotedly yours,

Georg Nicolai.

542.   From Max Wien

Jena, 18 May 1918

Esteemed Colleague,

Many thanks for both of your letters, which have reached me here. I am very sorry to have to inform you that I must insist on my refusal:[1] it really does not work, I cannot take on even more. Now I must spend the 1-$^1/_2$ vacation days at Pentecost almost exclusively at my desk, because all my private correspondence has been neglected.

I am very willing to assume the chairmanship sometime later when calmer times have returned. Some other uncommitted colleague would be much better suited for it now. The fact that I happen to be working in Berlin[2] has actually nothing to do with it, as I actually imagine the responsibility of a chairman of the German Physical Society to be broad; the management of the affairs of the Berlin "local chapter" would have to remain now as before in the hands of a Berlin member.[3] We could possibly discuss this once orally, of course.

With best regards, ever yours truly,

M. Wien.

543.   To Felix Klein

[Berlin, 19 May 1918]

Esteemed Colleague,

I shall be glad to send you my recent published notes.[1] On Thursday I submitted a fairly extensive exposition of my views on the energy law.[2] I shall send you the correction proofs, which I am probably going to be receiving in a few days.

I was very sorry that I could not attend the lectures of my dear colleague Planck.[3] My doctor has forbidden me all those sorts of things, unfortunately.[4] I shall be glad to register myself with the German Mathematicians Association, as you consider it appropriate.[5]

With amicable regards, yours truly,

A. Einstein.

Recipient's note: "Now the entire Levi-Civita material. Should I write a note on hypothesis B?"

544.   From Hermann Weyl

Zurich, 20 Schmelzberg St., 19 May 1918

Dear Colleague,

With my "correction" of sheet 15, I really have blundered![1] Although the statement at the top of p. 226 correct: on the sphere, h is an odd function of $x_4 = z$, just as in the fluid cap case, eq. (66), $h = \dfrac{b_0}{z}$.[2] What I had incomprehensibly overlooked, however, is this: that the boundary condition of vanishing pressure on both latitudinal spheres does not require the same value for h but rather the same value for f.[3] Now, the equation for Δ:[4]

$$(*) \qquad\qquad \frac{d\Delta}{dr} = \frac{rv}{2} h^3$$

yields the solution Δ, which is an *odd* function of z; hence the corresponding $f = \dfrac{\Delta}{h}$ is even. Because of the arbitrary additive constant in the solution of $(*)$, f can also be replaced by $f + \dfrac{\text{const.}}{h}$. If on both boundary spheres $r = r_0$, symmetrical to the equator, f must assume the same value $\boxed{f_0 = \dfrac{v}{\mu_0}}$, then this const. $= 0$

must be taken. The boxed equation produces a transcendental relation between μ_0 and r_0. This relation is formulated most elegantly in the following way.[5] If $r = a^*$ is the value for which (64) $\dfrac{1}{h^2}$ vanishes, and one sets

$$\gamma = \left(\frac{r_0}{a^*}\right)^3 : \left(1 + \frac{\lambda}{2\mu_0}\right),$$

$$\frac{1}{h^2} = \frac{1-x}{x} \cdot (x^2 + x + \gamma),$$

$$\rho = \frac{3\gamma}{2 + \gamma x^{-3}},$$

then $x = \dfrac{r_0}{a^*}$ satisfies the equation[6]

$$\rho + \frac{1}{h} \int_x^1 h\,d\rho = x^3;$$

given $\gamma(0 < \gamma < 1)$, it always has one and only one solution x. For an infinitely thin mass zone, an infinitely large mass density μ_0 results,[7] so the total mass contained in the zone assumes a specific finite value $\neq 0$; if I have not miscalculated: $4\sqrt{3} \cdot \dfrac{1}{\kappa\sqrt{\lambda}}$, whereas the mass, in the case that the space is filled uniformly at density λ, comes to: $\dfrac{\pi}{\sqrt{2}} \cdot \dfrac{1}{\kappa\sqrt{\lambda}}$.[8] This result, that the mass contained in the zone does not drop to 0 even when its thickness approaches 0, ought to meet with your approval.[9]

Regarding the "point mass solution" (65), a *zonal* horizon of mass can also be constructed:[10]

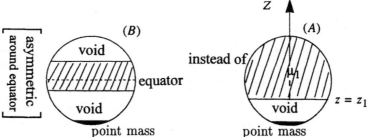

For the [spherical] mass cap (A), the density must be $\mu_1 < \lambda$, so that the pressure p is positive in the interior; for $\langle$if the radius $= 1$ is inserted:$\rangle$

$$\overset{\text{(pos.)}}{\underset{\text{(neg.)}}{\left(1 + \frac{p}{\mu_0}\right) \cdot \left(1 + \frac{\lambda - \mu_1}{\lambda + 2\mu_1}\frac{z - z_1}{z_1}\right) = 1}}$$

is valid. The continuous transition from case (A) to case (B) passes through [the case] $\mu_1 = \lambda$: Cap within which the pressure is everywhere $= 0$:[11]

By the way, the assumption of an elliptic space (identification of two diametrical points on the sphere) does seem far more natural to me now, after all, than a spherical one.–[12]

Once again, most cordial thanks for submitting my paper to the Academy![13] The corrected version was sent out to you a few days ago already. Your rejection of the theory weighs heavily on me; I know only too well how much closer a contact you have with reality than I. But my own brain still keeps faith in it. And as a mathematician, I absolutely must adhere to this much: My geometry is the true, local geometry [*Nahegeometrie*]; the fact that Riemann just arrived at the special case $F_{ik} = 0$[14] has merely historical reasons (development out of the theory of surfaces),[15] no substantive ones. If in the end you are right about the real world, then I would regret having to accuse God Almighty of a mathematical inconsequence.[16] I hope that in the near future I am given the opportunity of discussing this more thoroughly with you in person, here in Zurich. With sincere thanks and best regards, yours,

H. Weyl.

545.   Ilse Einstein to Georg Nicolai

[Berlin,] 22 May 1918

Dear Professor,

You are the only person to whom I can entrust the following and the only one who can give me advice, and that is why I ask you please to consider carefully what I am writing you now, and then let me know your view. You remember that we recently spoke about Albert's and Mama's marriage[1] and you said to me that you thought a marriage between Albert and *me* would be more proper. I never thought seriously about it until yesterday. Yesterday, the question was suddenly raised about whether A. wished to marry Mama or me. This question, initially posed half in jest, became within a few minutes a serious matter which must now be considered and discussed fully and completely. Albert himself is refusing to take any decision, he is prepared to marry either me or Mama. I know that A. loves me very much, perhaps more than any other man ever will, he also told me so himself yesterday. On the one hand, he might even prefer me as his wife, since I am young and he could have children with me, which naturally does not apply at all in Mama's case; but he is far too decent and loves Mama too much ever to mention it. You know how I stand with A. I love him very much;

I have the greatest respect for him as a person. If ever there was true friendship and camaraderie between two beings of different types, those are quite certainly my feelings for A. I have never wished nor felt the least desire to be close to him physically. This is otherwise in his case—recently at least.—He himself even admitted to me once how difficult it is for him to keep himself in check. But now I do believe that my feelings for him are not sufficient for conjugal life. I am just too afraid that then I might not be able to love him anymore and would even perceive him as a fetter. In the end, I would feel like a slave girl who has been sold. Maybe you think differently about this, but I have some doubts about it. I could imagine falling deeply enough in love with any other stranger to be able to live together with him, but my mind balks at the idea with A. I believe that the relationship between him and Mama that exists—in the alternate case, that *existed*—do not allow the sentiment necessary for connubial life to develop. I have become too used to regarding him somewhat as a "father." You will reply, this lies in the past, but I would be reminded of it daily by Mama's presence. You must admit, this is a somewhat unnatural thing and for our sensibilities nowadays, to my mind at any rate, is not quite clean either. (Although A. asserts that these are social prejudices.) By no means is A. trying to persuade me, since he does not want to assume the responsibility of binding such a young thing as me to himself. As long as I was here in the house,[2] it would not (in A.'s opinion) be a great difference for me whether or not I were married, at most it would merely be convenient, because I would always be an independent, free person. I would be valued just as much as Mama. And I am absolutely not jealous of all the public glamour that would descend on Mama. But it would be an entirely different matter if I were to marry someone, which I actually do want. A. told me that that would tear a hole in his life and that it would be painful for him to have to miss me. Now, this is a very sore point. I appear to myself like Heine's donkey.[3] On the one hand, I would like to live my lifelong with Albert, but to be married to him another kind of love is needed as well, I would think. Now you know exactly what the situation is between me and A. The third person still to be mentioned in this odd and certainly also highly comical affair would be Mother. For the present—because she does not yet firmly believe that I am really serious—she has allowed me to choose completely freely. If she saw that I could really be happy only with A., she would surely step aside out of love for me. But it would certainly be bitterly hard for her. And then I do not know whether it really would be fair if—after all her years of struggle—I were to compete with her over the place she had won for herself, now that she is finally at the goal. Philistines like the grandparents are naturally appalled about these new plans.[4] Mother would supposedly be disgraced and other such pleasant things. You know A. enough to know how much he cares about such empty words. I myself am too inexperienced to know whether Mama really would have any difficulties or even be harmed by a marriage between Albert and me. I wish to avoid such kinds of things in any case; Mama has suffered enough sadness and nastiness in her life.[5]

The question simply is, what is more beneficial for the happiness of the three of us, and especially of Albert? I think I have now discussed all the problems with you. Give me some advice. I can just rely on my instinct and do not know if in doing so I am hitting upon the right thing; this—my instinct—does tell me, though, that I ought *not* to become Albert's wife. It would be the bitterest for Albert himself if he had to see our good harmony, which until now has been only positive for me, transform into a fetter for me through the marriage. Marriage really is a devilishly silly affair! A. also thought that if I did not wish to have a child of his it would be nicer for me not to be married to him. And I truly do not have this wish. It will seem peculiar to you that I, a silly little thing of a 20-yearold, should have to decide on such a serious matter; I can hardly believe it myself and feel very unhappy doing so as well. Help me!

Yours,

Ilse.

Remark at the head of the text: "Please destroy this letter immediately after reading it!"

546.   To Mileva Einstein-Marić

[Berlin,] 23 May 1918

Dear Mileva,

Securities for 40 000 M is being transferred to the Swiss Bank Association, Zurich, for you one of these days.[1] I request, now, that you send the contract and file the divorce. The deposit in trust of 20 000 M., the interest of which you would be entitled to in the case of my death, will be taken care of tomorrow.[2] ⟨as soon as we are agreed on the contract.–⟩

I'm probably going to have to do without the trip to Switzerland this summer, considering that under these nasty traveling conditions the strain would be too great for me.[3] But I plan to go to a remote village by the Baltic Sea[4] for two months and would be pleased if Albert, or even Albert and Tete, could come. We would do a great deal of sailing there. I could certainly procure the travel permit. What do you think of this?

Kind regards, yours,

Albert

Kisses to the boys! We would just be sailing in a sheltered bay, not on the open sea.

547.   From Zionist Association of Germany

Berlin, 23 May 1918

Dear Sir,

The preacher at the Great Synagogue in Lodz, Rabbi M. Braude,[1] is coming to Berlin in the next day or so in order to instill interest in the *Institutes of Jewish Teaching* founded and directed by him with great success. Dr. Braude is the founder and director of the first Jewish secondary school [*Gymnasium*] in Poland, at which it is being attempted to provide Jewish pupils an education completely commensurate with a western European one on a consciously Jewish basis. The secondary school is composed of a section for boys and one for girls; the latter is still in the development stage, whereas the boys' section is releasing its first graduates this fall. Instruction is being carried out by first-class, exclusively academically qualified teachers in the Polish language and in a consciously Jewish spirit. The curriculum, which is based on very modern pedagogical principles, represents an excellent combination of the requirements a school must address in order to develop its pupils into well-rounded citizens of its country and, to the same degree, into individuals with a strong awareness of their Jewish national heritage.[2] A textbook publishing house is affiliated with the school through which a number of books necessary for instruction at the school have already appeared. The school enjoys a good reputation extending well beyond Poland's borders and maintains itself entirely with its own funds.

In continued pursuit of his purposeful cultural plans, Dr. Braude has now initiated the foundation of a *Jewish Teacher Training College* in Lodz which aims at qualifying its students as suitable primary and middle-school teachers working in this spirit, which has already proven so exceptionally fruitful at the secondary school in Lodz. The Polish Education Ministry has given permission to open this teacher training college, and thus the first Jewish teacher training college in Lodz will be inaugurated in the autumn of this year. Those who know about the terrible shortage of teachers and the resulting atrocious cultural poverty of the Polish Jews will duly acknowledge the eminent significance of this foundation. It is the purpose of Rabbi Braude's visit to attract interest in this college among broader circles of German Jews concerned about elevating the cultural niveau of Polish Jews.

We are counting on you, esteemed Sir, also to show your frequently established interest in this issue, which is of importance as much from the Jewish and Polish standpoints as from the German one, and permit ourselves accordingly to invite you to meet with Dr. Braude on

Tuesday, 28 May, at 8 o'clock sharp in the evening

in the conference room at our offices, 8 Sächsische St. We would be very grateful if you would pass on this invitation to your close acquaintance, among whom you could hope to find sympathy for this important cultural cause.

In utmost respect,

Zionist Association of Germany
Hantke,   O. Warburg.[3]

548.   To David Hilbert

[Berlin,] 24 May 1918

Esteemed Colleague,

I thank you belatedly for the extremely warm response you gave to my last letter with the circular.[1] I have since become persuaded that the psychical basis for my envisioned endeavor is lacking, in that just those persons on whom I would have depended do not harbor predominantly international sentiments either.[2] The whole endeavor was not supposed to satisfy any *political* purpose. Those persons of whom I would have expected to place the common cultural assets well above the current division were supposed to join hands. It would thus have been, to a certain extent, the antithesis to the notorious Manifesto of the 93.[3] This states: I am foremost a German and only in the second place a person of culture; we, on the other hand, are supposed to say: I am foremost a person of culture, and only in the second place a German or a Frenchman (insofar as it can be made compatible with the former!). But I gave up the scheme after receiving the responses.[4] If you do something in this sense later on, you can count on my participation, of course.

I very much appreciated your cordial invitation. However, in my current state, I cannot come to visit. The slightest dietary slip-up or the most innocuous of exertions can produce another relapse.[5] That is why I had to sacrifice going to Planck's talks, which would naturally have interested me very much, especially considering that a long-standing difference of opinion exists between us in this area.[6]

Yesterday I received a very interesting paper by Ms. Noether about the generation of invariants.[7] It impresses me that these things can be surveyed from such a general point of view. It would not have harmed the Göttingen old guard to have been sent to Miss Noether for schooling. She seems to know her trade well!

Cordial regards from yours truly,

A. Einstein.

549.   To Felix Klein

[Berlin, 28 May 1918]

Esteemed Colleague,

You may keep the correction proof, of course, and show it to whomever you wish.[1] I just sent it to you because I know that you are interested in this problem.[2] No comment on it is necessary, but I would be pleased if you read the paper when you have a chance. A mathematical gap in the train of thought remains.[3]

With all due respect,

A. Einstein.

550.   From Max von Laue

Würzburg, 29 May 1918

Dear Einstein,

I thank you heartily for your postcard. The Müller's Court Booksellers is in Karlsruhe (Baden). Sommerfeld thinks, not quite without justification, that the honorarium should not be waived, on principle; as soon as Warburg's and Planck's comments have also arrived, I intend to suggest 80 M to the publisher for the printer's proof.[1]

Planck wrote me, incidentally, it was now clear enough to him so that he could do without your and Warburg's manuscripts.[2] Please send yours to me here soon now, probably best by "registered," for safety's sake. I can then forward it to the publisher as soon as the negotiations with him have been concluded.

In order to report some science to you now as well, I would like to write that I examined a bit more closely the "electron gas" idea which haunts the literature on electrons in glow discharge. Mathematically, it involves the equation $\Delta \psi = e^{\psi}$ which can, however, by all means be integrated as far as is necessary for this. In two dimensions, its integration is almost as elegant as for $\Delta \varphi = 0$. The physical result is that a glowing body has an outer layer of electrons, which has a "surface pressure" (negative surface tension) still dependent on temperature. On curved surfaces, a "capillary pressure" is produced entirely as in capillarity, just of the opposite sign. The thermodynamics of this outer layer can also be developed easily.[3]

With cordial regards, yours,

M. von Laue.

551.   To Hermann Weyl

[Berlin,] 31 June [May] 1918[1]

Dear Colleague,

I am glad that you have put the zone affair in order, now.[2] The result of your calculation now corresponds completely with what had to be expected.[3] You probably already sent the relevant correction to Springer; I asked him to wait with the printing until your decision arrived.[4] Let's hope the abominable paper shortage does not delay the appearance of your book![5]

Now to the question: spherical or elliptic.[6] I do not think that there is a possibility of really deciding this question through speculative means. A vague feeling leads me to prefer the spherical one, though. For I sense that those manifolds are the simplest in which *any closed curve can be contracted continuously to one point*. Other people must have this feeling as well; since otherwise the case where our space could be Euclidean and finite would surely also have been taken into consideration in astronomy. The two-dimensional Euclidean space would then have the connectivity properties of an annulus. It is a Euclidean plane, on which every phenomenon is doubly periodic, where points lying on the same period grid are identical. In a finite Euclidean space there would be three kinds of closed curves not continuously reducible to one point. Analogously, an elliptic space, in contrast to the spherical one, possesses one sort that cannot be contracted continuously to one point; that is why it appeals to me less than the spherical one.

Can it be proved that the elliptic space is the only variety of the spherical space that can be obtained through the addition of periodicity properties? It seems to be so.

Now once again to your Academy paper.[7] Could one really charge the Lord with inconsequence for not seizing the opportunity you have found to harmonize the physical world?[8] I think not. In the case where He had made the world according to you, Weyl II would have come along, you see, to address Him reproachfully thus:[9]

"Dear God, if it did not suit Thy way to give objective meaning to the congruency of infinitesimal rigid bodies, so that when they are at a distance from one another it cannot be said whether or not they were congruent: why didst Thou Inscrutable One not then decline leaving this property to the angle (or to the similarity)? If two infinitely small, originally congruent bodies K, K' are no longer able to be brought into congruency after K' has made a round-trip through the space, why should the *similarity* between K and K' remain intact during this round-trip? So it does seem more natural for the transformation of K' relative to K to be more general [than] affine."[10] But because the Lord had noticed already before the development of theoretical physics that He cannot do justice to the opinions of man, He simply does as *He* sees fit.–

We are probably not going to see each other in Switzerland in the summer. With my "internal" sensitivity and the unpleasant traveling conditions, I shrink from taking the long trip. I am probably going to go to the Baltic Sea for relaxation.[11] Perhaps you will come up north sometime.

Cordial greetings, yours,

Einstein.

552.   From Felix Klein

Göttingen, 31 May 1918

Esteemed Colleague,

As promised, I am writing you before I interrupt my work on your latest papers for some weeks. Today now, first to your "cosmological" considerations.[1]

1. If ds^2 is defined in the way you did in 1917 I, in agreement with your view, find no obstacle to assuming the "space" as either spherical or elliptic, as one wishes.[2]

2. Not so, if one proceeds from a "world" of constant curvature (as I did, without noticing the difference with your assumption, in my Hedemünden elaboration and which is the basis of de Sitter's "Hypothesis B").[3]

In any event, for this purpose, it is simplest to set out from the "pseudo-spherical" form in 5 variables: $\xi^2 + \eta^2 + \zeta^2 - v^2 + \omega^2 = R^2$ and to define ds^2 as $d\xi^2 + d\eta^2 + d\zeta^2 - dv^2 + d\omega^2$.[4] The "elliptic" formulas then result in the known way,[5] in that

$$x = R \cdot \frac{\xi}{\omega}, \quad y = R\frac{\eta}{\omega}, \quad z = R\frac{\zeta}{\omega}, \quad u = R\frac{v}{\omega};$$

the projective metric with 4 variables is obtained, for which

$$x^2 + y^2 + z^2 - u^2 + R^2 = 0$$

is the absolute.– From the elliptic variables one returns to the spherical ones by "adjugating" the irrationality

$$\Omega = \sqrt{x^2 + y^2 + z^2 - u^2 + R^2}$$

and setting

$$\xi = \frac{Rx}{\Omega}, \quad \eta = \frac{Ry}{\Omega}, \quad \zeta = \frac{Rz}{\Omega}, \quad v = \frac{Ru}{\Omega}, \quad \omega = \frac{R^2}{\Omega}.$$

The question now is how to introduce the time t.

a) I find that the Schwarzschild–de Sitter[6] ds^2 is obtained by inserting[7]

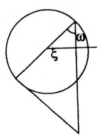

$$\xi = R\sin\vartheta\cos\varphi, \quad \eta = R\sin\vartheta\sin\varphi\cos\psi, \quad \zeta = R\sin\vartheta\sin\varphi\sin\psi$$

$$v = R\cos\vartheta\,\mathfrak{Sin}\,t, \quad \omega = R\cos\vartheta\cdot\mathfrak{Cos}\,t$$

$$\mathfrak{Sin},\ \mathfrak{Cos} = \text{hyperbolic functions.}$$

Hence inserting[8]

$$t = \arctan\frac{v}{\omega}.$$

And here arises the inconsistency with our tacit assumption about the time concept, which you point out in your note of 7 March,[9] namely, that t is undefined for those spatial points for which $v = 0$, $\omega = 0$.[10] This inconsistency is not eliminated even if we move over to an elliptic concept of space: at M_2, located in the space and which is of itself not special in any way, t has a singular region.[11]

b) We shall avoid this defect and generally have a natural point of departure if we set t in the spherical case equal to the perpendicular distance of some diametrical plane of the sphere, e.g., the plane $v = 0$:

$$t = R\cdot\mathrm{arc}\mathfrak{Sin}\frac{v}{R}.$$

In the elliptic case, this formula reads:

$$t = R \cdot \text{arc}\mathfrak{Sin} \left(\frac{u}{\sqrt{x^2 + y^2 + z^2 - u^2 + R^2}} \right).$$

And, if one wants to remain with the elliptic assumption, this is where the contradiction occurs that I wrote you about lately:[12] that past and future cannot be distinguished from one another.   For $+\sqrt{x^2 + y^2 + z^2 - u^2 + R^2}$ and $-\sqrt{x^2 \ldots - u^2 + R^2}$ are convex in the elliptic space; the sign of t necessarily remains undetermined.

On the other hand, the inconsistency disappears as soon as the mentioned square root is "adjugated," i.e., simply returning from the elliptic case to the spherical case. *I think, therefore, that the assumption of statement 2b can certainly be taken into consideration with the spherical interpretation.*–

With this I would like to close today. My entire letter is only supposed to be a more precise version of my earlier communication on the topic. If I now have provided certain formulas instead of geometric considerations, this is only because one can be more explicit with their aid; geometric considerations nevertheless do remain the source of the entire train of thought.

I do not know whether you will find anything new in this at all. In particular, I was not yet able to compare Weyl's developments.[13]

With best regards and wishing you a return to health, yours most sincerely,

Klein.

553.   To Arnold Sommerfeld

[Wilmersdorf, Berlin,] (5 Haberland St.) 1 June 1918

Dear Sommerfeld,

Yesterday evening you were elected Chairman by the Directors & Advisory Board as well as by a plenum of the German Physical Society,[1] and with visible enthusiasm at that. In the interest of the Society, I urge you to accept the nomination: No responsibilities, so to speak, will arise from this for you. When you happen to be in Berlin while the Society is meeting, you would take the chair; otherwise, one of the Board members residing here (Rubens in the first place) would substitute for you. Possible reservations you might have about accepting the nomination may be diminished by the following. We all shared the view that a non-Berliner ought to become chairman.[2] The choice first fell on

Mr. Max Wien because currently he is often in Berlin. Since this candidate has declined, owing to overburdening with military matters, no one else came into consideration whom we could hope to be present frequently at the meetings. So the choice fell spontaneously on you, without anyone else having been drawn into consideration. This unanimity deserves acknowledgment, and requesting that you soon convey this to the Society in the form of a ready

yes,

I am ⟨yours,⟩

A. Einstein

In addition, cordial private greetings.

554.   From Felix Klein

Göttingen, 1 June 1918

Esteemed Colleague,

Today two more questions, re your latest communication to the Berlin Academy.[1] I leave completely aside the case of the closed world.[2]

1. Why are there closed systems? Let us take the solar system as a whole, for instance. Then, at a great distance,[3]

$$\frac{\partial U_\sigma^4}{\partial x_4} = -\frac{\partial U_\sigma^1}{\partial x_1} - \frac{\partial U_\sigma^2}{\partial x_2} - \frac{\partial U_\sigma^3}{\partial x_3}$$

(including gravitational effect + radiation) would certainly become very small, but the integrals $\int \frac{\partial U_\sigma^4}{\partial x_4} dx_1 dx_2 dx_3$ would have to extend over a correspondingly large region, and I cannot see how their values would approach 0.

2. Furthermore, I do not understand (p. 4) how, with the methods of special relativity theory, it can be shown that J_σ is a four vector.[4] For the integrals, surely the upper index 4 is always singled out. I am missing some middle term there.[5]

Very truly yours,

Klein.

555.   From Arnold Sommerfeld

[Munich, after 1 June 1918][1]

D[ear] E[instein],

Although this note contains absolutely no secrets, its contents probably do not really fit in the official letter.[2]

You have probably read the new paper by Bohr.[3] His method of matching together wave th. and quantum th., using high quantum numbers, appears to me very effective, even if it does not offer any profound lessons. Certain concluding remarks of Bohr's coincide with a paper by Rubinowicz though, which meanwhile has been submitted to the *Phys. Z[ei]tschr[ift]* and about which I already spoke to you recently.[4] In the ms. to my Planck speech[5] [Laue's idea of the pamphlet is appealing, especially appealing in that in this way Pl.'s response and your researcher's dithyramb are also recorded.] I expanded on the matter a bit further than I could in the speech. Somewhat like this: It is not the atom, but the ether, whose métier it is to oscillate. It has this completely Maxwellian occupation, just as it must do in accordance with the amounts of energy and momentum of the atom. It has not yet been established definitely that the energy and momentum data determine the ether's oscillation. But there are so many confirmations of the interpretation already, with polarization in the Zeeman and Stark effects, and with the (since specified) quantum inequalities, that I do not doubt their soundness.[6]

For the last fortnight I have been writing a popular book on *Atombau und Spektrallinien* [*Atomic Structure and Spectral Lines*], the text itself for chemists, the supplements also for physicists.[7]

From Siegbahn I learned of a pretty confirmation of an initially unexpected finding on X-ray spectra.[8] K_β is a double line; the L ring expands at an electron's transition out of the M ring into the K ring, and in doing so reveals its double nature.

Could you not obtain from a Berlin benefactor a special grant for the presentation of talks at the G[erman] Ph[ysical] Soc.? As the first speaker, I suggest Siegbahn, for ex., or Bohr. As you see, I am already starting to meddle in the Society's affairs. Yours,

A. Sommerfeld.

556.   To Felix Klein

[Berlin, before 3 June 1918][1]

Highly esteemed Colleague,

I shall first answer your second letter.[2] If radiation exists in a system in the form of electromagnetic waves, then the U_σ^ν's do not vanish at the boundary to such an extent that the surface integral[3]

$$\int (U_\sigma^1 \cos nx + U_\sigma^2 \cos ny + U_\sigma^3 \cos nz) dS$$

extending over a remote surface would approach zero if the surface moved into infinity. In this case, one cannot say[4] that one has an isolated system in my sense of the term. I restrict myself in the paper to such cases where (with a suitable choice of coordinates) an isolated system can be identified. This agrees completely with the history of the energy concept; if approximately "isolated" systems did not exist, this concept could not have emerged.

The fact that J_σ, despite the asymmetrical way in which it is formed, can be a four vector (for linear transformations),[5] depends on the following: If the $T_{\mu\nu}$'s are tensor components, then the integrals[6]

$$\int T_{\sigma 4} dx_1 dx_2 dx_3 dx_4 = A_{\sigma 4}$$

$$\text{or} \int\int T_{\sigma 4} dV dx_4,$$

also are or, if one integrates over a piece of an isolated system's "world strand" [*Weltfaden*],[7]

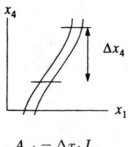

$$A_{\sigma 4} = \Delta x_4 J_\sigma.$$

Since this is a tensor component, and Δx_4 a vector component, it suggests itself that J_σ would be a four vector.

That this actually is so can be seen as follows. If a closed system is incompletely described, then the energy law for the described part reads

$$\sum \frac{\partial T_{\mu\nu}}{\partial x_\nu} = p_\mu,$$

where p_μ is the four vector of the force density acting on the system.[8] If this equation is integrated over a segment of a four-dimensional world strand, then one obtains on the right-hand side the time integral

$$\int dt \int p_\mu dV,$$

this is the total momentum and energy increase of the system, on the left-hand side, the expressions

$$\left| \int T_{\mu 4} dV \right|_1^2 = \Delta J_\mu.$$

Since the result of this integration has a vector character just as the integrand, the same is valid for ΔJ_μ, that is, for the increase that the momentum and energy experience on the segment of the world strand under consideration. Given this, it cannot be doubted that the J_μ's themselves have this characteristic.

The proof can be furnished, incidentally, also for the J_μ's themselves in that the $\Delta x_4 J_\sigma = A_{\sigma 4}$'s are components of a tensor $A_{\sigma\tau}$ ⟨whose components $A_{11} \ldots A_{33}$ vanish (as can easily be proven)[9]⟩. I do not want to go into this more formal proof here, though. So that out of the tensor character the possibility of describing it in the form $J_\sigma = E_0 \dfrac{dx_\sigma}{ds}$ follows, it is still necessary, however, that the momentum of a system at rest vanishes, which is not self-evident *per se*.[10] Physically, this means that a system at rest can be brought into another position through rotation without a finite exertion of force.–[11]

Your first letter[12] was very instructive for me; the consideration was new to me. From the physical point of view, I believe I can assert very definitely that this, because it is four-dimensionally uniform, mathematically more elegant conception of the world, does not correspond to reality. For the universe seems to be built in such a way that its finely distributed matter could remain at rest, with a suitable choice of the coordinate system.[13] This requires that $g_{44} = $ const. A physical interpretation of de Sitter's solution is easily obtained on the basis of de Sitter's own considerations. Namely, he finds that, through a choice of variables such that the $g_{\mu\nu}$'s become independent of t, ds^2 can be brought into the form[14]

$$ds^2 = -dr^2 - R^2 \sin^2 \frac{r}{R} \left(d\psi^2 + \sin^2 \psi d\vartheta^2 \right) + \cos^2 \frac{r}{R} \cdot c^2 dt^2.$$

This world can be obtained from the gravitational effect of a fluid which is concentrated at the "equator" (on a two-dimensional surface).

Two-dimensionally it is easy to draw a straightforward comparison with my conception (where time is omitted because of the field's static character):[15]

In my solution, the matter is distributed evenly over the entire surface of the sphere; for de Sitter, it is concentrated at the equator $\left(\dfrac{r}{R} = \dfrac{\pi}{2} \right)$.

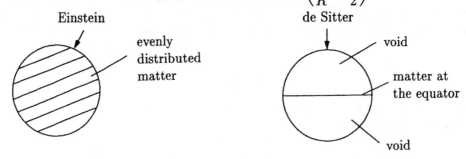

De Sitter incorrectly believes that his solution does not presuppose the existence of matter. Weyl has shown in his book[16] which will appear shortly, that de Sitter's case really is obtained as a limiting case of the more general

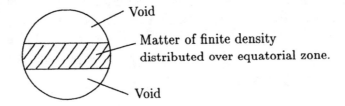

as I asserted without proof in my notice.[17]

If the world really were so, the fixed stars would have to have enormous velocities so that they could maintain their statistical distribution, owing to the tremendous differences in gravitational potential that would have to exist between the various points of such a world. The nonexistence of great stellar velocities compels us to believe that on a large scale matter is not all that unevenly distributed throughout the universe.

With amicable greetings, yours very truly,

A. Einstein.

557.   To Mileva Einstein-Marić

[Berlin,] 4 June 1918

Dear Mileva,

It is not without some gratification that I learn from your letter that you also have had problems with Anna Besso.[1] She has written me such impertinent letters that I've put an end to further correspondence[2] and am never going to be able to have dealings with her again. It must be granted her, though, that she is not quite normal. Michele, however, is independent of her to a great extent. All the same, if you prefer, speak with some other discerning man about our affairs. You can be sure that I have arranged the matter for your welfare, to the best of my knowledge and conscience. Prime securities at the current market value of 40 000 M have already gone out to you. This transfer is a great privilege granted to me personally.[3] I did it upon mature reflection. It is the best thing conceivable that I could do under the present circumstances to secure your and the children's future. In addition, 20 000 M are to be deposited here, the interest of which you'll receive in the case of my death.[4]

This summer I'm not going to come to Switzerland. I dread the strain and stress of the trip. I'm going instead to a little village by the Baltic Sea for two months.[5] I cordially invite Albert to come along. (Tete is maybe still too delicate.)[6] I could certainly get the traveling permission through, and for him the strain would be quite trivial. We would do a great deal of sailing there, but not on our own, because it is forbidden, unfortunately. Besides that, he could go swimming regularly in the sea. I leave it to you to consent to it; if not, there is no need for justification. Don't take me for a cruel father; it would not be sensible for me to embark on the trip under the existing traveling conditions.[7] Please do send me the corrected contract soon and file the divorce as soon as the opportunity arises.

Best regards, yours,

Albert.

Kisses to the boys.

558.   To Paul Ehrenfest

[Berlin, 5 June 1918]

Dear Ehrenfest,

Now poor you are sick as well;[1] maybe the thing has a purely psychological cause. It would be no wonder! I am curious what you are doing in political economics.[2] I always have the feeling that, owing to a lack of simple and rigorous chains of reasoning, everything in this field is too long-winded. But I find it tremendously interesting and understand how these problems can fascinate you. I have already argued with many about the question: Does money necessarily need backing (or some other fixing of its exchange value) or does it maintain itself? I've not been able to arrive at a firm conviction but tend toward the first view. What is your opinion? I am doing reasonably well; but I'm still unable to get along without constant care, since my stomach revolts at the slightest cause.[3] Why should it have less rights than my brain? From 22 June on I'm going away for two months to the Baltic Sea (Ahrenshoop near Stralsund).[4] I'm looking forward to sailing, my favorite amusement. You also should exchange wisdom for some harmless vegetating in nature for a while!

Warm regards to all 5 of you,[5] yours,

Einstein.

559.   From Anschütz and Company

Neumühlen near Kiel, 6 June 1918

To Professor Albert Einstein, Wilmersdorf, Berlin

Esteemed Professor,

We permit ourselves herewith to inquire whether you are inclined to provide us with a private evaluation of a patent matter.[1]

It concerns a patent for a gyrocompass which in our view

1) relies on an older patent of ours;

2) offers no inventive advantage over this older patent that could justify the issuance of an improvement or contingency patent.[2]

The case is relatively simple and burdened sparingly with documentation, so it would presumably take up only a little of your time, in spite of the fact that three members of the Complaints Department at the Patent Office did not succeed in penetrating to the core of the matter.

With best wishes for a complete recovery of your health, your very devoted servants,

Anschütz & Co.

560.   To Adolf Kneser

Berlin, 5 Haberland St., 7 June 1918

Highly esteemed Colleague,[1]

Thank you for sending me your original and inspired lectures.[2] In them *one* thing did pain me, though, which I do not want to pass over in silence. It hurts me when my name and my work are abused for chauvinistic propaganda,[3] as has been happening frequently in recent times. This is out of place even objectively. I am by heritage a Jew, by citizenship a Swiss, and by mentality a human being, and *only* a human being, without any special attachment to any state or national entity whatsoever. If only I could have said this to you before you held your lectures! You would certainly have had enough consideration for my feelings to omit the relevant utterances.

In utmost respect,

A. Einstein.

561.   To Felix Klein

[Berlin, 9 June 1918][1]

Esteemed Colleague,

It pleases me very much that you are going to give a talk on my energy paper.[2] I provide you now with the complete proof for the tensor character (for linear transformations) of J_σ.[3]

$\dfrac{1}{\sqrt{-g}}\mathfrak{U}_\sigma^\nu$ is a tensor for linear transformations, thus also

$$\int \mathfrak{U}_\sigma^\nu dx_1 dx_2 dx_3 dx_4 = A_\sigma^\nu, \ldots \tag{1}$$

extended over the region of integration

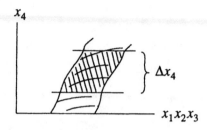

We consider specially[4]

$$A_\sigma^4 = \int \mathfrak{U}_\sigma^4 dx_1 dx_2 dx_3 dx_4 = \int J_\sigma \underbrace{dx_4 = J_\sigma}_{} \Delta x_4$$

$$\text{because } \frac{dJ_\sigma}{dx_4} = 0.$$

One has

$$J_\sigma = \frac{A_\sigma^4}{\Delta x_4}, \ldots \tag{2}$$

and the question arises, from which transformation law does this quantity follow?

In order to find this, I assign quantities Δx_1, Δx_2, Δx_3 to Δx_4 so that $(\Delta x_1, \Delta x_2, \Delta x_3, \Delta x_4)$ form a four vector. To arrive at this, I take Δx_4 to be large against the system's spatial dimensions, which can be indicated by the diagram.[5]

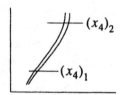

If now $(x_1)_1$, $(x_2)_1$, $(x_3)_1$, $(x_4)_1$ is a point within the system at time $(x_4)_1$, $(x_1)_2$, $(x_2)_2$, $(x_3)_2$, $(x_4)_2$ is a point within the system at time $(x_4)_2$, then the differences

$$\Delta x_1 = (x_1)_2 - (x_1)_1$$

etc.

are established up to relatively infinitesimal orders, and these quantities form a four vector.–

Now I introduce through linear transformation a second reference system K'. Then according to (2) for the transformation[6]

$$J'_\sigma = \frac{A'^4_\sigma}{\Delta x'_4} = \frac{\sum \frac{\partial x'_4}{\partial x_\alpha} \frac{\partial x_\beta}{\partial x'_\sigma} A^\alpha_\beta}{\sum \frac{\partial x'_4}{\partial x_\alpha} \Delta x_\alpha} = \frac{\sum \frac{\partial x'_4}{\partial x_\alpha} \frac{\partial x_\beta}{\partial x'_\sigma} J^\beta \Delta x_\alpha}{\sum \frac{\partial x'_4}{\partial x_\alpha} \Delta x_\alpha}$$

applies. This formula applies also when I choose the unprimed system in such a way that $\Delta x_1 = \Delta x_2 = \Delta x_3 = 0$.[7] In this case, it takes on the simpler form[8]

$$J'_\sigma = \sum \frac{\partial x_\beta}{\partial x'_\sigma} J_\beta \dots \tag{3}$$

Now, our proof would be finished if this formula were proven also for the transition from reference system K' to a third K''.[9] This is easily done. According to (3) we also have

$$J''_\sigma = \sum \frac{\partial x_\beta}{\partial x''_\sigma} J_\beta.$$

On the other hand, we have through the inversion of (3)

$$J_\beta = \sum \frac{\partial x'_\tau}{\partial x_\beta} J'_\tau.$$

Hence,[10]

$$J''_\sigma = \sum_{\beta\tau} \frac{\partial x'_\tau}{\partial x_\beta} \frac{\partial x_\beta}{\partial x''_\sigma} J'_\tau = \sum \frac{\partial x'_\tau}{\partial x''_\sigma} J'_\tau.$$

Thus the proof is complete.

With all due respect and greetings from yours truly,

A. Einstein.

[5] Recipient's label: "world strand."

[6] Recipient's comment: "Instead of this it ought to be possible to show that all A^α_β's, whose α's differ from 4, are zero. Yes!"

[7] Recipient's comment: "E. seems to compensate the first error with the 2nd. (I really do not have Δx_1, Δx_2, Δx_3 freely at my disposal anymore; they are given by the integration domain, and therefore $>< 0$ but very small relative to Δx_4.)"

[10] Recipient's comment: "What about the difference between cogradient and contragradient?"

562. Divorce Agreement

Berlin, 12 June 1918

Agreement.

1) Prof. Einstein in Berlin shall deposit in trust at a Swiss bank Forty Thousand Marks in securities,[1] with the stipulation that, in the case of a divorce of the spouses, this sum become the property of Mrs. Mileva Einstein née Marić.

2) From the moment of the opening of the trust account, Mrs. Mileva Einstein shall draw the interest. She shall not exercise power over the capital without the consent of Prof. Einstein (i.e., the securities shall neither be sold, nor mortgaged, nor exchanged without his approval).

3) Prof. Einstein shall send in quarterly installments to Mrs. Mileva Einstein an alimony sum which, including the interest from the gift sum just mentioned and including the interest from the Nobel Prize to be named under No. 4 of this contract, annually amounts to a total of Eight Thousand Francs,[2] about the disbursal of which Mrs. M. Einstein shall render no account.

4) Prof. Einstein shall instruct, in the event of a divorce and in case he receives the Nobel Prize, the principal of the above less Forty Thousand Marks, to become the property of Mrs. Mileva Einstein and shall deposit this capital in trust at a Swiss bank. Regarding this sum the following shall apply:

　　a) Mrs. M. Einstein shall have no authority over the capital without the consent of Prof. Einstein. She shall dispose freely of the interest, however.

　　b) In the case of the remarriage or death of Mrs. Einstein, the above-mentioned 40,000 Marks, or the above-mentioned 40,000 Marks along with the Nobel Prize ceded to Mrs. Einstein, less 40,000 Marks, shall go to the children, Albert and Eduard Einstein.

5) Prof. Einstein shall provide that Twenty Thousand Marks be deposited at a German bank,[3] the interest of which shall be paid out to Mrs. M. Einstein after his death, in the event that Prof. Einstein is not awarded the Nobel Prize.

6) Mrs. Mileva Einstein shall assume the care of the children and shall exercise all aspects of parental authority—obviously within the constraints of the legal regulations. She shall obligate herself to entrust the children to their father while he is sojourning in Switzerland during the school holidays.

7) Upon initiation of the divorce proceedings, Mrs. Einstein shall submit this Agreement to the judge for approval.

Albert Einstein.

(erroneously added! Einstein) $\left\{ \begin{array}{l} \langle \text{Zurich, the 7th of June 1918} \rangle \\ \langle \text{Mileva Einstein-Marić} \rangle \end{array} \right.$

563. From Hugo A. Krüss

Berlin, 13 June 1918

Esteemed Professor,

Attached you will find the letter by Eötvös. Its return was unfortunately delayed by my absence from Berlin.[1]

On Eötvös's comments I would like to add that a very essential aspect escapes his evaluation, namely, that taking into account the standing of the Potsdam Institute, the general scientific prestige and talent of its director is of critical importance, whereas Eötvös judges the matter predominantly, if not exclusively, from the professional standpoint.

It is only under this constricted professional aspect that, in my view, Krüger could come into consideration, as occurs in Eötvös's opinion.[2]

With best regards, yours very truly,

Krüss

P.S. Under the given state of affairs, I would consider Runge–Göttingen as the best solution.[3]

564. From Walter Dällenbach

Zurich, 15 June 1918

Report on scientif. research

Dear Professor,

This spring, after a 10-mo. stint of military service, I took up my d[octo]r[al] thesis again and would like to report on it to you before it is published.[1] Let a be any function of the world coord[inate], $d\tau$ the invariant volume element (I restrict myself in the following to the special theory of relativity, the drag of the grav. terms has no value). Then the mean value for a is

$$\bar{a} = \frac{1}{\int_G d\tau} \int_G a d\tau,$$

G a phys. sm[all] world domain. If one applies this averaging process to electro-magn. field equations, upon the basis of a model of matter consisting in uncharged molecules, convective charges & conduction current, then the resulting equations between the mean values are:

$\rho_0 =$ electricity density at rest

$u^\mu = \dfrac{dx^\mu}{ds} =$ four-component velocity

[*Vierergeschwindigkeit*].

$$F_{\mu\nu} = \frac{\partial \Phi_\mu}{\partial x^\nu} - \frac{\partial \Phi_\nu}{\partial x^\mu} \qquad (a)$$

$$\frac{\partial F^{\mu\nu}}{\partial x^\nu} - \frac{\partial M^{\mu\nu}}{\partial x^\nu} = J^\mu \qquad (b)$$

$$\frac{\partial F_{\mu\nu}}{\partial x^\sigma} + \frac{\partial F_{\nu\sigma}}{\partial x^\mu} + \frac{\partial F_{\sigma\mu}}{\partial x^\nu} = 0 \qquad (c)$$

$$M^{\mu\nu} = \overline{\frac{\rho_0}{2}(u^\nu x^\mu - u^\mu x^\nu)} \qquad (d)$$

An essential precondition for eq. (d) is that, for the phys. small interval of a molecule,

$$\int \rho_0 \mu^\mu d\tau = 0$$

applies.[2]

Derivation of eq. (d) & of the term $\dfrac{\partial M^{\mu\nu}}{\partial x^\nu}$ in (b) is but a transposition of the considerations by H. A. Lorentz in *Encycl. d. math. Wiss.* V 14, N° 27 and 28 from *space* into a *world*.[3]

If static matter $F^{\mu\nu}$ is identified as $\{\mathfrak{b}, \mathfrak{e}\}$ and $M^{\mu\nu}$ as $\{\mathfrak{m}, \mathfrak{p}\}$, then (b) (c) represent Maxwell's eq. & eq. (d) degenerates into the familiar expressions[4]

$$\mathfrak{p} = N \sum_{k=1}^{n} e_k \mathfrak{r} \qquad \& \qquad \mathfrak{m} = N \sum_{k=1}^{n} \frac{1}{2} e_k [\mathfrak{r}, \mathfrak{v}]$$

of electron theory. Φ_μ is a vector & scalar pot.

To determine the force density κ_σ, I take a straightforward course. Let G be a convex world domain that intersects arbitrarily many molecules. The force acting on the charges in G divided by the size of the domain itself yields the force density

$$\kappa_\sigma = F_{\sigma\mu}\frac{\partial F^{\mu\nu}}{\partial x^\nu} - \frac{\partial}{\partial x^\nu}(F_{\sigma\mu}M^{\mu\nu}) - \frac{\partial}{\partial x^\nu}\left\{\delta_\sigma^\nu\frac{1}{2}\int M^{\alpha\beta}dF_{\alpha\beta}\right\}$$

$$\qquad\qquad\quad 1)\qquad\qquad\quad 2)\qquad\qquad\qquad 3)$$

1) stems from the *true* current in domain G,
2) from the polarization current and
3) is the force exerted by all molecules attached to the exterior of domain G and cut by the surface of G. That means, all the countless rubberbands cut by a

surface element 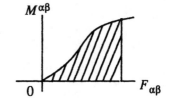 have as a resulting tension a simple pull

perpendicular to the element, regardless of how this element may be orientated, of the quantity $\frac{1}{2}\int M^{\alpha\beta}dF_{\alpha\beta}$. Since we are just considering isotropic, nondis-

persing, hysteresis-free media, each comp[onent] $M^{\alpha\beta}$ depends only upon its corresponding $F_{\alpha\beta}$. The integral should be extended from zero polarization until the actually existing value. The known rewriting of term (1) with the induction law yields

$$\kappa_\sigma = \frac{\partial T_\sigma^\nu}{\partial x^\nu} \quad\text{where the energy-momentum tensor is}$$

$$T_\sigma^\nu = -F_{\sigma\alpha}(F^{\nu\alpha} - M^{\nu\alpha}) + \delta_\sigma^\nu\frac{1}{2}\int(F^{\alpha\beta} - M^{\alpha\beta})dF_{\alpha\beta}$$

or let $D^{\alpha\beta} = F^{\alpha\beta} - M^{\alpha\beta}$ be the tensor "displacement" with the comp. $\{\mathfrak{h}, \mathfrak{d}\}$.

$$\boxed{T_\sigma^\nu = -F_{\sigma\alpha}D^{\nu\alpha} + \frac{1}{2}\delta_\sigma^\nu\int D^{\alpha\beta}dF_{\alpha\beta}}$$

Field & matter can be held strictly apart. Of more interest is T_σ^ν in comp. upon amalgamation according to[5]

$$T_\sigma^\nu =
\begin{array}{|c|c|c|c|}
\hline
\mathfrak{e}_x\mathfrak{d}_x + \mathfrak{h}_x\mathfrak{b}_x - \int(\mathfrak{d}d\mathfrak{e} + \mathfrak{b}d\mathfrak{h}) & \mathfrak{e}_x\mathfrak{d}_y + \mathfrak{h}_x\mathfrak{b}_y & \mathfrak{e}_x\mathfrak{d}_z + \mathfrak{h}_x\mathfrak{b}_z & -[\mathfrak{d}, \mathfrak{b}]_x \\
\hline
\mathfrak{e}_y\mathfrak{d}_x + \mathfrak{h}_y\mathfrak{b}_x & \mathfrak{e}_y\mathfrak{d}_y + \mathfrak{h}_y\mathfrak{b}_y - \int(\mathfrak{d}d\mathfrak{e} + \mathfrak{b}d\mathfrak{h}) & \mathfrak{e}_y\mathfrak{d}_z + \mathfrak{h}_y\mathfrak{b}_z & -[\mathfrak{d}, \mathfrak{b}]_y \\
\hline
\mathfrak{e}_z\mathfrak{d}_x + \mathfrak{h}_z\mathfrak{b}_x & \mathfrak{e}_z\mathfrak{d}_y + \mathfrak{h}_z\mathfrak{b}_y & \mathfrak{e}_z\mathfrak{d}_z + \mathfrak{h}_z\mathfrak{b}_z - \int(\mathfrak{d}d\mathfrak{e} + \mathfrak{b}d\mathfrak{h}) & -[\mathfrak{d}, \mathfrak{b}]_z \\
\hline
[\mathfrak{e}, \mathfrak{h}]_x & [\mathfrak{e}, \mathfrak{h}]_y & [\mathfrak{e}, \mathfrak{h}]_z & \int(\mathfrak{e}d\mathfrak{d} + \mathfrak{h}d\mathfrak{b}) \\
\hline
\end{array}$$

Energy cur[rent] $[\mathfrak{e}, \mathfrak{h}]$ + energy density[6] are correct. T_σ^ν degenerates for electrostatics to the expressions, as Cohn has found through variation of $\int \mathfrak{e}\, d\mathfrak{d}$.[7] However, tensor, as well as $T^{\mu\nu}$ is no longer symmetric. Momentum density

$$[\mathfrak{d}, \mathfrak{b}] = [\mathfrak{e}, \mathfrak{h}] + [\mathfrak{p}, \mathfrak{b}] + [\mathfrak{e}, \mathfrak{m}]$$

This looks as though these[8] certainly denoted momentum but not energy cur. This stems from that the electrical dipole at buildup of the electr. field $\mathfrak{e}$ in magnetic field $\mathfrak{b}$ obtains a force that is just equal to $\dfrac{\partial}{\partial t}[\mathfrak{p}, \mathfrak{b}]$. The corresponding energy current is free of divergence already for the self-contained individual dipole: The mean value for the momentum thus differs from zero, while the mean value for these energy currents[9] vanishes. $T^{\mu\nu}$ is thereby asymmetric and does not contradict the principle of the inertia of energy, after all.[10] T_σ^ν

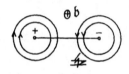

is the tensor that must be inserted in order to obtain the correct force density and actually has nothing to do with the mean value for the energy-momentum tensor in a vacuum.

κ_σ can be converted into the following simple form

$$\kappa_\sigma = F_{\sigma\mu} J^\mu - \frac{1}{2} \int \frac{\partial M^{\alpha\beta}}{\partial x^\sigma} dF_{\alpha\beta}.$$

This[11] $\dfrac{\partial}{\partial x^\sigma}$ is meant for a spatio-temp[orally] constant field. The integral has the same meaning as above. The result for matter at rest, divided into space & time is:

$$\begin{cases} \kappa_{1,23} = [\mathfrak{i}, \mathfrak{b}] + \rho\mathfrak{e} - \int \operatorname{grad} \varepsilon (\mathfrak{e}, d\mathfrak{e}) \\[2mm] \qquad\qquad - \int \operatorname{grad} \left(\dfrac{-1}{\mu}\right)(\mathfrak{b}, d\mathfrak{b}) \\[2mm] \kappa_4 = -(\mathfrak{i}, \mathfrak{e}) - \int \dfrac{\partial \varepsilon}{\partial t}(\mathfrak{e}, d\mathfrak{e}) - \int \dfrac{\partial \left(\frac{-1}{\mu}\right)}{\partial t}(\mathfrak{b}, d\mathfrak{b}) \end{cases}$$

$$\varepsilon = \text{Dielectr. const.} = \tan \alpha \qquad \mu = \text{Permeability}$$

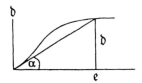

This, by & large, is the thesis, which I submitted today (I have already discussed it with Weyl). In the survey of the literature, I presented my view on the paper you wrote together with Mr. Laub and reiterate my comments here verbatim:[12]

"The authors determine the force on a volume element in space in this world in a way similar to mine; however, they require that its surface not intersect any dipole. Thus the pull originating from the internal bonds escapes their notice. Their force density in an electrostatic field {page 544, *Ann. d. Phys.* 26/1908}

$$\kappa = \mathfrak{e}\rho - \frac{1}{2}\operatorname{grad}\varepsilon\mathfrak{e}^2 + \frac{1}{2}\operatorname{grad}(\mathfrak{p},\,\mathfrak{e})$$

differs correspondingly from mine precisely through the pull $\frac{1}{2}\operatorname{grad}(\mathfrak{p},\,\mathfrak{e})$, which they are not in a position to interpret. I do not understand why, as is further asserted, this term should contribute nothing upon integration over a body existing in a vacuum. In accordance with the omission of the internal bonds, the above formula lacks the integral that always appears in my rendition as long as, aside from the one-to-one correspondence, no other condition is set about the dependence of polarization $\mathfrak{p}$ on the field $\mathfrak{e}$.

"The magnetic behavior of the bodies is represented by magn. dipoles rather than through moving electrons. That is the cause of the preference of vector $\mathfrak{h}$ over $\mathfrak{b}$. That is also why the objections to the term originating from Minkowski[13] [$\mathfrak{i}$, $\mathfrak{b}$] are unfounded, quite aside from that it absolutely never may be considered separated from the other terms. The *total* force density alone is decisive, which can be separated into the individual terms in a wide variety of ways, depending on the purposes needed. Thus, for example, my force density of the form

$$\kappa_{1,\,23} = [\mathfrak{i},\,\mathfrak{b}] + \rho\mathfrak{e} - \int \operatorname{grad}\varepsilon(\mathfrak{e},\,d\mathfrak{e}) - \int \operatorname{grad}\left(\frac{-1}{\mu}\right)(\mathfrak{b},\,d\mathfrak{b})$$

could give the impression that the displacement current $-\dfrac{\partial\mathfrak{d}}{\partial t}$ experienced a force in a magnetic field $\mathfrak{b}$ in a way quite different from the conduction current $\mathfrak{i}$. And yet, upon derivation of the force from the energy-momentum tensor, the temporal change in the momentum density for a practically constant magnetic field $\left[-\dfrac{\partial\mathfrak{d}}{\partial t},\,\mathfrak{b}\right]$ thoroughly corresponds to the term [$\mathfrak{i}$, $\mathfrak{b}$]."

"Although I have submitted the paper, the development of the asymmetry of $T^{\mu\nu}$ is not clear to me yet. I think it is connected to the fact that the substance in the world occurs in a threadlike manner (world lines).[14] When I have clarified this problem for myself, I shall write a postdoctoral thesis [*Habilitationsschrift*][15] and intend to habilitate at the Poly[technic] next fall on "phys. math. foundations of electrical engineering."

I am doing statistics together with Weyl & Besso.[16] We are reading your papers & the ones by Smoluchowsky on Brownian motion.[17] This I would like to continue until the spring of 1919 at the latest and then do practical work, if possible in France. I think that I am best suited for a lectureship that combines technology with science.

Despite your objection, I cannot shake off the great impression that Weyl's new theory has made on me.[18] What is certain is that, if it is accepted, *all* relations to observable fact are lost. Physics is then, no more than geometry, in an absolute world of its own which only Almighty God can operate. From there, some principle must exist in my consciousness that wanders along a world line, for inst., linking together the various space-time points of my experience. Or I must replace the present externally, dogmatically imposed coordinate arrangement of space-time points with something else that is firmly assumed as invariant against my stream of consciousness. You choose for this such complex things as *"rigid body"* or *"clock."*[19] This seems to me to be arbitrary. Perhaps much more profoundly based entities exist than these particular rigid bodies or clocks,[20] which we must pronounce over & over again as *the same*. For example, a unit for the quantum of action: Hence, although Weyl's theory is, for the time being, incapable of explaining the constancy of the chemical elements' spectral lines,[21] it does provide room for a new theory that would be in a position to do so, creates the purest foundation for it, and in this way does become the necessary transition point for a theory of matter, of the relation between the discrete & the continuous. But first we must wait until Weyl has found the electron. He has found the energy-momentum tensor. His Laue scalar is supposedly always strictly zero. This, as he told us yesterday, is what he is working on at the moment. He has also found the Newtonian gravitation eq. as a first approximation of his fourth-order equa[tion].[22] The skill he has in calculation is quite tremendous.

Mr. Besso has been telling me all the time how unsteady your health is.[23] I am very sorry about that. Also, I absolutely do not like your not coming to Switzerland. I would have been so pleased to see you again. But it could be that the long trip would have hurt you and that you can convalesce just as well during a longer stay by the sea as in the mountains.[24] At any rate, I wish you return to health as soon as ever possible.

I have had it up to here with what's happening in the world otherwise. And still no end is in sight.

Cord[ial] regards from your grateful student,

Walter Dällenbach.

Mr. Besso will be writing you in the next day or so and in the meantime sends his cordial greetings.

565.   To Walter Dällenbach

[Berlin, after 15 June 1918][1]

Dear Mr. Dällenbach,

It was very nice of you to send me an excerpt of your thesis.[2] I especially liked your fine expression for polarization, which seems to be new. The energy tensor is certainly correct. It has long been known that the values I had derived with Laub at the time are wrong; Abraham, in particular, was the one who presented this in a thorough paper.[3] The correct strain tensor has incidentally already been pointed out by Minkowski.[4] Its asymmetry arises from the fact that an anisotropic quasi-elastically bonded electron in an electrical field transmits a rotation moment to its molecule. Hence, $Y_z \neq Z_y$, etc. must be true.

Weyl's theory of electricity is wonderful as a concept, an ingenious theoretical achievement. However, I am convinced that nature is different.[5] Two infinitesimal bodies that were once congruent always remain so, irrespective of how different a turn their fates may take. Thus if two ds's lying at points a finite distance away from each other are measured as equal by *one* measuring rod, or by *one* measuring technique, then when measured otherwise they are again equal. This is a deeply rooted attribute of our world, which must find expression in the general foundations of physics. According to Weyl's conception, it would have to be expected that two originally identical measuring rods (at space-time point P_1) are no longer identical after passing through different world lines to reunite at P_2. Their difference would correspond to the value for the difference of both line integrals[6]

$$\int_{\text{curve 1}} \varphi_\nu dx_\nu - \int_{\text{curve 2}} \varphi_\nu ds_\nu.$$

This difference disappears, in general, only when there is no electromagnetic field lying between the world lines of both bodies. If, for inst., I place one of the bodies for a longer time T in a metal box charged to the potential φ and leave the other one outside, then at the end of time T, this difference has the value

$$\varphi T,$$

which is critical now to the relative quantities of both previously equal bodies.

I know very well that Weyl does not recognize this. He would say, clocks and measuring rods would have to appear merely as solutions; they do not occur in the foundations of the theory.[7] I find, though: if *ds*, measured with a clock (or a measuring rod), is an entity independent of the clock's prehistory, design, and material composition, then this invariant must as such also play a very fundamental role in the theory. But if real natural processes were not like this, there would be no spectral lines and no well-defined chemical elements.

Moreover, for Weyl there is *one* world line of a preferred nature, the geodesic. On the other hand, in nature there is *one* type of preferred world line, the trajectories of electrically uncharged point masses. It is hardly conceivable otherwise than that both are identical. The φ_i's now enter the equation of this geodesic line as moving forces, and this in a very dubious way.[8] You can figure this out yourself.

All in all: I am fully and firmly convinced that this theory does not correspond to reality. Formally as well, it appears unsatisfactory to me that the *angle* be regarded as invariant to displacement, but not the unit of length.–[9]

I was very pleased that you are no longer so militarily minded.[10] I hope you will advance well in your career. With your downright talent and your love of the subject, I should think so.

Accept my warm regards, yours,

Einstein.

In any case, I share Weyl's conviction that it must be possible to link gravitation and electricity to a unity; I just think that the correct link has not yet been found.[11]

566. From Felix Klein

Göttingen, 16 June 1918

Esteemed Colleague,

I discussed your May communication now variously with Hilbert and particularly with Runge[1] but, much as I understand your general train of thought, I cannot persuade myself of its correctness. In particular, I do not see how in the formula

$$A_{\sigma'}^4 = \sum \frac{\partial x_4'}{\partial x_\alpha} \cdot \frac{\partial x_\beta}{\partial x_\sigma} A_\beta^\alpha$$

you can replace A_β^α with $J_\beta \cdot \Delta x_\alpha$.[2] So for the time being probably nothing will come of a lecture by me on the subject, especially since in the coming weeks I may be occupied almost exclusively with the practical tasks attached to the 20th anniversary of the foundation of the Göttingen Association for Applied Physics and Mathematics.[3]

But last Tuesday I did report on your note re. de Sitter's ds^2.[4] I came to the result that the singularity you noticed can indeed simply be transformed away.[5] The formulas I prepared I had actually already written down recently.[6] I set

$$\left.\begin{aligned}
\xi^2 + \eta^2 + \zeta^2 + \upsilon^2 - \omega^2 &= R^2 \\
d\xi^2 + d\eta^2 + d\zeta^2 + d\upsilon^2 - d\omega^2 &= -ds^2
\end{aligned}\right\} \tag{1}$$

and arrive at de Sitter's line element[7] through the substitution:

$$\left.\begin{aligned}
\xi &= R\sin\frac{r}{R}\cdot\cos\psi, \quad \eta = R\sin\frac{r}{R}\cdot\sin\psi\cdot\cos\vartheta, \quad \zeta = R\sin\frac{r}{R}\cdot\sin\psi\sin\vartheta \\
\upsilon &= R\cos\frac{r}{R}\cdot\mathfrak{Cos}\left(\frac{c(t-t_0)}{R}\right), \quad \omega = R\cos\frac{r}{R}\cdot\mathfrak{Sin}\left(\frac{c(t-t_0)}{R}\right),
\end{aligned}\right\} \tag{2}$$

where $\mathfrak{Cos}$. and $\mathfrak{Sin}$. mean hyperbolic functions and t_0 is arbitrary. I do not need to write down the inverse of these formulas, which is self-evident; let me just note the value for t:

$$t - t_0 = \frac{R}{2c^2}\log\left(\frac{\upsilon+\omega}{\upsilon-\omega}\right). \tag{3}$$

Through this inverse, however, de Sitter's ds^2 is changed into $-d\xi^2 - d\eta^2 - d\zeta^2 - d\upsilon^2 + d\omega^2$, which on the "quasi-sphere" $\xi^2 + \eta^2 + \zeta^2 - \upsilon^2 + \omega^2 = R^2$ certainly has no singularity, q.e.d.[8]

In the last analysis, there are ∞^6 many legitimate ways of moving from equation system (1) to a de Sitter ds^2. This is because the "hyperplanes" $(\upsilon+\omega) = 0$ and $(\upsilon-\omega)$[9] are ultimately only two arbitrary tangential planes of the quasi-sphere's "asymptotic cone":

$$\xi^2 + \eta^2 + \zeta^2 + \upsilon^2 - \omega^2 = 0.- \tag{4}$$

It is amusing to picture how two observers living on the quasi-sphere and equipped with differing de Sitter clocks would squabble with each other. Each of them would assign finite time ordinates to some of the events that for the other would be lying within infinity or that would even show imaginary time values.–[10]

In order to give a physical turn to my letter after all, I note that de Sitter's ds^2 appears implicitly already in Schwarzschild's paper of 24 Feb. 1916.[11] One just has to set $\chi_\alpha = \dfrac{\pi}{2}$, $c = 2$, $R = \sqrt{\dfrac{\kappa\rho_0}{3}}$ in formula (35) there in order to have de

Sitter's ds^2.[12] Formula (35) relates, of course, to the *interior* of the sphere at rest considered by Schwarzschild of gravitating liquid of constant density. Formula (30) is thus applicable, which yields $p = -\rho_0$, hence a steady pull.[13]

Obliging regards from yours most truly,

Klein.

P.S. I think I can set your letter of 9 June[14] straight now. The A_β^α's are strictly equal to zero for $\alpha \neq 4$, the corresp. Δx_α's are arbitrarily small against Δx_4.[15]

567.   To Felix Klein

[Berlin, 20 June 1918]

Esteemed Colleague,

You are entirely right. De Sitter's world is, in and of itself, free of singularities and its space-time points are all equivalent.[1] A singularity comes about only through the substitution providing the transition to the static form of the line element. This substitution changes the *analysis-situs* relations. Two hypersurfaces

$$t = t_1$$

and $t = t_2$

intersect each other in the original representation, whereas they do *not* intersect each other in the static one.[2] This is related to the fact that, for the physical interpretation, masses are necessary in the static conception but not in the former one.[3] My critical remark about de Sitter's solution needs correction; a singularity-free solution for the gravitation equations without matter does in fact exist.[4] However, under no condition could this world come into consideration as a physical possibility. For in this world, time t cannot be defined in such a way that the three-dimensional slices $t = $ const. do not intersect one another and so that these slices are equal to one another (metrically).[5]

With sincere thanks for your fine and illuminating letter, yours truly,

A. Einstein.

Recipients note on the verso: "Did Weyl appear? Appointment regarding the building plans of the Math. Inst."

568.   From Anschütz and Company

Neumühlen near Kiel, 21 June 1918

To Professor Albert Einstein, Berlin W. 30, 5 Haberland St.

Esteemed Professor,

In receipt of your letter of the 9th inst., we thank you for your willingness and forward to you in the same post transcriptions of the patent applications we are challenging, as well as the complete set of documents.[1] The notification by the Examination Office, and the decision by Complaints Department I, reveal that the material we have presented was treated altogether superficially.

The same was the case for our oral presentations scheduled for 31 May. They culminated in the argument that the prominent advance made by our Patent Specification 241,637 consists in the revelation of the source of the stabilizer error and in the arrangement of the ancillary gyroscopes to augment the oscillation period as an axis along the north-south line.[2]

As is self-evident to an expert, the design corresponds thoroughly in its mode of operation to the familiar Schlick gyrostabilizer for ships.[3]

Even without the reference to Föppl[4] in Pat. 241,637, which at the bottom of page 223 terms the Schlick gyrostabilizer as a gyropendulum, the comparison with the nautical gyrostabilizer of gyrostabilization according to our Patent 241,637 was generally known in professional circles. In Föppl's book (*Vorlesungen über technische Mechanik* [*Lectures on Technical Mechanics*], Volume 6) which we are likewise submitting to you in the same post, the nautical gyrostabilizer is treated comprehensively. The explanations on pages 223 to 271 concern a single gyroscope with a vertical axis, thus the arrangement of the contested main application.[5]

On page 274 at the bottom, Föppl mentions an arrangement that resembles figure 4 of our Patent Specification 241,637.

We concluded from this that an expert who was prompted by the reference on page 1, line 54 of our patent specification to acquaint himself with Föppl's works required no inventive effort to discover from among "all cases" mentioned in our patent specification the special case of an ancillary gyroscope with a vertical axis. An inventive achievement would exist here only if mechanisms differing in principle from those in Patent Specification 241,637 or, in the case of similar operation, if additional special advantages were gained over our patent. Neither is the case, and even the alleged greater simplicity does not exist, for through the described arrangement of two gyroscopes according to our patent, the stabilizer error is completely eliminated. The third gyroscope, added for practical reasons not pertinent here (comp. offprint from the *Lehrbuch für den Unterricht in der*

Navigation [*Textbook for the Instruction of Navigation*], page 37),[6] has nothing to do with the stabilizer problem *per se*.

The reliance of Main Application G. 43359 on our Patent 241,637 was not deliberated on by the Nullification Department because it is not responsible for this. Both descriptions are explicit enough, so for the present we have nothing to add.

Finally, we mention that the process has been limited to Main Application G. 43359 so, in your evaluation also, you will want to consider the Supplemental Applications G. 44045 and G. 44067 only insofar as may perhaps seem useful to you in commenting on the substance of the main application.

In utmost respect, your very devoted servants,

Anschütz & Co.

569.   To Walter Schottky

[Berlin, 23 June 1918][1]

Dear Mr. Schottky,

I thank you heartily for the *great effort* you took for my sake and am glad that the information has come out to be so positive. It is being passed on immediately to my mother, who has asked me for it[2] (I made your signature illegible so that no indiscretion arise from it). If you ever are confronted with a mission with which you think I could help you, then please give me the opportunity to reciprocate.

Accept my best wishes for the next months and at the end of August or in September come again to visit your

A. Einstein.

Don't work too much but go on more walks!

570.   To Max Born

[Berlin,] 24 June 1918

Dear Born,

Tomorrow we *must* leave for a summer *resort* in Aarenshoop (at Mrs. Niemann's née Konow).[1] These lines in ceremonious farewell. A Greek gift also comes with it: I procured with Haber's help a travel permit to Finland for Nordström (at the General Staff). Now he wants to return to Holland,[2] and I am unfortunately no longer in a position to see to the matter. I beg you to settle the affair. It is urgent because Mrs. Nordström[3] is expecting to give birth to her child soon, and this, if possible, in Holland.

Wishing you and your little company pleasant days,[4] yours,

Einstein

I hope the 40 M, which I sent in a normal letter, have arrived.

571.   To Heinrich Zangger

[Berlin,] 24 June [1918][1]

Dear friend Zangger,

Now nothing will come of my trip to Switzerland after all, as sorry as I am about it. Although my health is reasonably good, I shrink from the long trip and other difficulties attached to it. The day after tomorrow I am going with Elsa and her children to a little village by the Baltic Sea for ca. 7 weeks (Aarenshoop, not far from Stralsund). You are now in the *fortunate* position of having to look for a successor to Sauerbruch.[2] I humbly turn your attention to a man here of approximately our age, whom I have come to know and value (also as a person) and who is looked upon as one of the best young local surgeons[3] but who also supposedly works successfully in other fields of medicine besides. This modest comment ought not to be interpreted as meddling.

Life goes on in quiet occupation without any special great new goals, only what developments draw along with them of their own accord. But for it, I have more peace to appreciate the ideas of others. I am painstakingly avoiding reading the newspaper and have few dealings with people.

In 1916 W. Kossel published in the *Annalen der Physik* a comprehensive article "Über Molekülbildung als Frage des Atombaues" ["On Molecule Formation as a Problem of Atomic Structure"], an almost exclusively qualitative paper, which I consider very lucid.[4] The author reported on it recently at our colloquium.[5] This you should not miss, but read it,[6] preferably with Besso. Have a good rest during the holidays; you must need it very much after all that has happened![7] We don't want to die without first seeing what the world's new state of equilibrium will look like!

If Weyl also leaves Zurich now, no professor of theoretical physics will be left there.[8] Then one ought to think also of Ratnowsky.[9] He is suited at least to the extent that he deserves a steady position at the university. His idea, which Guye . . . [stole] from him,[10] was certainly fine, and besides, he frequently produces more or less original ideas.[11] Before casting him completely aside, one should realize what one gets when the sense of inadequacy does not have enough force. There is presently a danger that he will work himself to pieces. Unfortunately, his wife[12] does not seem to be a good manager.

Now I have the opportunity of returning what I still owe you. It would be good if you gave me a complete itemization.[13] I am very glad now that I am able to secure my children's future financially so well that they will not be dependent on outside help if they want to study, as was the case with me.[14]

Cordial regards from your

Einstein.

572.   To Michele Besso

[Berlin, before 28 June 1918][1]

Dear Michele,

When I see your handwriting, I'm always very particularly pleased, for no one is so close to me as you and knows me so well and means so well. Don't worry at all about my experiences on paper with Anna; I don't want to bear a grudge against her for it and am grateful for her effort. But I must say one thing. Never before has anybody been so insolent to me, and I hope that no one ever will again in the future![2]

Weyl is an ingenious, fine fellow, but his conception of electricity is no good; I wrote Dällenbach about it in a bit more detail,[3] nothing new, by the way. I'm not going to try to convince you all but leave it to time. I would have liked to hear Weyl's lecture;[4] everything by him is so original and is elaborated so exquisitely. And he certainly can calculate!!

I have no patience for writing a statistical mechanics. Besides, Gibbs's book[5] is a masterpiece, although hard to read and with the main points between the lines. . . . It's a simple matter to recommend book-writing to a friend, you scoundrel, but where is *your* book then, which you were telling me about so earnestly last year? *Par nobile fratrum!*[6]

Now to the contract. I very much hesitate presenting Mileva with suggestions for new alterations or additions, now that she is satisfied.[7] She will surely handle the money carefully when I'm dead; so I don't want to burden her with any supervision, so let's leave aside the addendum to (2). The addition to (3) is inapplicable because the interest from the hypothetical Nobel Prize will not exceed Fr. 8 000.[8] Finally, I cannot expect my uncle to take care of my boys as well, indeed, I don't even want it. Who knows anyway, how much of his splendor this great era will leave him.[9] I'm very skeptical. Aside from this, what has now been agreed is that, upon my death, Miza receives Fr. 5 000; additionally, she has another Fr 10,000 in the bank (her dowry). If you add to this that her parents also have quite a bit,[10] then one must say that the children are provided for, if not magnificently, certainly quite decently, in any case incomparably better than I was provided for in my day.[11] So let's leave it at that . . . . .

Dear Michele! We'll be seeing each other again soon. My health is decidedly better than last year.[12] Elsa cooks me my chicken feed indefatigably every three hours, and I keep calm, am usually on my balcony. Everyone says I've never looked so well. Eventually there really will be peace again sometime, making it possible to travel more comfortably again. I'm very sorry that I can't see my boys. But I think that, despite Miza's associated lamentations, it is much the same to her either way.

Another thing. Tete has asked me for a handsome book (travel account). Sending this from here is very troublesome owing to censoring. So please buy one and send it to him in my name. Today I received my first letter from him. Please write me the price next time.

Vero wrote me a very nice letter in which he defended Anna. You would have had a chuckle from it. Send him my best regards, also to Anna.

Warm regards, yours,

Albert.

573.   To Eduard Einstein

[Berlin, before 28 June 1918][1]

My dear Tete,

I am very proud that now my second boy also can write already! Your letter delighted me. In return, I wrote straight away to Mr. Besso to have him buy a book of travels for you.[2] I hope you have it already. Sending one from here would be quite complicated, because a special permit is necessary. Go out very often on walks so that you get properly well, and don't read overly much, rather save some for when you are grown up. You are probably enjoying school[3] very much because you learn so easily and because it is fun to be with all those boys and because you are anyway a chirpy fellow who enjoys himself anywhere.

Kisses from your

Papa,

and write again soon.

574.   To Karl Scheel

[Berlin,] 29 June 1918

Dear Colleague,

I have so many pressing matters to settle today that I can no longer take leave of you personally. I wish you a pleasant vacation and herewith confidently leave the affairs of our Physical Society in your hands (during my 8 weeks absence).[1]

I did not yet reply to the letter by the Chemical Society. Also I am not sure whether note has already been taken of the two other attached communications.

With cordial regards, yours truly,

A. Einstein.

575.   To Max Born

[Ahrenshoop, after 29 June 1918][1]

Dear Born,

It is very kind of you to look after Nordström. You must simply write to the General Staff that N. has already had his trip out approved through the intercession of Haber. Then his return trip will surely be granted on the spot. He must return at the beginning of August, as I already wrote you.[2]

Here it is wonderful, no telephone, no responsibilities, absolute tranquillity. I really cannot understand anymore how one can bear it in the big city. The weather is splendid now as well. I am lying on the shore like a crocodile, allowing myself to be roasted by the sun, never see a newspaper, and do not give a hoot about the so-called world.

What you tell me about inertia in a crystal lattice is very gratifying. It clearly can only involve *electrical* energy though, since the potential energy of the other assumed forces does not contribute to inertia, according to the basic assumptions of mechanics. I am very eager to see your presentation of the matter.[3]

I am reading here, among other things, Kant's *Prolegomena* and am beginning to understand the enormous suggestive impact that has emanated and still is emanating from this fellow. If one just concedes to him the existence of synthetic judgments *a priori*, one is already ensnared.[4] I must tone down "*a priori*" into "conventional" in order not to have to contradict myself, but even then it does not fit in the details. Nevertheless, it is very nice reading, if not *as* fine as his predecessor Hume, who also had considerably more common sense.[5]

When I am back again, let us all squat together again cozily, so that you can gently introduce me back into the hustle and bustle of humanity, of which

I notice nothing now.   I hope that meanwhile you and your wife are returned
to good health. We are feeling well, the little harem also is feeding and thriving
sumptuously.[6] Cordial regards, yours,

Einstein.

Best regards also to your wife and your little ones;[7] more another time!

576.   To Hans Albert Einstein

Ahrenshoop, Pomerania. [after 29 June 1918][1]

My dear Albert,
    I was really sorry not to be able to take you along with me,[2] but was also
pleased besides that you miss me a little. You can easily imagine why I could not
come. This winter I was so sick that I had to lie in bed for over 2 months. Every
meal must be cooked separately for me. I may not make any abrupt movements.
So I'd have been allowed neither to go on a walk with you nor to eat at the hotel
without having to fear that the painstakingly gained recovery go to the devil
again.[3] How should I have done it? Even the trip itself was not without risk. I
could just have stayed at my sister's in Lucerne, where Mama only reluctantly
allows you to go. Added to this is that I had quarreled with Anna Besso,[4] that
I did not want to become a burden to Mr. Zangger again, and finally, that I
doubted whether my coming mattered much to you. How much simpler it would
have been for you to come to me if you had wanted. You are healthy, you know,
and it would have been very nice for you here. Furthermore, we would have been
undisturbed here, while in Switzerland I have many personal obligations to fulfill
which, added to everything else, deterred me only more. I can just travel to
Switzerland when my health has improved again adequately. So you see that you
are reproaching me unjustifiably; maybe later one day you'll think that it would
have been better if you had worried more about me at this time.–
    I'm very pleased that you feel so much at home with music already. I have
not been allowed to play violin anymore for half a year already, because of my
stomach, so you must take my place in this respect. I was pleased no less that
you must also engage yourself in the domestic arts. That's very healthy and keeps
you down-to-earth. I also reflected on many a scientific idea while I was taking
you out in your pram on walks!
    It is wonderful here at the Baltic Sea. But sailboats are not loaned out any-
where. I have to wait until the fisherman takes me along. The sea is magnificent
in stormy weather. You would certainly have liked it.

With best wishes for your holidays, heartfelt greetings from your

Papa.

I am living at: Old Customs House. Ahrenshoop (Pomerania) (until mid-August).
When do you have your fall vacation?[5]

577.   From Peter Debye

Göttingen, 2 July 1918

Dear Einstein,

Enclosed I am sending you a copy of a notice that I presented last Friday
to the local Sciences Society:[1] You are probably wondering about this mailing,
since you are, of course, used to receiving only finally printed items from me and
therefore suspect that this communication must have a very specific purpose.
This naturally is the case as well. In brief summary, the aim of the following
exposition is this:

a) I believe that the points treated in the notice have some bearing on problems
that ought to claim great interest.

b) For their further experimental treatment these points pose high demands with
regard to experimental means, such high demands that I cannot meet them without assistance.

c) In this situation, I turn confidently to you as director of the K. W. Institute
of Physics and ask for your support.

Now that I have unburdened myself and you need fear no more surprises, I
am surely permitted to give a more detailed report on what we intend.

Our investigations with X rays stemmed from considerations on the inner
atomistic cause of scattering. It is assumed that the electrons in an atom, brought
into motion by primary radiation, cause secondary radiation following classical
laws. If this is the case, then an interference effect of these electrons is present in
the scattering, which in observation is translated into a dependence between the
scattering intensity and the angle between the primary and scattered ray.

From the present notice you will draw the experimental bases in support of the
accuracy of this conception. This conception suggests that through correct evaluation of the intensity observations, inferences can be made about the number of
electrons and the distance between them. I already spoke with you once cursorily
about this subject but, as I recall, we did not go very deeply into it at the time,

so you probably will barely remember it. We have evaluated preliminarily the above type of considerations with the available apparatus

1) to conclude from the intensities of the Laue interferences the relative numbers of electrons and thus the orientation and size of the atoms' charges within the crystal lattice.

The experiments were performed a while ago already,[2] I told Sommerfeld and Born about it, as the occasion arose. Both have meanwhile been inspired by it to arrive, along other routes (piezoelectricity, residual rays), at figures for such charges.[3]

2) We could establish that, for diamond (which we chose because of its low thermal motion), a sharp drop in intensity against an increase in the angle between scattered ray and primary ray exists which cannot possibly be covered by present considerations. However, this drop is comprehensible if the dimensions of the C atom's system of electrons are not infinitesimal against the wavelength of Roentgen radiation. Its numerical evaluation hence leads to a direct determination of the size of that system.[4]

You understand that all of this is just the modest beginnings of a matter which, in my opinion, can be well developed. But for this we need *X rays of arbitrary wavelength and of sufficient intensity* and this the experimental equipment I presently have at my disposal does not deliver. It fails for short wavelengths owing to the $h\nu$-relation, and it is precisely there that conditions seem to me to be present which, as it seems, as far as can be concluded from the very incomplete and quite unreliable data, must point to a fault in classical electrodynamics. (I am thinking here of a quantization of the radiation emitted from a free electron.)

I initially opened contact with Siemens & Halske[5] and ordered a cost estimate there for the necessary apparatus. This estimate is available and amounts to 16 030 marks, the high total of which, naturally, is also determined by wartime conditions. Then I looked around regarding raising this money. Chance brought me into dealings with Mr. von Miller of the Deutsches Museum in Munich, who referred me to the Jubilee Foundation of German Industry.[6] This foundation has declared its willingness to support experiments on atomic structure with an amount of *up to* 5 000 marks, under the condition that corresponding contributions by other parties, particularly on the part of the government, be made toward covering the costs. A few days ago in Göttingen I had occasion to speak with his Excellency Schmidt, who referred me to you and the K. W. Society.[7] I obviously still cannot operate with the apparatus alone, and thus I come to the request of whether you, through the K. W. Society, were perhaps in a position to place at our disposal the funds for this apparatus.[8] I could then apply the grant by the Jubilee Foundation toward the running costs, which are likewise naturally not inconsiderable. The government also would certainly be willing to approve a smaller subsidy in support of the same.

My letter has become terribly long, so long that I do not want to add more, even though I have the feeling that some points among my reasons could be elaborated on further. But I hope to find your sympathy even so and would naturally be extremely pleased if you were in the position to regard the matter as a good one and release me from financial worries.

With best regards, yours,

P. Debye.

578. To Max Planck

[Ahrenshoop, after 2 July 1918][1]

Dear Colleague,

The letter speaks for itself.[2] I believe that there could be no better use of our money than by placing the desired apparatus at Debye's disposal (purchase and loan them out to him for as long as he would like to have them). Waiting until prices are more favorable is out of the question here, in m[y] o[pinion], because we have only *one* Debye, and his life span $< \infty$.

I asked Debye to send a copy of his preliminary notice on the subject to all members on the Board of Directors, so I think I do not have to send the manuscript along to you,[3] especially since the idea is explained clearly already in this letter. I ask you now please to convene a meeting of the Board of Directors as soon as possible and to confer on the matter. Then I request a brief report on the outcome, so that I can negotiate further with Debye.

With cordial regards, yours,

Einstein

(who is leading an enviable existence here in glorious nature.)
Address: Old Customs House, Ahrenshoop (Pomerania).

579. To Hermann Weyl

Ahrenshoop [3 July 1918]

Dear Colleague,

Your book has just come.[1] I already had a copy of it with me here on summer vacation (from Springer) and delight in your elegant and transparent deductions and calculations. Recently I gave a report at the Academy on your new theory, which Planck finds very interesting.[2] You are a genuine calculation wizard. This is revealed very particularly in the last chapters of the book as well, e.g., in the treatment of the static incompressible liquid,[3] also in the execution of the variation for the gravitational field.[4]

I am curious about whether you are going to remain committed to your explanation of electromagnetism.[5] Another comment on it: at all events preferred world lines do exist, the trajectories of uncharged bodies; on the other hand, geodesic lines exist, also according to your theory. Shouldn't both of these sorts of unique world lines be identical? There is little doubt that they should. Then, according to your theory, forces that are proportional to the electromagnetic *potentials* are exerted on uncharged bodies.[6] You ought to pursue this consequence more closely, especially since it is straightforward.

Cordial regards, yours,

Einstein.

cur[rently] at Mrs. Niemann-Konow,[7] Ahrenshoop.

580.   To Max Born

[Ahrenshoop, after 3 July 1918][1]

Dear Born,

This faithful confession by a fair soul born on stilts, for your edification.* How sweet it is to die, ...[2]

We are enjoying ourselves constantly and are vegetating like true good-for-nothings. To a happy reunion. Yours,

Einstein.

Show it also to Wertheimer![3]

Erbauung also meaning 'delight.'
[2]Author's footnote: "To be preserved for Philistine posterity!"

581.   From Felix Klein

Göttingen, 5 July 1918

Esteemed Colleague,

I still must write you about my Monday lecture.[1] I have the proof you had intended:[2] that the J_σ's form a contragradient vector, carried out under the restriction *that the world tube of the closed system can be contained within a cylinder of finite transverse dimensions.*[3]

The accompanying figure probably does not need any explanation (I choose the coordinate system in such a way that the x_4 axis is parallel to the cylinder's orientation; σ, σ' are any two points linked together by another parallel; they define the vector 0, 0, 0, Δx_4).– One obtains for the corresponding integrals[4]

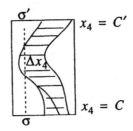

$$A_\sigma^\nu = \int \int \int \int U_\sigma^\nu dx_1 dx_2 dx_3 dx_4$$

the precise scheme:[5]

$$
\begin{array}{cccc}
0 & 0 & 0 & \Delta x_4 \cdot J_1 \\
0 & 0 & 0 & \Delta x_4 \cdot J_2 \\
0 & 0 & 0 & \Delta x_4 \cdot J_3 \\
0 & 0 & 0 & \Delta x_4 \cdot J_4
\end{array}
$$

Now I introduce, through some Lorentz transformation, instead of x, new $\bar{x}$ coordinates and define the new $\overline{A}_\sigma^\nu$'s by means of a cylinder section whose edges $x_4 = \overline{C}$, and $= \overline{C}'$, resp., again go through the points σ, σ'. The integration domain is thereby visibly different from the original one, but this difference bears less weight against the overall extension, the larger Δx_4 is relative to the cylinder's cross section. We conclude *that for* $\lim \Delta x_4 = \infty$, *the* $\overline{A}_\sigma^\nu$ *'s must be related*

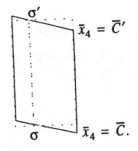

exactly to the A_σ^ν *'s as the* $\overline{U}_\sigma^\nu$ *'s are to the* U_σ^ν *'s.*[6] Simultaneously, $\overline{A}_\sigma^\nu = \overline{J}_\sigma \cdot \Delta \overline{x}_4$.[7] From this then follows explicitly the desired correlation between $\overline{J}_\sigma$ and J_σ,[8] first of all for Lorentz transformations, not[ably] such linear coordinate transformations for which the planes $\overline{x}_4 = \overline{C}$ cut the cylinder into "cross sections"; thus the $\overline{J}_\sigma$'s can be defined by integrals. For other linear coordinate transformations, by contrast, the values for $\overline{J}_\sigma$ are defined by the vector character of the J_σ's!–

As you see, I have gone through the same stages as in your letters, only that, owing to the restrictive precondition I placed at the start, everything could be formulated more precisely.[9] I wonder whether the restriction is necessary for the validity of the theorem?[10] In the case where the limitation is valid, do J_1, J_2, J_3 in the original coordinate system also vanish, hence does the vector J have the orientation of the cylinder enclosing the world tube?[11] I have not yet formed an opinion on either of these points.

I was able to hand your e[steemed] postcard on to Mr. Humm[12] right away on Monday. I am now reading Weyl with extreme interest.[13]

With best wishes for a pleasant stay in the country,[14] yours truly,

Klein.

582.   From Friedrich Adler

Stein-on-the-Danube, 6 July 1918

Dear Friend,

It has been almost one year since I wrote you, but I have been thinking of you incessantly, for I was occupied the whole time with relativity theory. Now the work is finally finished and I am very eager to know what you say to it.[1] I know that you are so convinced of the correctness of your foundations that you do not expect anything from further discussions of it. And yet I would like to burden you with the perusal of my book, for I imagine now having really caught Ariadne's thread, leading to a compelling derivation of the necessity for a preferred reference system from your transformation equations. The crux of the matter had long been clear to me, but it took quite some effort to work out all of its consequences in order to make it convincing to others as well. Today I would just like to know where I may send you the work, or for how long a mailing will still reach you in Berlin.

Now I have belatedly received the paper by E. Budde in the *Verhandl.* of 1914.[2] There the "Einsteinian optical clock" is repeatedly mentioned. Nowhere in your papers that I know about does such a thing appear, though. It would be very interesting for me if you could send me the relevant article or at least say where it has appeared.[3] Did you reply to Budde's paper, or was there otherwise any discussion attached to it? Considerable difficulties are naturally connected with my procurement of the literature, but I do believe that, bit by bit, I have received all the essential ones nonetheless. Please send me whatever you have published since the "Cosmological Considerations"[4] as well.

My wife told me with great delight about your visit last summer, but unfortunately also that you were not quite in good health.[5] I hope you are completely recovered again now. As far as I am concerned, I have absolutely no reason to complain. In virtually all respects, I am doing better here than in Vienna.[6] The air, above all, is so good here that all my acquaintances are astonished at how well I look. The diet I am provided with here is better than what the great majority of other people unfortunately have. And so my ability to work is maintained to such a degree that every day is too short for me. I am a $2\frac{1}{2}$-hour train ride away from Vienna, yet, I have had visitors from Vienna every week, up to now. In short, in this topsy-turvy world we now live in, it is in actual fact considerably nicer *intra muros* than *extra*.

With cordial regards, yours,

Fr. Adler.

583.   From Adolf Kneser

Breslau, 11 Hohenlohe St., 7 July 1918

Highly esteemed Colleague,

I deeply regret that some of my statements have evoked in you pained feelings;[1] however, even now neither can I look upon these statements as factually unjustified, nor even acknowledge that I should have refrained from these statements out of consideration for your personal views. Absolutely nothing is said, at the relevant places in my speech, about your views; but it obviously remains a fact that your brilliant discoveries were made in Germany during the war, where you are being granted protection and leisure for scientific research. So you must submit to having your researches credited to Germany and ranked as part of the German peace effort behind the frontline.

It sincerely pleases me that you did not take part in the exodus of many Swiss scholars, who left Germany as if abandoning a ship thought to be sinking. I believe I may conclude from this that you yourself felt, consciously or unconsciously, that Germany is the safest location for your scientific research. In enemy countries you would not by any means have been allowed to profess your convictions, set forth in your letter, without very serious harrassment and disruption in your professional life.

In expressing my utmost respect, yours truly,

A. Kneser.

584.   From Max Planck

Grunewald, 8 July 1918

Dear Colleague,

It is extremely welcome for the K. W. Institute of Physics to find such a fine opportunity to demonstrate its effectiveness.[1] But I believe that, to begin with, it is possible without a meeting of the Board of Directors (which, incidentally, I am not at all authorized to convene);[2] for the Board members will surely just be grateful if they are not imposed upon personally. As far as I can see, it would probably be most advisable to handle the matter just like Freundlich's, i.e., you agree on a *contract* with Mr. Debye in which the essentials are recorded, namely the amount and payment deadline for the funds contributed by the K. W. Institute. Care must also be taken to present the research formally as an endeavor by the Institute. Otherwise, there might perhaps be a problem with the Board of Trustees. Once the contract has been prepared, it can circulate, and the vote can take place in writing again, as in the Freundlich case.–[3] I am very pleased that you are feeling well. I must endure it here still until the middle of August.

Cordial regards, yours,

Planck.

585.   To Mileva Einstein-Marić

[Berlin, before 9 July 1918][1]

Dear Mileva,

I'll be happy to fulfill your wishes regarding the mode of money remittance, albeit only from January 1919 onward, since I have already instructed otherwise for this year and don't want to cause the bank so much trouble.[2] So for the time being, you can withdraw the interest, in the course of this half-year and exchange it when you please. Bad times for currency exchange are returning, I believe!

I hope you have received the guarantee document regarding the 20 000 M in trust.[3] Now I ask you to file the divorce right away[4] so that the affair is finally put in order.

It was virtually impossible for me to come to Switzerland. I don't want recklessness to be the cause of my having to lie in bed again for months on end.[5]

Amicable regards, yours,

Albert.

The report about Tete delighted me. You can be sure that I take the greatest interest in everything you tell me about the children. Every trifle is important to me.

586.   To Michele Besso

[Ahrenshoop, 9 July 1918]

Dear Michele,

I have received your letter with the original divorce advice—Till Eulenspiegel.[1] It's magnificent here, quite made for the likes of me. What a pity that you and our boys aren't here.[2]

An acquaintance of mine is coming to Zurich, a teacher (a Balt) by the name of Buek, who isn't known yet in Switzerland. I recommended that he look you up so that you can be of a bit of assistance to him. He is a really good, unaffected person, who deserves having his way facilitated.[3] He can report some things of interest to you as well.

Best regards, yours,

Albert.

587.   From Anschütz and Company

Neumühlen near Kiel, 12 July 1918

To Professor Einstein, Ahrenshoop. Old Customs House.

Esteemed Professor,

We acknowledge with most obliging thanks receipt of your letter along with the evaluation of the 7th inst. and are pleased with the resolute and clear form in which you expressed your opinion on the other application.[1]

Only in the next day or so can we report, after consultation with our patent attorney, whether any addition to the evaluation is necessary. In the meantime, we on our part just have the comment re the diagram under "C" (page 5) of the attached typed copy that the opponent[2] is expected to point to the omission in the directive force of the oscillation-impeding gyroscope. For depending on this gyroscope's size and revolving speed, the line a–a will not coincide exactly with the meridian but will include a specific, although invariable and easily ascertainable angle, because this gyroscope's directive force also is transmitted to the float k via the spring tied to it.

(If this one oscillation-impeding gyroscope is replaced by 2 of equal size but revolving in opposite directions, this deviation disappears, which case corresponds, by the way, to the diagram of figure 4 of our Patent Specification 241 637.)

In other words, with both gyroscope axes forming an angle with the meridian, your diagram thus again resembles the design example of our patent specification, the more so since it suffices for only one gyroscope to be equipped with a yielding link with the float.

Since it does not bear well to go into the details of these relations, it would perhaps be advisable to change the end of Section C a little.[3]

We reserve for ourselves to convey our patent attorney's opinion on this to you as well, and remain in the meantime, in utmost respect, your devoted servants,

Anschütz & Co.

588.   From Felix Klein

Göttingen, 15 July 1918

Esteemed Colleague,

I have succeeded in discovering the organic formation law for Hilbert's energy vector.[1] If the infinitesimal transformations of my January note[2] are not specialized to vanish at the edge of the integration domain, one obtains, in addition to the 4-fold integral (which must vanish for arbitrary p^σ), another triple integral taken along the edge (which, independent of the edge's shape, must correspondingly be zero):[3]

$$\int\int\int \sqrt{g}\{e^1 \cdot dw^2 dw^3 dw^4 + \ldots e^4 \cdot dw^1 dw^2 dw^3\}.$$

And here now is $e^1 \ldots e^4$, subject to changes one might want to attach owing to the field equations, or even the addition of a simple tensor, whose divergence vanishes identically, the sought vector.–[4] I hope very much to pave the way from here, also in conformance with my assumptions to your $\mathfrak{T}^\nu_\sigma + \mathfrak{t}^\nu_\sigma$.[5] But with other demands on me and reduced productivity, it is all just going very slowly.

Very truly yours,

Klein.

589.   To Felix Klein

[Ahrenshoop,] 22 July 1918

Highly esteemed Colleague,

It is very good that you want to clear up the formal meaning of t^ν_σ.[1] For I must admit that the derivation of the energy conservation law for the field and matter together seems unsatisfactory, from the mathematical point of view, making the formal characterization of t^ν_σ not at all feasible.

I would also like to note that the introduction of the Hamilton function $\mathfrak{G}^*$ for the $g_{\mu\nu}$-field against the invariant $\mathfrak{G}$ has another advantage apart from plain calculational simplicity.[2] In Schwarzschild's sphere problem[3] it was demonstrated that at the surface of the sphere the continuity of the $g^{\mu\nu}$'s, but not of the $g^{\mu\nu}_\sigma$'s, must be required, which latter quantities change discontinuously at the boundary. Hence $\mathfrak{G}^*$ remains *finite* throughout, even at the sphere's surface, whereas $\mathfrak{G}$ becomes infinite. Without the introduction of $\mathfrak{G}^*$ it might be difficult to justify being able to make do with requiring the continuity of $g^{\mu\nu}$ alone.

With best regards, yours very truly,

A. Einstein.

590.   From Hedwig and Max Born

[Berlin,] 28 July 1918

"The *bad*—this maxim is firmly fixed—
Is always the *good* from which one desists!"

Improve yourselves, you lazy Ahrenshoopers, and desist no more from the good, namely, from writing! Report rather on what pleasing thoughts the blue sky and "boundless" sea instill in you! And come back soon, plump as dumplings, to the delight of your

Borns.

Authors' caption on verso: "Vision by the lonely and pining Grunewalders."

591.   To Michele Besso

Ahrenshoop. [29 July 1918][1]

Dear Michele,

It's truly touching of you all to have dedicated so much time and care to my old papers.[2] But I do have to say that a good amount has been superseded, so it's not worth the effort. Particularly the opalescence paper—it is burdened by the superfluous Fourier expansion.[3] I am very pleased that Buek is finding a good guide in you. He is an excellent fellow but not a practical sort.[4] The book pleased my boy very much; he wrote that to me himself. Thank you;[5] I'll pay for it when I come to Switzerland, which I hope will be in the autumn. My mother is probably moving to my uncle (Jacob Koch) who lives in Zurich,[6] which provides a very welcome accommodation opportunity for me. It will not cause any difficulties, since I draw a very clear line of distinction between her and my boys. I'm very happy about the nice letters my boys are writing me; Mileva is also writing amicably. Your advice about my remarriage is well intentioned.[7] I'm not going to follow it, though. Because if I ever were to decide to leave a second wife as well, I wouldn't allow myself to be held down by anything. I would submit to any financial burdens that would arise out of it, as would be decent. For I count on her emerging from the war properly fleeced.[8] Besides, she is very good, so I really am content with her. I would have remained true to Mileva as well, if it had been bearable with her. In this regard, you have an entirely inaccurate opinion of me, that I know. I'm moderate and make no exaggerated demands of any sort on those around me. But Mileva was absolutely intolerable to me. When she's not my wife, I can like her quite well; she is all right to me even as the mother of my boys. I just can't live with her. Ever since she's been away I feel incomparably better, probably she does as well, apart from her illness, which you have all probably quite unjustifiably related to the separation.[9]

Here I have been pondering for countless hours about the quantum problem again, naturally, without really making any headway. But I no longer have doubts about the *reality* of quanta in radiation, even though I'm still quite alone in this conviction. That's how it will remain as long as no mathematical theory succeeds. I do want to present these arguments clearly sometime now, after all.

Warm regards, yours,

Albert.

If you ever get hold of a little book *Individualmoral und Staatsmoral* by Goldstein,[10] then read it. The conflict is described wonderfully in it.

In a paper on the principle of energy conservation in the gener. theory of rel. it is revealed that the total energy of a system is completely independent of the choice of coordinates (an *integral* invariant, which does not correspond to a *differential* invariant).[11] The energy of the whole world, in the case where the latter is closed with an even distribution of matter, is determined just by this matter alone; gravitational field energy and the energy contribution of the λ-term cancel each other out. Furthermore, it is not entirely uninteresting that the theory can easily be formulated so that λ appears not as a universal constant in the usual sense but as an integration constant, or as a Lagrangian factor. One only needs to say

$$\delta\{\int \mathfrak{H}d\tau\} = 0$$

must be satisfied for all variations that leave the volume, in natural measure,

$$\int \sqrt{-g}d\tau,$$

invariant. Such a formulation seems all the more natural since the vanishing of the variation of the Hamilton integral certainly cannot be postulated reasonably for *such* variations corresponding to a transition of the world to a neighboring smaller one that is also at equilibrium. There is nothing analogous to this consideration in standard mechanics, because there the masses and volume are not varied. I think I should publish this sometime when the occasion arises, because it cleanses the theory of a blemish.[12] I wonder whether one day other universal constants won't lose their painful character this way as well?[13]

592.   To Arnold Sommerfeld

[Berlin, between 1 August and 1 November 1918][1]

Dear Sommerfeld,

I am glad that you have given Mr. Usener's historical portrayal[2] the criticism it deserves.[3] There can be absolutely no doubt about his *mala fides*. I am completely informed about this because I had to draw up a brief private evaluation for Mr. Anschütz, in which Usener's portrayal of the circumstances behind the Van den Bos–Anschütz patents had to be taken into account.[4] Usener was formerly employed at Anschütz and is now cooperating with the competition.[5] In the book, he presented himself very cleverly as an impartial bystander but was intent on disparaging Anschütz's merits. You should let Anschütz give you the details; I was downright scandalized by the fellow! It is good that you uphold your statement in essence.[6] It is new to me that Martienssen was the first to have pointed to the application of long oscillation periods.[7] Was this really before Anschütz's first patent?

My suggested correction to the end of your notice has come out badly. I
believe, though, that—even if the long period by itself had already been suggested
by Martienssen—it must nonetheless assume this position, because (in my view,
at least) only the *combination*

$$\begin{cases} \text{effective damping} \\ \text{high eigenfrequency}^{[8]} \end{cases}$$

made success possible. Who knows when the matter would have come about
without Anschütz.–[9]

I also liked Weyl's paper[10] well, although I am convinced that the underlying
assumption cannot be physically correct.[11] The book is brilliant;[12] Weyl is a
highly talented and, added to that, a versatile man, very fine and likable as a
person as well. We may expect *great things* from him yet. It is quite natural
that you are even more interested in spectra; it has certainly become the most
promising field today. With your work on fine structure, you first drew it within
reach.[13] It certainly is a ruse to be able to live to see such a thing! I hope we
shall be seeing each other here again soon. Meanwhile, cordial greetings, yours,

Einstein.

593.   To Hedwig and Max Born

Ahrenshoop, 2 August 1918

Dear Borns,

The closer the homeward trip approaches, the more my conscience is stirred,
along with the fear of being scolded for my foul habit of not writing. But what
should a chap write who loafs around all day, sees no one, and who, at most,
totters about barefoot for a half-hour or so? If only we could bring about the
latter fine custom (voluntarily) in Berlin! I really relished the cloverleaf. It is
obvious that those three, brought together there in brotherly union, are inveterate
hobbyhorse-riders, two turned in upon themselves, one staring nonchalantly into
space.–[1] Recently I read that the population of Europe grew in the last century
from 118 million to almost 400 million . . . . a terrible thought that could almost
reconcile one to war!–

To a happy reunion! Yours,

Einstein.

[. . .][2]

594.   To Friedrich Adler

[Ahrenshoop,] 4 August 1918

Dear Adler,

Yesterday your manuscript arrived.[1] I have already studied the first chapter and am thus informed about all the essentials.[2] The basic physical assumptions are:[3]

a) no Lorentz deformation of moving rigid bodies,
b) no influence of motion on the running rate of the clocks.

Thus the geometrical and kinematic elements are entirely given, quite apart from the mathematical expression in coordinates.

The assumptions (a) and (b) are, *in principle*, directly verifiable, but not in *practice*. (a) leads immediately to Abraham's theory of motion of an electron, however, which is refuted empirically as soon as the nonexistence of the deformation of moving bodies is applied also to the electron.[4]

It must be considered, furthermore, that (a) leads straight away to a contradiction with the outcome of the Michelson experiment, if it is assumed that the law of the constancy of the velocity of light in a vacuum, which you are not likely to question, is valid relative to the preferred reference system K.

I find that a theory can be taken seriously from the physical standpoint only when it does justice to the following observational results:

1) Fizeau's experiment.
2) Motion of electrons in an electromagn. field.
3) Aberration law.
4) Michelson's experiment.

For it was these facts which compelled the formulation of the special theory of relativity. You have made no attempt to address these fundamental facts, however.–

I come now to the formal aspect. Basically, on making arbitrary assumptions about the behavior of measuring rods and clocks, one can use arbitrary transformation equations without coming into conflict with the logic or with experience.

If, however, as corresponds with your assumptions about the behavior of the measuring rods and clocks, it is assumed that rigid bodies at rest relative to one another follow Euclidean geometry regarding the positioning laws [*Lagerungsgesetze*], and that static clocks, relative to one another, run equally quickly, then it is certainly appropriate (although logically absolutely not necessary) to choose the coordinates in such a way that for all legitimate systems the following is valid:

The coordinates are directly measurable, obtained upon measurement with a standard measuring rod at rest relative to the relevant system. Time is measured directly with standard clocks at rest relative to the system. Standard measuring rods and standard clocks of all systems should all agree with one another when they are brought to the same place at rest relative to one another and are compared to one another.

If this is required, the quantities x, y, z, t are thus assigned physical meaning. Then arbitrary transformations are no longer admissible. Admissibility then depends on the physical laws. *Hence, in your case of the classical assumptions (a) and (b), no transformations other than the Galilean transformations are justified.* You arrive at other transformations just by denying that choice of coordinates suitable for a simple interpretation of the formulas.

These remarks were of a general nature. I now turn to the second chapter.[5] I have no alternative to making detailed comments on the text.

Re p. 41.[6] It is self-evident that I treat all times of the system as equivalent, since I set out from the postulate of relativity. I cannot be reproached for not making any use of your "zonal time." From the explanations at the bottom of p. 41, one sees that your *symmetry system* plays the role of Lorentz's ether.[7] You thus immediately encounter the difficulty springing from it for your system that our terrestrially observable space lacks electromagnetic "asymmetries." Your sentence "The error of Einstein's . . ." is completely incomprehensible to me. If a system's clocks are not set rel. to one another (in some way or other), their data are then incoherent and cannot serve for a time definition,[8] not even in a "symmetry system," p. 52.[9] It is not my view that "system time" corresponded to a kind of "higher time concept." My view is just that, in seeking to base time on identical clocks and signals between them, only a *system time* can be attained, but not a *universal time*. One could arrive at the latter only through instantaneous signals or through clocks whose rates are not influenced by motion. The existence of either of these things must be questioned in principle, though, *a priori.*– P. 46.[10] It is not correct that I based the time concept *exclusively* on that of simultaneity. *The concept of the standard clock is added to it as well*; the standard clock provides the *unit of time.* P. 47.[11] What you accuse me of here is incorrect.[12] I avoid an uncertainty regarding the time unit in the different systems by determining that, in all of them, standard clocks *of the same type* be used, i.e., standard clocks which, when placed next to one another at a state of relative rest, are identical (or at least identical running at the same rate). Re p. 50.[13] A pendulum clock is no "clock" in my sense, only a *pendulum clock with the gravitational Earth*; this also only if it is admissible to abstract from the spatial extension of this system. *I definitely contest your underlined assertion on p. 50.*[14] Likewise I contest the weight of the argument presented on this page that the method of identical clocks fails owing to the influence of motion on their rate; since for the determination of time (of a system) only clocks that are at rest relative to the relevant system are used. That is why you yourself surely cannot attach any weight to the remark at the bottom of p. 50.[15] P. 51.[16] The dial is a bit odd, why not the handle and the polish as well?[17] P. 52.[18] We imagine my standard clocks as having been produced identically somewhere and sometime by a clockmaker who enjoys a world monopoly and then having been brought to the different locations and into the states of motion of the various locations. I would like to see avoided as uneconomical and superfluous the establishment "by convention" of a unit of time independent of the standard clocks. I do not understand the sentence: "We thus do not need to acknowledge piously . . ."[19] Clocks are identical *by definition* if, after being brought side-by-side at relative rest to each other, upon observation they are found to be identical. This is clearly a harmless overdefinition, physically speaking; considering it physically inadmissible seems excessive skepticism to me! P. 53.[20] The poor dial again. P. 54.[21] I consider it to be the only constructive solution to select an identical "scale for the dials" of all the systems' clocks as well, since otherwise quite superfluous complications arise.[22] We make use of the same measuring instruments throughout. P. 55.[23] No, there is no gap here in my train of thought;[24] for I require that in every system the setting of the clocks be possible such that, measured by the system time, the velocity of light is c, *if in all systems identical clocks, in the above sense, are used.* (There follows a repetition of the unfounded objection regarding the unit of time.) The sources of inequality of the clocks indicated on page 55 are assumed by me to be *avoidable* and to have been avoided.[25] (This you actually do in your considerations as well.) I do not deny *the feasibility* of using different units of time, just the *practicability* of it for a consideration of the *principles* of space and time in physics.

The galvanometric method (pp. 59–60)[26] is no more than a new instantaneous signal of infinitely large propagation velocity.[27] It is plain that electrical current can achieve that no better than light. P. 61.[28] The concept of velocity relative to a system is clear if lengths and times are defined relative to this system. In my paper no special definition was needed;[29] only v_1 of your definitions could logically come into consideration.[30] Pp. 71–73 I agree with.[31]

Third Chapter.[32]

P. 79.[33] *"Moment of the onset of motion."* Re absolute simultaneity. The statement of the simultaneous coincidence of the partial dis[tances] of M and M' is meaningless without an indication of the frame of reference.[34] Again the illusion

of "galvanometric comparability." [P.] 80:[35] "This difficulty is insurmountable
if one wants to re[sort][36] to the (admittedly!) absurd assumption . . ."[37] The
difficulty vanishes immediately when the assumption of *absolute* simultaneity is
relinquished! At this point the Latinist says: "*Adventavit asinus, pulcher et
fortissimus.*" P. 82.[38] I had no cause to mention this state of affairs or to
conceal it, since it seems to be of little interest.[39] Everything until page 85,
incl., is materially correct.[40] P. 87:[41] "At the moment of separation, all clocks
are set to $\tau = 0$."[42] This definition of time relates to a *combination of two
systems moving relative to each other*, not to *one* system. It is therefore, in my
view, *extremely inappropriate* for the reflections; for one must then distinguish
between ∞^2 times, rather than only ∞, if "∞" is the number of all system times
in my sense. Even the "ideal clock" presupposes *two* systems in motion relative to
each other! The fact that, upon introduction of the combined time τ, the Galilean
transformation results between x and x' is uninteresting because τ, in particular,
cannot be defined from *one* system. On p. 94[43] the nonsensical assertion that,
"*at the moment of separation* (!) of the systems, $x = x'$ is required for reasons of
symmetry" is reiterated! This statement is just meaningless or false, depending
on how it is interpreted. P. 96[44] remains incomprehensible to me.[45] There is no
reason for further comment on the calculation up to p. 99;[46] I indicated above
the error underlying the entire consideration. Page 100, last paragraph[47] is a
fallacy. Namely, if

$$K_1 \text{ is the symmetry system of } S_1 \text{ and } S_1'$$
$$K_2 \quad " \qquad\qquad " \qquad\qquad " \ S_2 \text{ and } S_2'$$

then by no means can it be concluded that K_1 and K_2 are at rest against each
other. The "departure times" of S_2 and S_2' relative to K_1 are under no circumstances zero, you see, as it would have to be in order really to be able to draw the
conclusion formulated by you on page 101.[48] This error is of great significance
for the chain of reasoning, because you obviously wanted to construct a preferred
system out of it. P. 105.[49] The equation $x' = x$ in your meaning is simply not in
the least justified,[50] no more your setting out to force their validity in individual
cases through the choice of different scales. P. 106.[51] I do not understand what
is said under "Thirdly."[52] Your statement starting with "Under the condition
. . ." is false;[53] it also supports itself exclusively on the postulate $x' = x$, which
stems from the bias for absolute time. P. 107. Since the introduction of the
"preferred system" arises from a fallacy, as is shown above, no legitimacy can be
ceded to this concept in the following either. P. 111.[54] The condition quoted
by me must be satisfied in nature if the theorem of the constancy of the velocity
of light is allowed to be maintained.[55] I cannot concede any substance to the
assertion that "my system time were not isotropic"; it seems meaningless. Re

p. 116[56] "Measurements on the basis of zonal time," I cannot imagine how such measurements ought to be carried out. The statement, p. 116 below, is without grounds and incorrect.[57]

The rest of this chapter cannot be refuted in detail, because it is built upon earlier errors.–

Dear A[dler],

Now I have been sitting over your work for two and a half days and have studied a good half of it thoroughly.[58] I know your chain of reasoning and believe I can forgo studying the remainder for now, since it is probably based upon the foregoing deductions, which I do not view as correct. Now I expect from you some instructions on what I should do with your manuscript, or whether I should keep it. In any event, I would prefer if I were not invited by a third party to comment on it publicly. If you publish these reflections, I am unlikely to write a reply to it, because I have remained silent about a number of analogous publications already.–

Now I wish you continued fruitful researches in your asylum. Whenever a book falls into my hands that I think could give you pleasure, I shall send it to you. Prof. Goldscheid in Vienna published one recently, *Individualmoral und Staatsmoral* (or some similar title),[59] which could perhaps be made accessible to you.

With cordial regards, yours,

Einstein

From 15 August on: 5 Haberland St., Berlin.

595.   To Walter Dällenbach

[Ahrenshoop,] 8 August [1918]

Dear Dällenbach,

I congratulate you on your doctorate. Judging from my own experience, it facilitates dealings with people to a quite considerable degree. For the publication of your dissertation I recommend the *Vierteljahrsschrift* of the Zurich Scientif. Society. Publication in a German journal falls out of consideration because of the great paper shortage.–[1] There is no sign of influenza here at the Baltic Sea. But I am going home again to Berlin in a few days.[2]

Wishing you happy holidays, yours,

Einstein

Cordial regards to the Bessos.
Berlin address: 5 Haberland St.

596.   From Friedrich Adler

Stein-on-the-Danube, 9 August 1918

Dear Friend,

Cordial thanks for your letter, postcard, and the book by Weyl.[1] I have looked through the book and it appears to me to be very good as an introduction to the mathematical problems. However, he brushes off what seems to me to be the main physical question, namely, the time concept, just as lightly as do other explanations. You will see what I mean when you read my paper, which was sent out to you last week and which you probably already received in the meantime.[2] In the copy I prepared for you, a small emendation is still missing, as I see in retrospect. Namely, if the value for x from formula (77b)[3] is introduced into (48), the generally valid formula for the "starting time" results:[4]

$$t_a = \frac{x}{v}\left(1 - \gamma\sqrt{1 - \frac{v^2}{c^2}}\right)$$

and there

$$\frac{1}{v}\left(1 - \gamma\sqrt{1 - \frac{v^2}{c^2}}\right) = \frac{u_S}{c^2}$$

applies; so this starting time is always only a function of u, and the relation[5]

$$\frac{t_a'}{t_a} = \gamma\frac{u_S'}{u_S}$$

between the starting times corres[pondingly] results; hence a dependence on γ.

When you emphasize the limitation of theoretical possibilities as an advantage of the theory of relativity, I think that this limitation is not weakened by the proof of the necessity for the assumption of a preferred reference system. But I hope we shall understand each other more quickly on all these questions when you have read my work. For today I just want to say that my admiration for your achievement grew proportionately as I penetrated into the problem and as the immense amount of difficulties you were facing became clearer to me. And this admiration did not diminish in the least by the fact that it simultaneously became increasingly clear to me that your solution is an inadmissible generalization of a special case. What a pity that we cannot discuss this question orally, but I hope I have worked it out clearly enough in the book.

A note was inside the book by Weyl that you sent me, which you had forgotten to remove. As you may possibly be looking for it, I am sending it back to you herewith by return post.

What you say about the agreement with my papers[6] pleased me very much, although I knew from our last conversation in Berlin, of course, that we essentially take the same view, especially on the current questions.

Cordial regards, yours,

Fr. Adler.

597.   To Heinrich Zangger

[Ahrenshoop, before 11 August 1918][1]

Dear friend Zangger,

I am sitting here on a dismal rainy day and receive your impetuous postcard. I should come to Zurich and give a talk?[2] What do you want from an old wreck or empty eggshell like me? What useful things I had thought up are alive in fresher and younger minds, whereas I am plagued with a stomach that is constantly inclined to rebellion[3] and that takes offense at every form of excitement or exertion. The intellect gets crippled and one's strength dwindles away, but glittering renown is still draped around the calcified shell. You have Weyl in Zurich,[4] who is capable of presenting everything I could say much more reliably and clearly than I with my lack of command of [mathematical] form and with my bad memory. I am just right for the Academy, whose quintessence lies more in sheer existence than in activity.

But I do intend to travel to Switzerland this year, when my exacting stomach allows, somewhere around the boys' fall vacation.[5] I *must* take a breath of your fresh air, to assuage the feeling of isolation generated by the clashing mentalities.– My boys now write me heart-warming letters. Albert is already starting to think quite amusingly, oddly enough about technical questions. In the end, though, I am glad for any mental quickness, even if it clings to narrow-minded views. Perhaps he will realize the superfluousness of the many conveniences sometime, after all! I also was originally supposed to become an engineer. But the thought of having to expend my creative energy on things that make practical everyday life even more refined, with the goal of bleak capital gain, was intolerable to me.[6] Thinking for its own sake is like music! That is why I also never could take to Mach's principle of economy [of thought] as the ultimate psychological driving force.[7] Economy, correctly understood, may be one motive upon which intellectual aesthetics depends; however, the mainspring of scientific thought is not some external goal to be striven toward but the pleasure of thinking. When I do not have a problem to contemplate, I preferably rederive mathematical and physical laws that I have long been familiar with. So in this case there is absolutely no *goal*, only an opportunity to indulge oneself in the pleasant activity of thought.

That's the nice thing about my current existence, that I can occupy myself unperturbedly with thinking, without any worries about professional duties, whereas during my time as professor, giving courses was a greater burden to me than the Patent Office had been earlier, because the knowledge of having a lecture ahead of one keeps the mind in an uneasy state detrimental to quiet reflection. This is admittedly only the case with those who, like I, have the finished lecture neither in their heads nor in their notebooks.[8]

Warm regards, yours,

Einstein.

598.   From Heinrich Zangger

Zurich, [before 11 August] 1918[1]

Dear friend Einstein,

What makes you, at 40 years of age, call yourself a wreck, an egg after the hatching?[2] You have an instinctive need for breathing space—want to come this fall. Well: Off[icial] Adviser Mousson spoke with "His Excellency Naumann" and wants to speak with the President of the Education Council, and I am going to go to the Federal Councillor.[3] (I am given much credibility even, because I advocated, apart from Einstein, also Meyer and Debye,[4] and I am on the outside.) Thus you are faced on your own with a clear situation: either refreshing, healthy air and some lectures—when you are healthy—jointly for the Poly[technic] and the University—and are welcomed with joy, or the unhealthy and more comfortable option with the decrepit feeling.–

As a modern person you choose—unlike everybody else—what is more comfortable. Take time in deciding—a collision of values exists that is not immediately apparent. Consider many a *dixi et salutavi*, etc.[5]

Your boys are doing well. The younger one is a fine little fellow, with shy, bashful movements, not at all Einsteinian. Albert reportedly likes to write to you. Their mother had a serious throat and jaw infection. Besso is looking forward to Rome.[6] What a half year that was for me; well, dear Einstein, the loss of children gives one a hollowed out feeling[7]—one's eyes are constantly heavy and changed for life—so much is different. Now I have lost 4 pupils to influenza.[8]

This Descartian peace, as you imagine it, surely scarcely exists anymore—destiny gave you a time of creative labor.

Good-bye and good air. With kind regards,

Zangger

Edgar Meyer also sends his regards. Dällenbach is habilitating.[9]

A military mobilization would be the best for you.

599.   From Edgar Meyer

Zurich, 11 August 1918

Dear Mr. Einstein,

In the meantime you probably already received word that you are wanted here in Zurich again. I on my part also would like to do everything to help this plan become fine reality.

Prof. Zangger has probably already reported to you how exceptionally accommodating the Zurich government and the Polytechnic's Education Council demonstrated themselves to be.[1] This I can absolutely confirm as regards the Zurich government, because when I appeared there for the first time with the plan, I encountered such understanding and such efforts to clear away all possible difficulties that my breast was filled with the finest hopes. If you want to accept the offer, you can be sure that you can design your position here exactly as you wish, not only concerning the financial aspect but also the teaching reponsibilities. And with regard to the latter point, I would like to take the liberty of making a suggestion. Since, as I know, you do not like giving lectures, you could perhaps organize instruction in theoretical physics as follows: you read only an introduction of a sort and leave the individual matters after your introduction for private lecturers [*Privatdozenten*] to execute. Since in all probability Epstein will finally be here next semester,[2] furthermore, Meißner and Tank[3] just habilitated, you would have the help of 3–4 private lecturers and could therefore reduce your own lecturing hours to whatever minimum you please. Whatever you would like could be done so that you first have a good amount of time to recover thoroughly.

Obviously the conditions here are more modest than in Berlin and cannot be compared. But you would not be entirely without contacts in physics here either, and you would anyway have Besso, Weyl, Weiss, Epstein, Piccard, Ratnowsky, Dällenbach,[4] and others. Experimental physics is also not as equipped as you have it there, of course. But this I can guarantee you, that I and my entire institute will be at your disposal. Whatever you wish will be done, and I may also say with pride that the funds are considerably larger now than before, as I was able to inject approximately Fr 50 000 into the institute in the last $2\frac{1}{2}$ years.[5] And money is always obtainable here for scientific purposes, everything *can* be done here. Please do not take this as boasting, considering that since my presence here no major things have emerged from this institute yet; but the reorganization swallowed up very much time, and some quite fine papers are already in press.

Allow me to say one thing: I am obviously but a small part of physics in Zurich and it cannot be captured comprehensively enough in writing; do please come here sometime and speak with all who are a factor in the consideration. You will see *how much* you are wanted here. I shall do everything I can to satisfy all your wishes. Please come sometime, have a look at the new experimental conditions at the university, and let us talk about everything in detail!

Are you still coming during the holidays? I am always here in Zurich, and I urge you to stay with us then. We have enough room and shall do our best to make it comfortable for you.

With most cordial regards, to an imminent reunion, and with the profound wish that my hopes will be fulfilled, I am yours,

Edgar Meyer.

600.   From Theodor Rosenheim[1]

Schreiberhau i. R., 11 August 1918

Dear Professor,

How pleased I was with your letter. At the beginning of it, news of the best physical well-being possible for you; in the middle, an amusing description of your environment and its effect on you and, at the end, a profession of contentment with the world and the strongest affirmation of life. If ever there were preconditions for physical and mental reinvigoration, this time they appear to be present in your case, and that is why I am full of happy expectations of you.

I am almost filled with envy when I read that you are roasting in the sun on sand dunes.[2] I would like to do that one day as well; but here we had mostly cool days, often rain, and I only sparingly became privy to what I need: light and heat. Even so, since I am being boundlessly lazy, I also am fulfilling one portion of my mission that way and hope that upon my return to Berlin (end of August) I shall be more receptive and capable of working again than I was during the summer. Then it's back to the old grindstone again, and that's good—because since I do not have your cheerful optimism, I need steady occupation to distract myself and drown it out.

I am reading with particular interest the new book by Max Dessoir, whom you met once at our house, *Vom Jenseits der Seele* [*On the Other Side of the Soul*]—it contains many notable things, particularly for a medical doctor. All that is factually and critically stated about the frequently cited subconscience, hypnosis, spiritualism, occultism, etc., certainly deserves becoming generally known also among educated layman's circles.[3] Of course, he lacks a gripping descriptive style, he is a bit hard to read.

As we are discussing the subject of reading right now, I would not like to forget to make a request: In case you have available an extra offprint of your speech on Planck,[4] you would please me exceedingly if you sent it to me.

At the close of this epistle, I do not want to call myself so lucky, as you have done, but to express my contentment that up to now I have been often enough privileged with being allowed to enjoy fine things, to grasp problematic issues and, above all, to witness the profoundest, most wondrous, and complicated things man has ever encountered.

Cordial regards to you and your ladies, ever yours truly,

Rosenheim

My wife sends all of you her best regards!

Special thanks to you, esteemed Madam, for your kind note.

601.   To Heinrich Zangger

Ahrenshoop, 16 August [1918][1]

Dear friend Zangger,

You are a tormentor of moving kindness, throwing me once again into disarray with your letter.[2] You know my attitudes, albeit not the whole conflict inside me. But *one* thing is certain. In Berlin I am surrounded by so much personal kindness and solicitude, as concerns my colleagues and the authorities that, after all of that, I cannot think of going away from there. Planck and Warburg are so touching toward me, visited me constantly when I was lying sick in bed last winter. All this would be obvious to you if you had seen it as well. On the other hand, as before, I consider Zurich my true home and Switzerland as the only country to which I have an affinity. You know all of this very well. Now I have an alternative. I would like to hold frequent lecture cycles of about 6 weeks' length in Zurich; approximately twice a year, either during the holidays or during the academic term. From this the students gain as much as if I were there during the semester, and I can do this easily without having to be away from Berlin all too much. If you think so, we can start right away at the end of September or in

February. I ask for no more than the reimbursement of my expenses which, with my modest requirements, is of little consequence. I would take Else along, who cooks for me.

This solution seems to me a very favorable one and *it would be a great pleasure for me to carry out this plan.* Write me immediately what you think of it, that is, again to 5 Haberland St.

Cordial regards, yours,

Einstein.

602.   To Edgar Meyer

Ahrenshoop, 18 August 1918

Dear Mr. Meyer,

Yesterday a letter from my friend Zangger arrived, today yours, and both put me in quite a fluster. On the one hand, grateful devotion to Berlin, where my colleagues and the authorities are obliging me in every conceivable way, on the other hand, my dear Zurich, my hometown, to which I as a convinced democrat am attached now in these times more than ever.[1] You can *hardly imagine a worse* emotional conflict *than this one was.* But now I think I have a satisfactory solution for all sides.

I keep my positions in Berlin but come to Zurich twice annually for (about) 5 weeks each time. I would use each of these stays to give twelve two-hour lectures.[2] It is just unclear to me whether these courses should take place during vacation time or during the semester, furthermore, whether general introductory courses, covering the main fields of theory over two years, or courses on special topics would be more appropriate to your needs. I would be content if Fr 1 200 were awarded to me for such a course, in order to defray the costs for the trip and the stay, so your budget would thus be minimally burdened and *through this arrangement your professorship appointment options would be very little affected.* I do not doubt that the Academy and the K. W. Institute would kindly overlook such short absences, and yet in this way I could accomplish almost as much in Zurich as if I were constantly there.

Kindly consider this proposal with the authorities. Perhaps you will find it reasonable as well. For me also it would be a great pleasure to be able to teach in my home country again, without others being able to accuse me justifiably of some form of unthankfulness.– You cannot imagine how strange it seems to me that so much fuss is being made over me. I cannot do any differently than another person who would be happy to receive a modest teaching position. But reputation is everything . . . curious world!

Write me very soon to 5 Haberland St., Berlin, and thank the authorities in my name for their excessive esteem. When we are agreed, it can soon get started.

Cordial regards, yours,

Einstein.

603.   From Hermann Anschütz-Kaempfe

Neumühlen, Kiel, 19 August 1918

To Professor A. Einstein, Ahrenshoop, Old Customs House.

Esteemed Professor,

Thank you for your message of the 1st of this mo. and for the reprint of your cordial address for the Planck festivities.[1]

I am likewise indebted to you for your indication of the theoretical difference between the two constructions discussed in your evaluation, which I nevertheless consider, as you do, completely irrelevant practically.[2]

On the assumption that you may be interested in how, under the cloak of science, competitive jealousy operates, but primarily because your name is mentioned twice in it, I am enclosing for you a short paper on Usener's book *Der Kreisel als Richtungsweiser* [*The Pioneering Gyrocompass*].[3] I am obliged to you in advance for the return of the book itself, which I am sending to you in the same post.

With respectful regards, yours very truly,

Dr. Anschütz-Kaempfe.

Usener's book separately.
Article   )
D.R.P. 34513) *enclosed.*
D.R.P. 182855)[4]

604.   To Michele Besso

Ahrenshoop, 20 August [1918][1]
from 23 August back in Berlin, 5 Haberland St.

Dear Michele,

Unless you put the wrong date on it, your letter was en route for a full 13 days; I received it only yesterday. Unfortunately, I am not going to be able to come to Switzerland so soon, so we are unfortunately probably not going to be able to see each other this time. The following has happened and puts me in an awkward predicament. Zangger and Edg[ar] Meyer offered me a teaching position at the Univ. & Poly[technic] in Zurich,[2] and I really cannot split myself in two. In Berlin, everything conceivable is being layed at my feet . . . I want to sink into the ground with shame. (How happy I would have been 18 years ago with a measly assistantship.[3] But it's how Heine put it in verse:

If you have plenty, [then by and by]
Much more shall you receive besides

— — — — — — — —

Look up this exquisite little poem; it's one of the later ones.)[4] What to do? Days of dark brooding lie behind me. Proof: I dreamed I had cut my throat with a razor. Well, I made the following offer to Zangger and Meyer. I would keep my Berlin position but would come to Zurich twice a year for 4–6 weeks each time and give a lecture cycle of about 12 lectures.[5] As compensation for it, I would just take my expenses incurred by this traveling. With this kind of offering on the altar of my hometown, I am relieved of my oppressive feeling and may perhaps still be of some use, without having to act despicably toward my Berlin friends and benefactors. I'll endure it healthwise, because the matter has improved very much, thanks to the great amount of care. I haven't had any more attacks for over a quarter year. But I absolutely must avoid longer stretches by foot and rapid movements and generally only eat according to a diet.

I have confidence that Weyl not only is outstanding but also is a very delightful fellow personally as well. I'll not forgo any opportunity to meet him. He will come out of the relativistic dead end again, all right. His theoretical endeavor does not agree with the fact that two originally congruent rigid bodies also remain congruent, regardless of what destinies they follow.[6] In particular, it is inconsequential what value is attributable to the integral $\int \varphi_\nu dx_\nu$ of their world lines.[7] Otherwise, sodium atoms and electrons would have to exist in all sizes. But if the relative size of rigid bodies is independent of their prehistories, then there is a measurable distance between two (neighboring) space-time points.

Then Weyl's basic assumption is not correct on the molecular level in any case. As far as I can see, not a single physical [reason] speaks for it being valid for gravitational fields. The [argument against it], however, is that the gravitational field equations become of the fourth order,[8] for which there has been absolutely no indication in experience up to now, also that a somewhat plausible formulation for the energy law does not exist as soon as the Hamiltonian of the gravit. field contains derivatives higher than of the first order for the $g_{\mu\nu}$'s.

This leads me to the energy question. Your suggestion reveals to me that you also are of the opinion that an energy tensor for gravitation could be dispensed with.[9] But then the energy law immediately loses all value. The matter does satisfy [the] "energy law"

$$\frac{\partial \mathfrak{T}_\sigma^\nu}{\partial x_\nu} + \frac{1}{2}\frac{\partial g^{\mu\nu}}{\partial x_\sigma}\mathfrak{T}_{\mu\nu} = 0.$$

But the second term causes this equation to have *no* consequence of the form

$$\frac{d}{dt}\{\int dV\} = 0.$$

It is also intuitively immediately apparent that without a stress tensor for the static gravitational field, the Newtonian forces cannot be derived from an energy tensor. If the energy-momentum conservation concept [is not also] applied to the $g_{\mu\nu}$-field, it loses all physical value.–

What I wrote you about λ is worthless.[10] The reasons remain the following: Either the world has a center point, has on the whole an infinitesimal density, and is empty at infinity, whither all thermal energy eventually dissipates as radiation.

Or: All points are on average equivalent, the mean density is the same throughout. Then a hypothetical constant λ is needed, which indicates at which mean density this matter can be at equilibrium.

One definitely gets the feeling that the second possibility is the more satisfactory one, especially since it implies a finite magnitude for the world. Since the world just exists as a single specimen, it is essentially the same whether a constant is given the form of one belonging within the natural laws or the form of an "integration constant."–[11]

A priori, irreversible elementary laws can very well be expected. But closer observations up to now do not speak for it (particularly the quantum laws), no more so the fact of thermal equilibrium. Judging from all I know, I believe in the reversibility of elementary events. All temporal bias seems to be based on "order." You will counter with radioactivity. But I am convinced that the inverse process is only impossible practically.

Warm regards, yours,

Albert.

605.   To Felix Ehrenhaft

[Ahrenshoop, 20] August [1918][1]

Dear Colleague,

I studied both of your long papers with avid interest.[2] The energy with which you pursue your goals is admirable. *Now it can no longer be doubted that your particles are spherical and of the size you indicate.*[3] Considering the great care you devote to the subject, it would be desirable if you also proved stringently that the air in your condenser is not brought into motion by the electrical field.[4] For extremely weak movements already suffice to destroy the rationality. It is certain that the formula for the Brownian motion must be correct (because of its relation to mobility)[5] since, for this derivation, only the law of osmotic pressure is needed, which we know happens to be definitely valid for the smallest structures (of molecular orders of magnitude). Without having examined the calculations, I vaguely suspect a source of error from the circumstance that it is not the momentary position of the particle that is being registered, but some temporal mean value. The Brownian motion appears "smoothed out" to the observer. As a result, it could be merely an apparent shortening of the perceived paths. Practical testing immediately produces the $\sqrt{t}$-law. The Δ in the Brownian paths must be proportional to the root of the observation time. The limit $\sqrt{\dfrac{\Delta}{t^2}}$ for a large t yields the correct value.[6] Then one easily sees what large (numbers) times[7] are needed to circumvent subjectively determined errors.

Your negative photophoresis must be based on an as yet unknown type of momentum, i.e., a secondary process different from radiation, which is likewise directed and is immediately produced by the radiation.[8] For this phenomenon as well, it would be very desirable if it were shown that the light ray causes no horizontal motion in the gas, although you have already removed the thorn from this reservation with your magnificent proof of the pressure independence of the force of photophoresis and with your demonstration of the Schwarzschild-Debye maxima.[9] If you already can prove from your previous measurements that the motions of the gas could not have had a disturbing influence on this elementary quantum problem, please let me know the grounds for it. In your theoretical survey of the bases for an electron of constant size you, just as Mr. Konstantinowsky, failed to touch upon the strongest arguments.[10] Bohr's theory of spectral lines is conceivable only with an exactly constant e. The discovery by Franck and Hertz that electrons at discontinuous potentials cause the excitation of mercury vapor proves the equivalency of the elementary charges of electrons.[11] The constancy of e[12] likewise makes the constancy of both quantities at least plausible. Quite apart from all the particular theories, the type of lines from the luminescence in the gases proves to me that the molecular structures are sharply defined and do not vary with the individual mass limits. Your comparison of the theoretical arguments to the bundle of brushwood from the familiar fable is a good one![13] Your way of extricating the individual twig, however, and of showing that it can be snapped seems misleading to me. The value of a hypothesis lies in the multiplicity of its results. A hypothesis belonging within a theoretical complex can never be proven. Your experiments also are theoretically explained. You also recognize the admissibility of individual methods of determining B just by the fact that they provide concurring results:[14] hence along an indirect path. On the whole, it is thus with electrons as well. One more thing. Negative radiation pressures are impossible for substances that absorb light because such pressures would contradict the principle of reaction. Consider the following rigid system.

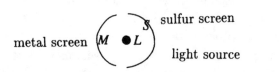

If the metal did not experience the same forces of pressure as the sulfur, the system would have to set itself into motion, if it was suspended in a vacuum.

Whatever may be the final result of your fascinating investigations, one thing is certain; none of your critics has offered anything convincing against your analyses.

With best regards, yours,

A. Einstein *m[anu] p[ropria]*

606.   To Hermann Anschütz-Kaempfe

Ahrenshoop, 22 August 1918

Esteemed Dr. Anschütz,

I read your explanations about the book by Usener with interest.[1] Now I really know why Usener sent me the little book a while ago![2] I do not doubt the dishonest intent of its author, who uses such sophisticated tactics, and I can imagine how offensive such conduct is. But everyone knows that it was you who first brought the gyrocompass into being. You arrived at this goal by reducing the horizontal stability. How obvious this solution was is a moot question; the point is to hit upon it!

In wishing that, from now on, you will not be plagued and injured anymore either by the *mala fides* of interested parties or by the limited understanding of impartial experts in the field, I am yours truly,

sig. A. Einstein.

(from the day after tomorrow, back at 5 Haberland St., Berlin)

I am returning in the same post both pat[ent] publications as well as the book with the Van den Bos sketch neatly modified.[3]

607.   To Michele Besso

[Berlin, 28 August 1918]

Dear Michele,

On rereading your last letter, I discovered something that downright annoys me: speculation allegedly had revealed itself to be superior to empiricism. In this regard you are thinking of the development of the theory of relativity. But I find that this development teaches something different that is almost the opposite, namely, that in order to be reliable, a theory must be built upon generalizable *facts*.

Old examples:

Main theorems of thermodynamics on the impossibility of the *perpetuum mobile*. Mechanics on the empirically explored law of inertia. Kin. gas theory on the equivalency of heat and mech. energy (also historically).

Special relativity on the constancy of the velocity of light & Maxwell's equations for the vacuum, which on their part are based on empirical foundations. Relativity with regard to unif. translation is an *observed fact*.

General relativity: *Equivalency of inertial and gravitational mass.*

———

No genuinely useful and profound theory has ever really been found purely speculatively. The closest case would be Maxwell's hypothesis for displacement current. But there it involved accounting for the fact of the propagation of light (& open circuits).

With affectionate greetings, yours,

Albert.

608.   To Paul Ehrenfest

[Berlin, 4 September 1918]

Dear Ehrenfest,

I congratulate you heartily on little Wassily.[1] Your will for life fills me with admiration. I also am full of hope for the future again, even if I don't give it concrete expression. Your observation about the entropy constant is absolutely correct. My paper to which you wanted to allude is probably the one from 1914,[2] not the one from 1916; there the number of possible quantum-like elementary states at absolute zero is indicated as

$$\Pi(n!),$$

as it obviously is.[3] Planck will not be talked out of his metaphysical probability concept.[4] When considering his type of inspirations, an irrational residue is left that I cannot assimilate (I then always have to think of Fichte, Hegel, etc.). At the time, you uncovered a false conclusion in my paper of 1914 (wrong application of your adiabatic hypothesis). I informed you once how a correction would be possible.[5] You never let me know whether you approved of this change in the chain of reasoning. Did you ever receive my message then?

Warm regards, yours,

Einstein.

609.   From Peter Debye

Maastricht, 14 Smede St., 4 September 1918

Dear Einstein,

Your letter, along with 2 copies of the contract, reached me here yesterday. Enclosed I am returning to you one copy of the contract, signed.[1]

As long as I was in Göttingen, I occupied myself during the waiting period with obtaining price estimates as well from the various suitable companies. Even after that, Siemens still seems to me the best supplier, though. Besides, I know some gentlemen there and I definitely think, as you likewise will suspect, that this must be taken advantage of.

Now I am at my parents' until 15 September, as you saw above, am doing as little as possible, and am going to do even the final ordering of the apparatus only from Göttingen. I hope we can then set up the apparatus soon and make some findings that please you: something that you could simultaneously perceive as an offering of thanks, which would suit me well.[2]

With best regards and cordial thanks, yours,

P. Debye.

610.   From Max Frischeisen-Köhler[1]

Halle-on-S[aale], 5 September 1918
cur[rently] Bad Leutenberg in Thur[ingia]. Guest house Zum Lamm.

Highly esteemed Professor,

May I permit myself to make a respectful request in the name of the editors of the *Kant-Studien*? You know what a lively interest philosophers take in the theory of relativity. Unfortunately this interest does not always match the understanding of this theory, which is the precondition for fruitful discussion of it. In particular, philosophers in general have not been in the position to follow the generalizations of the theory of relativity and of your newly developed theory of gravitation. Now, my inquiry and big request is, therefore, whether you, who has the first say in these matters, would not by chance have the time and inclination to report in the *Kant-Studien* on the state of the theory in a form that accommodates the philosophical statement of the problem. What interests us philosophers primarily is what the theory of relativity alters in the substance of the scientific worldview, whether we are dealing with a merely formal change in perspective, hence ultimately only with a new arrangement of the symbols with which we describe phenomena, or with a factual extension and reformation of our concept of nature. The other question occupying the epistemologist involves whether the system of categories, with which 18th and 19th-Century science as well as philosophy conceived of reality, has in principle been extended or shaken by the theory of relativity, or whether merely modifications are involved here that, as certain neo-Kantians (Natorp) or certain Positivists (Petzoldt) believe, can be comprehended within the framework of Criticism [*Kritizismus*] or Positivism, without encountering problems, or whether it is already provided for therein.[2] With such drastically contrasting opinions it would be of the greatest importance to us philosophers to become acquainted with the position of the author of this theory, which by now already has a rich history of development behind it. I am encouraged in making this request upon reading your contribution in *Kultur der Gegenwart*,[3] which seems to me to be a model of the form in which scientific questions can be treated, even without being clothed in the strict mathematical terminology of theoretical physics. I may remark, though, that the readership of the *Kant-Studien* is in part mathematically trained, at least to the extent that mathematical formulations would by no means have to remain excluded, as long as they are not too specialized and complicated. I also want to add that I would, of course, like to leave entirely to you the more precise formulation of the topic and how far you might like to address the specific philosophical issues under debate.

The matter naturally is not urgent. But I would be extremely pleased if you could send me a line informing me of whether you basically agree to my proposal. In case of an acceptance, all further questions (length, publication deadline, etc.) could be reserved for later discussion of the details.

In the hope that, in the interest of all philosophical circles, my request has not been in vain, I remain with all due respect, yours very truly,

Max Frischeisen-Köhler.

P.S. In the form of a footnote I may perhaps mention that, as editor of the *Philosophische Jahrbücher*, I have on one occasion already tried to present the theory of relativity to philosophical circles. At the time, Mr. Laue had the kindness to take on reporting about it.[4] But since that time, the theory has developed very much further, of course, and it would matter very much to me if, in particular, the author of the theory could decide to develop in the *Kant-Studien* those elements from it critical for our conception of nature and the world, and primarily free of superfluous trimmings.

611.   From Kurt Hiller[1]

Friedenau, Berlin, 9 Hähnel St., 7 September 1918

Strictly confidential.

Esteemed Professor,

The parliaments are letting us down. It is clear to many thinkers today that salvation cannot come from the official appointees but only from an association of intellectually oriented individuals who come together by force of voluntary resolve and impose themselves on the public at the critical "psychological moment."[2] Opinions diverge on whether this moment is already in sight. One feels tempted to answer the question in the affirmative upon reading letters like the one whose text I am enclosing herewith. The letter originates from a major at the front, commander of an engineer battalion in the Somme region.[3] The positive suggestion made by this officer coincides in essence with a suggestion by the writer Otto Flake in the June issue of the *Friedenswarte*[4] [*Watch-Tower for Peace*] (unfortunately not importable from Switzerland) and is in welcome agreement with what has been running through the minds of members of the immediate circle of activists for a long time now. The "League for the Cause" [*Bund zum Ziel*] which, according to plan, has unfortunately remained *in statu nascendi* since the Westend Resolutions of August 1917, would now perhaps have occasion to come fully to light and to do its first deed.[5]

The whether, how, and when of such a step will be the subject of the discussions, which are scheduled to begin on Saturday, 21 September, 4 o'clock in the afternoon, in the apartment of the undersigned and, if necessary, to continue on Sunday the 22nd.   The more significant your counsel is to the movement, the more welcome is your attendance! You will meet exclusively persons who can be assumed to agree with the basic mentality, whose reliability is beyond any doubt, and who are young in spirit. Should you be prevented from appearing, you would be of service to the gathering by explaining your position in writing.

With the humble request for a prompt response, in utmost respect,

Kurt Hiller.

612.   To Michele Besso

[Berlin,] Sunday, 8 September [1918][1]

Dear Michele,

My still not having been able to inform you when I'm coming to Switzerland is connected to the following affair.   Three weeks ago I received a call to the Zurich University and Polytechnic (letter by Edgar Meyer), which was obviously Zangger's enthusiastic doing.[2] I must say that I was tormented by this matter, and it was very hard for me to come to a decision. I don't need to say that the general conditions there appeal much more to me. But if you only saw what fine relations have formed between my closest colleagues and me (especially Planck) and *how* accommodating everyone was and still is to me here; moreover, if you realize that my papers gained influence only as a consequence of the comprehension they found *here*, then surely you'll understand that I can't decide to turn my back on this place. Added to this is that my *major* personal difficulties would persist if I pitched my tent in Zurich again, although it does seem very tempting to be able to be close to my children again; past experience from my visits to Switzerland gives me little encouragement in this. Here, everyone is close to me only to a certain limit, so life goes on almost without friction; this I have learned in life.

However, because I am very attached to Zurich, and because declining the offer was unutterably distasteful to me, I did something that I otherwise very much abhor: I resorted to compromise! I proposed that they permit me to hold two lecture cycles per year in Zurich, each of which should consist of 12 lectures (within 4–6 weeks).[3] This is not supposed to be considered a position but an extraprofessional activity for which I should receive only enough to cover my expenses arising out of it.–

Now you see why I cannot write you when I'm coming to Zurich. On condition that my proposal finds approval, I just want to travel to Zurich when the necessary dates are set so that I can link the trip to a lecture cycle.—Do you think that I did the right thing?

Your remarks about the role of experience and speculation in physics very much appealed to me.[4] I would just like to add that it doesn't fit to count Riemann's achievement as *pure* speculation. Gauss's accomplishment is the formulation of the positioning laws [*Lagerungsgesetze*] of little rigid rods on a given plane.[5] His *ds* corresponds to the little rod; without this experience-based structure, the entire consideration would of necessity not have been made. Riemann's generalization to the multidimensions admittedly is a purely speculative deed; but it likewise is based on Gauss's conception of the little measuring rod. When others later forgot the terrestrial origin of ds^2, that was certainly not an advance. In his fine book, Weyl rightly calls Riemann's theory the *geodesy* of multidimensional structures.[6]

Warm regards, yours,

Albert.

613.   To Kurt Hiller

[Berlin,] 9 September 1918

Esteemed Sir,

I read your brochure as well as your note of 7 September with great interest and do not want to omit thanking you and expressing my opinion.[1]

I would have liked to appear at your meeting. However, it is not appropriate for me as a Swiss to involve myself in local political affairs.

Your brochure is brilliantly written; your proposal, though, in the indefinite form it is, seems to me to contain little that lies within the realm of the possible.[2] If you summon together those local men who have distinguished themselves in their various fields by significant intellectual achievements, you will assemble a collection of power—or "pragmatic"—politicians who, if they are truthful, will agree on the tenet that private morals have no importance in the relations between states and nations;[3] that only the right of the strongest is decisive; that without war humanity would degenerate; concessions to pacifism should be considered merely as fine gestures for hoodwinking . . ., etc., etc.

These persons are simply the children of their time, their wishes (which are determined by their views) the product of a development of the mass psyche which is stamped with Bismarck's name.[4]

When a great scholar from a neutral country abroad was presented the mentioned political gospel with considerable eloquence, he said, "Perhaps the world is so, but then I do not want to live in it!"[5] I am firmly convinced that a "world teaming with enemies" is forged out of this reaction.

I see rapid and radical democratization according to the model of the western powers as the only way to save Germany. For only such a constitution—great as its faults may otherwise be—seems to guarantee such a far-reaching decentralization of the desire for power that a repetition of the line of action of 1914 would appear excluded. Nobody abroad trusts the current regime anymore.[6]

This, in a few words, is my conviction. As one who was a pacifist even before the war, I am authorized to air my views now.

Very respectfully.

614.   From Edgar Meyer

Zurich, 12 September 1918

Dear Mr. Einstein,

First of all, many thanks for your letter. I tried in vain to be able to reply quickly and still cannot report anything final to you now. I am doing the whole thing together with Mr. Zangger,[1] you see, and I have not been able to get hold of him the whole time. First, he was not well, and later he went away to recuperate.[2] In any event, though, we must speak with the authorities. Now, this is what I think of your proposal: It would be a crying shame if we could not get you to come here permanently, but if that is how it is, I am grateful to you for every hour you lecture at our institutions. Just provided that permanence is out of the question, I am very much for your proposal and shall do all I can to get it accepted. I believe quite firmly as well that the local government, which listens to anything reasonable, will consent to your proposal.

Specifically, I imagined that then your lectures would not be vacation courses but would rather take place during the semester. At the moment, that works out very well, of course, for as I heard, in Berlin you end your semester already at the end of January, whereas we do so only at the beginning of March. Specialized lectures would naturally be very nice.[3] I also quite definitely think that everything will be settled so that you can lecture in the winter term already.[4]

I would also like to say something briefly about your notice in the *Verhandlungen*.[5] I believe the 1 mm-wide dark stripe is brought about by a contrast effect.[6] Have a look at the image with a magnifying glass sometime and slowly cover the lighter part with a sheet of paper starting from the side. All this with the light shining through. The stripe then disappears. It is best to take a sheet of paper that reduces the light just as strongly as the dark part of the plate. Naturally, my view can be proven securely by analyzing the plate photometrically with a Hartmann microphotometer. Unfortunately we do not have such an instrument at the Institute, otherwise I would have done so. But there is certainly one available in Berlin. The whole thing incidentally reminds me of the observations by Haga and Wind (around 1900) on the diffraction of X rays by a conical slit.–[7]

I hope that I can send you favorable news about the lectures very soon. Until that time, I am, with the most cordial regards, yours,

Edgar Meyer.

615.   To Lise Meitner

[Berlin, before 14 September 1918][1]

Dear Miss Meitner,[2]

This afternoon I told Mr. Hopf[3] about our problem. I realized then that the execution of our fluctuation experiment would be of great importance after all, because Meyer's result is seen as secure and as virtual proof of the continuous nature of absorption processes.[4] They say it is somewhat like this:

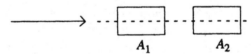

A_1 A_2

If the fluctuations of the ionizations in A_1 and A_2 reveal a partial coherence, this proves that the same elementary γ-radiation phenomenon (at least sometimes) affects not only A_1 but also A_2. This is conceivable only for a continuous-absorption property. Conversely, the independence of the ionizations in A_1 from those in A_2 would prove (albeit less stringently) that an elementary act can generate a secondary process only *either* in A_1 *or* in A_2, but cannot affect both simultaneously.

To achieve this goal we do not need such great precision, so we may safely give it a try. What do you say? Telephone me sometime about what you think of it. Even though, as I see it, the result is beyond question, just about all the other physicists do think otherwise about the matter.

Best regards, yours,

Einstein.

It suffices if we prove that the coherence in the ionizations drops rapidly with the relative distance between A_1 and A_2.

616.   To Lise Meitner

[Berlin, 14 September 1918]

Dear Miss Meitner,

I have performed the calculation again more exactly now, and found that the fluctuations *of both ionizations must be completely uncorrelated.* Thus our promising enterprise falls flat, except that Meyer's results definitely must be corrected.[1]

Even before the calculation, I recognized that in such cases statistical independence had to exist:

Let a ray S be split into two equally intense S_1 and S_2 through partial reflection. I contend that the fluctuations of S_1 and S_2 have to be independent of each other. For, since all rays of the same intensity (and frequency) must behave statistically identically, it must come to the same thing if I have S_1 combine with an equally intense, independently generated ray S_1', as if I had it combine with S_2. This is the case only when S_1 & S_2 are statistically independent of each other.

Cordial regards, yours,

A. Einstein.

617.   To Ernst Trendelenburg

[Berlin,] 16 September 1918

To the Administrative Office of the Kaiser Wilhelm Society for the Advancement of the Sciences. W. 9, 4 Voss St.

At your request by telephone yesterday, I am sending you some information about the K. W. Inst. of Physics, which could possibly be incorporated into the annual report.[1] Since I do not know exactly what details you need, I am not sure that what I am writing you here meets your wishes. The enclosed information does not regard the history of the Institute's creation nor its constitution, but rather its goals and activities up to now.

In great respect,

A. Einstein.

Kaiser Wilhelm Institute of Physics, W. 30, 5 Haberland St.

The Institute, which came into being on 1 October 1917, has as its exclusive mission the promotion of purely scientific, theoretical, and experimental research in the field of physics. Because the Boards of Trustees and Directors have found there to be no shortage of physical laboratories, no special building is being created for this Institute. In this way, the option is realized of assigning all the available funds completely for the benefit of scientific research. Ongoing activities: The astronomer Dr. Freundlich was taken on for the years 1918, 1919, and 1920 to perform astronomical analyses to test the consequences of the general theory of relativity.[2]

Furthermore, upon the application of Prof. P. Debye from Göttingen regarding his researches in the field of X-ray spectroscopy, the Board of Directors approved—for these investigations—the purchase of the necessary apparatus.[3]

618.   To Eduard Study

[Berlin,] 17 September 1918

Highly esteemed Colleague,[1]

I have been reading your little book on the epistemological foundations of geometry[2] these past few days and so much enjoyed your witty and unique explanations that I simply cannot refrain from *thanking* you personally *for the pleasure*. Much in the book is so aptly said and, at the same time, so humorously that often I had to laugh out loud in my quiet study.[3]

In general, any reader who has thoughts of his own will not able to be of the same opinion *on all points*. I liked exceedingly what you say about Kant and his successors[4] as well as your criticism of the offshoots of *axiomatics*, in particular, the requirement of an analytical foundation for geometry.[5] I am convinced that gradually, very gradually, this requirement will prevail.

In the hope of personally making the acquaintance of this inspired and truthful author sometime, I am yours sincerely,

A. Einstein.

619.   From Hermann Weyl

Samaden, Aola Pozzoli, 18 September 1918

Dear Colleague,

You have not heard from me for a long time. Throughout the summer I was not in the best of health and this is also the reason why I have just not come very much further till now with the consequences of my theory, which you are challenging.[1] But my faith in it is not yet shaken. To now it has only been by a complete penetration into the mathematical foundations of this theory. I am publishing an article about it entitled "Reine Infinitesimalgeometrie" ["Pure Infinitesimal Geometry"] in the *Mathem. Zeitschrift*, which is appearing soon.[2] There the natural development of geometry in the three stages of topology [*reinen "analysis situs"*] (1), affine (2), and metric geometry (3) is followed through. I say, now: God could have stopped at any of these 3 stages, namely, created either an empty world, or a world just with a gravitational field,[3] or the one familiar to us; but having once decided on the metric one, the most richly endowed, he had to adopt mine, the pure one.[4] I can perhaps best refute the objection you raised[5] with the following remark: if R is some invariant of weight -1,[6] e.g., the scalar of the curvature,[7] then the quadratic form $R(g_{ik}\, dx_i\, dx_k)$ is an absolute invariant;[8] by multiplying by this factor, an absolute normalization, so to speak, of the unit of length thus is achieved.–[9] When I sent you my note in the spring[10] I was already convinced that your gravitational equations follow in the 1st approximation from my theory for static fields, that is, *with* the cosmological λ term; yet in such a way that from the laws of nature it only follows that λ is a constant, without any specific value being prescribed for this constant (this value rather appears to be determined by the world's total mass).–[11] In my case also, a conservation law for the total energy and momentum is valid,

$$\frac{\partial \mathfrak{T}_i^k}{\partial x_k} = 0,$$

which results much more naturally than in your theory.[12] For the matter in the more restricted sense, $\mathfrak{t}_i^k$, the equation[13]

$$\frac{\partial \mathfrak{t}_i^k}{\partial x_k} - \Gamma_{is}^r \mathfrak{t}_r^s + F_{i\kappa}\mathfrak{s}^k = 0 \qquad\qquad (\mathfrak{s}\ \text{current})$$

applies; whereas for the current, the postulate $\mathfrak{s}^i = \rho u^i$ (u four velocity $\dfrac{dx_i}{ds}$, ρ charge density) is possible, $\mathfrak{t}_i^k = \mu u_i u^k$ contradicts the required invariance properties.[14] In consequence of this, the mechanical equations are still not clear to me; and as long as this is not settled, it is difficult to draw any consequences to be tested against experience (the influence of the electromagnetic potential on a uncharged point-mass, which you infer from the equation for the geodesic line,[15] is up in the air as long as the significance the geodesic line has in mechanics cannot be ascertained). This will be the next task, for me to arrive at some clarity here.[16] In addition, I have formulated the strict equations for a static, rotationally symmetric field; they are so complicated, though, that I do not know what to do with them at present; the constitution of the electron and the possibility of its existence ought to result from them.[17]

Hilbert was here for a few weeks and declared his unqualified support; I received a very approbative letter from Sommerfeld as well.[18] He wrote also that new, extremely careful measurements on Mt. Wilson did not yield any trace of astral redshift; what is the story there?[19]

At the moment I am recovering at Engadin; my state of health forced me to retract my acceptance of the call to Breslau.[20] If only we had you here now in Zurich again! You can imagine how happy I would be about that. But if it is not to be permanent, then I hope at least that you will be coming to us in the near future, for at least a few weeks.[21] With best regards, yours,

Herm. Weyl.

620.   From Friedrich Adler

Stein-on-the-Danube, 20 September 1918

Dear Friend,

Finally (on 14 Sept.) I received your long letter[1] after the short letter and the postcard had already arrived beforehand. I thank you heartily for all the effort you took, but to be honest, I am quite disappointed about your criticism. For you did not get to what seems to me to be the main issue. All your comments on chapter II are misunderstandings that arose from the fact that you made the comments before reading §24–35.[2] In my view, only the criticism to the 1st section of chapter III is legitimate, which I had meanwhile already become aware of on developing a 4th chapter, and which I am now busy changing. The statement on the observational bases that you miss is in the foreword, which I did not think necessary to send along to you. There it says that for all considerations, the validity of the Michelson & Fizeau experiment is *presumed*, it thus simply involves a choice between Lorentz and Einstein. My interpretation of these experiments, incidentally, becomes clearly evident in §34–35 as well.[3] For me it is not a matter of constructing a new theory, rather just showing what *possibilities* the observational foundations of Michelson and Fizeau leave open. There are ∞ many transformations that conform with *this* basis.[4]

$$x' = l\beta(x - vt) \quad y' = ly \quad z' = lz \quad t' = l\beta\left(t - \frac{v}{c^2}x\right) \qquad \beta = \sqrt{\frac{1}{1 - \frac{v^2}{c^2}}}$$

You want to prove that l can only be 1, but

$$l = \frac{1}{\gamma} \qquad or \qquad l = \gamma \qquad \text{where } \gamma = \sqrt{\frac{1 - \frac{u^2}{c^2}}{1 - \frac{u'^2}{c^2}}}$$

among others, conforms just as well to *experience*. Which hypotheses one adds for experience is merely a *matter of taste*. I therefore do not deny at all that the hypotheses can be chosen so that l has to be $= 1$, thus that one arrives at your transformation; I just contest (1) that *experience* virtually compels $l = 1$, (2) that the derivation in your first paper is correct.[5] For the problem of interest to me it does not help me in the least to avoid your first paper & to take another descriptive method as a basis for the discussion. I see in your first derivation an error that is admittedly somewhat difficult to recognize but that is completely clear to me now. I could question the correction of this consideration if you would demonstrate to me that there is an error in my deduction (§38, eqs. 114–126) as well. For only either[6]

$$t = \frac{a\left(t - \frac{u}{c^2}\xi\right)}{1 - \left[\frac{u^2}{c^2}\right]} \ (113) \qquad or \qquad t = a\left(t - \frac{u}{c^2}\xi\right) \ (126)$$

can come out; if my result (126) is right, then yours (113) must be wrong.– The case $l = \gamma$, which I treat in the 4th chapter, is interesting particularly because it determines *only* Lorentz deformations, but has no influence on the running speed of the clocks. (Albeit, Lorentz contractions in the direction of motion & perpendicular to it, which are in a relation of β to each other.[7])

Chap. II merely offers preparatory considerations for sections II–IV (chap. III), which are the main issue for me. There you can see that I did not overlook your "standard clock," rather that it is the actual point of departure of my reflections. I have been making a serious effort for a year now to understand it, but an *insurmountable logical contradiction* in your assertions remains for me:

a) The clocks U and U' are of the same construction.
b) The two systems in which these clocks are at rest are equivalent.
c) The two clocks display varying differences in hand position, when compared to each other at relative rest, when one has described a circular path relative to the other (comp. end of §4 of your 1st pap[er]).

For, according to the general laws of relative motion, U, seen from U', describes a circle when U', seen from U, describes one. Hence each of the two clocks must have lagged behind the other. There is just the logical possibility that, upon meeting up with each other again, a difference in hand position is the more probable situation (comp. §31–32 of my paper). This logical contradiction is admitted also by supporters of your theory (Berg, Petzoldt), but judged merely as a false consequence.[8] If you could clarify this point, very much would indeed have been gained, since then you would have to look differently at the entire problem of differences in the running rates. Something quite analogous to this applies with regard to the Lorentz contraction as well. You unfortunately stopped just before the doubly underlined question in §23: *Which of the two rods must be made shorter?*[9] This question was my starting point for all the considerations. It and the contradiction regarding the standard clocks, which on the one hand define the unit of time, on the other allow the ascertainment of the lapsing of a *differing* number of units of time, must continually reawaken doubts in any impartial person. But I do not want to pester you anymore right now; once it is in print, it will obviously be more comfortable to read and one or two details will have been improved. As concerns the manuscript, please, if it does not cause you too much effort, mail it to me in the same form that I had sent it out. Otherwise, I ask you simply to keep it but not to show it to anyone. Nobody was ever encouraged by *me* to ask you for information, and I am sorry if you were inconvenienced. I am sending you a transcription of your last derivation shortly,[10] so that this letter does not exceed the wartime standard. In principle, this derivation is, of course, identical with that in Laue's book, p. 38, except that one ray takes the place of the entire wave.[11] I would have had a great deal more to say, but for today space only remains to thank you very much once again for your effort. Cordial regards, yours,

F. Adler.

621. To Paul and Maja Winteler-Einstein, and Pauline Einstein

[Berlin, 23 September 1918]

My Dears,

Your telegram of yesterday awakes me from my indolent slumber and reminds me that you all don't know yet that I'm not coming to Switzerland at all, right now.[1] The reasons, the flu notwithstanding,[2] are the following.

As you know, I received a very attractive call to Zurich and at first didn't know what I should do. Well, although I declined the appointment, I offered to the Zurichers to lecture there twice a year for a month each time, in order thus to show my appreciation.[3] It's still not certain whether something will come of it, but it is very probable. Prof. Meyer in Zurich thought I should start lecturing there in *February*.[4] So it is understandable if I don't travel to Switzerland now, because it could otherwise become too much of a good thing. I would be very pleased if something came of it.

I'm doing extremely well, not only healthwise but also otherwise. I am coming back to life, now that my great, ardent wish is being fulfilled.[5] God's mills . . . [grind slowly but surely].

It must be quite lively at your place now compared to normally. I have a pretty good picture of what the conversations are like, I'm just a bit unsure of Uncle Jacob's role in them.[6] My health is extremely good, better than I could ever have dreamed. This I really do owe to Elsa's care, who cooks everything for me herself and spares no pains in procuring all the necessities for me.

Unfortunately I hear absolutely nothing of my boys again, despite many requests. Miza is supposedly in hospital, as I've learned from Dr. Zürcher.[7] How the children are managing is a mystery to me. I have to let it go on this way because I can't do anything from here and there is no one whom I could risk sending into the lions den either.

Edith wrote me that she wanted to join the German Phys. Society. This is strange, because she really gets nothing out of it, and it just costs money. Since she will not be dissuaded, I submitted her (and Janka's)[8] names there at their request. Their membership will be addressed at the next meeting. It's beyond me what use it is to her. For the present, tell her that their names are entered so that she does not think I had completely forgotten about the matter.

Affectionate greetings to all, yours,

Albert.

622. From Eduard Study

Bonn, 126 Argelander Street, 23 September 1918

Highly esteemed Colleague,

I thank you cordially for your so very kind letter to an author you do not even know personally.[1] I appreciate your verdict and attach great importance to it. You agreed on precisely those points closest to my heart.

The little book was given a very friendly reception elsewhere as well, despite its sarcasm, especially also among philosophers; but only very few also seem to have a sense of humor, like you do, for the topic. And since you are among these few, I may perhaps also tell you about a success that took me very much by surprise. Namely, advocates from *all* the philosophical schools I had attacked made an attempt to claim me as one of theirs.

I would give anything to know which points you are *not* able to agree with. Is it too forward of me to ask you, *provided it can be done briefly*, to commit some notes on this to paper? I really do not have the intention of dragging you into an exchange of correspondence that you might rightly fear. I also am not satisfied with all I had written at the time (1912). Above all, I certainly should have acknowledged and assessed your analysis of the time concept.

Now you have made such great advances in the meantime that it is difficult for someone who is fully engaged along other lines to follow. When the war broke out, I was so upset that I had to let mathematics lie for a while. I then took up biological studies, which I had been maintaining as a hobby until then, and am now working on a book about evolutionary theories and in particular mimicry which, I am happy to say, attracts the interest of the biologists here.[2] This does not mesh well with an orientation in new and complicated mathematical material, though. This occupation is more likely to lead me to Berlin sometime, where I would find all sorts of material that is lacking in my modest collection. But that is a tiresome question of upkeep and money, and maybe the war will thwart all my plans. If it becomes possible for me to come to Berlin, I shall give myself the honor of visiting you, and it will be a great pleasure for me to make your acquaintance.

With respectful greetings, I am yours very truly,

E. Study.

I am sending you a few more short papers. You probably already have my *Grundlagen der Kinematik*[3] [*Foundations of Kinematics*].

623.   From Hans Vaihinger[1]

Bad-Rothenfelde, 23 September 1918

Highly esteemed Colleague,

A few weeks ago the newspapers carried a report on a new philosophical periodical that is being founded by me in conjunction with Dr. Raymund Schmidt.[2] This report was essentially correct but premature and its content is not completely accurate. Since it is of importance to me that you, in particular, be correctly oriented in the new enterprise, I permit myself to forward to you a short two-page prospectus on the new journal, along with a general table of contents originating from the just published 3rd edition of *Philosophie des Als Ob*[3] [*Philosophy of As If*], as well as a short four-page program that will be printed as front matter to the 1st volume of the new journal, the exact wording of which is not yet definitely fixed, however.[4]

To this program is attached a typed addendum, constituting an essential extension of it.

From this latter addendum you will gather that a new idea is involved, namely, the thought that the journal attract not only philosophers in the restricted sense but also philosophically oriented and philosophically interested scholars from all the faculties and all the branches of knowledge of particular importance to philosophy. A number of them will also be mentioned right on the journal's cover page.

Thus the increasing valid fact that, on the one hand, philosophy inspires the individual sciences more than before, and on the other that they themselves consequently have a stronger effect than before on philosophy, should be lent expression, physically as well.

We have acquired as the representative of theology Professor Karl H. *Heim* at the Protestant Faculty of the University of Münster, who has distinguished himself with numerous philosophical writings in which he aligns himself with Avenarius.[5]

The Law Faculty is represented by Professor *Krückmann* in Münster, who already previously treated the problem of juridical fictions in a philosophical spirit.[6]

The Medical Faculty is represented by the famous physiologist *Abderhalden* in Halle, who is fully aware of the importance of thinking in terms of fictions for scientific, medical, and especially for physiological research and himself applies it in his papers.[7]

We have gained as our representative of mathematics the well-known *Pasch* from Giessen, the doyen of axiomatics, who aired his important ideas on the foundations of his science long before Hilbert and who is also already working on a paper for the 2nd volume of the *Annalen der Philosophie*.[8]

The inorganic sciences have their exponent in Professor *Volkmann* in Königsberg, who from his specialty, physics, has made valuable contributions to the epistemology of scientific thinking, which in part touch upon the ideas of the Philosophy of As If.[9]

For the organic sciences we have the botanist *Hansen* in Giessen, who has distinguished himself with his great work on Goethe's metamorphosis of the plants and in it already correctly treated as a fiction Goethe's idea of the protoplant.[10]

Art history and aesthetics find their advocate in Conrad *Lange* from Tübingen, who in his famous work *Das Wesen der Kunst* [*The Essence of Art*] described "unconscious self-deception" as the aesthetical fiction, as a main principle of artistic creativity and of art appreciation.[11]

The seven specialist scholars mentioned (all regular professors) will be named on the new journal's title page, as its sponsors, so to speak, its patrons, or however one would like to call it. The format for it will be the standard one, thus perhaps like this: "*Annalen der Philosophie, etc.* In conjunction with K.

Heim, P. Krückmann, E. Abderhalden, M. Pasch, P. Volkmann, A. Hansen, K. Lange, etc., etc., edited by H. Vaihinger and R. Schmidt."

As concerns philosophers in the proper sense, up to now I have negotiated with two German full professors about joining this editorial committee; one is *Cornelius* in Frankfurt on M[ain], who immediately consented, especially considering that he had independently expressed, even before I did, ideas that fit in the new journal's entire program.[12] The second is the Tübingener Karl Groos, who represents similar ideas in his game theory and his aesthetical and epistemological analyses.[13] He also agreed right away.

Negotiations are still underway with other philosophers.

I just found out now from Professor Frischeisen-Köhler[14] that you, highly esteemed colleague, mentioned that you have a close philosophical affinity to the direction advocated by me, and so I may perhaps venture to request your joining the above-described editorial committee and thus also forming a public link with the new philosophical line. I have long been following your highly significant publications attentively, and it soon became clear to me that a philosophical mind is behind them that coincides in essential parts with those convictions that I myself consider to be the correct ones. Through such an alliance our influence on each other would be mutually strengthened, and a fruitful collaboration would have to result in being useful to both sides. On the one hand, your new theory of relativity would be incorporated into the overall philosophical trend, on the other hand, this trend would be enriched by your ideas.

The first volume of the new periodical is appearing in a few weeks comprising approximately 45 printed sheets. A great deal of material has already been admitted for the second volume, which is projected to appear in 4 separate issues, and provisions have already been made for other volumes as well.

I am, naturally, happy to provide you with more information.

Those scholars whose names appear on the cover of the journal will receive a complimentary copy regularly from the publisher.

Should you, highly esteemed colleague, not have a personal copy of *Philosophie des Als Ob*, I should consider it an honor to submit to you a copy of the recently published 3rd edition of this book.[15]

There is no obligation to deliver papers, of course, but it would please the editors exceedingly to receive a contribution, esp. from [you] as well, now and then.

In requesting a prompt and, I hope, positive reply, and in expressing my great respect as a colleague, yours very truly,

Vaihinger.

Address until the first Oct. Bad Rothenfelde. Teutoburger Forest. Dietrich House; from 1 Oct. on, Halle-on-S[aale], 15 Reichardt St.

P.S. If you consent, the addition to the four-page program will be changed correspondingly, of course. You would probably have to be listed either as a representative of mathematical mechanics or of general mechanics, or as a representative of the inorganic sciences beside P. Volkmann. Please advise me on this.

Note at head of document: "Dictation. Confidential."

624.   To Eduard Study

[Berlin,] 25 September 1918

Highly esteemed Colleague,

Many thanks for the amicable letter.[1] I do not yet have your "Kinematics"[2] and would therefore be very pleased to have it. I am supposed to tell you my reservations?[3] By emphasizing them, it would appear that I wanted to pick holes in everything. But it is not so serious because I don't feel very comfortable or at home with any of the "isms." It always seemed to me as though such an ism was powerful only as long as it fed off the weakness of its counter-ism; once the latter is struck dead and it is alone on the deserted stage, it then proves to be just as weak-kneed. So *here goes with the cavilling!*

"The physical world is real." This is supposed to be the basic hypothesis. What does "hypothesis" mean here? For me, a hypothesis is a statement whose *truth* is temporarily assumed, *whose meaning, however, must be beyond all doubt.* The above statement seems to me intrinsically senseless though, like when someone says: "The physical world is cock-a-doodle-doo." It appears to me that "real" is an empty, meaningless category (drawer) whose immense importance lies only in that I place certain things inside it and not certain others. It is true that this classification is not a *random* one but . . . . now I see you grinning and expecting me to fall into pragmatism so that you can then bury me alive.[4] However, I prefer to do as Mark Twain, by suggesting that you end the horror story yourself, as you please.[5]

⟨Real and unreal seem to me to be like right and left.⟩

I admit that science deals with the "real" and am nonetheless not a "realist." This can be all the same to you, and I also do not feel like your opponent. For I would like to state the proposition:

If all the rubbish is cleared away from two arbitrary "isms," they then become alike.

That is why you *could* be mercifully accepted by the priests of all the isms. You *were* accepted because your grace appealed to them.

———

The positivist or pragmatist is strong as long as he fights against the opinion that concepts that were anchored in the "*a priori*" existed. If in his zeal he forgets that all knowledge consists in concepts and judgments, then this is a weakness lying not in the subject but in his personal disposition, likewise the senseless battle against the hypothesis; comp. the little book by Duhem.[6] In any event, the campaign against atoms is based upon this weakness. Oh, how bitter is this world for man; the path to originality leads through irrationality (in science), through hideousness (in art)—the one accessible to many, at least.

———

Your comment about Kant smuggling in psychological arguments is exquisite.[7]

———

My kinsman Cohen is not even palatable when he is served up by as skillful a gourmand as you, roasted on the spit.[8]

———

Confounded, I keep forgetting to gripe; *will do, immediately*!

———

Isn't your fight against pragmatism a bit demagogic? Is it not just a fight against *superficial* pragmatism? Shouldn't it be admissible to regard science from the biological point of view? What else is pragmatism, ultimately?

———

Isn't the concept "natural geometry" a little unnatural? Doesn't an idol really lie behind what you call "reality" of space? Pages 57–59 of your book gave me a bit of a stomach ache.[9] All in all, the discussion on the relation between geometry and experience seems to me to be the least felicitous aspect of your explanations.

Attached to this is the attack on Poincaré, which does not seem justified to me (except for the very legitimate criticism on page 116).[10]

It would be too lengthy to go into this in more detail, especially since what I would be telling you would not be new to you, essentially.

I am sending you a popular little book on the theory of relativity, in which I addressed the latter issue.[11] I know that I am certainly not going to win your applause with it because you will condemn my point of view, as it is presented there, as "empiricism." It seems to me to apply more to the presentation, though, than to the substance. But it would be nice of you if you read the relevant parts, so that we could take it up when we discuss the subject sometime in person.

With best regards, I am yours truly,

A. Einstein.

625.   To Paul Ehrenfest

[Berlin,] 27 September [1918]

Dear Ehrenfest,

You are such a kind person that I have no alternative but go visit you, despite the horrendous traveling conditions[1] and despite my mother lying so seriously ill (cancer) in Lucerne that I do not know when I'll have to go there again. I shall be glad to give a lecture. Send me an official invitation very soon and *take steps from there right away so that my entry is permitted immediately. The official way through the local embassy takes a minimum of 3 weeks*, as they themselves have informed me.

I am indescribably glad at the prospect of seeing you again and of taking in the refreshing Leyden atmosphere, now that the dreadful war no longer weighs upon everything.[2] With cordial regards to all of you, yours,

Einstein.

626.   To Hermann Weyl

Berlin, 27 September 1918

Dear Colleague,

I was just busy with your new paper[1] when your letter arrived.[2] It is a pleasure beyond words for me to read your scrupulously thought-out things. The division into the three theoretical themes is very effective and clear.[3] However, the point of sacrificing congruency while keeping similarity does not seem to work so naturally for me.[4] You know, of course, how I conceive the relation to reality; nothing has changed in this regard.[5] I know how much easier it is to convince people than to find truths, especially for someone who is such an incredible master of exposition as you are. But ultimately, far be it from me to be presumptuous; I can be mistaken here, as I have been countless times before. In a couple of years it will be obvious whose eyes saw more clearly. And I know precisely that the only goal either of us has is to see the truth.

From what you wrote me, I was very impressed by the fact that the statics comes out right, and this with a λ still at one's disposal, also that an energy equation, $\sum \dfrac{\partial \mathfrak{T}_i^k}{\partial x_k} = 0$, results.[6] This does indeed mean more to me than the results I already knew about previously. I am looking forward to the opportunity of studying this more closely. The expression $R g_{\mu\nu} dx_\mu dx_\nu$ for the measured length[7] is, in my opinion, by no means acceptable, however, when the curvature invariant is taken for R, because R is very dependent on the mass density. A very small change in the measurement path would have a very pronounced influence on the integral of the square root of this quantity.[8] Furthermore, although it is correct that you did not assert anywhere that your geodesic line was the path of an uncharged point-mass,[9] nonetheless this does appear necessary to me without extra proof. For the only type of world line preferred in reality must surely correspond to the world line preferred by the theory; I at least cannot think of any other arrangement. If this is so, however, we encounter a contradiction with your theory with the energy principle, in the following manner. The vector potential $\varphi_1 \varphi_2 \varphi_3$ would act as a moving force upon a static point; a permanent magnet generates closed lines $dx_1 : dx_2 : dx_3 = \varphi_1 : \varphi_2 : \varphi_3$.[10]

Of course, I know very well that the state of the theory, as I have set it up, is not a satisfactory one, disregarding the fact that matter remains unexplained. The unrelated juxtapositioning of the gravitation terms, electromagnetic terms, and λ-terms is undeniably a product of resignation. I share your firm conviction that this must and will be otherwise. I just do not believe that the route you have embarked on is the right one, as finely thought-out as it is. In the end, the answer must include that the action densities do not have to be stuck together additively. Many ideas occurred to me as well, but I always had to drop my head in resignation again.[11] The Lord did not make it easy for us!

I also have heard about the American measurements and discussed them with Freundlich.[12] They do not seem to prove anything yet.   No flawless measurements have been carried out yet on terrestrially generated lines; the electrical arc used up to now is unsuitable. We are now in the process of procuring or begging for the funds for an electrical furnace that generates flawless lines *thermally*. Only in this way will secure results be attainable. In a few years the verdict will be in.[13]

You probably know that I received a call to Zurich on the instigation of my friend Zangger. After agonizing vacillation, I declined.[14] I am so much indebted to my colleagues here that it would not have been right to go away, as fond as I am of my Zurich. Now that you write me that you are staying in Zurich, I find that it would be completely superfluous for me to go there as well. I am even more sorry though that it is not free will but a concern for your health that forces you to stay in Zurich.[15] For Zurich, on the other hand, I am glad that you are not going away. Young people can learn a great deal from you, not from me, since I personally am incapable of anything. I am one of those who ponders a lot but has learned nothing. I suggested that I come to Zurich twice each year for a month in order to hold lectures there. Now that you are continuing to teach there, my proposal seems ridiculous to me; I am going to write to Zangger in this vein as well.[16] Now that the semester has started,[17] I cannot easily come to Switzerland. However, if my always somewhat unstable health allows me, I shall go there at the beginning of February for at least four weeks; I am very much looking forward to being able to speak with you.

Cordial regards, yours,

Einstein.

627.   From Eduard Study

Bonn, 126 Argelander Street, 27 September 1918

Highly esteemed Colleague,

Your second engaging letter amused me very much as well.[1] Many thanks for it, likewise for kindly mailing me your little book, which I read immediately.[2] I knew most of it already, but still must study it more thoroughly later, of course. But why did you fear offending me? Any decent author is surely glad for reasonable criticism, from which he can learn. And such a very kindhearted one as yours, at that. I possibly did not deserve it at all, since I myself have sometimes been a downright malicious critic. This, at least, is the view of many of my colleagues in the field. *Vox populi, vox dei.* The majority is *always* right!

I abandon the statement that mathematics belongs within logic.[3] I myself do not understand how I could have written that. What I actually meant, I cannot say, I appear not to have kept any proofs of this paper. Presumably I only wanted to emphasize the contrast to empiricism.

However, I can*not* set the statement that the external world is real = cock-a-doodle-doo. If this statement makes no sense (as you say), then the antithesis: the external world is unreal, a *chimera*, also makes no sense. I do think, though, that this statement is understood by anyone who says: "The devil does not exist, the devil is a chimera." The content of such statements cannot be *explained*, but at most *clarified* (with familiar examples). You use such an explanation nonetheless in that you, as a pragmatist in the style of Poincaré (forgive me, but no one can avoid the isms!), set out from the correct comment that more can be done with the "real" than with the "unreal," and then want to apply it to the *definition* for the real. At least, that is how I understand you. But then you have the biological advantage as a criterion for truth or as a basis for proof, and with it the entire litany of pragmatism. What you call "offshoots" become, in my view, *the correctly drawn consequences* of your premise; I do not believe that this premise can be admitted and the consequences rejected.

And now I'll turn the tables. I did *grin*, as a matter of fact, when I read at the top of your page 7 "that orbits of themselves *do not exist*."[4] What does *"exist"* mean? Oh, it's senseless! Likewise p. 42, when you speak of "absolute physical reality." *I* allow you this, but *you* should not allow yourself this! Otherwise I'll come and say cock-a-doodle-doo!

I do not understand what you have against my concept of natural geometry and some other things as well. I shall save up your letter, of course, and when I come to Berlin sometime, I'll bring it along. Corresponding about such a thing, which after all is not urgent, I would perceive as an abuse of your kindness. I would prefer to chat with you now about your book.

Our difference of opinion, which is indeed not great, can be illustrated completely with the statement on p. 7 already commented on above. I find that the thesis of the theory of relativity is (1) *somewhat carelessly* written, (2) *dogmatically*, (3) *unclearly.*

1) All physical relations *that we know* of are relative. Having made this completely clear and having said what significance such a statement has is your great discovery. But then you assert something more: that an absolute place in space, in which the physical laws (known to us) do not cease to hold, also "does not exist." As long as radioactive bodies were not found, anyone who would have

wanted to suspect such a thing could, with about equal legitimacy, have been confronted with the statement that such a thing obviously "does not exist."

2) Even given that it was certain that no fact could *ever* be discovered forcing us to attribute physical importance to an absolute place in space, surely it cannot be forbidden to represent the facts in such a language—which is possible; therefore, presenting the new theory of relativity in approximately the same way that *Lorentz* had presented the old one.[5]

3) The assertion: *There is* no such a thing as an absolute place (or space), really appears senseless to me. How can one presume to make a statement about something about which we admittedly know nothing?

These are my basic reservations; I do not regard them as serious, especially considering that in the history of science it has rarely been the case that discoveries of great consequence were not exaggerated. I find that in your representations, as far as I understand them, *what counts* is in the best of order. Not even elegance do I miss in your style. I absolutely do *not* agree with Boltzmann, however, concerning cobblers and tailors.[6] The "elegance" of the formulas in the theory of invariants, for example, is a *characteristic of natural laws* whose existence must not be ignored. To me, Boltzmann is unreadable. If he has a symbol for x, he then needs another one for πx and one for $\pi^2 x$ and a third one besides for $\frac{3}{2}\pi x$. That is simply sloppiness and totally insufferable. Nobody has the right to publish papers in the draft stage.

I also have a few misgivings about the conclusion to your exposition.[7] In a [quasi-]spherical space, the [approximate][8] function $\frac{c}{\sin^2 r}$ would have to take the place of the function $\frac{c}{r^2}$. Or am I wrong? If this were the case, though, if furthermore gravitation needs time for its propagation, and third of all, if no screening effect is present, then very strange consequences would result from the convergence of the force lines at the antipodes.[9] A body moving in such a way would have its mirror image at the antipodes, of which only the gravitational field would be verifiable. This mirror image would have a second in the proximity of "the body itself," this one a third image again at the antipodes, etc., *ad infinitum*.[10] Could one then also arrive at knowledge of the physical laws? Your little essay contains no mention of these difficulties (to which others are added in an *elliptic* space), and even less is said about how you come to terms with them. Shouldn't this be clarified? If the spherical hypothesis were tenable, the flare-ups of new stars could probably be explained with it.

Who is *Schlick*? A physicist or a philosopher? I find his little book very charming, but he treated me miserably in it. All that the reader learns about my book—which could, of course, have remained completely unmentioned—is that I had *naturally* misunderstood Riemann and Helmholtz.[11] It is completely beyond him that he could have misunderstood various things himself. For inst., he thinks that the world lines are the shortest (rather than the longest) lines. I was very annoyed about this gentleman and was thus doubly pleasantly surprised upon receiving such an amicable letter from you.

I am also sending you my lecture on kinematics.[12] But you really are on my mailing list. Should you happen to find the other copy, please be so kind as to return one. I hope that you will take a glance inside, it is one of my best works, the content *perhaps* even useful in physics, if only in the broader consequences.

So, now I really have written you a long letter, and you will be glad that it is now coming to an end.

With best regards, yours very truly,

E. Study.

628.   To Friedrich Adler

[Berlin,] 29 September 1918

Dear Friend,

Many thanks for the nice and concise letter.[1] The good part about these grim times is that they compel brevity. So let's go! The reflections of §38[2]—apart from the introductory criticism—is, in my view, thoroughly correct including the resulting final equation (126). But you are not allowed to offer the choice between either equation (126) *or* (113). For, since the constant a can still depend arbitrarily on u, both equations are *entirely equivalent*. Your criticism at the top of page 221 would be valid only if the ξ'''s signified lengths *measured in S*, which is not the case, however: $\xi' - \xi'_0$ is the difference between the two abscissas measured in K, relative to a specific time τ of K (that is, a time common to both points).[3] I fully maintain my earlier consideration, although I must admit that it was a bit more impractical.

Now the remaining main issue is the determination of the factor l in the equations

$$\left.\begin{aligned} x' &= l\beta'(x - ut) \\[2mm] y' &= ly \\ z' &= lz \\ t' &= l\beta\left(t - \frac{u}{c^2}x\right) \end{aligned}\right\} \tag{1}$$

You unjustifiably contest the accuracy of my rod consideration. It is completely correct when the two assumptions

1) principle of relativity
2) isotropy of the physical space

are taken as a basis. I do seem to have forgotten, though, to emphasize in the first explanation that the principle of relativity must be applied again at this place. The consideration is best clothed in this form:

A sphere that when measured at rest has the radius R must always have the same shape & size, measured from a coordinate system K relative to which it is moving at the velocity u, independently of the choice of system K. The static sphere relative to K' with the equation

$$x'^2 + y'^2 + z'^2 = R^2$$

must therefore have the same shape and size, seen from K, as the sphere at relative rest to K with the equation

$$x^2 + y^2 + z^2 = R^2,$$

seen from K'. This, translated into formulas with the aid of equations (1), necessarily requires $l = 1$. The consid. yields $l^2 = \dfrac{1}{l^2}$, whereby l should be positive.)

The situation naturally is different if you do not want to presuppose the principle of relativity or the law of isotropy. Then the problem reads as follows: Are there known observations, or are observations conceivable that can clarify this? In any event it is clear that the choice of l is not merely a matter of formal convention but is a realistic hypothesis. This hypothesis determines, for ex., the form of the electron in connection with velocity and thus also the dependence of electromagnetic mass on velocity. Thus, for instance, for a while Bucherer advocated a theory that boils down to another choice of l.[4] But now that the laws of motion of the electron have been verified with great precision, a different choice of l is out of the question today.[5]

A decision between Lorentz and Einstein is impossible, anyway, since *factually* Lorentz's theory agrees entirely with the special th. of rel.; it is just a more specialized (exclusively electromagnetic) theory.–

Now to the clock problem. This paradox is solved, from the point of view of special relativity theory, as follows. If U is permanently at rest relative to a Galilean reference system while U' is describing a circle relative to it, then U' lapses behind U even though the clocks are of the same construction and even though—looked at kinematically—U thus likewise describes a circle relative to a system K' that is rigidly connected to U'. Only if systems K and K' were both *justified* systems in the sense of the special principle of relativity would there be a contradiction with the (special) principle of relativity; they are not both justified, however, *because K' is not a Galilean (acceleration-free) reference system.* Only when general relativity is taken as a basis are both frames of reference equivalent. In this case, the difference in the rates of the clocks is explained by the combined effect of the influence of the velocity and *the gravitational potential.* Nowhere is a contradiction evident (Berg and Petzoldt are thus mistaken).[6]

Finally, to the galvanometer experiment.[7] Here your bias for absolute time or for the instantaneous signal is exposing itself. *Under no circumstance* will a current flow if the contacts made are so brief that their duration is small against the propagation time of light or of an electrical wave between the two contact points. The negligence of the continuously distributed capacity, which determines the finite propagation velocity of the electrical waves along the rods, must not be disregarded here, in principle. However, your example can be replaced with a similar one. Two rods of equal length when observed at rest have an orthogonally projecting arm at both ends which are each rotatable around the rod's axis. At the middle of each rod is the push-button of a signaling device that can make the two arms rotate in such a way that, from the rod's point of view, both arms are simultaneously rotatable. The paradox then occurs that "at the instant of meeting" each rod can prove, by means of its arms, that it is longer than the other. The paradox disappears immediately when one considers that—judging from the point of view of the other rod—the motions of a rod's two arms are not simultaneous.

With cordial regards, yours,

Einstein.

Author's marginal note: "I am keeping the manuscript for the time being, so that you can refer to it again."

629.   To Friedrich Adler

[Berlin, 30 September 1918][1]

Dear Friend,

The editor of the *Naturwissenschaften*[2] is pestering me for lack of manuscripts, saying I should give him something to print in his journal. So I have in mind to write an article in dialogue form on paradoxes in the theory of relativity.[3] You have provided a particularly suggestive form for one paradox with your example of the rods equipped with contacts;[4] may I make use of this, whereupon I would refer to your forthcoming work?[5]

I hope you have received my letter, which was sent out yesterday.[6] With best regards, yours,

Einstein.

630.   From Felix Ehrenhaft

Vienna, 3 October 1918

Dear Colleague,

Upon returning to Vienna a fortnight ago from an excursion to Switzerland, I discovered your kind letter postmarked 20 August[1] as well as your postal card stamped 23 August 1918. I beg your pardon for having neglected for so long to answer the questions posed in your letter, but I was very busy.

The fact that the air in the condenser is not brought into motion by the electrical field is elucidated by the circumstance that the velocity of fall of un-charged test specimens remains precisely the same when a random voltage is applied to the condenser plates or when the test specimen is allowed to fall with the condenser short-circuited. We performed such experiments already in 1910 in Vienna. Hence not the slightest motion in the gas itself arises from the electrical field, so the conclusions about the rationality or irrationality are not obscured by it in any way. Thus your hypothetically proposed gas motions from the electrical field play absolutely no role, because they do not occur.

As far as the nonagreement you mentioned between theory and experiment for Brownian motion is concerned, we on our part have made clear (: Konstantinowsky, 1915 *Annalen* :) that the $\sqrt{t}$-law is confirmed by the experiments.[2] By contrast, we have had no success until now with the mobility calculation from Brownian motion, as you can gather from my recent papers of 1918 as well as from the paper by Dr. Parankiewicz (: *Annalen* 53, issue 15, p. 564 :).[3] That is why I also wrote on p. 71 of my paper "Über die Teilbarkeit" ["On the Divisibility"], "this course will require further investigation."[4] This investigation is now in progress. We have succeeded, in particular, in subjecting one and the same test specimen to a series of measurements at various gas pressures. You write in your letter, "The Brownian motion appears 'smoothed out' to the observer. It appears to me that it is not the momentary position of the particle that is being registered, but only a certain temporal mean value." I do not quite understand this concept of smoothing out, since the theoretical formula is, of course, valid only for the visible Brownian motion. Perhaps your time will permit you to inform me of this in somewhat closer detail; for further clarification in this regard I would also like to refer to *Annal. der Phys.*, vol. 46, p. 288, where for the same test specimen observed successively over various distances (: dropping distances :), the same $\overline{\lambda^2}$, and therefore the same mobilities, resulted.[5]

With respect to photophoresis, you yourself note that the phenomenon's independence of gas pressure and the unique correlation between the velocity of fall, color, and photophoretic velocities provide the best proof that the observed phenomenon is not disturbed by gas currents. The fact that the light ray introduces no horizontal motions in the gas is also supported by the existence of optically neutral particles of matter.

Recently, we succeeded in measuring *the same* particle at a variety of gas pressures, whereby a complete constancy of the light-positive or light-negative forces, and thus independence of the gas pressure, resulted. Furthermore, new analyses have demonstrated that the effect is also completely independent of the gas's chemical properties. Thus the photophoretic energies obtainable from the following table[6] [*Vide* 6th page.] resulted for light-negative red selenium particles in argon, nitrogen, and hydrogen. (: If gases are seen as atomistically structured, it also seems to me very implausible anyway that the closer proximity of the test particle, which is at most of the order of magnitude of the molecules' mean length of path at low degrees of evacuation but is already much smaller than the mean length of path, should exert directed forces on the test specimen. :)

The proofs I presented seem to me to have determined unequivocally that a direct effect of the radiation on matter is involved. As concerns the motion of the plates, this has not yet been analyzed but is on my program for the immediate future; only after performing such experiments do I want to draw the relevant conclusions.

Now like you, esteemed colleague, I would like to return again to the electron. Regarding the system of hypotheses you have put forward for the electron, up to now I only went so far as to say that all these argumentations contain no direct charge determinations that could conflict with the charge determinations for the single test specimen.[7] Also, the entire system contains no proof against individual test specimens, as they are encountered at measurement, having arbitrarily small charges. Once one has clarified the existence of such small charges, the question of whence these grouped charges come will have to be raised in the future, provided that grouped charges are thought to have been identified in various phenomena in nature.

Incidentally, your proposed theoretical arguments against the relevant theory are eliminated, since in order to maintain e and h, Maxwell's electrodynamics would have to be abandoned, without being able to replace them yet with new ones.

Concerning the paper by Dr. Bär, which should be appearing soon in the *Annalen* in a somewhat more complete text, I had the opportunity of studying closely the experimental arrangement in Zurich[8] and have come of the view that it is far from precise enough to have a say in such questions, owing to a deficiency in their optics, and a deficiency in the voltmeter readings (: the voltmeter and the particles are handled by the same observer, which seems completely unacceptable because, for an objective judgment, the observer must attend exclusively to the particle while another observer must operate the voltmeter :).[9] If my assessment is correct, then the intellectual author of this paper, Professor Edgar Meyer, who incidentally already conceded in this paper, among other things, my and Konstantinowsky's principles with regard to the narrowing down,[10] seems to have adopted my view because he intends to have the experiments taken up again with one adapted to my arrangement, which might perhaps be facilitated a bit in that Dr. Bär has asked me to arrange an opportunity for him initially to study the matter closely in Vienna. Another result of the discussion in Zurich was that the published series of measurements claimed by Dr. Bär also do not speak nearly as consistently and flawlessly for such a simple integer view as might appear at first reading. The remaining unpublished records have been promised to me. A transition to the appropriate optics will clarify the whole situation there as well. You recall that in 1914 I indicated the proportions for the charges that are even simpler than all of Bär's series of measurements;[11] I remember the series

$$+2 : +3 : +2 : +1 : -2 : -3 : -2 : -1 : -2 : -1,$$

and yet they are meaningless, because the jumps from one particle to the next are varied and because a more accurate narrowing down requires more complicated considerations.

In Zurich I spent some very stimulating hours together with Mr. Besso, from whom I received a ⟨lengthy⟩ postal card on physics today as well.

If I may write an honest word about all these affairs, in the electron problem of today almost the most interesting aspect for me is repeatedly seeing how people so easily prove what they wish to see fulfilled. The most important principle for the experimental physicist is objectivity, though, otherwise all his effort is worthless.

Accept my very cordial regards, yours truly,

Ehrenhaft.

Selenium

Gas	Color	Vel. of Fall $\cdot 10^3$ cm/sec	Radius $a \cdot 10^6$ cm	Photophor. Veloc. 10^3 cm/sec	Moving $B \cdot 10^{-7}$	Photophor. Force $\mathfrak{P} = \frac{\mathfrak{B}}{B} \cdot 10^{10}$ Dyne
Argon	rot	1.53	15.30	11.72	2.42	4.85
Nitrogen	rot	1.91	15.26	14.37	3.05	4.70
Hydrogen	rot	5.16	15.15	38.68	8.09	4.78

631.   To Pauline Einstein

[Berlin, 8 October 1918]

Dear Mother,

Now you had to get the loathsome flu as well. I hope it's over with now. Margot also is lying in bed with backache & fever; she probably has it as well.–[1] I remind you now of our conversation at Prager Place[2] two years ago, and think about whether I hadn't seen correctly after all! And we're not out of the woods yet. There was nothing special to say about myself except that my health is good. The day after tomorrow I am starting my lecture course.[3] I'm producing nothing to speak of at present, without loafing about either. Recently I made the acquaintance of the chess master Lasker, a small, fine little man with a sharply cut profile and a Polish-Jewish, yet genteel manner. He has been world champion in chess playing for 25 years and is a mathematician and philosopher to boot. He stayed contentedly seated until 12 o'clock, even though a great tournament awaited him the next day.[4] Rathenau was also there and sparkled both in wit and eloquence. The last essay ("Charakter") of his most recent little work to appear *An Deutschlands Jugend*[5] [*To Germany's Youth*] is well worth reading.

Affectionate regards to all, yours,

Albert.

632.   From Friedrich Adler

Stein-on-the-Danube, 12 October 1918

Dear Einstein,

Many thanks for your letter of 29 September and the postcard of the 30th.[1] I received the postcard of 3 September as well, but only very late because you wrote the address very unclearly and it therefore wandered about at other places.

The same must have happened with the relativity brochure announced in it, for it has not come in, to this day.[2] I suspect it constitutes a third edition of the writing from the Vieweg collection. Although I have the first edition, I naturally would be very interested to know what new things were added.

If you can make use of my example with the contact rods, it would please me very much.[3] I just want to add that if the rods lie in two equivalent systems and, seen from their "symmetry system"[4]—with which is meant a system that is purely kinematically *symmetrical* to them—the contact takes place *simultaneously* for both ends, which is immediately clear from the consideration without need for calculation, of course, then both rods are subject to *the same* Lorentz contraction, seen from the symmetry system, hence are "equally long," which is naturally confirmed by the calculations. The intersection of the [two][5] edges x_2 and x'_2 occurs *earlier, later, and simultaneously*, just as for the edges x_1 and x'_1, depending on whether observed from system S or S' or from symmetry system K. The question is, from which system does the simplest description result? From the symmetry system, we can retain the old theory of closed currents. For the system times in S and S', we are impelled to a theory of unclosed currents. There is a good analogy to this from the spatial perspective. We have three equally long rods a, b, c which we arrange parallel to one another on a plane so that the distance between any two is d. Then for the observer who stands at a, c is shorter than b, and this latter is shorter than a. For the observer at c, on the contrary, c is the longest rod and a the shortest. For the observer on the symmetrical plane at b, a and c are equally long. These facts from the spatial perspective are "real" and incontestable. Nevertheless, I can list a series of *physical* experiments, according to which the three rods are of equal length. This is entirely analogously valid for the consideration of your theory of relativity from the *temporal perspective*. The propagation velocity of electrical waves can be taken into account without altering the crucial symmetry argument, although it does get a bit more complicated. It then depends on the theories from which one sets out. Let us very simply assume, for inst., that an electrified particle is moving from the positive pole of the element to the negative one at velocity c; the "circuit" must thus be closed for at least the time $\dfrac{3L}{2c}$ so that this particle can pass by both contacts. (Where L is the length of the rods.) If D is the width of the contact, then necessarily $D = \dfrac{3L}{2} \cdot \dfrac{v}{c}$, hence the contact width is of the order of magnitude $\dfrac{v}{c}$, whereas the shortening owing to the Lorentz contraction is of the order of magnitude $\left(\dfrac{v}{c}\right)^2$. It can now be shown that, seen from the symmetry system, a current is formed with this minimum contact width but not when seen from S and S', since through the Lorentz contraction the time of contact became too brief. It is also my opinion that a decision on constant l in the transformation equations is, in principle, possible through experience; by no means is a mere convention involved here. What I maintain is just that the experiments by Fizeau and Michelson cannot bring about this decision but rather leave open the quantity of l. Therefore the question is whether *other* experiments can decide the question. Now, it would interest me very much to find out *which* experiments you view as the finally decisive ones on the laws of motion of the electron.[6] For, as far as my knowledge of the literature goes, nowhere did I find any statement about a final decision. It would be possible, though, that another publication I do not know about succeeded Hupka (1910).[7] Hupka says (*Annalen, 33*, p. 402): "I would like to emphasize again that the last word has not yet been spoken on this subject." Likewise, Laue says in his book (2nd ed., 1913, p. 18),[8] that the "opinions on the conclusiveness of the experiments by Bucherer, Ratnowsky,[9] and Hupka are so divided that the theory of relativity has certainly not yet obtained unconditionally reliable support from this quarter." Lorentz likewise says in Hinneberg, *Kultur der Gegenwart* (Physics, p. 332), thus still in 1915: "Various experiments along this line have been performed; however, final and concurrent results have yet to be obtained."[10] So it would interest me greatly if you could indicate to me on which papers you base your opinion that the problem is solved, and specifically, I would be grateful for an exact bibliographic reference, since the local library unfortunately does not have the *Annalen* and I consequently cannot research it.

Unfortunately there is not enough space anymore today for the other issues; I shall thus seek to show you another time that you have two meanings for ξ' when applying it in the derivation of equation (113).[11] And on the other hand, you should clarify for yourself sometime the fact that you do not maintain the clock paradox in the way it appears in the *first* paper,[12] since *there* it is deduced from two *justified* systems! Cordial regards, yours,

Fr. Adler.

633.   To Edgar Meyer

[Berlin, after 12 October 1918][1]

Dear Mr. Meyer,

I thank you sincerely for the kind news. Obviously, I did not expect any reply to my proposal yet, because the holidays had already arrived in the meantime.[2] At all events, I would be pleased if I were bound by real ties to dear old Zurich schoolmastership again like that, so that I do not have to feel like a renegade deserter.

I am writing you today about something else, however. Namely, I still have little stomach for your recent result on the fluctuations of ionization by γ-rays, and for theoretical reasons I am *firmly convinced* that the statistical dependencies you have found in the two cases

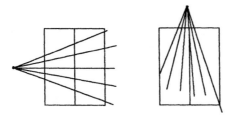

were based on secondary β or γ-rays that had entered from one chamber into the other.[3] The theoretical basis for my view is the following. The energy that appears during elementary γ absorption is of the same order as the energy an electron must have in order to generate the γ or Roentgen impulse through deceleration. Therefore, each radioactive elementary decay process probably corresponds to a single absorption process. The decay of *one* radioactive atom can thus never correspond to an elementary absorption process *in both of the two chambers*. So any sort of statistical relation between the elementary processes of absorption in both chambers is out of the question, whatever the particular theory behind it may be.

I would now like to ask you very earnestly to take up experimenting on the subject again;[4] you would really do a service for a good cause. There are—as far as I can see—two ways open. Either one varies distance a between the two chambers and examines whether the fluctuation dependence changes with a,[5]

aperture

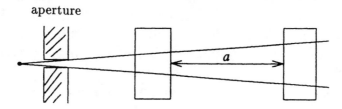

or one compares the case concerning the dependence

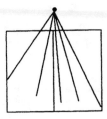

with the case

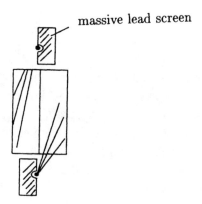

massive lead screen

where each chamber is irradiated by a separate preparation.

To me the method of varying a seems most natural, because no screens are needed, which produce secondary γ-rays. But obviously I know very well that it would be ludicrous for me to presume to give you suggestions on the *method*. It is only to tell you concretely what observations would be needed.–

The experimental resolution of the problem of whether each decaying atom really does correspond to *one* act of γ-absorption could, in principle, also be settled observationally with fluctuation measurements using γ-rays. It would be magnificent if you ventured to tackle this as well.

The paper by Bär on the deceleration potentials interested me very much.[6] This muddle really ought to be put straight sometime once and for all. (Elimination of this physical Bolshevikism).

Cordial regards, yours,

Einstein.

634.   To Hans Albert Einstein

[Berlin, 17 October 1918]

My dear Albert,

Prof. Zangger's information was out-of-date. Now I am very happy with you because you are writing to me more often.[1] How is it that the holidays shriveled up into two miserable days? The confounded flu is on the loose again;[2] it's uncannily rampant here as well. Unfortunately, the money was not sent for so long through the fault of the bank.[3] I have complained at the bank now. If you all don't have anything left, you can withdraw interest from the 40 000 M in the meantime.[4] By the time you get this postcard you will have received the money, I hope.

Now I am definitely coming in February, provided I am healthy. From now on, I am probably going to be in Zurich for two months every year, in order to give regular lectures.[5] Then we'll see each other more than before. That will be nice! But I am going to have a lot of work to do each time, since I shall have to read three lectures a week. I hope Tete does not catch the flu. How satisfied are you all with his health? What do you two want from me for Christmas? Now there will be peace, I hope. Then a Wall of China won't exist between us anymore and letters will arrive quickly again.

Warm regards to you all, yours,

Papa.

635.   To Nobel Committee for Physics of the Royal Swedish Academy of Sciences

Berlin 4. 30, 5 Haberland St., [before 18 October 1918][1]

Roy. Academy of Sciences.
Nobel Committee for Physics.
Stockholm.

With reference to the inquiry you sent me regarding the Nobel Prize in Physics for the year 1919, I suggest herewith that the prize be awarded to Prof. Max Planck for his achievements in the area of thermal radiation, and more specifically, for the two papers: "Ueber das Gesetz der Energieverteilung im Normalspektrum" ["On the Energy Distribution Law in the Normal Spectrum"] (Drude's *Annalen*, pp. 553–563, 1901);[2] "Ueber die Elementarquanta der Materie und der

Elektrizität" [On the Elementary Quanta of Matter and Electricity"] (Drude's *Annalen*, pp. 564–566, 1901).[3]

With this work, the named author not only provided the first exact determination of the absolute size of atoms[4] but, in particular, also laid the foundations for quantum theory, which only recently was revealed to be so eminently fruitful for the whole of physics. Planck's theoretical accomplishment also forms the basis of *Bohr's* theory of spectra and thus constitutes the most highly significant advance made in recent years in our knowledge of atomic structure.

In utmost respect,

A. Einstein.

P. S. I do not have the mentioned articles at hand. However, I hope that the validity of my proposal does not suffer from the fact that I cannot enclose them herewith, especially considering that the referenced articles are available at any physics library.

636.   To Friedrich Adler

[Berlin,] 20 October 1918

Dear Adler,

Three recent papers exist that have *securely proven*, so to speak, that the relativistic equations of motion of the electron are applicable (e.g., in comparison with Abraham's).[1]

1) One by Cl. Schaefer and one of his students[2] [very precise measurements] according to the method by Bucherer (Published in the *Annalen*).[3]

2) One, where the applicability is indirectly deduced from certain anomalies of spectral lines (Sommerfeld and one of his students).[4]

3) One by Guye and Lavanchy following Ratnowsky's method (*Extr. des Archives* . . . Geneva, 1916, last quarter).[5]

The last article was the only one I could find among the chaos of my things! If mailing were not so damnably inconvenient, I would send you the separate. My popular little writing, which I wanted to send you, also came back.[6]

I did not use the example with the contacts in my little essay;[7] it would have become too lengthy. On the other hand, the clock paradox was discussed in detail, also from the point of view of the general theory of relativity. Substituting the current with an electron moving at the speed of light in your example is thoroughly acceptable.[8] You will certainly discover, though, that the condition relating to the moment of passing is independent of the choice of coordinates.

Philipp Frank in Prague will certainly be glad to provide you with the above articles if you write him a card.   They are worth reading *per se*. It is too complicated from here.

I energetically contest that in my first paper[9]I had used two different meanings for ξ'.[10] $\xi' = \xi - \nu(\tau)$, nothing else![11] [The consideration there is the formulation of the statement: The light signals with which the observer of the "moving" system registers their simultaneity, have—observed from the system "at rest"—propagation velocity c.] Hence a defined quantity seen from the system "at rest." I likewise entirely uphold what I wrote about the clocks in my first paper along with everything else I wrote there. It is a pity that we cannot discuss these things orally! I am curious who between us will be the first to manage to come and see the other. Who can know? Could I perhaps visit you, if I came to Vienna?[12] For the time being it is out of the question, for when the semester is over I shall be visiting my boys.

Cordial regards, yours,

Einstein.

637. From Edgar Meyer

Zurich, 20 October 1918

Dear Mr. Einstein,

Many thanks for your letter.[1] I just want to report quickly how matters stand here. Your proposal was before the faculty yesterday and was accepted, thus you will be with us every semester for about 5 weeks![2] I cannot tell you how pleased I am. But that goes for all of us here. Acceptance by the Education Director is self-evident, of course.[3] I also requested that it start this semester already, and I think that February will suit you well. At that time Debye also will be giving two lectures at the Physical Society,[4] so it will become a February of physics that could even bear comparison with scientific activities in Berlin. Since, furthermore, Weyl is remaining with us,[5] it really could turn out to be very nice indeed.

Thank you very much for the suggestion regarding the fluctuation measurements.[6] That has long been one of my future projects. The energy relations obviously cannot be understood from the couplings. But I intend to proceed somewhat systematically. You see, if the fluctuation measurements really are to be very clean, we would have to be able to take absolute measurements because, with all that relativization [*Relativerei*] (that is, not yours), things do occur which until now have not been completely clear. For this reason, this problem is *already in progress* here as well. Namely, Weyssenhoff is now examining the method systematically, first.[7] Although he is not very far yet, I do consider my ideas about it good ones. As soon as everything about the method has been cleared up, the experiments you desire should be performed immediately. Weyssenhoff's research is being undertaken with this in view as well because I have long been planning to analyze the relation between the number of incidences of atomic disintegration compared to γ-absorption; I have even received substantial support for this plan from the Solvay Institute. The work was drawn out first by the war and then by my many Zurich responsibilities. I do believe, though, that I told you once about the plan (with the application of β-ray fluctuations). In any case, I very definitely hope that when you are here we shall have a decent thing or two to tell you about.

I am very pleased that you were interested in Bär's paper. The complete communication must be in one of the next issues of the *Annalen*.[8] I think you will like it because all doubts about the "law of integers" are accordingly absolutely impossible. As a consequence of the preliminary communication in the *Archives*, Ehrenhaft was also here to see us.[9] We had much to discuss. I believe we are on the right track but are proceeding very cautiously and systematically as well. I have bought Ehrenhaft's entire exact arrangement, so that every last objection to another apparatus is eliminated. I am firmly convinced that the Brownian motion has played a nasty trick on him because of his method of narrowing down.

With warmest regards, I am yours,

Edgar Meyer.

638.   To Felix Klein

[Berlin,] Tuesday. [22 October 1918]

Highly esteemed Colleague,

I have already studied your article[1] thoroughly, and with sincere admiration, at that. You have cleared up this difficult issue altogether. Now everything is wonderfully transparent. I am very eager to see your new study; in my study[2] I did have to leave open some important proofs, as far as the closed world is concerned. It is obvious that your paper can be published by our Academy; it will be my pleasure and an honor to present your paper.[3]

The only thing I was unable to grasp in your paper is the conclusion at the top of page 8 that ε^σ was a vector.[4] Perhaps you will be so good as to fill in this gap for me, when you have an opportunity.

Best regards, yours very truly,

A. Einstein.

I was also especially pleased that you do not reject the t_σ^ν's.[5] At first, I had some trouble in understanding your equation (6). The trick is just that, with the variation method you prefer,

$$\delta\left(\frac{\partial g^{\mu\nu}}{\partial x_\sigma}\right) = \frac{\partial}{\partial x_\sigma}\underbrace{(\delta g^{\mu\nu})}_{p^{\mu\nu}} \text{ is true.}^{[6]}$$

639.   From Hans Mühsam[1]

Liège II. Military Base Hospital II/126, 24 October 1918

Esteemed Professor,

As you see from the dateline, I have joined the general movement and am concentrated at a strategical position of retreat. How long we are going to stay here in Liège[2] also depends less on the opportunity to practice medicine as on Wilson. Isn't it despicable that thousands of young human lives depend on the haggling of the ruling powers, on their willingness to respond more or less quickly, and on their double-dealing intrigues?–[3] This wretched word "justice" has caused all this misfortune, the world over. One person defines the word from his point of view and another, who has a different worldview, defines it otherwise. And both sides perhaps do not even have the goodwill simply to define it subjectively correctly. If justice means equal rights for all, that is cause enough for the implementation of such a principle necessarily to give rise to quarreling and confusion. For the natural law of organized nature claims preferential rights for the strong. Although the Darwinian assumption of unlimited genetic variability cannot be upheld, the thesis of the survival and expansion of the better-adapted variety and, within this variety, of the fittest individuals, is undisputed. The Norway rat has supplanted the house rat, in violation of hereditary right, and among other animal species we see such phenomena by the hundreds. Thus also the intellectually less astute dark peoples from Oceania and America were driven out by the white race and more or less deliberately exterminated. This is a pity, but it is in keeping with natural law. The stronger, fitter, better adapted must come to his preferential rights; in the long run it cannot be withheld from him without causing severe conflicts. If the League of Nations, now being planned,[4] is to encompass all peoples or nations on Earth, in my view, it would be a mistake with severe consequences if all were supposed to be given the same rights such that none could rise above the general standard, win recognition of its superiority, and make gains, that is, draw from it the material conditions for its continued development. That would be a sin against nature and for that reason could not last. In my opinion, the League of Nations should restrict itself to the protection of the weak but should go only so far as to guarantee them an economic existence. Would it be right to grant Ecuador or Siam, for instance, the same—absolute or relative—amount of raw materials necessary for industry, in the form of colonies as, shall we say, Germany or Belgium? And as infeasible as it is to establish equal rights for all among the nations without taking into account their biologically diverse viability, simply in that the natural preference for the superior is reduced while the lesser rights of the inferior are arbitrarily increased, I can just as little accept the same social rights for all within a nation. The Social Democrats' attempts to even out differences go against nature. Protection of the weak also at the individual level must be limited to guaranteeing his physical subsistence. If every citizen is granted by the state the minimum of life's necessities, being food, housing, and clothing, regardless of whether or not he is working, if educational opportunities (schools, trade and art colleges, etc.) were made equally available to all, the state will have gone to the limit to which it may advance in protecting its weaker members without hindering the strong ones from assuming the superior place they deserve. It is unnatural and therefore senseless to give the same salary to the minister as to his porter, as the Bolsheviks are doing now. The superior must advance. Nature is aristocratic, not democratic.–

Obviously, I consider our autocratically feudal system, which has fortunately been eliminated now, just as nonsensical as the equalization sought by the Social Democrats. The nobility's tacitly made assumption, "our forefather was a valiant fellow; valor is hereditary, *ergo* I also am justified to assume a higher rank in the state than the masses," is nonsense, of course. Everyone must first personally provide proof of his superiority. But those who furnish this proof ought not to be artificially barred from social advancement.–

I believe that the world will not progress in steady and healthy development before the social fabric of the states and the coexistence of the nations are placed on scientifically secure foundations. Wars will not be preventable sooner either; no international agreements and guarantees, no matter how cleverly conceived, will guard against their return if their intention violates the natural law of higher rights for the more powerful.–

Now to something personal. Please do not take it amiss, esteemed Professor, if I meddle again, without invitation, in the treatment of your indeed happily withstood ailment.[5] Should your treating physician not already have pointed it out to you, I urgently recommend to you two things: (1) once a month— after 3 meatless days—have your stool examined for what is called occult blood. Prof. Boas[6] has the most experience in the detection of the slightest traces of blood in stool; his laboratory (Trantenau Street, at the Clinic) should be the most competent in this. (2) Have the functioning of your stomach checked roentgenoscopically every 2–3 months.—These are merely precautionary measures in order to determine in time whether the scar is not impairing the cardia's functioning[7] to the extent that the stomach is forced to exert itself in a way it cannot sustain over long periods.   When detected in time, the problem is easily dealt with.   If one waits until the stomach does not function properly anymore, however, and causes its owner to feel unwell, then the recovery lasts considerably longer.—Since such additional complications are very rare, I hope and confidently assume that normal conditions will be found every time.   As absolutely no discomfort is involved for you in either of the two examinations, I would consider it negligent if you did not have them done regularly anyway.-

Ca. 3–4 years ago I began to test your law of friction[8] ($\eta' = \eta\left[1 + 2.5\dfrac{v'}{v}\right]$) on real solutions, in order to check whether with its aid one can determine whether hydrated crystalline salts release their water into the aqueous solution or whether it is retained in the molecule. The experiments were conducted so that a measured weight of the relevant salt was added to a specific volume of distilled water and, using a Traube stalagmometer, the drip speed of pure water was compared against that of the solution.—These experiments cannot claim great precision because the volume could just be determined exactly to 1 cc and the temperature could not be held constant either. Furthermore, in part only approximate and in part conflicting data were available to me on the molecular and specific weights of some salts.   As a consequence of my so terribly limited time in Berlin, only a very modest number of experiments resulted; a portion of them I did not perform myself but had them conducted at my laboratory.-

I shall first indicate the observed figures, then the molecular volume calculated in the following way:

If 1 mole of a salt of molecular weight M contains . . . a molecules, then 1 g of this salt contains $\dfrac{a}{M}$ molecules and n g: $\dfrac{n \cdot a}{M}$ molecules.

Now, if these n g take up the space of b cc, then 1 molecule fills the space of $\dfrac{b \cdot M}{a \cdot n}$ cc. b is the observed volume or volume calculated from the experiment, resp.; n the number of grams of the salt contained in the solution.—I then calculated the quotient $\frac{b}{n}$ for each experiment.–

There is, first of all, Glauber's salt $Na_2SO_4 + 10\,H_2O$, about which I have arrived at the view, along other routes, that it releases its water of crystallization—which according to Werner's view is, as you know, bound to the central atom or to a lateral chain,[9] thus belonging to the molecular bond—entirely to the solvent water. It comes to the molecular weight of Glauber's salt 322.32, of which 142.16 goes to Na_2SO_4 and 180.16 to the 10 H_2O's.– I found the following data on solubility: 100 cc water dissolve at 0°: 12.17 g $Na_2SO_4 + 10\,H_2O$ and *5.02* g anhydrous Na_2SO_4.– Now, 12.17 g $Na_2SO_4 + 10\,H_2O$ contain .... 5.4 g Na_2SO_4 and 6.77 g water. If the latter are added to the 100 cc water used as a solvent, we would have 5.4 g Na_2SO_4 in 106.77 cc water or *5.05* g anhydrous Na_2SO_4 in 100 cc, which agrees very well with the above specification.

For 11°67 I found specified as the solubility:

$$Na_2SO_4 + \quad 10\,H_2O \ . \ . \quad 26.38 \text{ g} \quad \text{in 100 g water}$$
$$\text{''} \qquad\qquad \text{anhydrous} \quad \textit{10.12}\text{g} \quad \text{''} \quad \text{''} \quad \text{''} \quad \text{''} \ .$$

26.38 g $Na_2SO_4 + 10\,H_2O$ contains 11.64 g Na_2SO_4 and 14.74 g water. If the crystallization water escapes from the molecular bond and is added to the solvent, then 11.64 g Na_2SO_4 are contained in 114.74 g water, thus *10.14* g anhydrous Na_2SO_4 in 100 g water, which corresponds to the observed solubility.—The calculation yields accordingly at 32°73: 50.72 g the observation 50.65 g, and at 40°15: 48.89 g and 48.78 g Na_2SO_4, resp.– At 32°73 the Glauber's salt melts into its crystallization water and from this temperature on its solubility drops as well. At this temperature the saturation limit is reached at 322.32 g $Na_2SO_4 + 10\,H_2O$ to 100 g water. This number is just equal to the molecular weight of Glauber's salt. Therein, $\dfrac{322.32 \cdot 142.16}{322.32} = 142.16$ g Na_2SO_4 and $\dfrac{322.32 \cdot 180.16}{322.32}$ = 180.16 g water; hence 142.16 g anhydr. Na_2SO_4 in 280.16 g water = 50.72 g to 100 g water, which corresponds to the saturation value of anhydrous Na_2SO_4.

The behavior of the specific weight agrees with the same assumption. I found in this regard the following data: Specif. wt. of a

10%	solution	of	Na_2SO_4	$+10\,H_2O$	= 1.040
20%	''	''	''	''	= 1.082
10%	''	''	Na_2SO_4	anhydrous	= 1.093

In the 10 g $Na_2SO_4 + 10\,H_2O$ of a 10% Glauber's salt solution, 4.4 g Na_2SO_4 are contained and 5.6 g water. If the latter are added to the 90 g solvent water, then 4.4 g Na_2SO_4 exist in 95.6 g water. Since the specif. weight of the 10% Glauber's salt solution is 1.040, 100 g of such a solution thus has the volume $\frac{100}{1.040} = 96.15$ cc. Of this, 95.6 cc is water, so for the 4.4 g Na_2SO_4, 0.55 cc remain. The specif. wt. of anhydr. Na_2SO_4 would therefore be $\frac{4.4}{0.55} = 8.-$

The specif. weight of a 10% solution of Na_2SO_4 determined from this quantity is 1.095, which agrees well with the observed value 1.093.

Likewise, the determination of the specif. wt. of a 20% solution of Na_2SO_4 + 10 H_2O, interpolated in the same way from the 10% solution, yields 1.083; the observed sp. w. is 1.082.

The friction observations performed at my instigation by a chemist at the military base hospital, then on active duty as a soldier, refuse to agree with this at all, however. The data on the change in volume for the solution already are striking. Upon dissolving 10 g of Glauber's salt into 90 g water (at 15°), the total volume of the solution is supposed to have amounted to only 92.5 cc, whereas according to the specific weight, (1.040) it would have to be 96.15 cc.—Since the table leaves an unreliable impression otherwise as well and since it was not possible for me to come up with any kind of system behind it, I shall refrain from reproducing it.–

The behavior of the similarly structured *Epsom salt* $MgSO_4$ + $7H_2O$ is very interesting; specif. wt. of the dry substance 1.685, molar wt. 261 (of this 135 to $MgSO_4$ and 126 to 7 H_2O).—At 13°5 η for water 50.4 s.

2 g	Epsom salt in	98 cc	Water: vol of solution	ca. 99 cc	$\eta' =$	53.3 s.	Temp. 16°5	
4 "	"	96 "	" "	ca. 98 "	"	56.6 s.	"	"
6		94		ca. $97^1/_4$		59.1	"	"
10		90		ca. $95^1/_2$		64	"	"
14		86		ca. $93^1/_2$		69.5	"	"
16		84		ca. $92^3/_4$		75	"	"
20		80		ca. $91^1/_4$		85	"	"

If the difference in volume is entered into your formula as V', the observed values for η' come out quite well, if K is set equal to 5 instead of to 2.5.–[10] The factor for the molecular volume calculated under the same assumption yields

| | According to | |
	Friction	Vol. Increase
2%	0.56	ca. 0.5
4%	0.6	ca. 0.5
6%	0.55	ca. 0.54
10%	0.5	ca. 0.55
14%	0.5	ca. 0.54
16%	0.56	ca. 0.55
20%	0.62	ca. 0.56

Furthermore, according to these specif. weights of dry $MgSO_4 + 7 H_2O$: 0.59, and according to the specif. weight of a concentrated (44.44%) solution (1.243) 0.56.

Since all these figures, which agree among themselves to a certain extent within the probably considerable experimental error, have the same underlying assumption that the 7 H_2O's in the molecular bond are retained and are not added to the solvent water, as is the case for Glauber's salt, it would be very interesting to compare with them the relevant Glauber's salt figures on solubility and specif. wt. of solutions, which figures, unfortunately, are not available to me.

$MgSO_4$ *anhydrous*: spec. wt. ?– mol. wt. 135.–

Drip time for water at 15°–16°: $47.5^{\underline{s}} = \eta$.

2 g	$MgSO_4$ in	98 g	Water: vol. of solution	ca. $98^3/_4$.	η' :	50.4 s.	15°–16°.
4	"	96	"	ca. $97^1/_2$	"	54.5 s.	"
6		94		ca. $96^1/_4$		58	"
10		90		ca. $93^3/_4$		67	"
14		86		ca. $91^1/_4$		82.7	"
16		84		ca. 90		97	"
20		80		almost 86		114.5	"

The molecular volume obtains the numerical factor (for $K = 5$)

| | According to | |
	Friction	Vol. Increase
[2%]	0.6	0.375
4%	0.75	"
6%	0.71	"
10%	0.77	"
14%	0.96	"
16%	1.17	"
20%	1.2	0.3

Thus, whereas the molecular volume has very exactly the same quantity for all concentrations—except for 20%—when calculated from the increase in volume of the solvent, it is very dissimilar when calculated from the friction.–

If the specif. wt. of anhydr. $MgSO_4$ is calculated from that of Epsom salt, as for Glauber's salt, the result found is *4.7*, hence the molecular volume at 0.21; whereas a spec. w. of *2.67* would result from the solvent's increase in volume for this solution.—Whichever volume is inserted in the formula for V', neither with $K = 2.5$ nor with $K = 5$ do we succeed in obtaining the observed numerical values for η'. Well, I performed the calculation for the case that the volume of the $MgSO_4$ in solution is apparently enlarged by the entry of H_2O. The resulting behavior of the volume increase would then correspond to the addition of 2 H_2O's to the $MgSO_4$ molecule (molecular volume 0.38), the friction, however, to the addition of 4 H_2O's, or 7 H_2O's, resp.:

Solution % of $MgSO_4$	Upon Addition of	η' Calculated ($K = 5$)	η' Obs.
2%	4 H_2O	50.9 *s*	50.4
4%	"	54.7	54.5
6%	"	58.6	58
10%	"	66.3	67
14%	4 H_2O	74.8	
	5 "	79.5	
	6 "	84.5	82.7
	7 "	89.1	
16%	7 H_2O (Epsom salt)	94.5	97
20%	"	111.3	114.5

It would naturally be of great importance to come to know the spec. wts. of the Mg sulfates of differing crystallization-water content, and furthermore to know which sorts crystallize out of the solutions.

The fact that the high concentrations 16% and 20% exhibit a friction equaling the quantity that the dissolved amount of $MgSO_4$ would have as Epsom salt, agrees with the behavior of Epsom salt solutions, of which it had to be assumed, of course, that the crystallization water remains within the molecular bond.

Copper chloride $CuCl_2 + 2 H_2O$: Spec. wt. 2.4. Mol. wt. 170.8 (thereof $CuCl_2$ 134.5 and 2 H_2O 36.03). According to the one result, I found Cu 63.6, according to the other, 63.12; likewise for Cl, 35.45 on the one hand, 35.18 on the other. That is why I took as a basis for the calculations the total mol. wt. 170, of this $CuCl_2$ 134.

CuCl$_2$+ 2H$_2$O	Water	Vol. of Solution	Spec. Wt. of Solution	Drip Time of Solution	Temp	Drip Time of Water
2 g	98 cc	ca. 99 cc		55 s	15.5°	54
4 g	96 "	ca. 98 "		55.4	16°	53
6 g	94 "	ca. 97 "		55.8	16.5°	51
10 g	90 "	ca. 93.5 "	1.092	60	"	"
14 g	86 "	ca. 91 "		65	"	"
16 g	84 "	ca. 89.5 "		67.3	"	"
20 g	80 "	ca. 87 "	1.222	75.3	"	"
30 g	70 "	ca. 81	1.362	89	"	"
40 g	60 "	ca. 74.5	1.528	104.7	18°	50.3
50 g	50 "	ca. 68$^3/_4$		148.5	"	"

The specific wts. of the solutions are taken from data in the literature.

If η'_r means the friction, which is calculated out of V' = change in volume; η'_s is the one calculated out of $V' = \dfrac{\text{weight}}{\text{spec. wt. (2.4)}}$, then results, for $K = 5$,

%	η'_r	η'_s	$\eta'_{obser.}$
2	56.7	54.2	55
4	58.3	57.5	55.4
6	58.9	57.6	55.8
10	60.7	62.7	60
14	65.0	67.3	65
16	66.6	70	67.3
20	71.4	75.5	75.3
30	85.7	91.8	89
40	99.1	106.0	104.7

On the whole, there is thus no great difference between whether the volume is calculated from the change in volume or from the specific weight of the dry substance. In both instances, the result agrees sufficiently with observation. From this it would probably follow that also in solution the crystallization water remains within the molecular bond.–

The values I find indicated for the specific weights of the solutions I can reconcile neither with the dry substance's specific weight nor with the volume increase upon dissolving the corresponding weighed amounts of copper chloride.–

The molecular volume results from the dry substance's specific weight at 0.41, from the solution's increase in volume at 0.34–0.5 (at a mean of 0.40), from the friction at 0.3–0.415 (at a mean of 0.36), whereas from the spec. wt. of the solutions at 0.16; 0.09; 0.11; 0.14.

Anhydrous CuCl$_2$ yields the following figures: η for water 49$\underline{s}$.

Temperature throughout 15–15.5. The specif. wt. of anhydrous $CuCl_2$ is indicated as 3.05; the sp. w., derived from $CuCl_2 + 2\,H_2O$ in the same way as for Na_2SO_4 from $Na_2SO_4 + 10\,H_2O$, comes out to be 3.85.– From the former specif. wt. figure follows a molecular volume of: 0.328, [from the latter, 0.286].[11] But from the increase in volume for the solution follows a mol. vol. of only 0.12–0.23 (an average of 0.17). From the friction ($K = 5$), 0.35–0.5 (on average 0.41) results.

g $CuCl_2$	Water	Solution Vol.	$\eta'_{obs.}$
2	98	ca. 98	51
4	96	ca. 96.5	52.7
6	94	ca. 95	55.0
10	90	ca. 91.5	59.6
14	86	ca. 88.25	64.7
16	84	ca. 86.75	68.5
20	80	ca. 83.75	75.0
30	70	ca. 75	97.3
40	60	ca. 69	151.0

If the volume resulting from the specif. wt. is inserted for v' in the formula, the η''s then become substantially smaller than the observed values; this is even more the case when the observed volume differences are used.

However, if one makes the assumption that, in solution, $CuCl_2$ adds 1 or 2 water molecules, resp., then the η' figures following from the thus enlarged volume correspond very much better with observations. Depending on whether one proceeds from the specif. wt. of $CuCl_2 + 2\,H_2O$ (2.4) or from that of anhydr. $CuCl_2$ (3.05), the resulting spec. wt. for $CuCl_2 + H_2O$ is either 2.87 or 2.67. The following table contains the figures found in this way, whereby $\eta'_{2.67}$ incorporates those volumes following from the assumption of a $CuCl_2 + H_2O$ of spec. wt. 2.67.

%	$CuCl_2 + H_2O$		$CuCl_2 + 2\,H_2O$	
	$\eta'_{2.67}$	$\eta'_{2.87}$	$\eta'_{2\,H_2O}$	$\eta'_{obs.}$
2	51.1	51.0	51.7	51.0
4	53.3	53.1	54.4	52.7
6	55.6	55.2	57.3	55.0
10	60.3	59.8	63.2	59.6
14	65.5	64.5	69.6	64.7
16	68.1	67.1	73.0	68.5
20	74.0	72.5	79.9	75.0
30	90.7	88.2	100.9	97.3
40	109.3	105.8	124.5	151.0

For 2–20% the observed friction is in very tolerable agreement with the calculation following from the assumption of the addition of 1 H_2O molecule; at 30% the observed η' lies closer to the figure following from the addition of 2 H_2O molecules. η' at 40% is completely out of place; perhaps some material remained undissolved there.

From the specif. wt. of $CuCl_2 + H_2O$, 0.37 and 0.35 result, resp., as the mol. vol., which agrees very well with 0.40 following from η' for the 2–20% solutions; although the mol. vol. of $CuCl_2 + 2\,H_2O$ agrees with it even better: 0.41.

I had chosen copper chloride not only because it occurs both with and without crystallization water but, above all, because both types are, at the same time, soluble in alcohol.

The 99.8% alcohol used for the solution had $\eta = 75.5\ s$.

g CuCl$_2$ + 2 H$_2$O	cc Alcohol	ccm Solution	η'
2	98	98	80 s
4	96	almost 97	88
6	94	" 96	97.5
10	90	93.5	112
14	86	90.5	129 .

The mol. vol. would thus, according to the increase in volume, be ca. 0.3–0.35 but according to the friction ($K = 5$) ca. 0.96.

Unfortunately, I do not have the table for the alcoholic solutions of anhydr. CuCl$_2$.–

The molecular volume results consequently are

From Spec. Wt. Dry	Vol. Increase Hydrous	Friction (Hydrous)	Spec. Wt. of Hydr. Solutions	Vol. Increase Alcoh.	Friction Alcohol	For CuCl$_2$+ 2H$_2$O
0.41	0.40 } mean	0.36 } mean	0.125	0.32	0.96	CuCl$_2$
0.33	0.17	0.41	—	—	—	anhydr.

(0.36   CuCl$_2$ + 2 H$_2$O)

Although the differences existing in these figures are not substantial, considering the wide margins of error, one can nevertheless certainly conclude from them that CuCl$_2$ + 2 H$_2$O does not release all of its water, at least, to the aqueous solution and that water is added to the anhydr. CuCl$_2$ compound.

—o—

The important questions to ask in advance of all such experiments and conclusions would be, of course, whether the assumptions on which you, Professor, based your friction formula also apply to real solutions, furthermore, whether it is admissible to set $K = 5$ in this formula.

The questions regarding all these solutions interest me so particularly because I always think that only after a complete clarification of all aspects of the problems regarding inorganic solvents can there be some prospect of success in attempting the study of complicated organic liquids and because surely at least part of the immunity puzzle is of a physical nature. That is why I also beg you not to take offense, if I turn to you with a request for your advice and support.–

—o—

Unfortunately the institutes at the local university are closed; neither are any lectures being given, nor are any positions being made available. So I am forced to spend my generous amount of free time uselessly.–

Here in Liège there are no signs of the war. Everything can be bought— naturally, at horrendous prices. All the bars are overcrowded. There is much playing, dancing, and living. The Belgians are confronting us more and more impudently and provocatively by the day, so we have already been advised always to take our guns along when we go out in the evenings.—Do you have any requests? Should I procure for you coffee, ham, or anything else? The pound of butter sent off from the outskirts of Couvin[12] at the beginning of October has arrived safely at your home, I hope.–

I wish all goes well with you; please give my respects to your esteemed family, and accept my best regards to you personally, yours very truly,

H. Mühsam.

640.   From Max Planck

Grunewald, 26 October 1918

Dear Colleague,

Upon receiving your friendly letter this morning, I, in warm gratitude for your having thought of me, pondered long and deeply about whether I should subscribe to the public declaration you had enclosed.[1] You know that I am in full agreement with its content, at least in substance. But this is not what is at issue, of course, but whether the publicity will have the effect that the signers wish and imagine. And this question I cannot by any means answer positively with the conviction necessary to move me to overturn my resolution, made in the fall of 1914, of absolute restraint in future public statements about this war.[2] I could very well imagine, for ex., that such a publication could serve as a stimulus for our opponents to raise their demands against us even more than they are going to anyway; whereas on the contrary, there is the consideration that, by being pushed to the limit, we could inflict quite some damage on them and possibly have the effect of dampening somewhat their confidence in a victory. I lack knowledge of the actual circumstances to see this clearly enough. Only one thing is entirely certain: namely, that this step will elicit an agitated reaction on the other side, which could spoil the plans of our current government again which, in my opinion, is presently following a promising course.[3]

And I believe I can see clearly another thing as well and would gladly be prepared to do my part in being involved: it is that it would be extremely fortunate, a saving deed, if the bearer of the crown voluntarily renunciated his rights.[4] But in the word "voluntarily" already lies the impossibility for me to work in this direction; for first I think of the oath I took, and second I feel something that you admittedly will not be able to understand at all (if the Kjellén has not had an influence on you in the meantime), namely, a reverence for and an unshatterable solidarity with the State to which I belong, about which I am proud—and especially so in its misfortune[5]—and which is embodied in the person of the monarch.

I have written you more openly here than I tend to do otherwise and surely may ask you preferably not to let this letter fall into other hands. I am eager to speak with you about Kjellén's book sometime but also about all sorts of other matters remote from depressing politics.

Cordial greetings, yours,

Planck.

641.   To Felix Klein

[Berlin, 28 October 1918]

Highly esteemed Colleague,

Thank you very much indeed for your extremely friendly note. I still do not see the vector character of ε^σ.[1] For if the invariance of[2]

$$\int\int\int \sqrt{g}\begin{vmatrix} \varepsilon^{\mathrm{I}} & \varepsilon^{\mathrm{II}} & \varepsilon^{\mathrm{III}} & \varepsilon^{\mathrm{IV}} \\ dw^{\mathrm{I}} & \cdot & \cdot & \cdot \\ d'w^{\mathrm{I}} & \cdot & \cdot & \cdot \\ d''w^{\mathrm{I}} & \cdot & \cdot & \cdot \end{vmatrix}$$

were proven for any *arbitrary* hypersurface, then that conclusion could immediately be drawn. But as far as I understand it, proof is only given for *closed* hypersurfaces.[3] I thus certainly can see that

$$\frac{1}{\sqrt{g}}\frac{\partial(\varepsilon^\sigma\sqrt{g})}{\partial x_\sigma}$$

is an invariant,[4] but not that ε^σ itself is a vector.–

With amicable regards, yours very truly,

A. Einstein.

642.   To Lise Meitner

[Berlin, 29 October 1918]

Dear Miss Meitner,

Bär's notice is on its way to Regener[1] in the same post, but Meyer wrote me that he is already working on the revision of his fluctuation researches.[2] It's good that we did not get started.[3]

Cordial regards, yours,

Einstein.

643.   From Paul Bernays[1]

[Göttingen,] 2 November 1918

Esteemed Professor,

Forgive me for only today acknowledging receipt of your kind postcard as well as of the returned book. I wanted to add a couple of remarks at the same time and was not able to come to it until now.

It is quite a shame that Nelson's *Methodenlehre* [*Methodology*] appealed so little to you.[2] I believe I understand, to some extent, why it goes against the grain for you. It is surely that merely apparent lack of presuppositions in the method of introducing the ethical problems, whereby one is basically steered toward a rational ethics from the outset, without any indication of an even preliminary view toward life as a whole, by means of which one could orientate oneself intuitively.

Although I concede this to be an essential flaw in the work, I nonetheless cannot refrain from expressing my opinion, in view of my familiarity with Nelson's philosophical personality, that it would be doing an injustice to Nelson just to be willing to grant him subordinate talent and routine. It is true that a certain over-rating of dialectic argumentation often finds expression in his explanations. Yet such overestimation is found in many others, even among the greatest philosophers.

Besides, the talent for logical keenness and clarity (disregarding the fact that it really is very desirable in an academic teacher and is by no means particularly prevalent nowadays among philosophers) does not just have a bootblack's role in philosophy, perhaps, but rather, I believe that essential philosophical tasks exist that only a logically trained philosopher can accomplish. Such a task is, for example, the conceptual fixing of the manner of evaluation, which forms the basis of what is called, in the more restricted sense, the moral judgment of actions (as is exercised more or less consciously by an educated person).

This problem, which Kant was the first to pose in a precise form, was first solved satisfactorily, as it seems to me, in Nelson's ethics on the basis of Kant's approach (which had frequently been completely abandoned owing to the obvious shortfalls in Kant's more detailed expositions).

Surely you will not misunderstand me, Professor, in the sense of my wanting to convert you with my statements. It is just that, as much as I feel myself to be a pupil of Nelson's, I could not content myself with allowing your frankness to go unanswered.

Now I would like to ask you something else concerning physics. You put forward, in objection to Weyl's theory, the fact that one and the same type of crystal always occurs only with a very specific density.[3]

In a corresponding way, couldn't the fact be advanced against your theory that certain substances in a crystalline state always take on the same forms (as regarded in mineralogy), the defining elements of which do not by any means only occur in Riemann's geometry but also include rectilinearity in the Euclidean sense, parallelisms, etc.? Could it not also be argued that: if physical structures in the limited (Euclidean) sense have no invariant significance then, in general, the forms into which a crystal cleaves must be dependent on its prehistory, its momentary state notwithstanding?

Shouldn't the correct refutation of this last argumentation perhaps also provide, at the same time, a refutation of your objection to Weyl?

With my incompetence in the field of physics, I draw very much into consideration the possibility that I am speaking nonsense here. In any event, I would be very grateful if you would clear up this question for me sometime.–

Surely you also are hoping for an expeditious armistice?

Most amicable greetings, yours,

Paul Bernays.

644. To Edgar Meyer

[Berlin, 4 November 1918]

Dear Mr. Meyer,

I am very glad that the lectures are now materializing.[1] So I plan to be with you all on February 1st. I have already informed Planck, who assured me that no one has anything against it here. No less am I pleased that the statistical analysis of γ rays is already on your program.[2] Your handsome experiment on the generation of cathode rays appealed to me very much.[3] It is also good that Mr. Bär has taken verification of the Viennese results firmly in hand.[4] Ehrenhaft is still defending himself heftily.[5]

Cordial regards, yours,

Einstein.

645.   From Felix Klein

Göttingen, 5 November 1918

Esteemed Colleague,

The vector character of my η^σ (formula (11) of the new note[1]) can be derived through calculation, by working from the *way* the invariant K is *constructed*.[2] As follows (whereupon I note in advance that Dr. Vermeil[3] helped me very considerably with the complicated calculations):

1) The element Kp^σ in (11) is in and of itself a vector component and should be disregarded henceforth.

2) $\dfrac{\partial K}{\partial g_{\rho\sigma}^{\mu\nu}}$ is, because of the special nature of K, a term dependent only on the $g^{\mu\nu}$'s and hence a mixed tensor which may be called $K_{\mu\nu}^{\rho\sigma}$.[4]

3) In the third term of η^σ—which is now called $-K_{\mu\nu}^{\rho\sigma}p_\rho^{\mu\nu}$—we replace $p_\rho^{\mu\nu}$ with the corresponding covariant derivative:

$$A_\rho^{\mu\nu} = p_\rho^{\mu\nu} - (\Gamma_{\tau\rho}^\mu p^{\tau\nu} + \Gamma_{\tau\rho}^\nu p^{\tau\mu}).$$

The term that emerges, $-A_\rho^{\mu\nu} \cdot K_{\mu\nu}^{\rho\sigma}$, again has the desired vector character.

4) The component of η^σ remaining to be analyzed is:

$$\left(-\frac{\partial K}{\partial g_\sigma^{\mu\nu}} + \frac{1}{\sqrt{g}}\frac{\partial(\sqrt{g}K_{\mu\nu}^{\rho\sigma})}{\partial w^\rho}\right)p^{\mu\nu} - K_{\mu\nu}^{\rho\sigma}(\Gamma_{\tau\rho}^\mu p^{\tau\nu} + \Gamma_{\tau\rho}^\nu p^{\tau\mu}),$$

or arranged otherwise, by carrying out the differentiation:

$$p^{\mu\nu}\left\{\frac{\partial K_{\mu\nu}^{\rho\sigma}}{\partial w^\rho} + K_{\mu\nu}^{\rho\sigma}\frac{\partial \log\sqrt{g}}{\partial w^\rho} - K_{\tau\nu}^{\rho\sigma}\Gamma_{\rho\mu}^\tau - K_{\tau\mu}^{\rho\sigma}\Gamma_{\rho\nu}^\tau - \frac{\partial K}{\partial g_\sigma^{\mu\nu}}\right\}.$$

Here, now, according to Vermeil's calculation, the expression enclosed in braces is nothing but the covariant derivatives summed over ρ:

$$K_{\mu\nu,\,\rho}^{\rho\sigma} = B_{\mu\nu}^\sigma.$$

5) Now $B^{\sigma}_{\mu\nu}p^{\mu\nu}$ again has the desired vector character, which thus brings the proof to an end.–

I chose the designations $A^{\mu\nu}_{\rho}$ and $B^{\sigma}_{\mu\nu}$ in order to allow the close relation with formulas (8), (9), and (10) of Hilbert's first note[5] to stand out.—Hilbert's results agree with mine so long as $K + L$ are substituted for his H.[6] However, he provides these results in his note even before he specifies H in this way. There are strong doubts about whether the results are correct in this generalization— whether Hilbert had not originally developed them only for $K + L$ and had just put them in the wrong place later during revision.

Of course, I am eager to hear what you say to these considerations. My form of proof is admittedly not pretty, because the execution of (4) requires too much mechanical computation. But a scoundrel offers more than he has.

Very truly yours,

Klein.

646.   To Felix Klein

[Berlin, 8 November 1918]

Highly esteemed Colleague,

Thank you very much for the transparent proof, which I understood complete-ly.[1] The fact that it cannot be performed without calculation does not detract from your entire analyses, of course, since you do not make use of the vector character of ε^{σ}.–

In the whole theory, *one* thing still disturbs me formally, namely, that $\mathfrak{T}_{\mu\nu}$ must necessarily be *symmetric* but not $t_{\mu\nu}$, even though both must appear *equivalently* in the conservation law. Maybe this incongruency will disappear when "matter" is included, not only *superficially* as it has been up to now but *really*, in the theory. The logically so fine approach by Weyl, which would achieve this, unfortunately does not seem to me to provide the solution, for *physical* reasons.[2]

With thanks and my best regards, yours very truly,

A. Einstein.

647. To Mileva Einstein-Marić

[Berlin, ca. 9 November 1918][1]

Dear Mileva,

I agree fully to the suggested procedure. From the first of January 1919 on, you should receive Fr. 1 600 at the beginning of every quarter year, and the balancing of accounts should take place at the end of each year.[2] I sent you Fr. 1 800 now in the hope that you pay the life insurance with it as well. (The 2 000 M you claim are currently only Fr. 1 400,[3] hence there is nothing at all secure about it; it would therefore probably be better for you if the amount were fixed in fr[ancs] not in m[arks].) All will go well if my income is not substantially reduced by the consequences of the war. But I'll do all I can to ⟨maintain⟩ fulfill my obligations toward all of you completely. Be very careful with Tete so that his health is not undermined again. I would prefer him somewhere else for two months than at regimental Pedolin, though;[4] the atmosphere there did not appeal to me. Do make a few inquiries about alternatives. See that you accelerate our divorce so that the 40 000 M are transferred to your name;[5] that may possibly be crucial. Here also the flu is very potent and virulent; I have been spared until now, though.[6] I am glad that Albert has an intense interest in something. On what it is directed is less important to me, even if it is engineering, by God.[7] Children cannot be expected to inherit mentalities. I am going to be in Zurich throughout the whole of February and will give a lecture series. This will happen twice a year from now on. The negotiations on this are essentially closed.[8] I am just accepting the travel expenses for it, since the undertaking is supposed to be an acknowledgment on my part for an appointment at Zurich issued in the summer, which I refused out of consideration for my friends here.[9] Tell the boys that I would be very pleased to receive a few lines from them.

Best regards, yours,

Albert.

It would be better if we left open the question of what kind of certificate ought to be procured from the 2 000 M, since the conditions will all have been changed decisively by then.

Censor's note: "Please inform send[er] that henceforth letters without his adr. information on the envelope will not be delivered. Censor."

648.   From Heinrich Zangger

[Davos, ca. 10 November 1918][1]

Dear friend Einstein,

It was with a kind of joy, a kind of astonishment, perhaps a kind of almost jealous admiration that I thought so often of the passage in the postcard addressed to Davos where you noted in all thankfulness that the essentials in life are coming out extremely fortunately for you—because all things your easily satisfied nature can scarcely hope for are being realized at your hands with the same force as in the creative world of the development of knowledge.

The likes of us try to liberate what we have in ourselves during a few weeks of vacation. I read Weyl's *Kontinuum*[2]; saw him. Now multiple death has burst in on the student and doctor communities; yesterday we buried the 4th member of the university hospitals since my becoming dean,[3] others are seriously ill. The telephone is constantly ringing every day, from morning to evening. All cars have been requisitioned for transportation of the sick, the military barracks that we have requisitioned do not suffice at all.

As dean, the great responsibility falls to me; should I call for having the clinics closed as in the summer, against the rector's office's wishes, and dismiss the students as practical help?[4] If it goes on like this, we shall have a death rate of 10 % of the population in a single year.–[5]

Meanwhile, I myself had the flu, pneumonia pleuritis, and was sent to Davos (Davos Place—Villa Regina, English quarter). Besso came along; I still feel miserably sick and weak.

Regarding Teddy, at the *sanatorium, a healthy diet is guaranteed* right now,[6] otherwise it is very difficult because many guesthouses, etc., are closed. Thanks for the postcard.

649.   From Michele Besso

[Davos,] 10 November 1918

Our friend Zangger is urging me to write more. My trip was delayed repeatedly[1] and now I have followed my old wish of, for once, spending some days with our dear friend[2]—not without worries, since I left Anna still unwell and still a bit feverish and with shooting pain in her legs at every attempt to get up. Also, circumstances are critical even in Zurich, these days. We will probably have to count on a frenetic state of shock everywhere.[3] The economic leadership had counted much too much on a continuity in the state of affairs and had not the least perception that it was based on psychological grounds that had become increasingly unstable. People of the likes of Godin, Ruskin, Friese, Merton, and Abbe, who strove with all their might toward new grounds[4]—as was necessary within the foreseeable future for a worker-dominated population, even with tolerable management and without the convulsions brought on by war—remained solitary voices.   Today one would be prepared to sacrifice anything in that line, of course: but not only does it seem to be late for that, there is also a complete lack of psychological preparation. One is just as little acquainted with the functions that had sufficed for the foregoing social organs to be able to attend to the emotional conditions for a new creation and for a recognition of things in the making. This was certainly the case everywhere to almost the same degree—the advantage that West and East had with regard to directness in human relations, as well as the West's advantage with regard to the development of the communal will through the counsel of equals, are probably hardly adequate to prevent the encroachment of this movement. Perhaps, if we are still alive in three or four years, the time will have come when social and political development will be seeking intellectual crystallization centers and there will be something for us to do as well. In the meantime, erstwhile, like Archimedes, we are plagued by the missing red shift and, on the small scale, by the still open explanation of Ehrenhaft's experiments.[5]

Affectionately yours,

Michele

On the Ehrenhaft experiments: I initially thought I understood clearly that as soon as the velocities coming into consideration in Brownian motion—maybe $\dfrac{\overline{mv}}{M}$, not $\dfrac{\sqrt{\overline{mv^2}}}{M}$—become large against the directed motion, the mobility would rapidly become smaller. But I cannot succeed in fixing the idea quantitatively: it seems to be something more deeply seated than that.

16 November 1918. After a few days of stress, I am happily back in Zurich. May the regrouping in Germany come out well, which really can be of immense significance to the whole of mankind![6]

In your family home all is well and in good order; the children are in excellent shape. The day before yesterday little Albert came over for Vero's birthday.

17 November 1918: I just saw little Albert and Tete—in good health.

650.   From Felix Klein

Göttingen, 10 November 1918

Esteemed Colleague,

Your postcard of 8 November has just arrived.[1] Meanwhile, with Miss Noether's help, I understand that the proof for the vector character of ε^σ from "higher principles" as I had sought was already given by Hilbert on pp. 6, 7 of his first note,[2] although in a version that does not draw attention to the essential point. The specific structure of K is not used there,[3] rather K can be any invariant formed from the arguments coming into consideration. In the following manner:[4]

1) At the start, we cut away from ε^σ or η^σ, respectively, all components that recognizably have the desired vector character. There remains

$$\zeta^\sigma = (\quad)p^{\mu\nu} - \frac{\partial K}{\partial g^{\mu\nu}_{\rho\sigma}}p^{\mu\nu}_\rho$$

for investigation of its nature.

2) In any case, $\dfrac{\partial \zeta^\sigma}{\partial w^\sigma}$ is an invariant.[5] The highest term (a term with the highest differential quotients for $p^{\mu\nu}_\rho$) to appear in it is: $-\dfrac{\partial K}{\partial g^{\mu\nu}_{\rho\sigma}}p^{\mu\nu}_{\rho\sigma}$.

3) Now we have the obvious statement: $p^{\mu\nu}_{\rho\sigma}$ transforms (under arbitrary transformation of w) as a tensor, *if the terms with lower differential quotients are disregarded.*

This is enough to conclude that $-\dfrac{\partial K}{\partial g^{\mu\nu}_{\rho\sigma}}$ is a tensor, which now, as in my previous letter, will be called $K^{\rho\sigma}_{\mu\nu}$.[6]

4) Once we had made this finding, we set, again as in my earlier letter, in making use of the process of covariant differentiation

$$p^{\mu\nu}_\rho = A^{\mu\nu}_\rho + \Gamma^\mu_{\tau\rho}p^{\tau\nu} + \Gamma^\nu_{\tau\rho}p^{\tau\mu},$$

where $A^{\mu\nu}_\rho$ is now a tensor.

5) Accordingly, $-K^{\rho\sigma}_{\mu\nu}A^{\mu\nu}_\rho$ is a regular vector, which we eliminate. There remains: $\Theta^\sigma = (\quad)p^{\mu\nu}$ to be analyzed, which we immediately want to abbreviate, $= B^\sigma_{\mu\nu}p^{\mu\nu}$.

6) Then it's the same game anew as in (1), (2) . . . . :

a) In any event, $\dfrac{\partial \Theta^\sigma}{\partial w^\sigma}$ is an invariant.[7]

b) The highest term to occur in it is $B^\sigma_{\mu\nu}p^{\mu\nu}_\sigma$.

c) Except for the lower terms (which just contain factors $p^{\mu\nu}$), $p^{\mu\nu}_{\sigma}$ transforms like a tensor.

d) Hence $B^{\sigma}_{\mu\nu}$ is a tensor.

e) Hence $B^{\sigma}_{\mu\nu}p^{\mu\nu}$ also is a regular vector.

Finished!

This really is beautiful, isn't it, and seems to contain the principles of many more far-reaching considerations, which I have not thought through yet, though.

Best regards, yours,

F. Klein.

651.   To Pauline Einstein

[Berlin,] 11 November [1918]

Dear Mother,

Don't worry. All has been going smoothly, impressively even, up to now. The current leadership really seems to be equal to the task.[1] I am very pleased with the matter's development. Now I'm beginning to feel really comfortable here. The bankruptcy has done wonders.

All of us are well. We are healthy, and Haberland St. is peering[2] half curiously, half fearfully into the world. I'll be writing you more often now so that you don't worry. Among the academicians, I am some kind of high-placed Red [*Obersozi*].[3]

Heartfelt regards, yours,

Albert.

652.   To Paul and Maja Winteler-Einstein

[Berlin,] 11 November [1918]

My Dears,

The great event has taken place! I had feared some sort of breakdown in the order. But up to now the action has been taking a truly impressive course,[1] the greatest public experience conceivable. And the funniest thing of all is: people are adjusting themselves remarkably well to it. That I could live to see this!! No bankruptcy is too great not to be gladly risked for such magnificent compensation. Where we are, militarism and the privy-councillor stupor has been thoroughly obliterated.

By the way, 1st Maja, many happy returns of the day.[2] I am going to send you a little something as well, if it's possible.

All of us have been doing well until now, also healthwise.

Best regards, yours,

Albert.

Ilse, who is as red as they come, got into a little shoot-out and took to her heels.

653. To Leo Arons

[Berlin, 12 November 1918 or later][1]

Esteemed Colleague,[2]

Since I know that as an intrepid champion of free speech[3] you certainly cannot feel offended by the following, I may be completely frank.

Professors revealed during this war that *from them we can learn nothing* in political affairs, on the contrary that it is imperative that *they* learn *one* thing, namely: To

> **Shut up!**

That is why I cannot support your suggestion.[4]

In the hope of soon making your personal acquaintance, I send you my best regards . . .[5]

654. To Svante Arrhenius

[Berlin,] 14 November 1918

Esteemed Colleague,[1]

It is unusually thoughtful of you to send me greetings in these eventful days, without knowing me personally. For me as an old, convinced democrat and republican and a person almost fanatically fixated on rights, joy drowns out all other reactions of anxiety, especially considering that I am Swiss. I hope Europe recovers soon again somewhat from the terrible blow and the more refined feelings and interests soon will make themselves felt again so that life regains its more cheerful character. Then the economic pressures will not be able to make any decent person overly sad. It is quite remarkable how flexibly most have adjusted to such entirely new circumstances here now already; it is the most surprising experience of all the surprises.[2]

In heartily returning your greetings, I am yours truly,

A. Einstein.

655.   To Ludwig Quidde

[Berlin, 15 November 1918, 10:36 a.m.][1]

New Fatherland League strives toward realization of Socialism through Democracy. Wants to propagate[2] this idea among all population groups. Next goal is convocation of a constitutional National Assembly.[3] Request consent by telegraph to nominating you for representative Executive Committee.[4]

New Fatherland League, Berlin, 53 Unter den Linden, by proxy
Count [von] Arco, Professor Albert Einstein, Prof. August Gaul,
Wilhelm Herzog, René Schickele.[5]

656.   From Ludwig Quidde

Munich, 16 November 1918

To the "New Fatherland" League [*Bund "Neues Vaterland"*] Berlin.

I did not wire a response to your amic. telegram of yesterday[1] because I could not so concisely express my reservations, which quite precluded my simply consenting.

"The realization of Socialism through Democracy" is much too ambiguous a formulation for me to be able to agitate for an association whose entire agenda is summarized in these words. I do *not* want the uninhibited installation of Socialism in the sense of the program of Social Democracy; and to the extent that I do want the materialization of Socialist ideas, the safeguarding of personal freedom, that is, the preservation of individualistic needs, is to me of equal importance beside it.[2]

In the meantime, this morning I had occasion to read the newly founded Democratic Party's platform, whose appeal for support is signed among others also by Professor Albert Einstein, and by Helmuth von Gerlach besides.[3]

If "the realization of Socialism through Democracy" must be understood in the sense of this appeal, then I do not understand what business the "New Fatherland" League still has outside of this party. Then, programmatical reservations I would consequently not have, but the more so, doubts about an unnecessary and possibly damaging fragmentation.

Respectfully.

657. From Hermann Weyl

Zurich, 20 Schmelzberg St., 16 November 1918

Dear Colleague,

Here now follows my extensive reply,[1] which I hope will arrive in your hands soon despite the political upheavals and demobilization.[2] My own faith in my theory was also shaken, since I could not straighten out the problem of the mechanical equations for weeks and weeks on end.[3] My present *certainty* was thus not achieved without a struggle. I am in suspense about what position you will take now. After all the trouble that the 1st note has given you,[4] can you bring yourself to present this 2nd communication to the Academy as well? I would be very grateful to you for it. I chose the form of a letter to you because, while writing it, I actually did constantly have the feeling that I was speaking to you and was presenting my opinion to you. It would be preferable to me if, for simplicity's sake, you (or whomever you could entrust it to in Berlin) read the correction proofs directly as well. I hope the length does not exceed 1 printer's sheet!

Regarding my book *Space, Time, Matter*,[5] Springer informed me that it is nearly sold out (if there are people beating the publicity drum *so very* much in the *Naturwissenschaften* . . . it is ultimately no wonder);[6] I accepted his suggestion of arranging for another unaltered printing right away in small volume (600 copies),[7] although the last chapter is in dire need of improvement and additions.[8] But thereupon, as I hope, a completely revised edition will follow within a short time.[9]

I am very much looking forward to your coming in February!–[10] As far as can be judged from here, the signs do look good that Germany will be spared from chaos and terror. May it be so! Cordial regards, yours,

Herm. Weyl.

658.   To Arnold Berliner

[Berlin, before 19 November 1918][1]

Dear Mr. Berliner,

I shall be glad to give a report on Nordström.[2] The figures are incorrect. I shall sketch the correct figures with the request that they be incorporated into the two columns in which the two illustrations of the clock paradox are described.[3]

Best regards, yours,

A. Einstein.

659.   From Paul Bernays

Göttingen, 43 Nikolausberger Way, 22 November 1918

Esteemed Professor,

Thank you very much indeed for your kind letter in answer! In view of my too imprecise argumentation, I hardly deserved such a thorough reply.[1]

I am answering your letter only today because I first wanted to allow the substance of your explanations to run through my head for a while. These considerations I have been engaging myself in were not able to assist me with a certain difficulty, however, with regard to the concept of the rigid body, and I would like to take the liberty of interrogating you again on this account. It involves the following:

As far as I understand it, a body can count as rigid only when it is in "the same" state of rest at all times, given a suitable division of space and time, whereby "the same" means that *every* establishable property of the body that *can be expressed as invariant* remains unchanged. (Otherwise I would not know how, in the case where a Euclidean determination of mass possibly occurs twice at separate times, a return to the same state should be guaranteed.)

Now, the invariantly expressible determinations of a Euclidean body, of a cube, for instance, obviously include not only the angles and edge lengths but also include, for ex., that every pair of points of an interface can be linked together with a geodesic line lying completely on this surface (where "geodesic" relates to the three-dimensional space); furthermore, that the geodesic linkings of three points on the interface form a geodesic triangle with the angular sum π.

Now, it seems to me very doubtful, though, that for any given metric field the division of time and space can be executed in such a way that, in the relevant space (in which the rigid body would then have to be in a form of rest), the mentioned conditions can be realized.

If this is not satisfied, I do not see how the "rigid body" ought to be defined accordingly.–

Then I also have another question on my mind with regard to the facts of Weyl's theory.[2] Would it not be possible that, instead of $\int \sqrt{\sum_{\mu\nu} g_{\mu\nu} dx_\mu dx_\nu}$, the "length" were given by an expression $\int \sqrt{\kappa \cdot \sum_{\mu\nu} g_{\mu\nu} dx_\mu dx_\nu}$, where κ means an invariant function of x_1, x_2, x_3, x_4 against coordinate transformations expressible by $g_{\mu\nu}$ and φ_ν, which function is constant in the case where $g_{\mu\nu} = \delta_{\mu\nu}$ and which, upon introduction of Weyl's λ-factor, changes into $\text{const} \cdot \lambda^{-1} \cdot \kappa$? Then, of course, all length relations would be invariants in Weyl's sense, and we would have the usual length measurement for Euclidean metrics.[3]

Given, for inst., the world were composed in such a way that in it a point O could be chosen so that the geodesic lines starting from O covered the world completely and simply (except for O itself, of course), then the expression $e^{-\int_O^P \sum_\nu \varphi_\nu dx_\nu}$ in which the integral extends over the geodesic link between O and P, would be a function κ of the coordinates of P, which has the required properties.[4]

Hilbert, from whom the idea of the introduction of a κ factor originates, has indicated the equation

$$\sum_\nu \frac{\partial}{\partial x_\nu} \left\{ \kappa \cdot \sqrt{g} \cdot \sum_\mu g^{\nu\mu} \cdot \left(\varphi_\mu + \frac{1}{\kappa} \cdot \frac{\partial \kappa}{\partial x_\mu} \right) \right\} = 0$$

to determine such a function[5] which, interpreted as a differential equation for the κ to be determined, is invariant against coordinate transformations and, in addition, remains unchanged if

$g_{\mu\nu}$ is replaced by $\lambda \cdot g_{\mu\nu}$, φ_ν by $\varphi_\nu + \dfrac{1}{\lambda} \dfrac{\partial \lambda}{\partial x_\nu}$, and κ by $\lambda^{-1} \cdot \kappa$.

Perhaps through invariant constraints a solution dependent on the $g_{\mu\nu}$'s and φ_ν's could be chosen for this differential equation such that in the case $\varphi_\nu = 0$ ($\nu = 1, \ldots, 4$) it is equal to the constant 1 and, upon introduction of a λ-factor, is multiplied by λ^{-1}.

I would appreciate hearing your opinion on these matters.–

Regarding the Nelson topic, I thoroughly understand your point of view.[6] I gladly accept the prospect of discussing it with you orally when the occasion presents itself.–

Currently, the political events are presumably at the center of your interest. What turn destiny will take "lies in the lap of the gods."[7]

Most cordial regards, yours,

Paul Bernays.

660.   To Carl Heinrich Becker

Berlin W. 30, 5 Haberland St., 25 November 1918

To Privy Councillor Becker, Ministry of Culture

Highly esteemed Sir,[1]

A few days ago Mr. Born told me that some uncertainty regarding my position here stood in the way of his appointment to Frankfurt and Mr. Laue's call to Berlin.[2] That is why it should be of interest to all concerned if I explain to you in the following how I reacted to a call to Zurich. Last summer Prof. Edgar Meyer (Zurich) informed me on behalf of the faculty that the Zurich University and Polytechnic jointly wanted to offer me a professorial chair in case I were inclined, in principle, to return to Zurich. As I am thoroughly satisfied with my position here and I particularly did not want to leave my excellent colleagues here, after brief deliberation I declined with thanks. But in order not to seem ungrateful to my kind countrymen, I promised to give two short lecture cycles annually in Zurich at the university,[3] which should be possible without neglecting my responsibilities here.

I hope that the affair is adequately clarified by the above statement but am otherwise very willing to report at the Ministry of Culture, if that appears desirable to you.

In utmost respect,

A. Einstein.

661.   To Hermann Weyl

[Berlin,] 29 November [1918][1]

Dear Colleague,

Yesterday I wanted to submit your paper to the Academy.[2] It could not be done, though, because—something I had forgotten—an earlier decision had already been reached that papers by nonmembers exceeding a length of 8 printed pages may no longer be accepted under any condition in the *Sitzungsberichte*. I therefore believe it would be best if I put it aside until you have dealt with it otherwise. You can inform me by letter or telegram what I should do with the manuscript. The above decision arises from hard necessity—paper shortage and unaffordable printing costs; *necessity* is anyway the grinning spook we face everywhere.[3]

I have studied your paper but am more than ever convinced that you have gotten onto a very dubious track[4] which is regrettably costing you your valuable energy. I believe you are placing too much importance on the beautiful conservation laws to which the gauge invariance[5] leads.[6] It is not so surprising that the form of Maxwellian electromagnetism comes out of it, since it is known *a priori* that the Maxwell equations satisfy the gauge invariance.[7] The question is, however, whether the other parts of the action function are gauge invariant as well. The existence of spectral lines (electrons of a specific size), etc., speaks very much against it, as I already said earlier.[8] But now I would like to present you with another counterargument that suggests itself from your latest considerations.

I make the following preamble. If we want to retain the common units for mass and length and define the infinitesimal displacement in your geometric theory, we must write, instead of $1 + d\varphi$,[9]

$$1 + \gamma d\varphi,$$

where γ is a universal constant. We must do this if we set

$$d\varphi = \varphi_\nu dx_\nu,$$

where φ_ν signifies the four vector of the el[ectric] potential *in standard units* (cm, gr).[10]

In the Hamilton function given on page 8 of your paper, which should be valid at least outside of electrons and atoms, we must then write

$$\gamma^2 \varphi_i \varphi^i \text{ instead of } \varphi_i \varphi^i,$$

and on page (9), as the equation for the electromagnetic field,

$$\frac{\partial f^{ik}}{\partial x_k} = -\frac{3}{2} \frac{\sqrt{g}\gamma^2}{\lambda} \varphi^i.$$

In this connection γ means the gravitational constant. ($\sim 10^{-27}$.)[11] This field equation is easily integrated to determine the field outside of an electron, whereby it is legitimate to select $g_{\mu\nu} = \begin{matrix} -1 & 0 & 0 & 0 \\ 0 & -1 & 0 & 0 \\ 0 & 0 & -1 & 0 \\ 0 & 0 & 0 & 1 \end{matrix}$. One is then simply led to the equation[12]

$$\Delta\varphi = -\text{const }\varphi,$$

the solution of which is $\dfrac{e^{-\frac{r}{l}}}{r}$.[13] For φ_4 to come out practically $= \dfrac{\text{const}}{r}$,[14] $l \sim \dfrac{\sqrt{\lambda}}{\gamma}$ must be a *very large* length (approximately the world radius?) (measured in cm, a very large number indeed!). The order of magnitude of $\frac{1}{\gamma}$ is hence in any case well over $\frac{1}{\sqrt{\lambda}} \sim 10^{13}$. This universal constant has the dimension of an electricity quantity and must be of immense size if the theory is not to violate the known laws of electrostatics. Not the slightest hint of this can be found in the known laws of nature. It therefore seems downright insane to introduce such a thing for the sake of the gauge invariance! This surely is a very serious objection.

Moreover, I absolutely cannot concede that your theory leads to the equation of motion

$$\frac{d(mu_i)}{ds} - \frac{1}{2}\frac{\partial g_{\alpha\beta}}{\partial x_i}mu^\alpha u^\beta = 0$$

because this expression, in which m is supposed to be constant, of course, is invariant neither against coordinate nor gauge transformation.[15] In your consideration you cannot assume a priori that the φ_i's do not yield any moving forces. The special role of the electrically uncharged point-mass consists in that its path does not depend on the electrical field f^{ik}. I am convinced, now as before, that from the aspect of invariance and gauge invariance, the geodesic line is the only one that can come into consideration at all.–[16]

Furthermore, I do not see how the introduction of a gauge according to $F = \alpha$ is possible, even though one can make free use of a gauge factor.[17] It may very well be possible, for inst., that any attempt at imposing such a gauge would bring about singularities.–

My first objection seems to me by far the most substantial: the gauge invariance leads to a modification of the laws, which not a single fact supports, and also leads to a new natural constant that is meaningless, according to current knowledge. Added to this is that the existence of spectral lines, i.e., of clocks independent of their prehistory, makes the only natural way seem to be treating ds as an invariant from the beginning.

I am already looking forward to being able to chat—and argue—with you in February. Until then, cordial regards, yours,

Einstein.

662.   From Arnold Sommerfeld

[Munich,] 3 December 1918

Dear Einstein,

For a particular purpose (a popular book on atomic models) I need a simple description of the foundations of quantum statistics.[1] For that I must make plausible that $\prod_1^f dq_k dp_k$ (f—the system's degrees of freedom) measures the *a priori*-probability of the phase $q\ p$. The *a priori*-p[robability] means: I know nothing about the motion and also do not know its energy. Liouville's law is cited here unjustifiably, however. Now, L's law just states that domains in phase space *that are transformed into one another by motion* have the same probability (carry an equal amount of system points).[2] If this result is expanded to *any* two domains of the same size in phase space, then the law is extended in a way that leaves nothing of the original meaning. This extension is what is specifically needed, though.

Let us take, for ex., Planck's resonator. L's law refers to two domains like A and B. They are completely unimportant to me, though, because I combine them within the same elementary domain R anyway. What I rather need are two domains like 3 and 1. They are never transformed into each other by motion, however; hence Liouville says absolutely nothing about them.

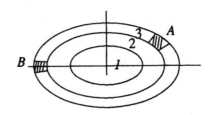

Thus if one does not want to cheat, one can actually only say the following: L's law tells us that certain *specifically situated*, equally large domains in phase space have the same probability. This statement is extended *without any compelling reason* to domains of the same size in phase space *regardless of their location*. As far as I can see, nobody has tried to support this extension—probably because it is groundless. The ergodic hypothesis is of absolutely no use for this since it just relates to the points of the *energy surface* in the phase space, thus on no account to *all* the points in the phase space. Ehrenfest and Gibbs[3] tell us nothing about *a priori*-probability in the above sense. And yet it is needed for the derivation of the H theorem as well as for the explanation of quantum theory. For the latter it would be more honest to leave Liouville out entirely and to proceed immediately with the elementary domains R in the (q, p) plane.

Do you approve of this? Or is there anyone who can dispel my doubts? So much has been written on these general matters that one would think they had been finished off somewhere.

I hear from Kossel[4] that you have confidence in the new times and want to participate. May God preserve your faith![5] I find everything contemptible and idiotic beyond words. Our enemies are the greatest liars and scoundrels, we the greatest imbeciles. Not God but money rules the world. I hope you are healthy at least. I would be very grateful for a reply.

Cordially yours,

A. Sommerfeld.

663.   To Michele Besso

[Berlin, 4 December 1918][1]

Dear Michele,

A major thing really has been achieved. The military religion has vanished. I think it will not return anymore. Nothing has taken its place, of course. While southern Germany promises to develop more after the Swiss model, here the Russian example is disquietingly prevalent.[2] Runaway slaves without a true sense of solidarity and without a general overview. The government, which is dependent on the common herd, of course, is struggling assiduously but with minimal success against the economic crisis caused pitilessly by the vicious circle of forced wage increases—accelerated bill printing—currency devaluation[3]—forced wage increases . . . . . In spite of everything, most people feel relieved after the drastic cure, even those who know very well that they will have to say good-bye once and for all to their money purses. The Ac[ademy] meetings are amusing now; the old folks are for the most part completely disoriented and dazed. They perceive the new era as a sad carnival and are mourning over the bygone state of affairs, whose disappearance means such liberation to the likes of us. When I see Treitschke in the university quad with his ostentatiously arrogant posture and his solid belly of brass, he seems to me like a mammoth from the Bismarckian ice age.[4] Long ago it was, ..., that is, about four weeks, when he was still a god beside the higher god Bismarck, who now all of a sudden appears to many in the same light that he had always appeared to *us*. If only they survive the drastic cure; some symptoms are a bit too suspect. But I will not be dissuaded from my optimism. I am enjoying the reputation of an irreproachable Socialist [*Sozi*];[5] as a consequence, yesterday's heros are coming fawningly to me in the opinion that I could break their fall into emptiness. Funny world!

But by prowling about Zurich for so long you are a veritably abominable institute father. I would give a lot to accompany you to the scene of your duty,[6] for which many a loden-knight here now pines. (I always console them by telling them that failure brings reconciliation.) Yet I would spend time with the kind Levi-Civita in Padova;[7] perhaps you will do the same?

Weyl sent me an exceptionally fine and brilliant paper for the Academy, which cannot be printed there, however, because the Academy had to close its *Berichte* to nonmembers owing to a far too great congestion as a consequence of a paper shortage. I am very firmly convinced that Weyl's gauge invariance does not apply to nature and recently imparted to him support for the objections I have.[8] But I know that someone who has stayed infatuated with one idea for more than half a year can no longer be saved from its spell, at least not by others.

The business with my divorce is an amusement to all who know about it. I should have been depositioned about it now at a local court but received the summons too late. Meanwhile, the files have been sent back to Zurich![9] It's touching how much trouble Zürcher[10] is taking in the matter.

Greetings to Vero and your gruff regency (Anna),[11] as well as Zangger and Dällenbach. Affectionate regards, yours,

Albert.

664.   To Paul Ehrenfest

[Berlin,] 6 December 1918

Dear Ehrenfest,

Hearty thanks for your invitation,[1] which I nonetheless cannot accept at the moment. In the next few days I am going to be traveling via Switzerland[2] to Paris in order to ask the Entente to save the famished population here from starvation.[3] After so much falsehood, it is difficult to help give credence to the bitter truth. But I think they will believe me if I give my word of honor. Besides, I must say that the people here have borne the collapse with composure and calmness, once they had been somewhat enlightened about the truth related to the causes of the war.

In February I shall be giving a lecture cycle in Zurich for students; this is supposed to take place twice annually from now on. I am doing this to show my gratitude for a call I did not accept.[4] I could not decide to leave here under the prevailing circumstances. But who knows if bare necessity will demand it in the end?

I was very pleased about Burgers's dissertation and the news of his appointment.[5] Your researches on adiabats are much appreciated here generally.[6] I hope de Sitter recovers soon;[7] my criticism of one of his papers was only partially applicable, which I regret now especially.[8]

I am feeling well enough, particularly since I am being exceptionally well cared for and monitored in matters of health. It is hardly probable that I shall ever be completely healthy again; but the main thing is that one is content, and that I really am. I am particularly thankful to fate that my great and ardent wish has now been fulfilled.[9] The flu is less virulent here than elsewhere. We have been spared until now.   What do you say to Ehrenhaft's negative light-pressure?[10] Nobody has been able to raise any sound objection to the interpretation of the experiments yet.

Cordial greetings to all of you, the Nordströms, de Sitter, and Lorentz, yours,

Einstein.

665.   To Arnold Sommerfeld

[Berlin,] 6 December [1918][1]

Dear Sommerfeld,

In my view it is not at all arbitrary for elementary domains of the same size to be treated *a priori* as equally probable.[2] For the following is evidently a neat consequence of mechanics. If a system S is in contact with a second system of relatively infinitely great energy (interaction of the systems with additive energies), then

$$dW = \text{const.}\, e^{-\frac{E}{kT}} dq_1 \ldots dp_n$$

is valid for the elementary probability (frequency) of the state of S. The essential thing there is that the factor "const" be independent of $q_1 \ldots p_n$. Disregarding the temperature factor, equally large elementary domains are hence equally probable (equally frequent). (In the limiting case $T = \infty$, equally large domains in phase space are in fact equally probable, whereas it really would be conceivable *a priori* that another factor $\varphi(q_1 \ldots p_n)$, *e.g.*, $\varphi(E)$ were added.) Computationally, this state of affairs is usually taken into account by assigning *a priori* equally high probability to equally large domains. It must also be mentioned that the equation for dW from the Liouville equation is deduced for the (∞ large) total system, so although the usual form of expression is imprecise, it is not exactly false.–

It is true that I have hopes for these times, despite the many ugly things it is bringing forth in the details.[3] I see the political and economic organization of our planet advancing. If England and America are sensible enough to agree, wars of some consequence will not be able to occur at all anymore. The military economy so repugnant to me will also virtually disappear. If the period of transition becomes quite grievous for us, in particular,[4] in my opinion it is—frankly speaking—not entirely undeserved. However, I am of the firm conviction that culture-loving Germans will soon again be able to be as proud as ever of their fatherland—with more reason than *before* 1914. I do not believe that the current disorganization will leave permanent damage.

Cordial regards, yours,

Einstein.

666.   From Zionist Association of Germany

Berlin, 9 December 1918

To Professor Albert Einstein, Schöneberg, Berlin W., 5 Haberland St.

Esteemed Professor,

With reference to our discussions, I am respectfully submitting to you in the attached a notice intended for the *Jüdische Rundschau*.[1]

At the same time, I am sending you a draft of the invitation for the little meeting which is to take place on Thursday, 19 December, 8 o'clock in the evening in the Yellow Hall of the "Rheingold" at Potsdamer Place.[2]

We formulated the topic in a way that points especially to Jewish problems in Germany. The Jewish notion of nationality in relation to the idea of internationality will be discussed thoroughly in the general theoretical introduction to the report, however.

I would be very obliged to you if you could inform us as soon as possible of your approval of the drafts. The return of the drafts is not required.

In great respect, sincerely,

Rosenblüth[3]

2 drafts
1 response card.

667.   To Hans Albert and Eduard Einstein

[Berlin,] 10 December 1918

Dear Children,

Your long letters pleased me very much. Today the bank was given instructions to remit the money for you. Unfortunately the rate of exchange is getting worse and worse[1] (do you know what that is, d[ear] Albert? Have it explained to you; this is a very interesting affair.). For Tete I have a big, handsome book to read, which will be sent out immediately, but for Adn I have nothing but to wish him a Merry Christmas. I'll be coming to see you in 6 weeks[2] and am very much looking forward to it.

Your strike was just a weak reflection of the revolution here.[3] This one began in Kiel. The sailors there refused to fight on, ousted their officers, and then took control of the public buildings and the administration of the city.[4] After that they sent delegations to the soldiers in other cities, who followed their example. The workers everywhere joined in, and everywhere workers' and soldiers' councils were formed, which took the place of the previous government. Finally the coup was carried out in Berlin, the Republic was proclaimed, parliament was dismissed, the emperor and crown prince were forced to flee,[5] the generals and officers were removed, and a Socialist government was installed which is accountable to the representatives of the Workers' and Soldiers' Councils, what is called the Executive Council. A normal National Assembly is supposed to convene shortly, elected by all men and women in Germany.[6]

In the interim, quite wild things have been going on, though. Many are trying to take advantage of the confusion. The funds and other properties of the state are being used recklessly. Paper money is being printed in vast quantities, so money is constantly dropping in value. Workers are demanding enormous wages even though there is nothing to do at the factories. All factories are on the verge of going bankrupt.

Today the soldiers marched in. That was a cheerful stir. Many boys like you drove and rode around in the streets with the soldiers and everyone was up and about in order to welcome the men who had had to endure the so very long and hard ordeals and dangers. But Tete should not go in for vile soldier games. I don't like that in the least.

Affectionate regards from your

Papa.

Amicable greetings to Mama.

668. To Moritz Schlick

[Berlin,] 10 December 1918

Dear Mr. Schlick,

I find your exposition excellent, like your earlier one.[1] You will understand the few minor suggested corrections I have made, without any need for explanation.

You are a true artist of exposition.

Best regards, yours,

A. Einstein.

669.   From Hermann Weyl

Zurich, 10 December 1918

Dear Colleague,

I am a little disappointed that you cannot present my paper to the Academy, but much more so that you do not want to have anything to do with the matter.[1] This worries me very much, of course, because experience has shown that your intuition can be relied upon;[2] although I must admit, your counterarguments so far have not been very illuminating for me. Thus I am caught between faith in your authority and my view. I am now in a really difficult position; through my upbringing so conciliatory by nature that I am almost incapable of discussion, I must now fight on all fronts; my attack on analysis and attempt at a new foundation of it[3] is encountering much fiercer rejection among mathematicians engaged in these logical things than yours is in my efforts in theoretical physics. What should I do now with the revision of *Space, Time, Matter*?[4] Certainly: I shall construct the infinitesimal geometry as in the *Math. Zeitschrift*[5] so that Riemann's appears as a special case of geometry with [action at a] distance [*fern-geometrisch*]; I simply cannot do otherwise if I do not want to trample on my mathematical conscience (I can present one or two details there more clearly). [Incidentally, you must not believe that I came via physics to introducing the linear differential form $d\varphi$ in the geometry alongside the quadratic form; rather, I really wanted finally to remove this "inconsequence,"[6] which had always been a thorn in my side, and then noticed to my own astonishment: it looks as if this explains electricity. You are throwing your hands up in the air and are exclaiming: But physics is not done that way! And I do, in fact, very well understand your anger, which I had probably only aggravated with the triumphant ring in my last note,[7] when you see dreamers and speculators beginning to go their own way along paths that you, who had always regarded reality so seriously, had pointed out.[8] I just have to say in my defense, I believed in all earnestness that you would now abandon your resistance and take my side.][9]

To your counterarguments I can reply this much, in brief:

1) If you set the measuring units in your theory rationally: unit of length = world radius; velocity of light and gravitational constant = 1, then you obtain an immensely large unit of charge,[10] that charge whose "gravitational radius" = the world radius, and just the one for which you are rebuking me.[11] Its "meaninglessness" is based on the fact that the world radius is a "physically meaningless" quantity in the sense that, up to now, it has not been possible to draw any upper limit for it from observation. The fact that the electron's charge is a tiny fraction

of this unit is always the same "pure number of enormous size" mystery.

2)
$$\frac{d(mu_i)}{ds} - \frac{1}{2}\frac{\partial g_{\alpha\beta}}{\partial x_i} mu^\alpha u^\beta$$

really are, whether m is constant or not, the components of a *coordinate-invariant* vector (which reduce to $\dfrac{d(mu_i)}{ds}$) in a coordinate system where $\dfrac{\partial \gamma_{\alpha\beta}}{\partial x_i} = 0$.[12]

3) I do not understand what you mean when you say that, in my derivation of the equations of the mechanics, I "assumed the φ_i's do not yield any moving forces."

4) You repeat without any backing your assertions about the geodesic line and the dependence on the prehistory; I think I have given reasons refuting them and cannot say more than that on it;[13] maybe also just this: You yourself made me aware of the following: If the geodesic line is valid, a contradiction arises with the energy principle; but the energy principle is valid in my theory (it is precisely from it that the equations of the mechanics are derived); *ergo* . . .

Ergo, you will conclude: he is beyond help! Perhaps you will make one more attempt, even so, in February when you are here. And what should happen now with my script? I think in its existing form[14] it was suited only for publication in the Academy *Berichte*; so please send it back to me! I shall then try to find a place for it elsewhere in a revised form.[15]

Do we have a right to worry about the law governing the remotest worlds when that of the nearest is in the process of such tumultuous change, menaced on all sides by peril and calamity?

Even though the war hatchet between us has been dug up, I give my regards in sincere respect. Yours,

H. Weyl.

670.   From Heinrich Mousson

Zurich, 12 December 1918

To Prof. Einstein, Berlin, 5 Haberland St.

The Philosophy Faculty II of the University of Zurich is petitioning to have a teaching assignment at our university in the field of theoretical physics issued to you already for the 2nd half of the current winter semester, namely, in the form of twelve two-hour lectures which, subject to more specific fixing with the university Rector's Office, are to be interspersed throughout a period of ca. 5 weeks.[1]

The University Commission and Education Board welcome it if in this way your valued efforts can again be channeled into the service of the university and give their consent to such an arrangement. We request that you let us know whether and, if yes, under what conditions you would be inclined to take on an assignment toward this end. Repetition of the assignment in the coming summer semester will be reserved for further negotiation.[2]

So as to be able to make the necessary arrangments for the current semester, we should be particularly obliged to you for a prompt reply.

Director of Education:
Mousson.

Note by recipient: "Therma Fabr. for el. heating, Schwanden, Glarus. Gustav Maier without mentioning names. 20 000 fr. Zuoz 35."

671.   From Zionist Association of Germany

Berlin W. 15, 8 Sächsische Street, 12 December 1918

To Professor Albert Einstein, Berlin. 5 Haberland Street

Esteemed Professor,

I extend to you my most respectful thanks for your commitment to attend the meeting on the 19th of this mo.[1]

We have decided to refrain from using your name on the invitation.[2] It would have been particularly important to us to have the invitation issued by a committee that did not consist predominantly in individuals long known as leading members of our organization. Time is too short to interest twenty, or in any event a substantial number, of new prominent members in such a committee. We shall therefore just invite in the name of our organization.

In great respect, devotedly,

Rosenblüth.[3]

672.   To Mileva Einstein-Marić

[Berlin, mid-December 1918][1]

Dear Miza,

The reasons I am trying to accelerate the divorce are purely superficial. I'll explain this to you more precisely when I come to Zurich again; it would be too involved in writing.[2] I am coming around February 1st in order to give lectures in Zurich.[3] It's genuinely shocking what the exchange rate is now: according to the current rate, the Fr 8 000 are 14 000 M![4] Who knows, one fine day I might be faced with the necessity of leaving here because I couldn't drum up the money for you otherwise.   I'll wait it out as long as possible, though. *Because here I am received with a kindness that is truly moving.*

I have made sure that this time you'll receive the money on time.[5] So don't worry about it.

I wish you a Merry Christmas, yours,

Albert.

673. To Hermann Weyl

Berlin, 16 December 1918

Dear Mr. Weyl,

I am sending your manuscript to you right away.[1] I can only say to you that everyone I have spoken to talks about your theory *with the deepest respect*, from the mathematical point of view, and that I also admire it as an edifice of ideas. You do not need to fight, least of all against me. It really is out of the question that I be angry: genuine admiration but disbelief, those are my feelings toward the subject.

With my objection I wanted to say (1)[2] that the φ-additional terms to the Riemann tensor [3] certainly cannot contribute anything toward an explanation of the electron. (2) Your equations for the motion of the point are not gauge-invariant, indeed not even coordinate-invariant, or am I not seeing the wood for the trees?[4]

I do not want to argue but suggest one more thing to you. If you consider it proven that the additional φ-terms in the Riemann tensor can play no role in the electron problem,[5] then this problem really does seem to take on a very simple form. In my view, this is where you must set to! When we are together in Zurich in February[6] we shall come to terms, with or without gauge invariance. You will see that I am not obstinate, rather, I am gladly prepared to follow through any chain of thought. No events of the day could distract my interest either in remote ideas; the yonder delivers us from slavery to the passions.[7]

In happy anticipation of seeing you again soon, let this little be enough; all technical points orally.

Cordial regards, yours,

Einstein.

674. To Heinrich Mousson

[Berlin,] 17 December 1918

To the Director of the System of Education for the Canton of Zurich.

With reference to your letter of the 12th of this month,[1] I declare herewith that I gladly accept the teaching assignment kindly offered to me. If you and the university agree, I shall be in Zurich on 1 Feb. 1919 and then begin the lectures immediately. (3 two-hour lectures per week.) Unless I receive specific suggestions on the subjects I should be treating, I shall lecture on relativity theory.

My idea of arranging regular lectures in Zurich stems from the wish to show my gratitude toward my fellow citizens for the honorable call they issued to me last summer.[2] In accordance with this motive, I would have liked to give the lectures completely gratis, had I been in the position to do so; however, the more difficult circumstances created by the war necessitate me to have my extra expenses reimbursed. I therefore suggest to you that for the cycle I be given a travel and accommodation allowance of Fr. 1 200.[3]

In great respect.

It is my great pleasure, thanks to your kind cooperation, to be able to take part again in the intellectual life at our university.

675. To Fritz Haber

Dahlem, Berlin, [before 20 December 1918][1]

Dear Haber,

I come with a small request for the German Phys. Society. We have decided, despite the present difficult circumstances, to attempt nevertheless to scrape together a fund from relevant industries in support of academic physics instruction (laboratories).[2] Scheel[3] and I are supposed to take care of the matter. Everything is ready to be sent out. But now it also occurs to us that if our appeal does not arrive in the hands of the right *persons*, our glory will definitely find its way unread into the various wastepaper baskets. But there we lack information about the important leading individuals at the firms to whom we could turn. (We would like to call on the accessible ones personally as well.) *In this predicament we beg you to look through the enclosed lists and to give us tips with regard to the addresses.* Please send the lists—with your emendations—to Mr. Scheel.

Cordial regards, yours,

A. Einstein

P. S. *Please, do not under any condition let the matter wait.* If you have no time for the affair, then please send the lists to Mr. Scheel unchanged or send them with the same request to another person who would be ready and in a position to advise us.

Recipient's note: "To Dr. Arnold Berliner with regards & request to execute, Haber."

676.   Deposition in Divorce Proceedings

Schöneberg, Berlin, 23 December 1918

⟨Public Hearing of the Royal⟩ The District Court.
In re divorce matter⟨s⟩
Einstein *versus* Einstein

Present:
   D[istrict] C[ourt] Justice Jacobi **as Magistrate,**
   Sol[icitor] Hildebrandt **as Court Clerk.**
In the original with attachments.
Respectfully returned to Regional Court Section II in Zurich.
Schöneberg, Berlin, 23 December 1918. The District Court Sec. 10.

Jacobi.

Appeared:
   Upon producing the writ of summons, Professor Einstein of ⟨Schöneberg, Berlin, 5 Haberland St.⟩, proved his identity.[1]

He requested to be heard today already, as he was unable to come on January 3rd. He was cautioned to speak the truth and declared:

My name is Albert, **I am** 39 years **old.** ⟨**Religion,**⟩ dissenter.

O[n the] m[erits of the case]:[2]

1. It is correct that I committed adultery. I have been living together with my cousin, the widow Elsa Einstein, divorced Löwenthal, for about 4 $1/2$ years and have been continuing these intimate relations since then.[3] My wife, the Plaintiff, has known since the ⟨spring⟩ summer of 1914 that intimate relations exist between me and my cousin. She has made her displeasure known to me.

2. I do not wish to claim separate grounds for a divorce against my wife.

3. I approve of the Agreement of 12 June 1918 regarding the proprietary consequences of the divorce and the assignment of the children to the custody of my wife. I also agree to the children's visiting rights in such a manner that they be entrusted to me over the school holidays during my sojourns in Switzerland.   r[ead out,] a[pproved,] s[igned]

Albert Einstein.
Jacobi.   Hildebrandt.

Added note: "*Urgent!*"

677.   To Felix Klein

[Berlin,] 27 December 1918

Highly esteemed Colleague,

First of all, belatedly, my heartiest congratulations on the beautiful jubilee you celebrated recently.[1] When we look back on the 50 years of successful work you have lived through, it is a splendid spectacle for us all; for you yourself, such a retrospective view must be an experience of purest satisfaction.

I thank you furthermore for your elegant and fine proof of the vector character of ε^p.[2] One more comment about the field equations. Your analyses completely clarified the relations

$$\frac{\partial(\mathfrak{T}_\mu^\nu + \mathfrak{t}_\mu^\nu)}{\partial x_\nu} = 0$$

formally. It is also important, though, that it be possible to bring the field equations into the form

$$\mathfrak{T}_{\mu\nu} + \mathfrak{t}_{\mu\nu} = \text{Div}.$$

For these relations are the physical expression of the fact that the total energy of a system determines the outward directed flow of force. It would be nice to know whether this relation also is independent of the special choice of the Hamilton function for the gravitational field.–

What prompts me to write today, though, is a different matter. Upon receiving the new paper by Miss Noether,[3] I again feel it is a great injustice that she be denied the *venia legendi*. I would very much support our taking an energetic step at the Ministry.[4] If you do not consider this possible, however, I shall make the effort on my own. Unfortunately, I have to go away on a trip for a month.[5] But I beg you to leave me a short message by the time I return. If something should have to be done beforehand, please avail yourself of my signature.

With cordial regards, yours truly,

A. Einstein.